Lecture Notes in Networks and Systems

Volume 369

AF806718

The series "Lecture Notes in Networks and Systems" publishes the latest developments in Networks and Systems—quickly, informally and with high quality. Original research reported in proceedings and post-proceedings represents the core of LNNS.

Volumes published in LNNS embrace all aspects and subfields of, as well as new challenges in, Networks and Systems.

The series contains proceedings and edited volumes in systems and networks, spanning the areas of Cyber-Physical Systems, Autonomous Systems, Sensor Networks, Control Systems, Energy Systems, Automotive Systems, Biological Systems, Vehicular Networking and Connected Vehicles, Aerospace Systems, Automation, Manufacturing, Smart Grids, Nonlinear Systems, Power Systems, Robotics, Social Systems, Economic Systems and other. Of particular value to both the contributors and the readership are the short publication timeframe and the world-wide distribution and exposure which enable both a wide and rapid dissemination of research output.

The series covers the theory, applications, and perspectives on the state of the art and future developments relevant to systems and networks, decision making, control, complex processes and related areas, as embedded in the fields of interdisciplinary and applied sciences, engineering, computer science, physics, economics, social, and life sciences, as well as the paradigms and methodologies behind them.

Indexed by SCOPUS, INSPEC, WTI Frankfurt eG, zbMATH, SCImago.

All books published in the series are submitted for consideration in Web of Science.

More information about this series at https://link.springer.com/bookseries/15179

Hrvoje Glavaš · Marijana Hadzima-Nyarko ·
Mirko Karakašić · Naida Ademović ·
Samir Avdaković
Editors

30th International Conference on Organization and Technology of Maintenance (OTO 2021)

Proceedings of 30th International Conference on Organization and Technology of Maintenance (OTO 2021)

Springer

Editors
Hrvoje Glavaš
Faculty of Electrical Engineering, Computer Science and Information Technology Osijek
Josip Juraj Strossmayer University of Osijek
Osijek, Croatia

Marijana Hadzima-Nyarko
Faculty of Civil Engineering and Architecture
Josip Juraj Strossmayer University of Osijek
Osijek, Croatia

Mirko Karakašić
Mechanical Engineering Faculty
University of Slavonski Brod
Slavonski Brod, Croatia

Naida Ademović
Faculty of Civil Engineering
University of Sarajevo
Sarajevo, Bosnia and Herzegovina

Samir Avdaković
Faculty of Electrical Engineering
University of Sarajevo
Sarajevo, Bosnia and Herzegovina

ISSN 2367-3370 ISSN 2367-3389 (electronic)
Lecture Notes in Networks and Systems
ISBN 978-3-030-92850-6 ISBN 978-3-030-92851-3 (eBook)
https://doi.org/10.1007/978-3-030-92851-3

This Springer imprint is published by the registered company Springer Nature Switzerland AG
The registered company address is: Gewerbestrasse 11, 6330 Cham, Switzerland

Preface

This book contains papers presented at the 30th International Conference on Organization and Technology of Maintenance (ORGANIZACIJA I TEHNOLOGIJA ODRŽAVANJA – OTO). The International Conference on Organization and Technology of Maintenance represents an opportunity to become familiar with new approaches and maintenance methods for the purpose of direct exchange of experience by experts from all fields of maintenance aimed at exploring and analyzing the implementation of new methods and procedures. The conference seeks to raise the level of maintenance knowledge, taking into account the continuous advancement of engineering and technology in all spheres of economy, infrastructure, and public services. Maintenance relates to different scientific disciplines and professional activities with a synergistic effect. Some disciplines are required for the successful maintenance, and the implementation of maintenance procedures promotes other activities. Communication between scientists and experts in maintenance practice raises the level of expertise and introduces new methods and maintenance procedures into everyday practice. Furthermore, the majority of the papers include contributions from Master's and Ph.D. students, providing them with a unique chance to meet, network, and exchange their research ideas and findings with world-class programmers, students, and researchers.

The conference covers the following topics: organization and management of maintenance, maintenance technologies, quality management in system maintenance, information systems in maintenance, product life cycle management, design for maintainability, material and structure properties, reliability of technical systems and environmental safety, diagnosis and prognosis of failures and operational malfunctions, design optimization for maintenance, maintenance in technical systems, analysis of efficiency and cost effectiveness of maintenance, influence of maintenance on the environment and employee safety, maintenance legislation, and education for maintenance.

The book brings 36 original papers which underwent a review process by the international review board members. The articles were written by authors from eight Croatian faculties originating from four Croatian universities, as well as from nine countries abroad.

The most recent developments in the field of maintenance in a high-quality manner are provided. It contains theory, applications, and various viewpoints on the state-of-the-art and future developments relevant to maintenance strategies in various domains.

Throughout the book, all areas and subfields of maintenance, as well as new challenges, with topics including wireless communication systems, mobile applications, soft computing methods, software simulations, design, as well as questionnaires and statistical analyses implemented in a wide range of areas that cover electrical, mechanical, and civil engineering, architecture, industry, economy, production, and even education are covered

The first paper is dedicated to the history of operation and development of the scientific conference "Organization and Technology of Maintenance" (OTO) over three decades from a regional scientific and professional conference to an international scientific symposium.

The topics of papers in the field of electrical engineering and energy address various issues ranging from product life cycle assessment (LCA), which was analyzed on an example of two household appliances and showed a significant impact of proper material selection at the early stage of construction on reducing the impact of production and products on the environment and human health, to the optimization of hydropower operations, also from the design and production forecasting phase to the verification and updating of the plan. The presented articles also deal with motors, both in the field of frequent testing and/or measuring the condition of electric motors, in order to determine the correctness and reliability in operating conditions, and simulations and experimental measurements of electrothermal induction motor models for predicting motor temperature in different working conditions. As the most important components of substations, simulations are covered in the article on fault diagnostics and protection of power transformers and the article on the model of the battery storage system for the possibility of active and reactive power management. In addition, one paper considers of a new method for cyber-maintenance of distributed power sources, given that IT devices used for remote control usually have a shorter life span than the rest of the system, and if exposed, there is a possibility of malicious attacks.

Different areas of civil engineering and architecture are covered by about a dozen papers. They dealt with up-to-date topics, from maintenance of buildings to extreme weather events and climate change. The need for all stakeholders to join in and with coordinated efforts finds a better way to adapt maintenance of agricultural drainage systems to the new climate conditions is presented in a case study of hydrotechnical maintenance of agricultural drainage systems.

An overview of maintenance measures for the main building elements, from control to repairs, and eventually to replacement is provided. The availability of financial resources dictates how and when these measures will be conducted. In this respect, the costs of measures for maintenance and improvement of building properties are analyzed in line with the new requirements.

In light of the devastation of structures in the Zagreb and Petrinja earthquakes, interesting research regarding the influence of state preservation and maintenance on the seismic vulnerability of buildings in two rural settlements in Eastern Croatia was presented, hoping to raise awareness of important research in the field of earthquake engineering. Two articles provide an analysis of the importance of proper building design to withstand earthquake activities and, in addition to the determination of the bedrock or the strict ground conditions in the bedrock, the effect of insufficient cover thickness on the structural performance of reinforced concrete buildings.

Proper modeling of the durability of cementitious composite elements is necessary to schedule maintenance and servicing of civil engineering structures. With this in mind, the authors presented a review of the latest research on the modeling of the durability of cementitious composites with different admixtures using different machine learning algorithms. Machine learning techniques were also applied to assess material consumption in the construction of prestressed concrete road bridges.

By using software simulations of a masonry wall, some authors showed how the construction systems affect the comfort of the users and how the obtained results are valuable for the maintenance of the building as they foresee future problems and guide the technician in the resolution of architectural encounters. By using the post-occupancy evaluation (in the form of a questionnaire), the authors explored the architectural aspect of buildings (site, function, interior design) and the methods in which they should be evaluated by users.

Other areas covered by dozens of papers are related to the industry, economy, production, and even education with a connection with maintenance.

The importance of the successful business of microenterprises was presented by the results of empirical research on the presence of continuous education of employees and their use of modern information technology in maintaining and measuring the success of microenterprises.

Machine learning algorithms are also applied in the analysis of surface defects in real-life ceramic tile production. A wireless communication system to achieve remote monitoring and/or process control is set up and adjusted in a local wood processing industry. Wireless devices are used for constant monitoring of environmental parameters and wood structural moisture in wooden planks during the naturally aspired drying process phase as well as the forced drying phase in a controlled environment.

A comprehensive summary of the evolution of winery monitoring systems that helps to maintain wine quality during its production is provided in one of the articles. To ensure the demanded quality for malting barley, each season several quality indicators have to be determined and evaluated. Bearing this in mind, a five-year continuous study was analyzed that can provide mean values for the tested indicators important for growers, maltsters, and brewers.

During a three-year period, maintenance of the quality parameters of homemade "kulen" (a dry Croatian sausage) was observed and analyzed. Homemade "kulen" is a dry fermented sausage with specific sensorial characteristics (smell and taste), which originates from the use of high-quality pork, specific spices, smoking (use of open fire), and a long ripening time (at least four months).

The papers presented in the book best reflect the interdisciplinary approach to the topic of maintenance. The editorial board tried to rank them in accordance with the subject reviews of individual authors, taking into account the characteristics of individual professions of the authors.

While editing, the reflections of the authors and the suggestions of the reviewers were taken into account, and care was taken to meet these requirements.

We would like to thank all the authors for their contributions, as well as all experts involved in the reviewing process and members of conference committees. We owe a debt of gratitude to all for raising awareness of the importance of maintenance.

Conference Editorial Board.

Contents

Three Decades of OTO Conference Contributions to Organization Practices and Maintenance Technologies

Milan Ivanović(✉)

Panon – Think Tank for Strategic Studies, Osijek, Croatia

Abstract. History of operation and development of scientific conference "Organization and Technology of Maintenance" (OTO) during the three decades from regional scientific-professional conference to international scientific symposium is presented. The paper provides an overview of the number and structure of published papers, number of authors and their professional affiliation, sponsors and venues, and analyzes the structure of published papers. The conclusion points to the contributions of OTO papers to the practice of organization and maintenance technology in the region of Slavonia i.e. Central and Eastern Europe, and gives recommendations for better cooperation of scientific institutions with companies and public affairs institutions in the OTO area.

Keywords: Industry · Equipment maintaining · Organization of maintaining · Technologies of maintaining

1 Introductory Remarks

It has been 30 years since the first regional conference of OTO (Organization and Maintenance Technology) at which multidisciplinary issues of organization and technology of exploitation and maintenance of industrial equipment were discussed. Several generations of experts from industry and academia - in difficult socio-economic and political conditions - have invested extraordinary efforts and social enthusiasm in the organization of this conference and accompanying events; and, no state-political or professional body ordered that these activities be carried out, nor were they paid for that work, nor was any private non-transparent interest realized. So, it is time to give an assessment of the OTO rallies held over the past three decades.

1.1 Techno-Economic Characteristics of the Region's Economy

Techno-economic characteristics of the economy of the region of Slavonia (see Fig. 1) summarize the following three excerpts, with graphs and tables, from our previous papers:

"In 1979, in the area of the Slavonia region, 184742 people were employed in the social sector, of which 85702 in industry. In following years, employment grew so that

H. Glavaš et al. (Eds.): OTO 2021, LNNS 369, pp. 1–20, 2022.
https://doi.org/10.1007/978-3-030-92851-3_1

in 1987, 221326 and 104890 workers were employed in industry [1]. This shows that the region of Slavonia was not an explicitly agricultural region (as was often assumed) but had a developed industrial production (metal processing, wood, textile, food, chemical, rubber, leather and foot-wear industry)." [1]; Fig. 2, 3, 4 and Table 1.

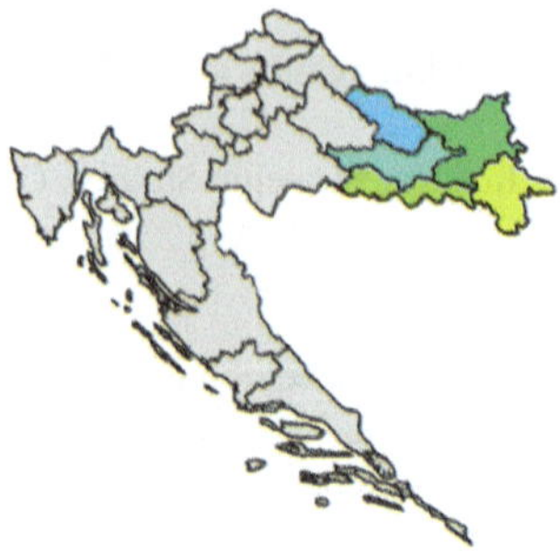

Fig. 1. Croatia region of Slavonia in 1983 (14 then municipalities).

In socialist Yugoslavia, economy was divided into the social and private sectors. Only the number of employees in social sector is shown in Fig. 2. This means that private craftsmen, farmers and free professions (approximately 30000 employees) are not included [1].

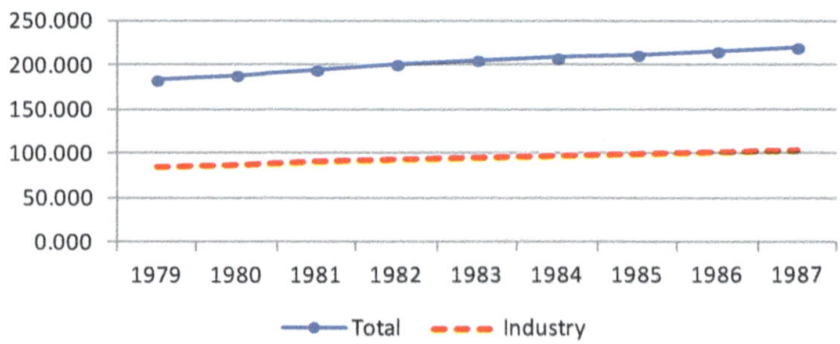

Fig. 2. Number of employees in social sector of economy and industry of Slavonia in period 1979–1987 [1].

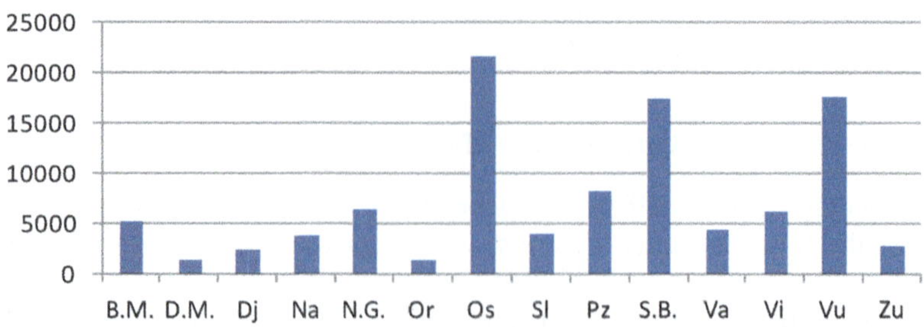

Fig. 3. Number of employees in industry of region Slavonia by municipalities in 1987 [1].

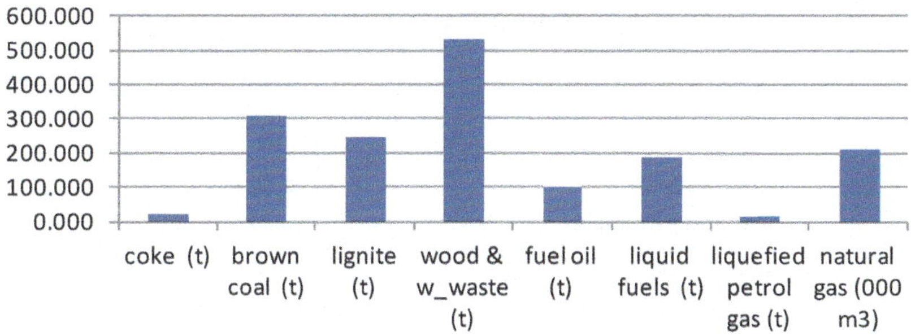

Fig. 4. Structure of substitutable energy consumption in Slavonia industry in 1984 [2].

In the area of Slavonia region in 1984, 25217 t coke, 159 t hard coal, 309318 t brown coal, 249072 t lignite, 532785 t wood and wood waste, 102669 t fuel oil, 88000 t liquid fuels, 16952 t liquefied petroleum gas and 213.5 million m^3 natural gas were consumed (see Fig. 3 and 4) [2, 3]. The region had a dozen large energy consumers that were high on the list of energy consumers in Croatia (see Table 1).

Table 1. Slavonian industrial capacities on the ranking list of the largest energy consumers in Croatia (1981–1983) [2].

On	Company name	Headquarters	Rank in Cro
1	Kombinat Belišće	Belišće	5
2	Tvornica šećera IPK Osijek	Osijek	6
3	Kombinat Borovo	Borovo	11
4	Tvornica cementa	Našice	13
5	Tvornica šećera PPK Županja	Županja	15
6	Tvornica šećera PIK Belje	B. Manastir	17
7	SOUR Đuro Đaković	Sl. Brod	21
8	DIK Đurđenovac	Đurđenovac	26
9	Slavonija DI	Sl. Brod	28

In those decades, 12 large industrial-agricultural companies with developed agricultural production and food industry operated in the region (see Table 2).

Table 2. Industrial-agricultures companies in Slavonia region (1983) [1].

On	Company name	Headquarters
1	Agro-kombinat Jasinje	Slavonski Brod
2	IPK Osijek	Osijek

(*continued*)

Table 2. (*continued*)

On	Company name	Headquarters
3	PIK Belje	Darda
4	PIK Đakovo	Đakovo
5	PIK Nova Gradiška	Nova Gradiška
6	PPK Slavonska Orahovica	Slavonska Orahovica
7	PPK Kutjevo	Slavonska Požega
8	PIK Đuro Salaj	Valpovo
9	PIK Vinkovci	Vinkovci
10	PIK Vukovar	Vukovar
11	PPK Županja	Županja

Eleven large industrial and agricultural companies in Slavonia in those years were on ranking list of 140 largest companies in Yugoslavia (see Table 3).

Table 3. Slavonian companies on ranking list of largest companies in industry, mining, agriculture, forestry and construction in Yugoslavia - according to total income [4, 5].

Company	Headquarter	Ranking position		Number of employees
		1983	1989	1989
1. Kombinat Borovo	Vukovar	16	35	21.557
2. IPK Osijek	Osijek	33	22	16.520
3. PIK Belje	Darda	57	–	8.710
4. SOUR "Đuro Đaković"	Sl. Brod	76	–	16.619
5. PIK Vinkovci	Vinkovci	87	102	4.338
6. Kombinat Belišće	Belišće	121	...	5.038
7. Elektroslavonija	Osijek	–	89	2.500
8. Saponia	Osijek	–	104	2.839
9. PPK Đuro Salaj	Valpovo	–	126	2.473
10. PIK Đakovo	Đakovo	–	131	3.579
11. PPK Kutjevo	Kutjevo	–	135	2.669

Unfortunately - after the democratic social changes in the 1990's in the Slavonian region (as well as in the whole Republic of Croatia) three (for the region) objective difficulties significantly reduced the potential of industry: a) war aggression on the Republic of Croatia caused great human and material losses and broken social relations which has set back economic and social development; b) the concept and implementation of privatization in the Republic of Croatia during the same period also made huge losses in the financial and social capital of the region - for which incorrect implementation there is a very significant part of the responsibility of local politicians; and c) the processes of globalization have changed the economic environment which, with large food imports into the country, has done additional damage to deveopment. It should be added that during this period there were a number of subjective weaknesses of local politics and science. [6] Thus, dozens of industrial companies in the region disappeared, and in hundreds of companies, production - as well as the number of employees - was decimated. According to Statistical reports, in 2008 there were 525 industrial companies in the Slavonia region, employing 42,183 workers [7].

1.2 Establishment of the Association of Industrial Equipment Maintainers of Slavonia and Baranja

Developed industrial capacities and mechanized agricultural production have conditioned development of strong equipment maintenance services with numerous professional staff in a number of engineering and technological professions. Consequently, experts founded their professional organization – Association of Maintenance Osijek (industrial equipment in Slavonia), which was founded in Osijek on October 27, 1983.

The Association brought together hundreds of engineers and experts of other profiles from thirty cities and municipalities of the region, who at that time had a professional need to exchange experiences, expand their knowledge of new technologies, develop specific solutions in operation and maintenance of industrial equipment. At that time, members of the Association participated in a series of seminars, lectures and visits to large industrial complexes, both in the Slavonia region and in Croatia, as well as in other republics of Yugoslavia. Also - a number of experts participated in their work at professional meetings in Croatia. The Association was programmatically oriented towards manufacturing companies in industry, agriculture and construction, as well as the transport sector. Exchange of experiences and processes of continuous education (today it would be called lifelong learning) were basic tasks of Association. Thus, a number of technical exhibitions, study visits, courses, forums and round tables were organized during the operation. In the late 1980s, need to hold a regular annual regional meeting on organization and technology of equipment maintenance was articulated. Namely, developed industrial production in the region, capacities with high energy consumption, and introduction of new technologies and crisis of the socialist business model (which indicated need for stronger market respect) imposed a number of new organizational and technological problems, among others, in sector of exploitation and maintenance of industrial equipment [8]. At the beginning of 1990 - on the initiative of prof. dr. sc. Branimir Kovačić (Faculty of Electrical Engineering Osijek) the Association started by organizing regional professional-scientific meeings on topic of maintenance of industrial equipment. That's how it was on April 20, 1990 the first scientific colloquium on topic "Organization of Maintenance in New Conditions" was held at the Faculty of Electrical Engineering in Osijek [9].

2 Conference on Organization and Maintenance Technology

Due to the aggression against the Republic of Croatia and the bombing of Osijek in 1991, the conference was not held, and next two conferences were also held during the war on 12 December, 1992 and March 12, 1993 in Osijek under the title "Maintenance of Industrial Equipment" (In the fall of 1993, the scientific colloquium "Informatization of Maintenance" was also held in Osijek). The next two conferences were held: March 4, 1994. ("Maintenance Organization and Technology") in the city of Đakovo and January 27, 1995 ("Management and Maintenance Organization") in the city of Našice. The seventh conference was held on February 16, 1996 in Slavonski Brod under the title "Organization and technology of maintenance - OTO 96." and since then all subsequent conferences are held under that title, and OTO (abbreviation from the Cro name "Organization and technology of maintenance") becomes the brand of the conference [9]. In the past 30 years OTO conferences have been held in 11 cities in the Slavonian region and in Kutina (Croatian city outside territory of the Slavonia) Table 4.

Table 4. Cities and number of OTO conferences held [8–38].

1	Osijek	14
2	Požega	3
3	Đakovo	2
4	Kutina	2
5	Vinkovci	2
6	Županja	2
7	Donji Miholjac	1
8	Kneževo	1
9	Našice	1
10	Slavonski Brod	1
11	Virovitica	1
12	Vukovar	1

2.1 Number of Papers and Structure of Authors

In 30 years 1321 authors have presented at OTO conferences (and published in proceedings) a total of 630 papers [10–40]. In the first years of conferences (which were during the war) number of papers is smaller (about 10), but in later years it grows (20 per conference) and from 2020 to over 40 papers (see Fig. 5).

The analysis of the number of authors and published papers shows:

- In the first two decades (1990–2010) there were the most papers with one author (224), followed papers by two authors (61), with three authors (33), with four athors (6) and by with five author (4);

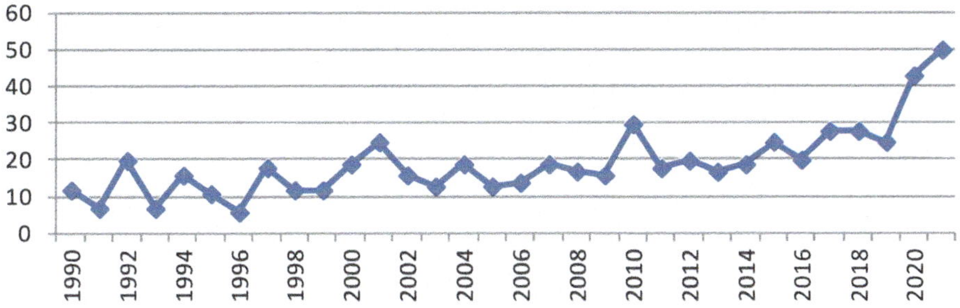

Fig. 5. Number of papers in OTO proceedings from 1990 to 2021 [10–40].

- Since 2012, the number of authors per paper has grown significantly; this is mostly result of more complex research on specific technical issues that are published, alhough it should not be excluded "friendly" addition of some "authors" (according to this author at least 15% of papers) who thus get the necessary points for re/election in scientific and teaching positions (see Fig. 6 and 7).

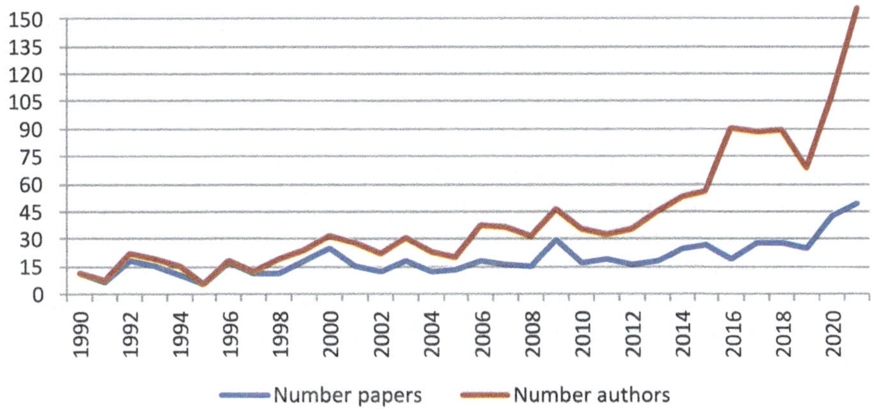

Fig. 6. Number of authors and number of papers in OTO proceedings 1990–2021 [8–38].

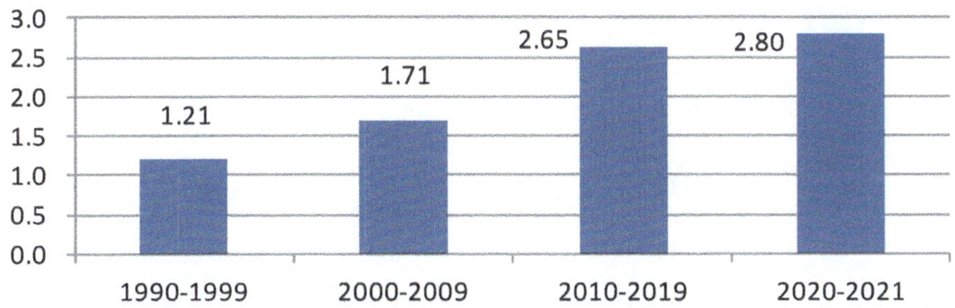

Fig. 7. Average number of authors per published paper in OTO proceedings 1990–2021 [10–40].

Analysis by author profession shows that most of authors are in field of electrical engineering (power engineering, electromechanical engineering, computer science, and telecommunications; 31%), and mechanical engineering (26%), followed authors in economics (16%), agronomy (11%) and construction (11%) and others (law, socology etc. 4.0%), see Fig. 8.

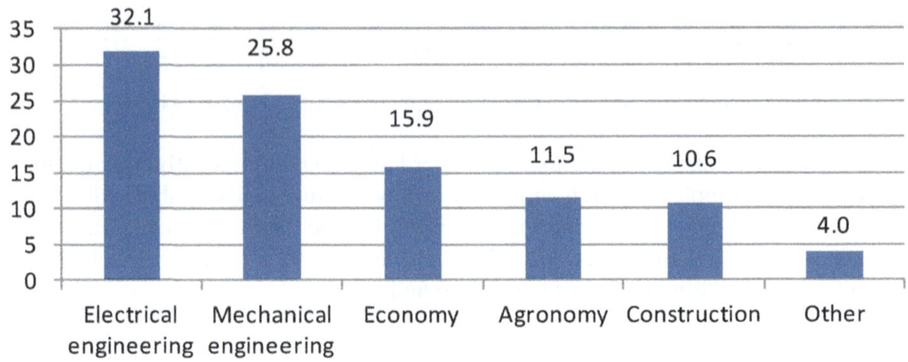

Fig. 8. Authors of papers at OTO conference 1990–2021 by professions (%) (Total = 100) [10–40].

It should be emphasized here that over 98% of authors have a university degree, or that over 65% of the first authors have the academic degree of master or doctor of science. In multi-author papers, more than 50% of these author teams are interdisciplinary.

Looking at the domicile institution of the authors - most authors are in group of technical faculties (TF), followed: authors from faculties of social sciences and coleges (SFHS), large companies in business sector (LBS), public institutions and local government (LSG), small enterprises (SME) and others (unemployed professionals and students), see Fig. 9.

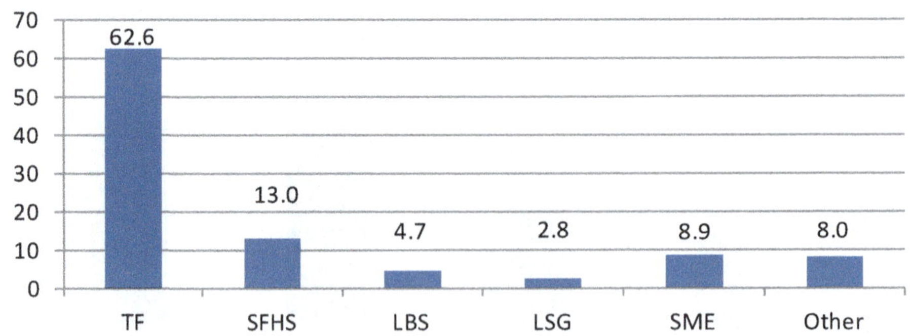

Fig. 9. Authors of papers in OTO proceedings 1990–2021 by employment institutions [10–40].

According to place of residence of (first) author (in team work) in first 20 years, the most represented authors are from Osijek (225), followed by: Slavonski Brod (25), Požega (17), Zagreb (16), Kutina (12), others cities in the region of Slavonia (9), other cities in Croatia (28). There were few papers from abroad (13). Figure 10 shows the number of authors by decades according to country of residence in the period 1990–2021.

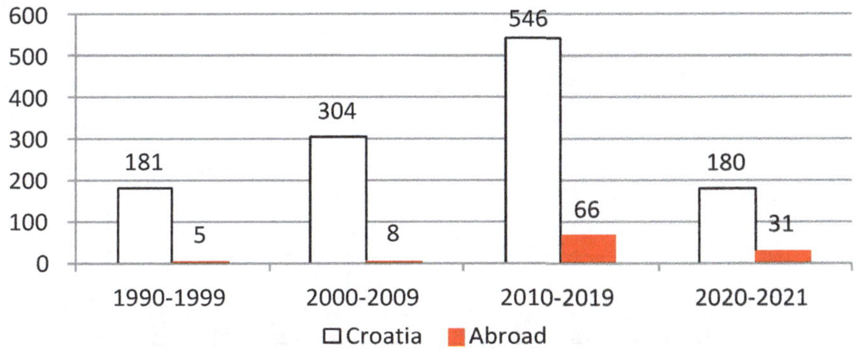

Fig. 10. Authors of papers in OTO proceedings 1990–2021 by country of residence [10–40].

Most authors outside Croatia are from Serbia (71), Bosnia and Herzegovina (17), Slovenia (8), and a smaller number of papers are from Austria, Germany, Hungary, Kosovo, Poland, Spain, and Turkey.

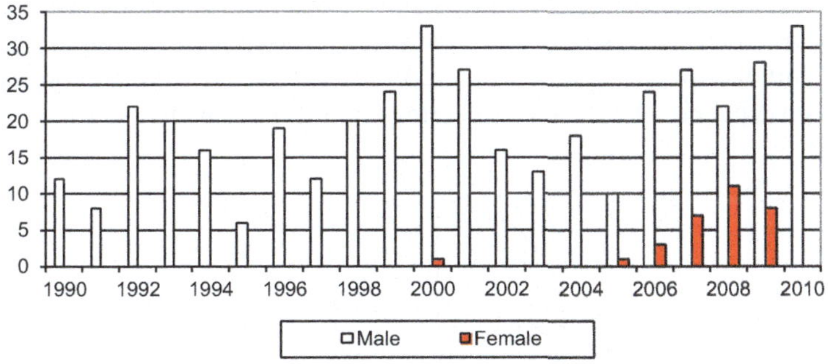

Fig. 11. Number of authors in OTO proceedings 1990–2010 by gender [7].

It is especially important to point out gender structure of papers authors: in the first 15 years in OTO proceedings, except four female authors, all authors are male. Only since 2006 have papers of female authors been present - as shown in Fig. 11. Looking at the total number of authors in the first 20 years of the OTO conferences female authors was 10% (see Fig. 12). Since 2011, female authors have been regularly reprsented at OTO in increased number compared to previous period (see Fig. 13 and 14).

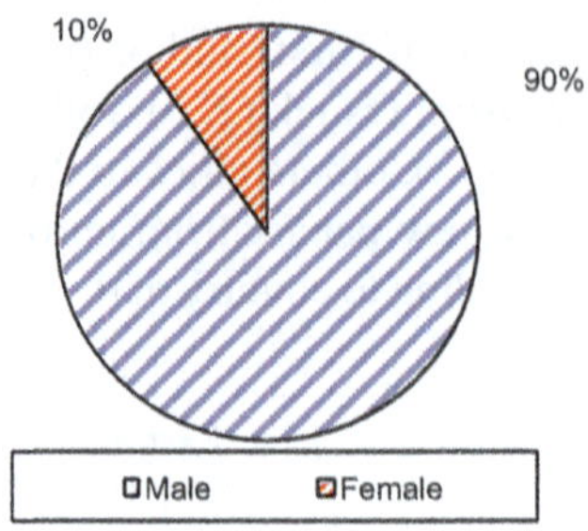

Fig. 12. Percent of authors in OTO proceedings 1990–2010 by gender [7].

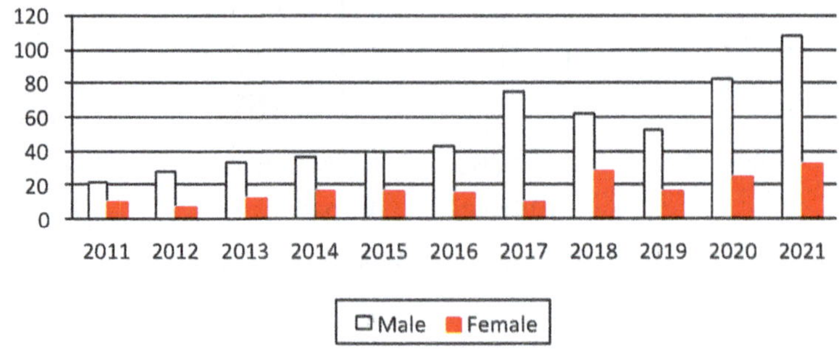

Fig. 13. Number of authors in OTO proceedings 2011–2021 by gender [31–40].

Increased representation of female authors is a result of changes in material prduction technology and increased share of tertiary sector (services and education) in the structure of economy and society and strong development of informatics and computing - in which sectors a significant percentage of women professionals - as well as representation of women in construction, electrical engineering and mechancal engineering.

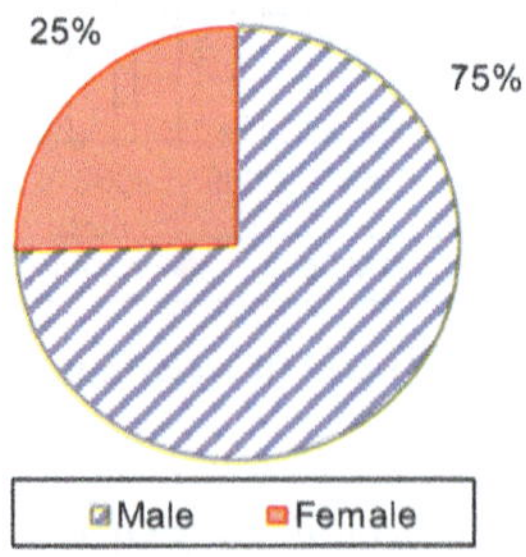

Fig. 14. Percent of authors in OTO proceedings 2011–2021 by gender [31–40].

3 Thematic Framework of Published Papers

The analysis of the thematic framework of published papers at OTO conferences [10–40] was made for two periods; a) 1990–2010 and b) 2011–2021, and in each of the period on two levels:

1. Thematic framework of papers; the papers are grouped into three groups: (1) "Maintenance organization", (2) "Maintenance processes and procedures" and (3) " Maintenance technologies";
2. Analysis of papers according to area of purpose; There are five sets of intended use of paper: (1) " In principle models", (2) "Intended use for industry", (3) "Intended use for agriculture", (4) "Intended use for construction", and (5) "Intended use for the non-economic sector".

3.1 Period 1990–2010

According to the thematic framework, most papers were published in the group "Maintenance processes and procedures" (MPP), followed by thematic groups "Maintenance organization" (MOR) and "Maintenance technologies" (MTH), see Fig. 15.

The analysis of works by purpose shows that most works were published in group "Principle models" (Pri_mo), followed by: "Purpose industry" (P_ind), "Purpose agrculture" (P_agr), "Purpose_ construction" (P_con) and "Purpose non-economic sector" (P_nes), see Fig. 16.

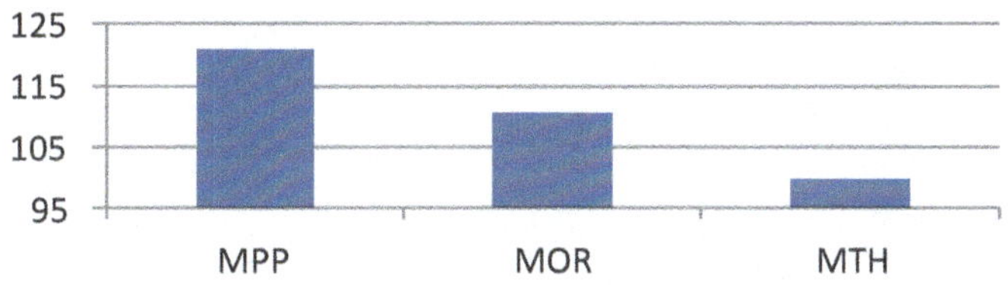

Fig. 15. Thematic framework of papers in OTO proceedings 1990–2010 [9].

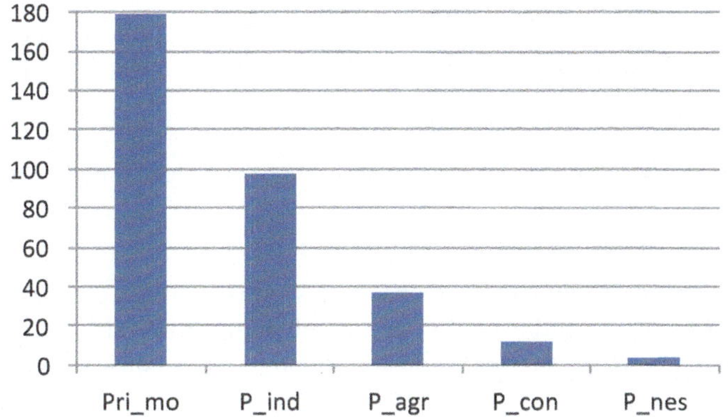

Fig. 16. Papers in OTO proceedings 1990–2010 according to purpose [9].

In thematic group "Maintenance organization" according purpose most papers cosider Economic aspects of maintenance (Ecn), followed by: Macro (Mac) and Micro (Mic) models on maintenance organization (Orm), Informatics in maintenance (Info) and Safety at work (SaW), see Fig. 17.

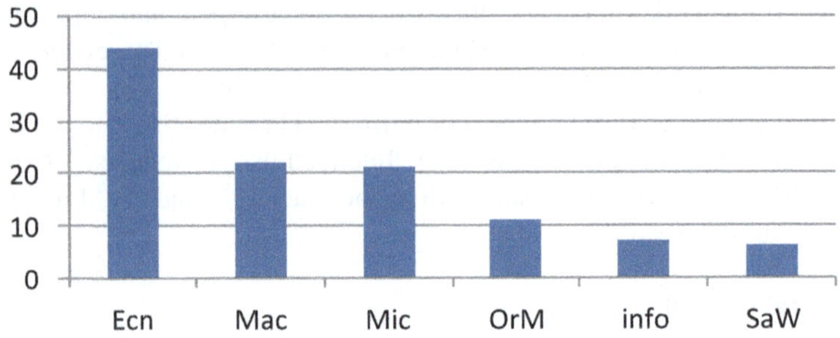

Fig. 17. Papers in OTO proceedings 1990–2010 according to field of papers purpose [9].

In thematic group "Maintenance processes and procedures" most papers are from field of mechanical engineering (Mc_eng), followed by: electrical engineering (El_eng), informatics (Info), models (mod), agriculture (Agr), industry (Ind) and costruction (Con), see Fig. 18.

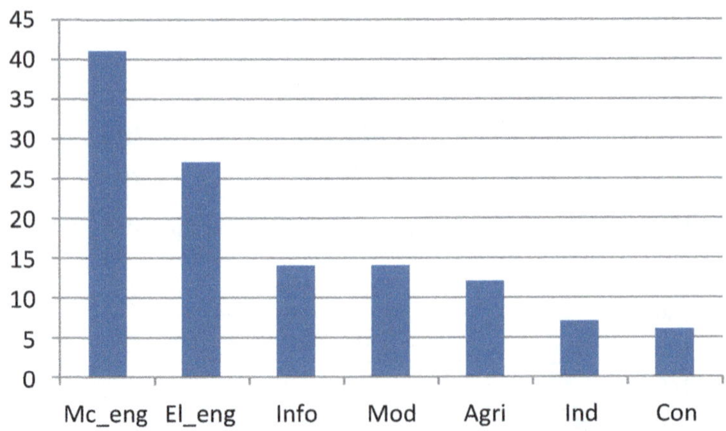

Fig. 18. Papers in field "Processes and maintenance procedures" in OTO proceedings 1990–2010 [9].

In thematic group "Maintenance Technologies" most papers are in field of mechaical engineering (M_eng_t), followed by: electrical engineering (El_en_t), informatics (Info_t), Chemistry (Che_t), food industry (Food_t), and construction (Con_t), see Fig. 19.

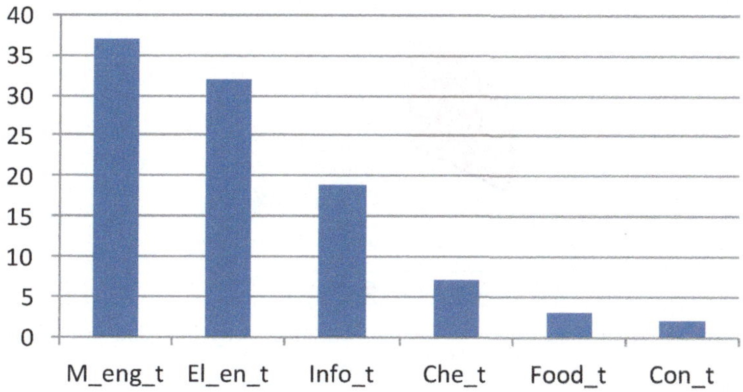

Fig. 19. Papers in "Maintenance Technologies" in OTO proceedings 1990–2010 by technology [9].

Analysis of papers according to area of purpose based on five sets of intended use of paper: "In principle models" (81%), followed by groups: "Intended use for indutry", "Intended use for agriculture", "Intended use for construction", and "Intended use for the non-economic sector". Basic analysis of purpose shows in Fig. 20; a detailed analysis is published in the book [9].

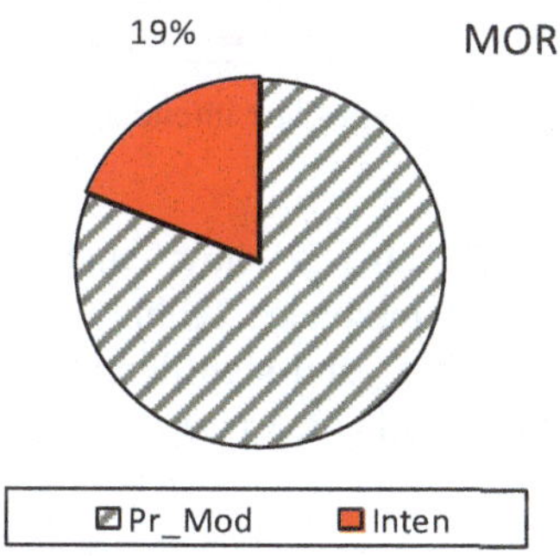

Fig. 20. Papers in "Maintenance Technologies" by intended in OTO proceedings 1990–2010 [9].

The category "Principle Models" in purpose of paper means that this paper discusses professional issues in this area and that this article is in principle useful and applicable in all sectors of maintenance - but in all other sectors of human activity – e.g. from fisheries, firefighting to activities to popular music). Namely, we want to emphasize that: (a) in such a category of works there is too much retelling of other people's works and empty dosing, and what knowledge can be found in hundreds of other places (books, magazines, Internet), or (b) that there is little works related to the specific maintenance practice of a specific business entity. In other words - it would be far more useful for all participants in OTO conferences and the professional public if most of these papers (content) were presented on example of a specific business entity; theoretical and principled considerations are certainly needed, but not like this - by retelling other people's models - and in a large number of papers.

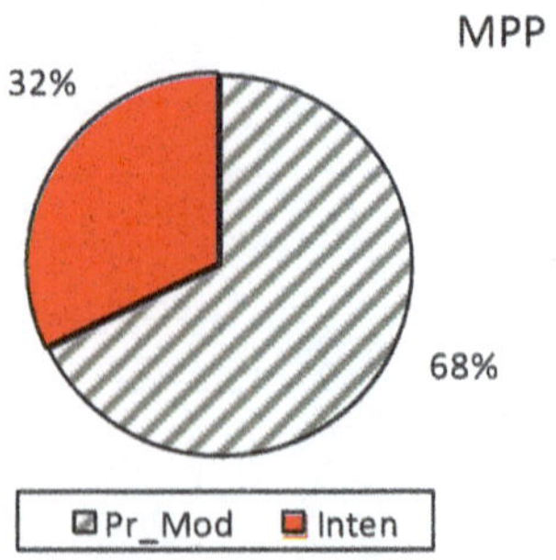

Fig. 21. Papers in "Maintenance processes and procedures" by intended in OTO proceedings 1990–2010 [9].

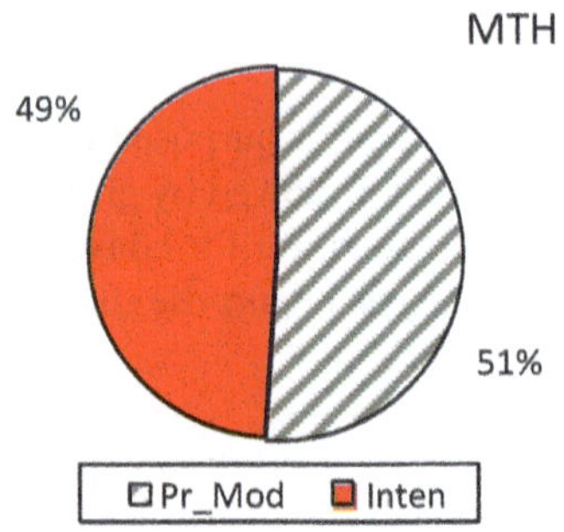

Fig. 22. Papers in "Maintenance Technologies" by intended in OTO proceedings 1990–2010 [9].

A comparison of Figs. 20, 21, and 22 shows that in the areas of "Maintenance Processes and Procedures" and "Maintenance Technologies" there are far fewer papers in the category of "Principle Models" and significantly more in the field of concrete application in industry, agriculture and construction.

3.2 Period 2011–2021

Thematic framework in this period have similar structure; most papers were published in group "Maintenance processes and procedures" (MPP), followed "Maintenance organization" (MOR) and "Maintenance technologies" (MTH), see Fig. 23.

Major changes in the structure of papers published at OTO conferences in this period occurred in terms of content, namely: a) the topics covered and b) the purpose of the papers. Papers in the group of "principle models" are much less represented, i.e. over 90% of papers are considered specific issues in the field of specific organizational and technical procedures and technology (see Fig. 24).

Also engineering areas (civil engineering, energy, informatics, machine and electrical engineering, computing etc.) are far more represented (see Fig. 25).

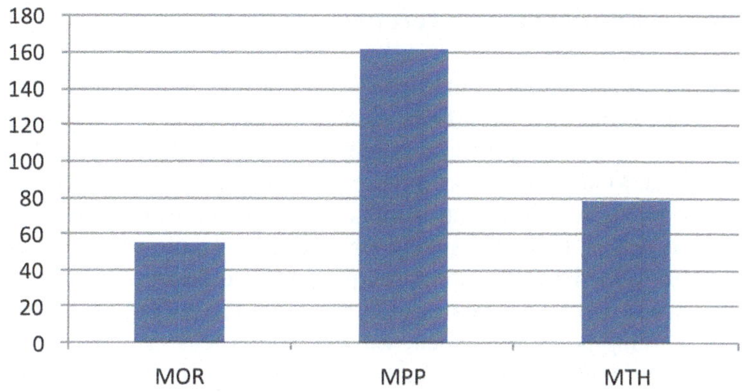

Fig. 23. Thematic framework of papers in OTO proceedings 2011–2021 [31–40].

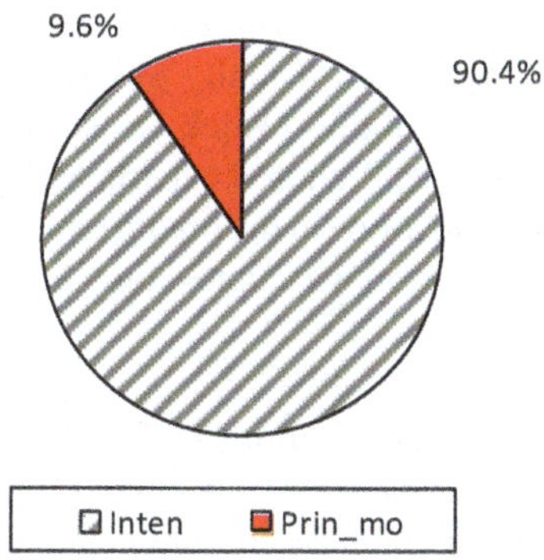

Fig. 24. Thematic framework of papers in OTO proceedings 2011–2021 [31–40].

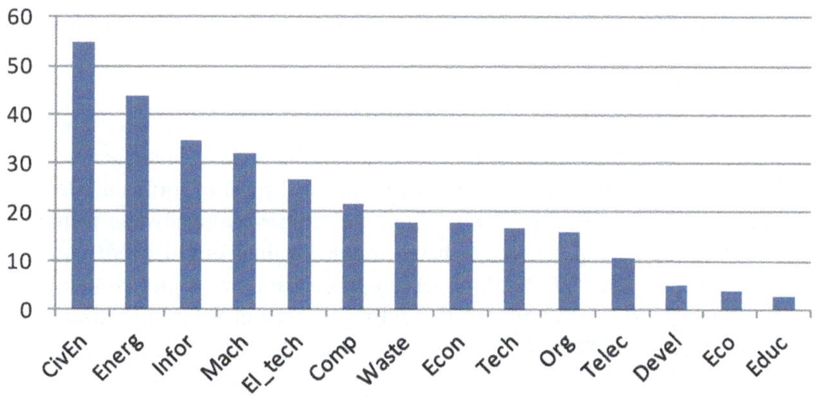

Fig. 25. Thematic area framework of papers in OTO proceedings 2011–2021 [31–40].

4 Organization of OTO Conference

The organization of a scientific conference, as well as the publication of proceedings, is complex projects that involve a large number of executors and require significant financial resources that exceed the capabilities of a non-profit professional association (such as the Osijek Maintenance Society) especially in the war and postwar years.

However, with the self-organization and correct commitment of dozens of experts from all over the region, these demanding jobs are always realized with quality and on time; starting from the first call, conference program and posters, through the holding of the conference to the UDC classification and technical editing of the proceedings, thousands of working hours were spent without paying fees. Neither the authors nor the reviewers were paid. Conference halls and printing of proceedings were paid by the sponsors of the conferences - large companies, chambers of commerce or higher education institutions from the region.

The first 15 conferences were organized by the Osijek Maintenance Association (OMA) with the co-organization of the Faculty of Electrical Engineering Osijek (EEF). The organizers and co-organizers of later OTO conferences were: Faculty of Agriculture Osijek (FAO), Polytechnic Požega (PP), Faculty of Civil Engineering Osijek (FCEO) and Panon, think tank for strategic studies Osijek - as shown in Table 5.

Over the past thirty years, several hundred experts and professors have volunteered and enthusiastically performed functions in the organization of OTO conferences - many of them for ten years. On this occasion we cannot state their names - this is published in Book 1 (see Fig. 26).

- Members of scientific and program committee of the OTO conference were profesors of Osijek faculties until 2015, when OTO became international, and in addtion to scientists from Croatia, members of these committees were also scientists from Bosnia and Herzegovina, Hungary, Slovenia and Serbia.
- The members of the Organizing Committee were from: OMA, Osijek faculties and representatives of sponsors - large industrial companies.
- The editors of the proceedings were professors from Osijek faculties.
- Promotional and other organizational activities were carried out by employees of the Faculty of Electrical Engineering Osijek.

On average, over a hundred participants attended OTO conferences; within each meeting participants visited production processes in the industrial system - host of the meeting, and with each meeting was organized an exhibition of tools and devices used in the maintenance of industrial plants. The proceedings were always printed before the OTO conference - so participants had a book so they could more easily follow authors'presentations. Until 2015, all papers in proceedings were in Croatian, and from 2016 to 2020, the papers (at choice of the author) were published in either Crotian or English. As of 2021, the OTO proceedings are complete in English.

Table 5. Organizer and Co-organizers of OTO conferences [10–40].

Year	Organizer	Co-organizers		
1990	OMA	EEF	–	–
1992	OMA	EEF	–	–
1993	OMA	EEF	–	–
1994	OMA	EEF	–	–
1995	OMA	EEF	–	–
1996	OMA	EEF	–	–
1997	OMA	EEF	–	–
1998	OMA	EEF	–	–
1999	OMA	EEF	–	–
2000	OMA	EEF	–	–
2001	OMA	EEF	–	–
2002	OMA	EEF	–	–
2003	OMA	EEF	–	–
2004	OMA	EEF	–	–
2005	OMA	EEF	–	–
2006	OMA	FAO	–	–
2007	OMA	FAO	PP	
2008	OMA	FAO	–	–
2009	OMA	FAO	–	–
2010	OMA	FAO	–	–
2011	OMA	FAO	–	–
2012	OMA	EEF	FCEO	FAO
2013	OMA	EEF	FCEO	FAO
2014	OMA	EEF	FCEO	FAO
2015	OMA	EEF	FCEO	FAO
2016	OMA	EEF	FCEO	FAO
2017	EEF	–	–	–
2018	EEF	–	–	–
2019	PANON	EEF	FCEO	–
2020	PANON	EEF	FCEO	–
2021	PANON	EEF	FCEO	–

Fig. 26. OTO 2020 proceedings.

5 Conclusion

The analysis organization of OTO conferences in period 1990–2021 and analysis of published papers in OTO proceedings indicate successful work of a large group of industry experts from Slavonia region and professors from Osijek faculties in first two decades, and in the third decade and experts from several neighboring countries. Dspite the war aggression against Croatia and the difficult transition and post-war coditions, the OTO conference has been held continuously for the past 30 years.

Thirty OTO conferences were organized in 12 different cities, with sponsorship of about thirty companies and several higher education institutions from the region of Slavonia.

At thirty OTO conferences, a total of 1321 authors presented (published in the proceedings) over 630 professional and scientific papers on the following topics: (1) Maintenance organization, (2) Maintenance processes and procedures and (3) Tecnologies used in maintenance.

Most of papers are in the field of technical sciences (mechanical engineering, eletrical engineering, construction, and petrochemical technology), followed papers in field of economics and organization and agronomy. All papers for the OTO confeence were reviewed by two reviewers, and since 2006 the papers have been categorized.

In OTO activities, autopoiesis organization and self-denying and volunteer work of dozens of experts from companies and higher education institutions from the region prevailed.

OTO conferences have contributed to (a) a high level of expertise in the operation and maintenance of industrial plants in Slavonia region, and (b) the development of industrial culture in the region's economy and public services. Thus, a modern form of dissemination of professional technological and organizational knowledge that croses the borders of the Slavonian region was realized.

In 2021, the OTO conference outgrows the regional framework of Croatia and neighboring countries of Southeast Europe and becomes an international European conference.

References

1. Ivanović, M.: Science and regional energy - research on energy development and energy use in Slavonia and Baranja; (in Croatian - Znanost i regionalna energetika – istraživanja o razvoju energetike i korištenju energije u Slavoniji i Baranji); Elektrotehnički fakultet Osijek, Osijek (2006). ISBN 953-6032-50-3
2. Institut za elektroprivredu: Possibility and justification of replacing liquid fuels with alternative domestic fuels, study. Zagreb (1985). (in Croatian - Mogućnost i opravdanost zamjene tekućih goriva alternativnim domaćim gorivima, studija), Zagreb (1985)
3. INA Razvoj i istraživanje: Perspective of energy consumption in the Community of Osijek Municipalities for the period 1980–1985, Study,). (in Croatian – Perspektiva potrošnje energije u Zajednici općina Osijek za razdoblje 1980–1985, Zagreb (1982)
4. Business weekly (Ekonomska politika): special editions - ranking list of the 200 largest companies in Yugoslavia 1980, Belgrade (1980)
5. Business weekly (Ekonomska politika): special editions - ranking list of the 200 largest companies in Yugoslavia 1989, Belgrade (1989)
6. Ivanović, M., Požega, Ž.: Economic Development of The Slavonia and Baranja - contributions for macroeconomic analysis of regional development in Croatia (in Croatian - Ekonomski razvoj Slavonije i Baranje - prilozi za makroekonomsku analizu regionalnog razvoja u RH), 2nd International Conference, Vallis Aurea: Focus on Regional Development, Požega, 3 October 2010. DAAAM International Viena i Veleučilište u Požegi; Procedings, pp. 475–486 (2010). ISBN 978-953-98762-7-0
7. Statistical reports 1376/2009
8. Ivanović, M.: Two decades of professional and scientific contributions to the practice of organization and maintenance technology - OTO 1990 – 2010 (in Croatian - Dva desetljeća stručnih i znanstvenih doprinosa praksi organizacije i tehnologije održavanja - OTO 1990–2010), pp. 7–18, Proceedings OTO 2011, Osijek (2011)
9. Ivanović, M.: Two and a half decades of autopoiesis - scientific-professional conferences of OTO between 1990 and 2015 (in Croatian - Dva i pol desetljeća autopoiezisa – znanstveno-stručni skupovi OTO između 1990 i 2015 godine), Panon, Osijek (2015). ISBN 978-953-7973-08-7
10. Proceedings of the Conference, Organization of Maintenance in New Conditions 1990, Osijek, 4 April 1990
11. Proceedings, Maintenance of Industrial Equipment 1991/1992, Osijek, 12 December 1992
12. Proceedings of the Professional Colloquium Maintenance Computerization, Osijek, 12 February 1993
13. Proceedings of the Professional Colloquium Maintenance Organization and Technology 1993, Osijek, 12 November 1993. ISBN 953-6032-06-6
14. Proceedings of the Professional Colloquium Maintenance Organization and Technology, Đakovo, 3 March 1994
15. Proceedings of the Professional Colloquium Maintenance Organization and Technology Našice, 27 January 1995
16. Proceedings of the Professional Colloquium Maintenance Organization and Technology - OTO 1996, Slavonski Brod, 2 February 1996

17. Proceedings of the Professional Colloquium Maintenance Organization and Technology - OTO 1997, Virovitica, 2 February 1997
18. Proceedings of the Professional Colloquium Maintenance Organization and Technology - OTO 1998, Vukovar, 26–27 March 1998
19. Proceedings of the Professional Colloquium Maintenance Organization and Technology - OTO 1999, Požega, 9 April 1999
20. Proceedings of the Professional Colloquium Maintenance Organization and Technology - OTO 2000, Županja, 14 April 2000
21. Proceedings of the Professional Colloquium Maintenance Organization and Technology - OTO 2001, Kneževo, 20 April 2001
22. Proceedings of the professional colloquium Organizacija i tehnologija održavanja - OTO 2002, Osijek, 19 April 2002
23. Proceedings of the Professional Colloquium Maintenance Organization and Technology - OTO 2003, Kutina, 24–25 April 2003
24. Proceedings of the Professional Colloquium Maintenance Organization and Technology - OTO 2004, Osijek, 16 April 2004
25. Proceedings of the Professional Colloquium Maintenance Organization and Technology - OTO 2005, Đakovo, 14–15 April 2005
26. Proceedings of the Scientific-Professional Conference Maintenance Organization and Technology - OTO 2006, Vinkovci, 7 April 2006
27. Proceedings of the Scientific-Professional Conference Maintenance Organization and Technology - OTO 2007, Požega, 12 April 2007
28. Proceedings of the Scientific-Professional Conference Maintenance Organization and Technology - OTO 2008, Osijek, 9 May 2008
29. Proceedings of the Scientific-Professional Conference Maintenance Organization and Technology - OTO 2009, Osijek, 8 May 2009
30. Proceedings of the XIX. Scientific-Professional Conference Maintenance Organization and Technology - OTO 2010, Kutina, 23 April 2010. ISBN 978-953-6331-82-6
31. Proceedings of the XX. Scientific-Professional Conference Maintenance Organization and Technology - OTO 2011, Osijek, 6 May 2011. ISBN 978-953-6032-76-1
32. Proceedings of the XXI. Scientific-Professional Conference Maintenance Organization and Technology - OTO 2012, Županja, 20 April 2012. ISBN 978-953-6032-77-8
33. Proceedings of the XXII. Scientific-Professional Conference Maintenance Organization and Technology - OTO 2013, Osijek, 25–26 April 2013. ISBN 978-953-6032-79-2
34. Proceedings of the XXIII Scientific-Professional Conference Maintenance Organization and Technology - OTO 2014, Požega, 24–25 April 2014. ISBN 978-953-6032-84-6
35. Proceedings of the XXIV. Scientific-Professional Conference Maintenance Organization and Technology - OTO 2015, Donji Miholjac, 17 April 2015. ISBN 978-953-7973-07-0
36. Proceedings of the XXV. Scientific-Professional Conference Maintenance Organization and Technology - OTO 2016, Osijek, 22 April 2016. ISBN 978-953-7973-13-1
37. Proceedings of the XXVI. Scientific-Professional Conference "Maintenance Organization and Technology - OTO 2017", Osijek, 22 April 2017. ISBN 978-953-6032-92-1
38. Proceedings of the XXVII. Scientific-Professional Conference Maintenance Organization and Technology - OTO 2018, Osijek, 22 April 2018. ISBN 978-953-6032-97-6
39. Proceedings of the XXVIII. Scientific-Professional Conference Maintenance Organization and Technology - OTO 2019, Vinkovci, 12 December 2019. ISSN 2706-4131
40. Proceedings of the XIX. Scientific-Professional Conference Maintenance Organization and Technology - OTO 2020, Osijek, 12 December 2020. ISSN 2706-4131

Design and Optimization of Lattice Supporting Structure by Direct Optimization Method

Mirko Karakašić[1(✉)], Karlo Vrančić[2], Ivan Grgić[1], and Hrvoje Glavaš[3]

[1] Mechanical Engineering Faculty, University of Slavonski Brod, Slavonski Brod, Croatia
mirko.karakasic@unisb.hr
[2] University North, Varaždin, Croatia
[3] Faculty of Electrical Engineering, Computer Science and Information Technology, Josip Juraj Strossmayer University of Osijek, Osijek, Croatia

Abstract. In the design process, new and creative design solutions have a goal to increase competitiveness compared to similar design solutions from other manufacturers. Material saving, easier montage and transport, by the mass reduction of the construction in design process, are a significant factors, which affect on the lower price and easier product maintenance, especially for steel made lattice prefabricated constructions. With the development of the numerical methods and computer tools, the optimization process, in the modern design process, has an increasingly important role. By clearly defining the objective function, design optimization process, makes design solutions significantly more competitive. This paper presents design process of a floating bed frame. After the design of the initial solution, this solution was optimized, in order to reduce the mass, respectively material saving. Optimization process was performed by Direct Optimization Method (DOM). Total mass of the structure, compared to the initial design solution, was reduced by 70%. According to the results obtained by the optimization process, a design solution of the floating bed frame was made.

Keywords: Design process · Optimization process · Direct Optimization Method (DOM) · Floating bed · Maintenance process

1 Introduction

The task of the designer in the process of a new product design consists of the requirements mapping [1] into functional and physical structures that define a new product prototype [2]. During product development process, designers encounter limitations such as physical, ergonomic, time, production, and product cost [3]. Using different optimization methods, it is possible to influence on the numerous constraints. Thus, mass reduction affects on the saving of material required to make a design solution, or affects on the cost of the product. Also, mass reduction has a direct impact on a certain design properties of the product.

Modern design process uses application of various optimization methods in order to solve the set objective function. These methods are part of the optimal design process [4, 5].

H. Glavaš et al. (Eds.): OTO 2021, LNNS 369, pp. 21–34, 2022.
https://doi.org/10.1007/978-3-030-92851-3_2

In the process of optimal design, it is necessary to formulate design process using a mathematical model. Depending on the type of numerical method, mathematical models have a certain peculiarities. What they have in common is the definition of the one or more objective functions (multi-objective optimization problem) [6], the constraint functions and the variables [4]. Application of Direct Optimization Method (DOM), especially Response Surface Method (RSM), is of great importance in the modern process of design optimization. RSM is a combination of statical and mathematical techniques utilized to develop, improve and optimize processes of industrial production [7, 8]. This method is using in the construction of complex systems such as individual parts of aircraft subsystems, such as landing gear systems [9] or planet lander [10, 11]. RSM method, by 3D diagrams, allows representation of the interdependence of independent and dependent construction variables (parameters).

This paper presents the process of design and optimization process of a floating bed frame. The specificity of the construction is reflecting in the request derived from the requirement list. This request defines a structure that has no vertical supports (legs). In practice, there are numerous versions of the floating bed that use a vertical support placed on the central part of the structure. This design completely excludes a vertical support (see Fig. 1). Development of this conceptual variant, has several goals: material saving, mass reduction, easier montage and maintenance, and reduction of positions in the hierarchical structure of product. Position reduction and transferring of their functions to the other positions of the product hierarchical structure, is an important factor in a design for maintenance and design for assembly process [12, 13].

Fig. 1. Conceptual solution of floating bed [14].

The construction of the floating bed frame consists of a steel frame, which, depending on the design, is bolted at one of its ends to a fixed support (wall). Conceptual design of the structure and its fixture is presented on Fig. 2.

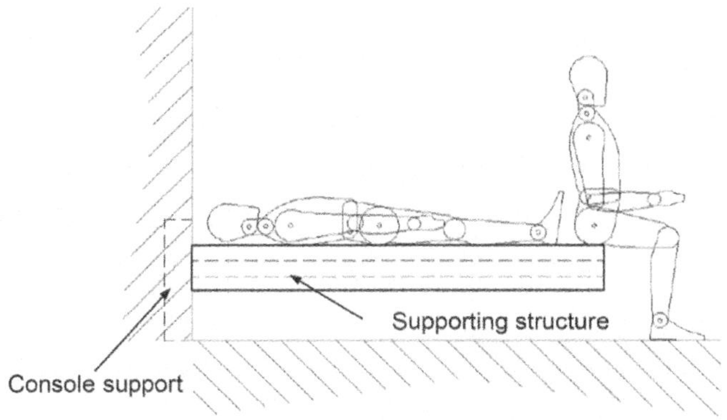

Fig. 2. Conceptual solution of console support.

It is important to abstract some of the more significant requests defined in the requirement list. These requests are: minimum mass, structure stiffness (due to the load) needs to ensure a displacement reduction, structure shape needs to enable installation of side panels and floors, frame structure needs to be in two parts and structure needs to have following dimensions 1800 × 2000 mm. In order to simplify maintenance, montage and transport, since it is a lattice prefabricated structure, mass reduction is an important requirement which is therefore selected as objective function of the optimization process of the floating bed lattice structure.

2 Design and Modeling of Floating Bed Structure

Structure of the floating bed consists of two basic positions: supporting structure and console support (see Fig. 3). Supporting structure is made by welding steel pipe profiles 80 × 40 × 3,2 mm. Material is steel S235 (norm EN 10027-1). Structure is lattice. It consists of one left and one right supporting structure. These constructions are connected by a bolt connection. Connection is made with six bolts M16 × 85 (ISO 7412).

Function of the console support is reflecting in taking the load of the supporting structure (see Fig. 3). Console support, by vertical supports, rests on the base and is placed on the wall. Supporting structure is connected by a bolt connection to the console support. The structure is welded and reinforced by horizontal reinforcements.

Using eight M16 bolts and four M16 × 100 bolts (ISO 7412), console support is fixed to the wall and base (see Fig. 4).

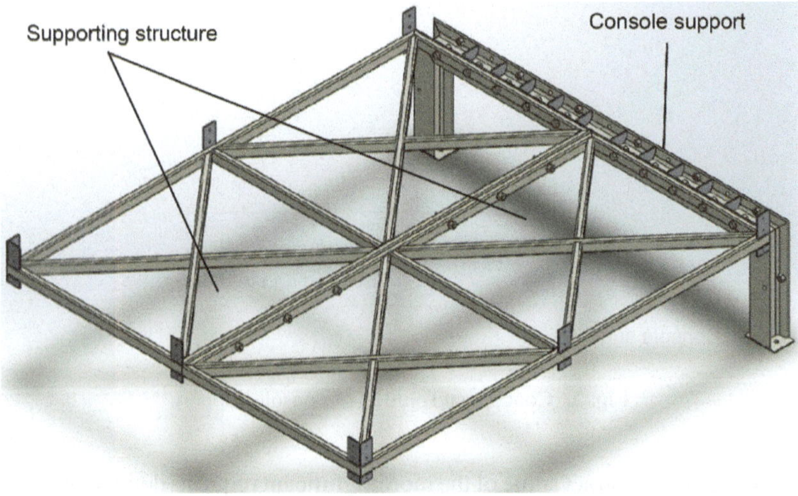

Fig. 3. Lattice structure of bed.

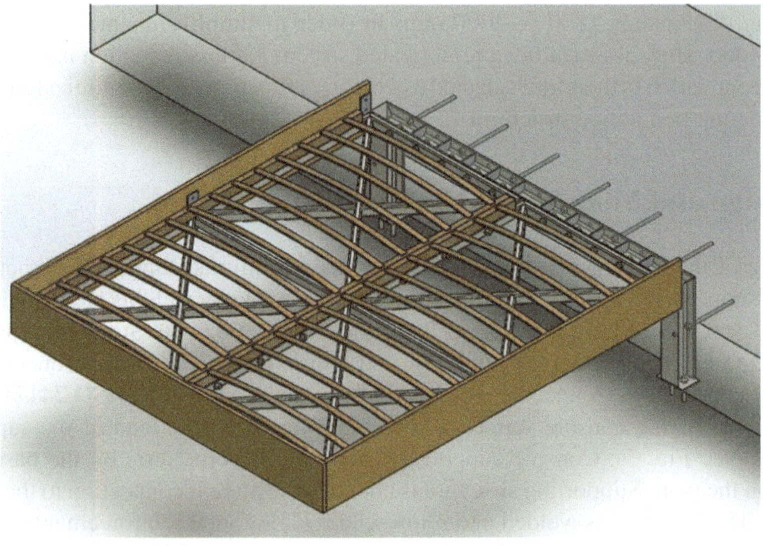

Fig. 4. CAD model of floating bed structure.

3 Initial Numerical Calculation of the Supporting Structure

Initial calculation of the structure includes calculation of the stress and deformation (deflection) of the supporting structure. Steel S235 with the following properties was selected for the material: density $\rho = 7{,}850\ \text{kg/m}^3$, Young's Modulus of Elasticity $E =$ 210 GPa, Poisson's Ratio $\nu = 0{,}3$, Shear Modulus $G = 81$ GPa, Tensile Yield Strength R_e = 250 MPa and Tensile Ultimate Strength $R_m = 460$ MPa. Tetrahedral finite elements with mid-nodes were selected for the mesh, size 15 mm. At the joints of the lattice structure, the finite elements were reduced to 5 mm. Total number of finite elements is 109752, and the total number of nodes is 279404. Mass of the supporting structure is 113,9 kg. Load amount on the structure is 2000 N. Amount of the load arises from the request that the bed needs to be able to carry two adult persons whose individual mass is 100 kg. The area on which the load acts is 3,6 m^2. Therefore, the pressure on the supporting structure is 55,55 Pa (Fig. 5).

Supporting structure is fixed to the surface, which is bolted to the console support. After numerical calculation was performed, the following results were obtained: $a =$ 0,54982 mm (Total Deformation, see Fig. 6) and $\sigma_{\text{ekv}} = 33{,}884$ MPa (Equivalent Stress, see Fig. 7).

Results of the numerical analysis, presented by the initial calculation, are satisfactory, since Maximum Equivalent Stress, $\sigma_{\text{s,dop}} = 160$ MPa.

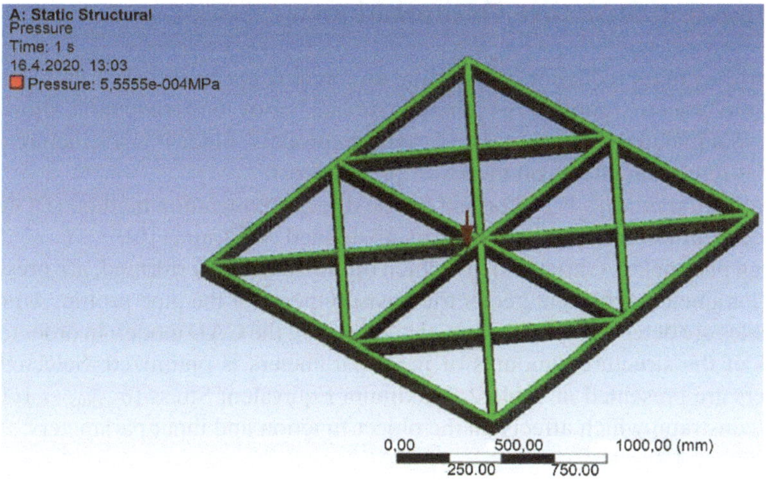

Fig. 5. Load on the supporting structure.

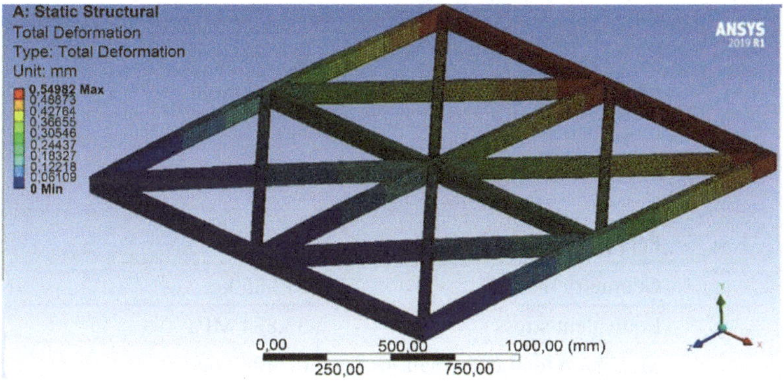

Fig. 6. Total deformation of the supporting structure.

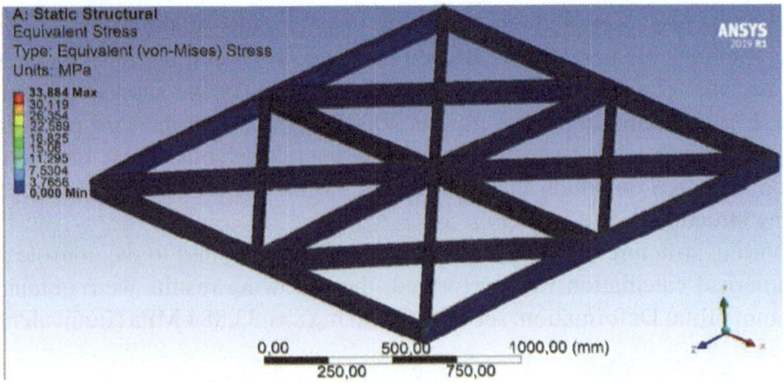

Fig. 7. Equivalent stress of the supporting structure.

4 Supporting Structure Optimization

In objective of mass reduction, supporting structure is optimized. Therefore, the same name of the objective function is defined. Advantage of mass reduction allows easier structure manipulation in transport and montage process. Since it is a structure made of steel, mass reduction affects on the lower product cost.

Optimization process is performed by the direct optimization method (DOM). This method determines multiple objectives using defined constraints [6].

Design parameters (variables), by which optimization is performed, are presented in Table 1. Parameters represent geometric measurements of the pipe profile. Amounts of the variables are determined by the initial geometry of the CAD model. In order to reduce the mass of the structure, amounts of input parameters is optimized. Selected output parameters are presented in Table 2. Maximum Equivalent Stress ($\sigma_{s,dop} = 160$ MPa), presents constraint which affects on the object function and input parameters.

Table 1. Input variables.

Profile	Parameter name	Amount
P1	DS_Wall_Thickness	3,2 mm
P2	DS_Tube_Width	40 mm
P3	DS_Tube_Height	80 mm

Table 2. Output variables.

Parameter name	Amount
Geometry mass	113,85 kg
Equivalent stress	33,884 MPa
Maximum total deformation	0,54982 mm

Amounts of lower and upper value of the pipe profile input variables presented are in Table 3.

Table 3. Interval values for input variables.

Input parameters	Lower bound, mm	Upper bound, mm
DS_Wall_Thickness	1	3
DS_Tube_Height	20	80
DS_Tube_Width	10	40

During optimization process by DOM in the software package Ansys, program increases interval values for every input parameter from lower to upper value. This process presents iterative optimization process.

Changes in the values of the input parameters, are presented in Figs. 8, 9 and 10. By changing values of the input parameters, value of the objective function (mass reduction of the support structure) changes. This change is presented on Fig. 11.

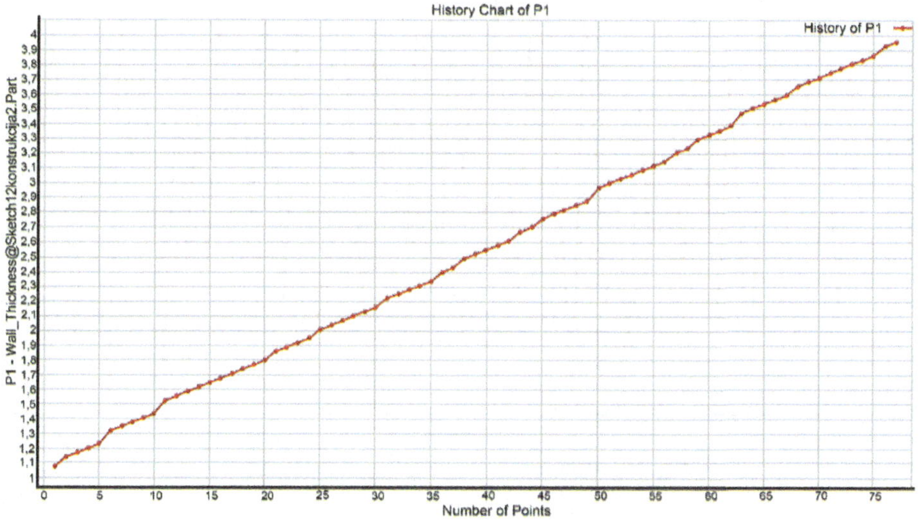

Fig. 8. Parameter value change – wall thickness.

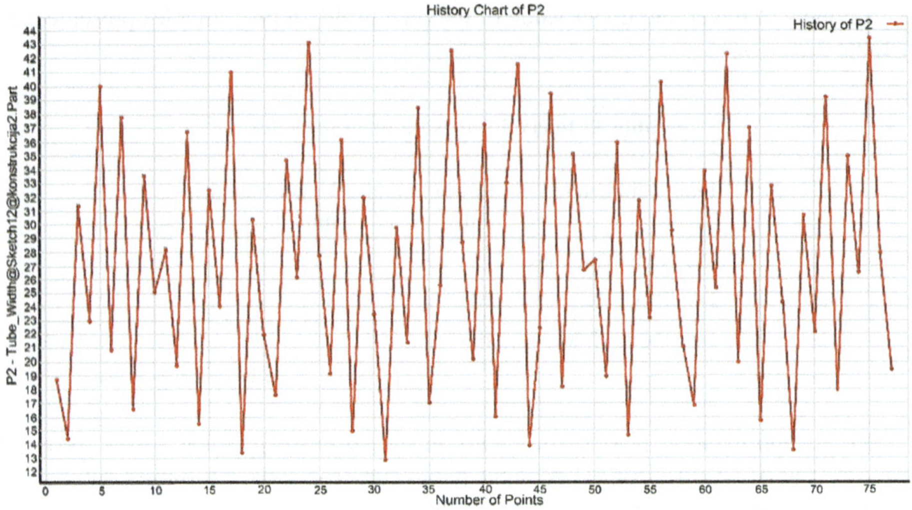

Fig. 9. Parameter value change – tube width.

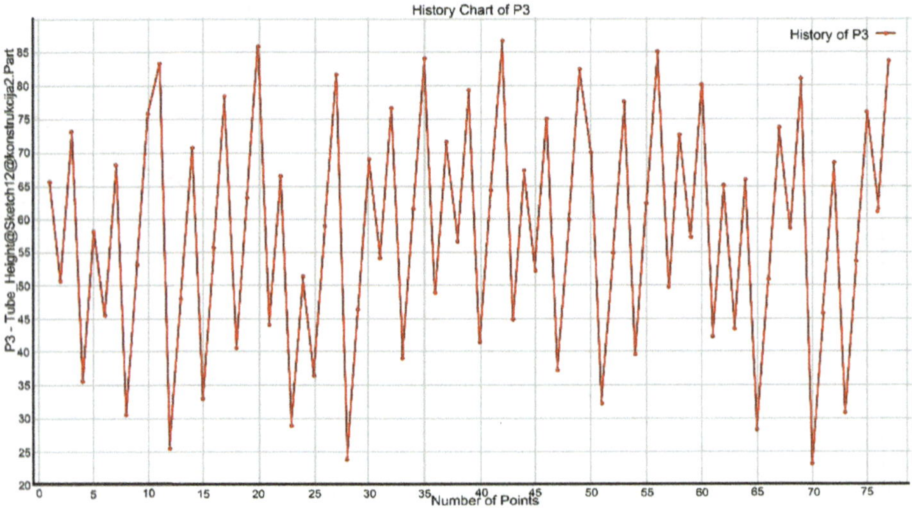

Fig. 10. Parameter value change – tube height.

On Fig. 12, also is presented change of the Maximum Equivalent Stress. Interdependence of the mass parameter in the comparison with Maximum Equivalent Stress is presented on Fig. 13. From the picture it can be concluded (despite individual cases to the contrary) that Equivalent Stress decreases with mass increasing.

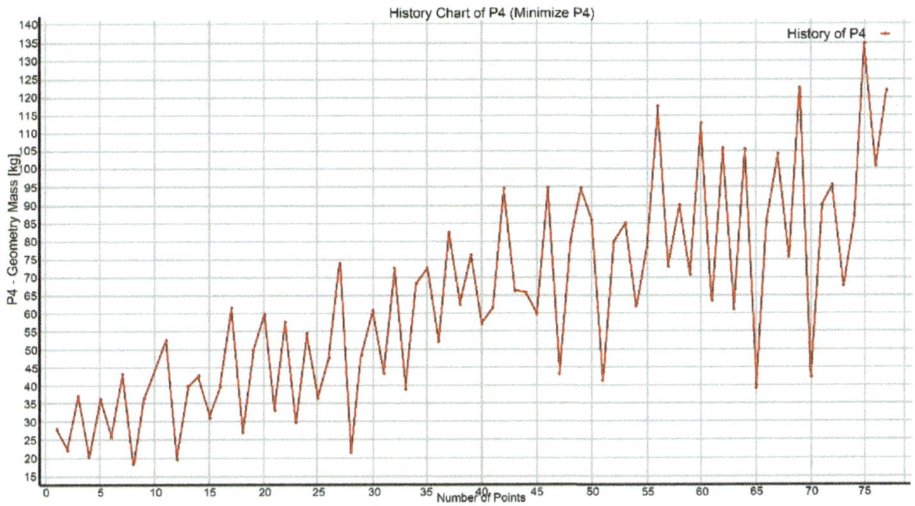

Fig. 11. Value change of objective function – mass reduction.

Interdependence of the output parameters (Geometry Mass, Maximum Equivalent Stress and Maximum Total Deformation) is presented on Fig. 14. Mass increasing, reduces Equivalent Stress and Total Deformation.

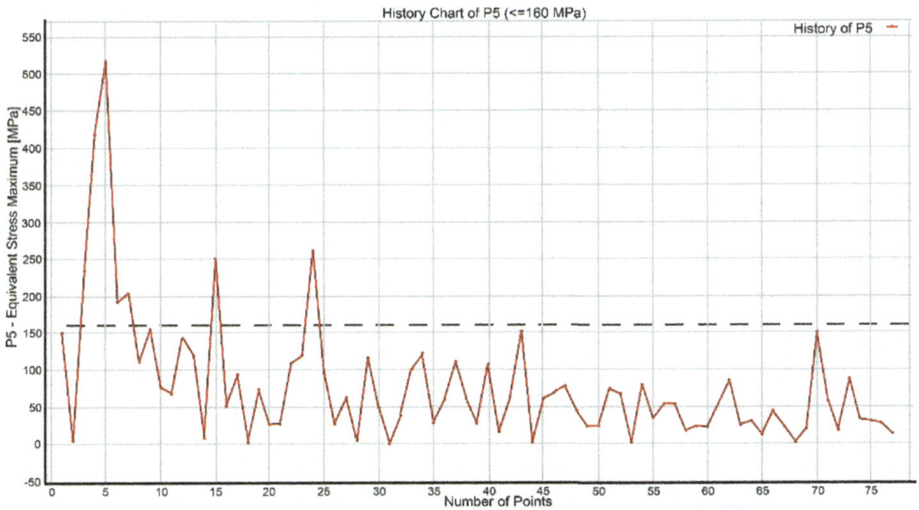

Fig. 12. Value change of constraint – maximum equivalent stress.

Sensitivity of the objective function to the change of input parameters can also be determined from the solution (see Fig. 15). Wall thickness of the tube profile has the greatest influence on the mass of the structure. Width of the tube profile has the least impact on the mass. The most significant impact on the Equivalent Stress and Total

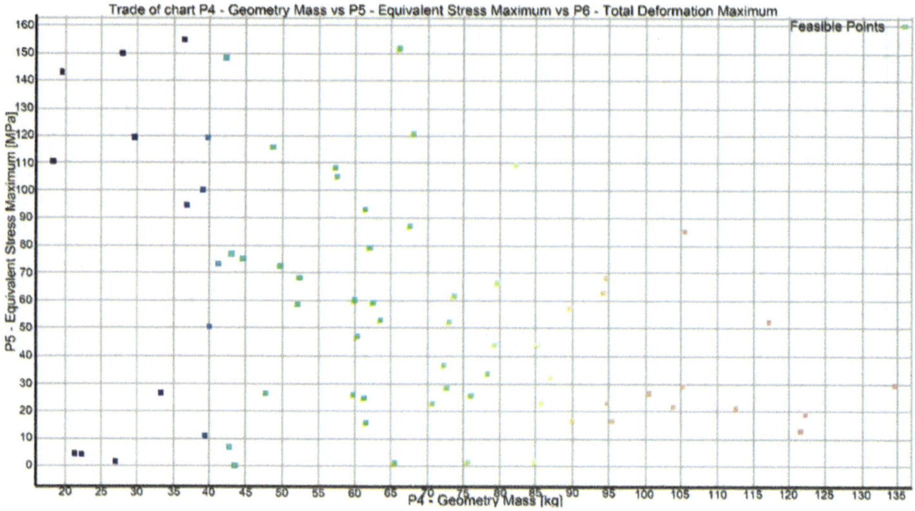

Fig. 13. Optimization process – 2D view.

Deformation has width of the tube profile. Wall thickness and profile height have less impact on the Equivalent Stress and Total Deformation (see Fig. 15).

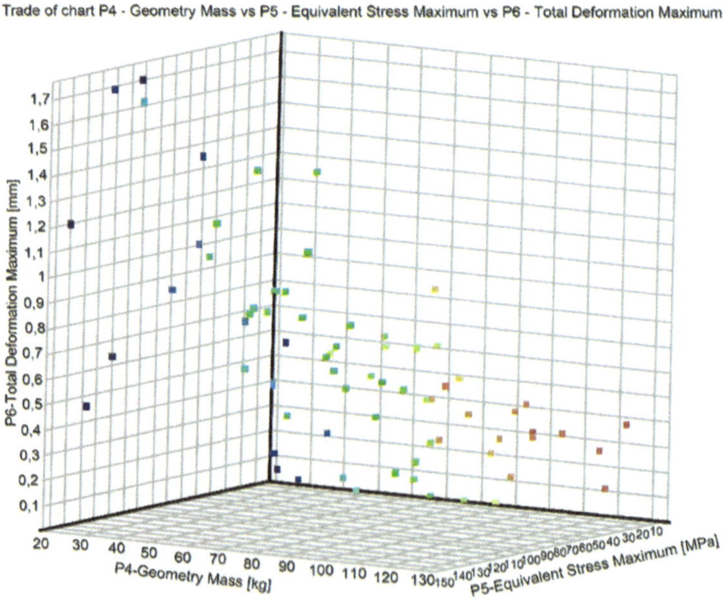

Fig. 14. Solution of the optimization process – 3D view.

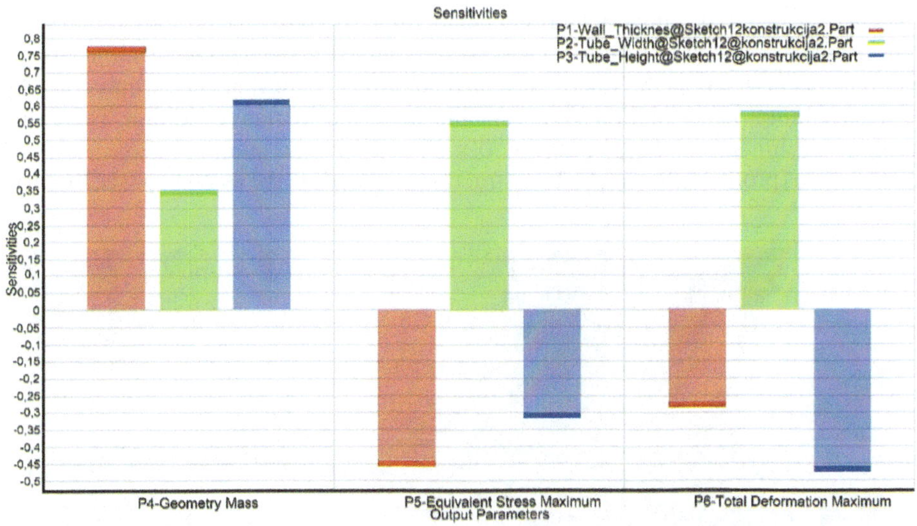

Fig. 15. Sensitivity of the objective function on the input parameters.

After the using DOM and analysis of the results shown in Figs. 8, 9, 10, 11, 12, 13, 14 and 15 three the best solutions were selected. These solutions are presented in Table 4. Solution 1 best solves the objective function, since the mass of the structure is 18,295 kg (see Table 4).

Table 4. Selection of the best solutions by the DOM.

Variables	Input variables			Output variables		
Solutions	Wall thickness (mm)	Tube width (mm)	Tube height (mm)	Geometry mass (kg)	Maximum equivalent stress (MPa)	Maximum total deformation (mm)
Solution 1	1,375	16,545	30,414	18,295	110,25	0,6326
Solution 2	1,555	19,733	25,377	19,466	142,94	1,1961
Solution 3	1,135	14,42	50,562	22,244	3,9964	0,058159

Problem that arises when choosing this solution is certainly the non-standard dimension of the pipe profile (see Table 4). Second solution has the largest amount of Equivalent Stress. Third solution has the lowest amount of Equivalent Stress and the lowest amount of Total Deformation.

Since the material suppliers supply standard dimensions of pipe profiles, dimensions of the profiles obtained by solution 3 best correspond to the standard dimensions, so this solution was selected as the finally accepted solution. The difference in mass of 4 kg between solution 1 and solution 3 is acceptable due to the previous condition. Therefore, the standard dimension of pipe profiles 50 × 20 × 2 mm was adopted.

Using this profile, a floating bed construction was made in the workshop (see Fig. 16).

Fig. 16. Supporting structure montage.

Supporting structure is made as a lattice structure, and the assembling of individual segments is achieved by welding technology (see Fig. 17). The final appearance of the manufactured floating bed is presented on Fig. 18.

Fig. 17. Supporting structure welding and grinding.

Fig. 18. Floating bed.

5 Conclusion

Using DOM, the optimization process of the supporting structure of the floating bed is presented in this paper. The objective function includes mass reduction of the structure. Optimization of the input variables of the pipe profile was performed, in order to reduce the total mass of the structure.

After optimization, three design solutions were obtained that best solve the objective function. Of these three solutions, the solution that did not best solve the objective function was selected. The reason stems from the fact that the optimized variables with their amounts were closest to the standard values of the dimensions of the pipe profiles. By choosing a standard profile 50 × 20 × 2 mm, the construction of a floating bed was made. The total mass of the structure, made with standard profile, is 34,1 kg. Compared to the optimized solution (solution 3), the mass is increased by 53%.

Since the construction solution with standard profiles is the optimal solution, its mass represents a reduction of mass by 70% compared to the initial construction.

Future development of an optimized design solution, within development of smart furniture, would aim development of a smart bed. Such a solution, by applying movable design elements such as s spindle with a trapezoidal thread, would give the possibility of vertical displacement of the structure upwards or downwards. The application of such a design solution would be useful in small living spaces.

References

1. Srinivasan, V., Chakrabarti, A.: An integrated model of designing. J. Comput. Inf. Sci. Eng. **10**(3), 1–10 (2010)

2. Pastović, M.: Procedure for product development of modular and variant products in the conceptual design. Doctoral thesis, University of Slavonski Brod, Mechanical Engineering Faculty, Slavonski Brod, Croatia (2020)
3. Dörner, D.: Gruppenverhalten im Konstruktionsprozess. VDI – Berichte 1120. VDI Verlag, Düsseldorf (1994)
4. Ravindran, A., Ragsdell, K.M., Reklaitis, G.V.: Engineering Optimization – Methods and Applications, 2nd edn. Wiley, New Jersey (2006)
5. Arora, J.S.: Introduction to Optimum Design, 2nd edn. Elsevier, Amsterdam (2004)
6. Heidari, M., Hosseini, S.V., Parvaz, H.: Modelling and optimization of surface roughness and specific tool wear in milling process. Tech. Gazette **28**(5), 1626–1633 (2021)
7. Myers, R.H., Montgomery, D.C., Anderson-Cook, C.M.: Response Surface Methodology: Process and Product Optimization Using Desgined Experiments. Wiley, New York (2016)
8. Bae, J., Kim, H., Park, S., Kim, K.S., Choi, H.: Parametrization study of electrospun nanofiber including LiCl using response surface methodology (RSM) for water treatment application. Appl. Sci. **10**(20), 1–16 (2020). https://doi.org/10.3390/app10207295
9. Shengyong, G., Xingbo, F., Xiaohui, W.: Parametric analysis on landing gear strut friction of light aircraft for touchdown performance. Appl. Sci. **11**(12), 1–18 (2021). https://doi.org/10.3390/app11125445
10. Ding, Z., Wu, H., Wang, C., Ding, J.: Hierarchical optimization of landing performance for lander with adaptive landing gear. Chinese J. Mech. Eng. **32**(1), 1–12 (2019). https://doi.org/10.1186/s10033-019-0331-0
11. Zheng, G., et al.: Parametric design and analysis on the landing gear of a planet lander using the response surface method. Acta Astronaut. **148**, 225–234 (2018). https://doi.org/10.1016/j.actaastro.2018.05.020
12. Boothroyd, G., Dewhurst, P., Knight, W.A.: Product Design for Manufacture and Assembly, 3rd edn. CRC Press, Boca Raton (2011)
13. Ben-Daya, M., Kumar, U., Prabhakar Murthy, D.N.: Introduction to Maintenance Engineering. Modeling, Optimization and Management, 1st edn. Wiley, Hoboken (2016)
14. Vrančić, K.: Konstrukcijska razrada i optimiranje konstrukcije konzolnog kreveta. Graduate thesis, University North, Varaždin, Croatia (2020)

Design Optimization of Induction Motor for Efficiency and Reliability Improvement

Vasilija Sarac(✉)

Faculty of Electrical Engineering, Goce Delcev University, Stip 2000, North Macedonia
vasilija.sarac@ugd.edu.mk

Abstract. The paper analyses the induction squirrel cage motor type 5AZ 100LA-4 product of company Konačar. As a matter of fact two version of the same motor from Končar are analysed. The old one, known under commercial label 2AZ 155-4 and the new one, labelled as H5AZ100LA-4 (IE3 efficiency class). The computer model for calculating the parameters and operating characteristics is set in software program, based on the exact cross section of laminations of iron core from the producer. The motor is calculated on the base of these data, resulting in parameters and operating characteristics, obtained as output data from the computer model. First modification is realized by changing the number of conductors per stator slot and the length of the air gap. The improvement of efficiency from 78.4% to 78.8% is observed in the first optimized model with significant improvement of power factor from 0.8 to 0.85. Second step in motor optimization was to increase the slot fill factor. Therefore the stator slot was optimized i.e. changed which resulted in bigger slot fill factor and lower core saturation. Besides the stator slot modification, the stator outer diameter was increased as well as the motor length. The final optimized model has efficiency 86.5% and power factor of 0.87 at full load.H5AZ100LA-4 has efficiency of 86.8% and power factor of 0.76 at full load.

Keywords: Induction motor · Optimization · Efficiency · Power factor

1 Introduction

Electrical motors are one of the largest electricity consumers and they are responsible for almost 45% of consumption of electrical energy. The strict efficiency regulation had come into force in EU from January 2017 that impose that all of the motors released in the EU market in power range from 0.75 to 375 kW must have efficiency class IE3. This covers efficiency levels of above 80%. On the other hand, the induction motors due to their principle of operation and construction are limited with respect to achieving high efficiency and power factor. Some of the drawbacks of the asynchronous induction motors are high magnetizing current that implies poor power factor, induction of the current in the rotor winding, usually of squirrel cage type that adds additional copper losses located in the rotor winding, lowering the motor efficiency. Usage of high quality materials such as steel with low specific losses can improve the efficiency but also contributes to the higher production costs. Power factor is another issue that is important for operation of the induction motors. Low power factor is followed by the increase of the line current and consequently with increased copper losses in the stator winding. All

H. Glavaš et al. (Eds.): OTO 2021, LNNS 369, pp. 35–49, 2022.
https://doi.org/10.1007/978-3-030-92851-3_3

losses in the motor contribute to the motor heating and potentially shorten its operational life. Therefore a careful and detailed analysis regarding all design aspects of the induction motor is needed. The analysis concerning distribution of the rotor bars in squirrel cage winding and their influence on the generation of the rotor slot harmonics is described in [1]. Interesting idea regarding efficiency improvement of the induction motor with small design modification such as introduction of stator and rotor slits can be found in [2]. An important construction detail in operation of the motors is the slot fill factor and its optimization [3, 4]. The high slot fill factor of the stator slots can significantly contribute to the higher efficiency of the motor. Yet, the accurate determination of all types of losses including stray losses and friction and windage losses is very important for accurate determination of efficiency [5, 6]. Various authors propose the method of Genetic Algorithms or the improved version of it for finding the optimal values of different design parameters [7, 8]. Each computed optimized model of the motor needs a validation with experiments. The comparison of computed and measured results allows a deep insight on the capability of the mathematical model to accurately reproduce the phenomena, and consequently of predicting the device performances and in particular the power losses under different operating and supply conditions [9, 10]. Following a major constrain in optimization, output power of the motor to remain unchanged, two optimized models of the three phase induction squirrel cage motor in this paper were derived. Starting point is the old version of Končar motor 2AZ 155-4, available for testing in faculty laboratory, along with all necessary constructing details available from motor manufacturer. Based on the data from manufacturer (the motor dimensions, constructional details of steel laminations, winding type etc..), the computer model of the motor is modeled which gives as output, the motor parameters and operational characteristics (BM). Optometric analysis is run by varying the number of conductors per slot (CPS) and air gap length in order to find the best combination of CPS and air gap length that allows the best operating characteristics. The first optimized model (OM1) has improved power factor from 0.8 to 0.85 and slightly improved efficiency from 78.7% to 78.8%. The comparison is done with the starting model (BM). The real Končar motor has power factor of 0.8 and efficiency of 77.8% for 87% of the rated load, measured at the faculty's laboratory. Second optimized model (OM2) is derived by varying the CPS, air gap length along with the modified stator slot that allows improvement of slot fill factor, increased outer diameter of the stator and the motor length. This second and final optimized model, obtained by optometric analysis as well, has achieved considerable improvement of efficiency i.e. 86.5% and power factor 0.87 while the motor output power is kept unchanged-2.2 kW. Obtained results are compared with Končar motor of IE3 class-H5AZ100LA-4. The numerical models, (FEM) models of the motors, were derived as well. The OM2 model has lower flux density in stator yoke, compared to OM1 and BM. Finally the transient characteristics of speed, torque and current are obtained for OM2 model for acceleration with the rated load, showing satisfactory dynamic behavior of the optimized model. The presented optimization methodology allows obtaining relatively simple and cost effective optimized model of the motor with increased efficiency and power factor, reduced total losses and core saturation and consequently longer service life and reliability of the motor. Moreover the proposed model has even larger efficiency at 75% load of 87.3%. At 50% load the model achieves efficiency of 86.6%.

2 Methodology and Results

2.1 Analytical Models

The available data from motor manufacturer allow exact modeling of the motor regarding its dimensions and the materials' properties in the computer program Ansys that allows obtaining the motor parameters and operating characteristics. The obtained output results i.e. parameters and characteristics are compared with the Končar motor, type 2AZ 155-4, i.e. the old type of the motor, as the available data from the manufacturer were for this type of the motor. This comparison presented in Table 1 shows good correspondence between computer model and real motor, thus the starting model (BM) is verified as sufficiently accurate.

Table 1. Comparison between BM and motor type 2AZ 155-4.

Parameter/Characteristic	BM	2AZ 155-4
Output power [W]	2.2	2.2
Rated speed [rpm]	1353	1410
Rated torque [Nm]	15.5	14.9
Current [A]	5.2	5
Power factor [/]	0.8	0.81
Efficiency [%]	78.4	78
Locked-Rotor Torque Ratio [/]	2.55	1.8
Locked-Rotor Current ratio [A]	4.3	4.2
Break-Down Torque Ratio [/]	2.65	2.2
No-Load Stator Phase Current [A]	2.9	2.4
No-Load Power Factor [/]	0.07	0.12

The equations for determining the operating characteristics are presented in (1)–(7). The electromagnetic power P_m, or air gap power, is computed by:

$$P_m = 3 \cdot I_2^2 \cdot R_2/s \tag{1}$$

I_2 is the rotor current, R_2 is the rotor resistance and s is the motor slip. Electromagnetic (air gap) torque is:

$$T_m = P_m/\omega \tag{2}$$

Where ω is synchronous speed in rad/s. The output mechanical torque T_2 is:

$$T_2 = T_m - T_{fw} \tag{3}$$

T_{fw} is the friction and windage torque. The output power is:

$$P_2 = T_2 \cdot \omega_2 \tag{4}$$

where $\omega_2 = \omega(1 - s)$ is the rotor speed in rad/s and s is the motor slip. The input power is:

$$P_1 = P_2 + P_{cu2} + P_{Fe} + P_{cu1} + P_s \tag{5}$$

P_{cu2}, P_{Fe}, P_{cu1} and P_s are rotor copper losses, iron-core losses, stator copper losses and stray loss respectively. The power factor is derived from:

$$\cos\phi = P_1 / m \cdot U_1 \cdot I_1 \tag{6}$$

m is the number of phases. The efficiency is computed by:

$$\eta = \frac{P_2}{P_1} \cdot 100 \tag{7}$$

Optometric analysis is set in order to find the optimal number of conductors per slot and air gap length. For that purpose variation range of CPS is defined to be from 100 to 130 and the air gap length from 0.2 to 0.4 mm. The BM model has 115 CPS and air gap length of 0.3 mm. After optometric analysis is run the optimal number of CPS is found to be 118 with air gap length of 0.2 mm, which gives the best result regarding efficiency and power factor i.e. efficiency of 78.8% and power factor of 0.85. This optimized model which is a result of first optometric analysis is called OM1. More detailed data about this model are presented in Table 2. If we compare the efficiency of BM (Table 1) and efficiency of OM1 a very slight increase can be observed. Therefore the second optometric analysis is run but in the motor model, the stator slot was modified in order to obtained bigger slot fill factor. The good slot fill factor is important for the motor efficiency. The bigger the slot fill factor is, the lower the copper losses are. The fill factor depends of the wire shape. Round shapes can only be used with limited efficiency. Other cross sectional shapes such as a rectangular one can bring improvement, but have limited applications due to its high cost and unsuitability for any but heavy gauge wires [4]. In our case the round magnetic wire is used. The usual slot fill factor is between 60–70%. The limit for the slot fill factor in the computer program is set to be 75%. In this second optometric analysis apart from the modification of the stator slot, the CPS are varied again in interval from 95 to 130, the air gap length from 0.2 mm to 0.4 mm. Additionally, two more variables in OM2 were added, motor length and outer stator diameter i.e. the motor length was varied from 100 mm to 105 mm and the outer stator diameter from 152 mm to 163 mm. The output result of this second optometric analysis is second optimized model OM2 with 92 conductors per slot, the length of air gap 0.2 mm, stator outer diameter of 163 mm and motor length of 105 mm. Data are presented in Table 2. Considerable improvement of efficiency is observed from 78.8% to 86.5% and power factor of 0.87. The slot fill factor was increased from 62.8% at OM1 to 71.3% at OM2. The data of the last optimized model were compared with the latest version of the Končar motor, type H5AZ 100LA-4 (IE3), in order to verify the second optimized model [11]. The comparisons between OM1, OM2, H5AZ 100LA-4 (IE3) as well as the newer version of Končar model 5AZ 100LA-4 (IE1 class) are presented in Table 2.

Analytical calculation of the flux density in the constructing parts of the motor was done as well. Obtained results for BM, OM1 and OM2 are presented in Table 3.

Table 2. Comparison between OM1, 2AZ 155-4 (IE1), OM2 and H2AZ 155-4 (IE3).

Parameter/Characteristic	OM1	5AZ 155-4 (IE1)	OM2	H2AZ 155-4 (IE3)
Output power [kW]	2.2	2.2	2.2	2.2
Rated speed [rpm]	1346	1410	1418	1445
Rated torque [Nm]	15.6	14.9	14.8	14.5
Current [A]	4.9	4.8	4.4	4.8
Power factor [/]	0.85	0.83	0.87	0.76
Efficiency full load [%]	78.8	79.7	86.5	86.7
Efficiency 75% load [%]	81.7	/	87.3	86.3
Efficiency 50% load [%]	82.3	/	86.6	86
Locked-Rotor Torque Ratio [/]	2.55	2.2	4.4	3.5
Locked-Rotor Current Ratio [A]	4.3	5.2	7.9	7.1
Break-Down Torque Ratio [/]	2.65	2.7	4.6	3.8
CPS [/]	118	/	92	/
Air gap length [mm]	0.2	/	0.2	/
Motor length [mm]	100	/	105	/
Stator outer diameter [mm]	152	/	163	/
Slot fill factor [/]	62.8	/	71.27	/
Stator copper losses [W]	252.6	/	125.56	/
Rotor copper losses [W]	253.31	/	128.16	/
Iron core losses [W]	26.69	/	28.6	/
Frictional &windage losses [W]	20	/	22.25	/
Stray losses [W]	39.6	/	39.6	/

Table 3. Comparison between BM, OM1 and OM2.

Flux density	BM	OM1	OM2
Stator teeth flux density [T]	1.3	1.26	1.63
Rotor teeth flux density [T]	1.35	1.31	1.6
Stator yoke flux density [T]	2.03	1.98	1.26
Rotor yoke flux density [T]	0.86	0.84	1.12
Air gap flux density [T]	0.56	0.55	0.67

From the presented results in Table 3 it is evident that the flux density in stator yoke of BM and OM1 is high which causes the core saturation and machine heating. The increase of the diameter of the stator has decreased the flux density in stator yoke and the machine heating. In terms of the flux density distribution the OM2 model is designed better than previous models of the motor. The complete distribution of the flux density in the machine cross section will be presented in the numerical models and calculated by the aid of Finite Elements Method (FEM). FEM models are created in Ansys program as well. The program allows besides analytical calculation of motor models and their optometric analysis also creation of FEM models and calculation of magnetic flux density inside the models. The cross section of BM and OM2 is presented in Fig. 1.

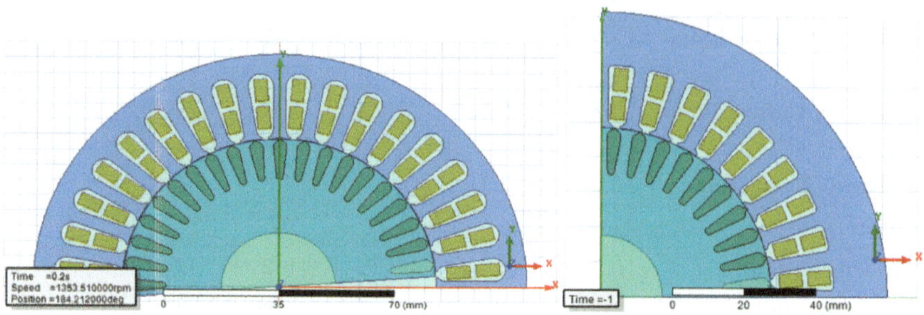

(a) BM (b) OM1

Fig. 1. Cross section of motor models

In Fig. 2 is presented efficiency of BM, OM1 and OM2 versus output power. The key points of efficiency for 100%, 75% and 50% load are presented for OM2. The Figures should support the data in Table 2 and more clearly illustrate the operation of all three models. In Fig. 3 is presented the power factor versus output power for all three models. Finally the output torque versus speed is presented in Fig. 4. Increased efficiency in OM2 model compared to BM and OM1 is a result of the increased fill factor, due to a slot modification, the increase of power factor and consequently the decrease of the current and the copper losses. Yet, it should be noted that the air gap length in both optimized models is decreased compared to the original motor from 0.3 mm to 0.2 mm. Decreased air gap increases the power factor and efficiency but worsens the heat dissipation and motor cooling. Initially, the length of the air gap was calculated according to [12]:

$$\delta = (0.1 + 0.012 \cdot \sqrt[3]{P_n}) \cdot 10^{-3}\text{m} \approx 0.25 \cdot 10^{-3}\text{m} \tag{8}$$

During optometric analysis it is chosen to vary the air gap length in interval (0.2–0.3) mm in steps of 0.1 mm. As the target of optimization is efficiency the length of the air gap was reduced in the optimized model to 0.2 mm. The OM2 model was calculated also for the air gap length of 0.25 mm and satisfactory results of efficiency and power factor at rated load were obtained i.e. 86.3% for efficiency and 0.85 for the power factor. Since several variables are varied within certain limits simultaneously it is very useful to run optometric analysis in order to determine the best combination of all variables which gives the best performances of the motor. The increased efficiency of OM2 comes with the increased weight of the motor and the material consumption. Similar conclusion can be reached for Končar motors where IE3 class of the motor 2AZ 155–4 has increased weight of almost 5 kg compared to IE1 class of the same motor [11]. The OM2 model shows very good operating characteristics also at partial load (Table 2).

One of the important factors for good efficiency is good slot fill factor. In all optimized motor models, the wire diameter is automatically adjusted in order not to exceed the limited slot fill factor of 75%, set in the computer program. Yet, the operation of the motor highly depends on many design details such us good slot fill factor, air gap length,

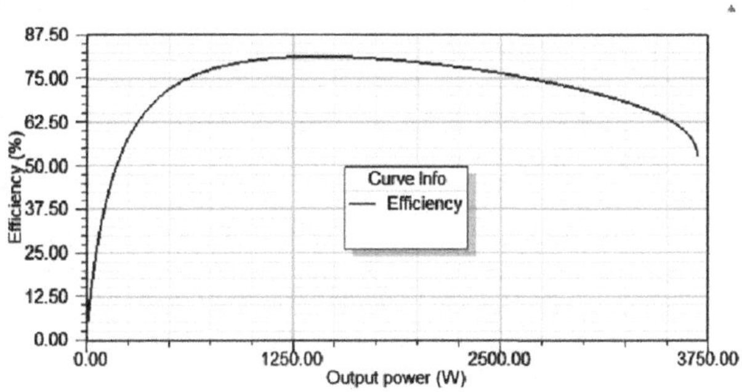

(a) BM

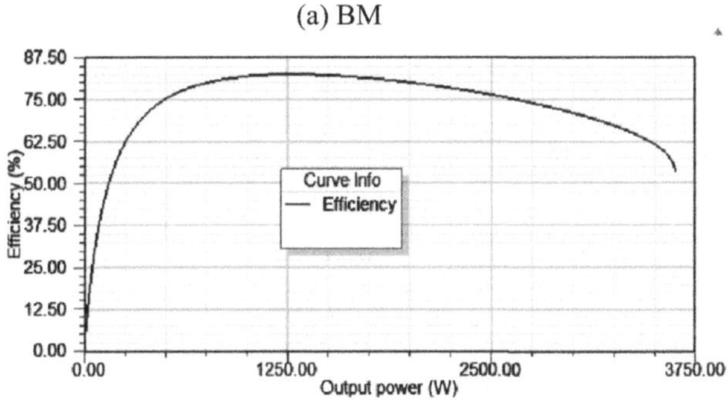

(b) OM1

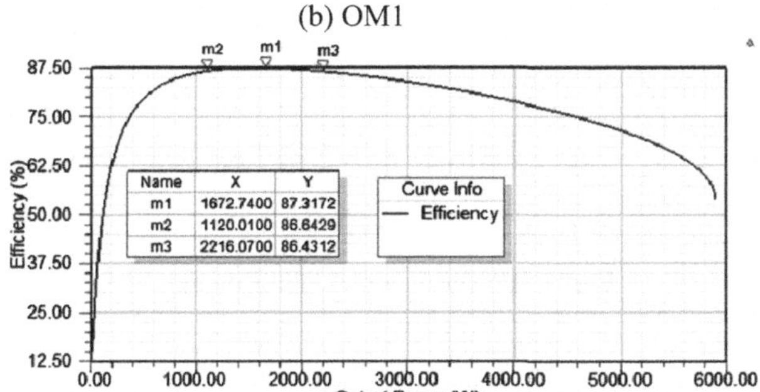

(c) OM2

Fig. 2. Efficiency of motor models

good cooling etc.... Many of these details are modified during production which affects the final operating characteristics.

The good power factor is very important for the motor operation and for the industry as electricity bills are reduced and the low power factor penalty is avoided. The system capacity is increased as additional loads can be added without overloading the system. Improved system operating characteristics are also due to the low line losses and voltage drop as a result of the lower motor current. In that respect the model OM2 shows promising results.

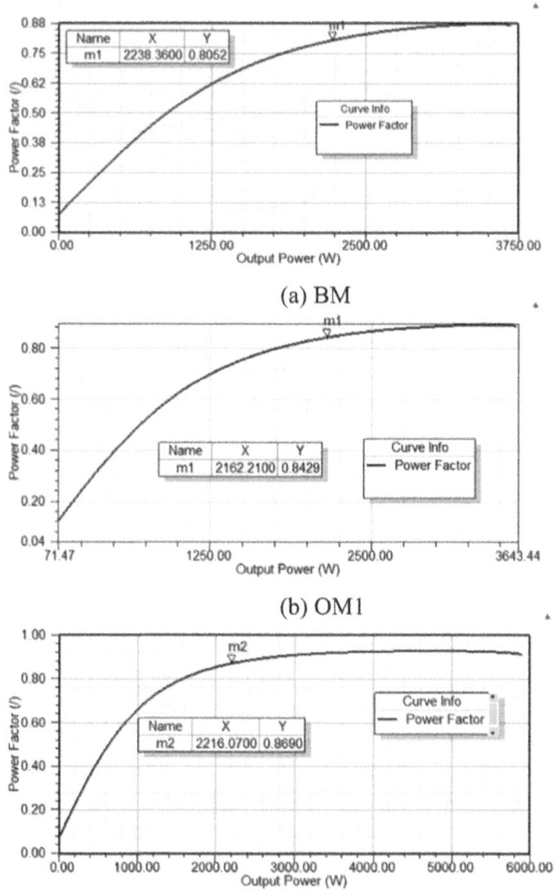

(a) BM

(b) OM1

(c) OM2

Fig. 3. Power factor of motor models

As there is no significant difference in torque characteristics of BM and OM1 in Fig. 4 is presented torque versus speed only for BM and OM2.

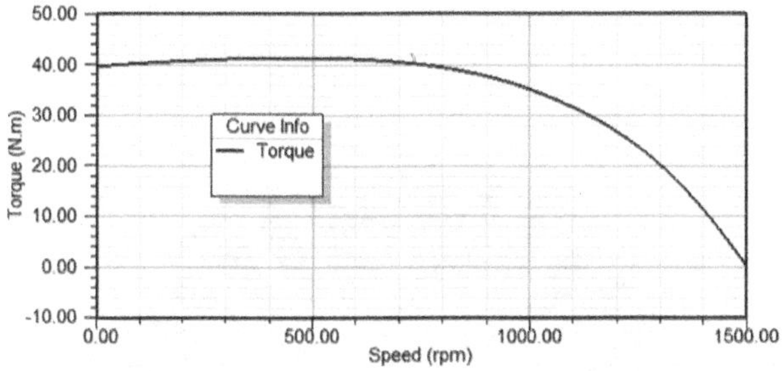

(a) BM

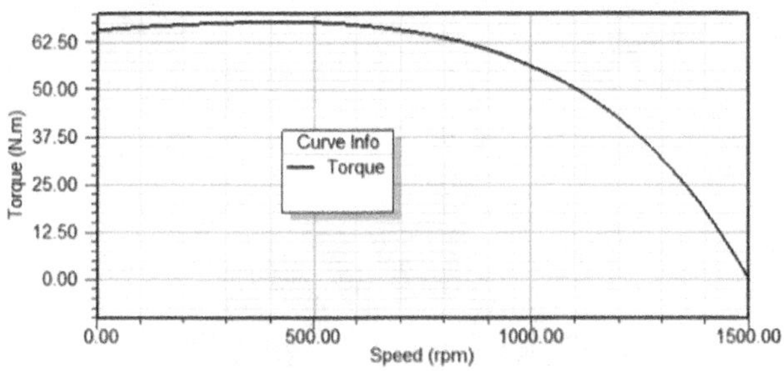

(b) OM2

Fig. 4. Torque of motor models

Both optimized models OM1 and OM2 have sufficiently large starting and breaking torque. The overloading capability (break-down torque ratio) is sufficiently large and in good correspondence to IE1 and IE3 class of motors from Končar. Efficiency of the machine is highly affected by accurate determination of frictional and windage losses and stray losses. In the analyzed case of optimization, the frictional and windage losses are calculated as 1% of the motor output power while stray losses are calculated as 1.8% of the output power according to [11] and [7]. The practical measurements may differ from these assigned values.

2.2 Numerical Models

The design of electrical machines is verified usually by numerical models which determine the distribution of flux density inside the entire cross section of the machine. The Finite Elements method (FEM) is useful for detection of points of the magnetic core with high flux density which cause the core saturation and machine heating. In the standard design of the induction motor generally the flux density in the stator core may be assumed varying between 1.2 to 1.4 T. The flux density in stator and rotor teeth is limited bellow 1.8 T. According to [12] the air gap flux density for a four pole machine is recommended to be 0.65 T to 0.78 T. Yet, the final calculation of the flux density is performed by finite elements method where the motor cross section is divided into numerous elements and the magnetic vector potential is calculated in each element and consequently the flux density. In Fig. 5 is presented the flux density distribution in all three models, calculated by FEM.

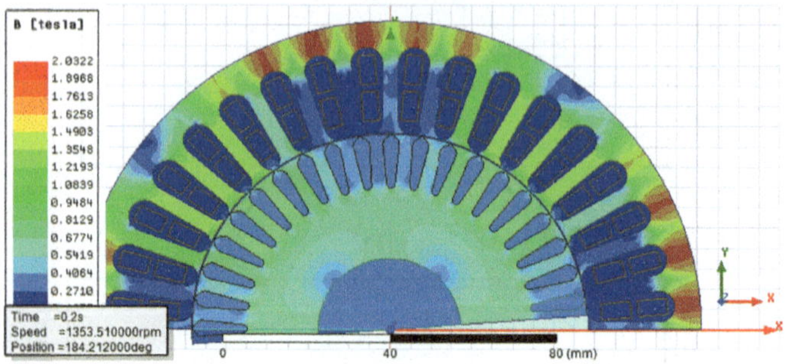

(a) BM

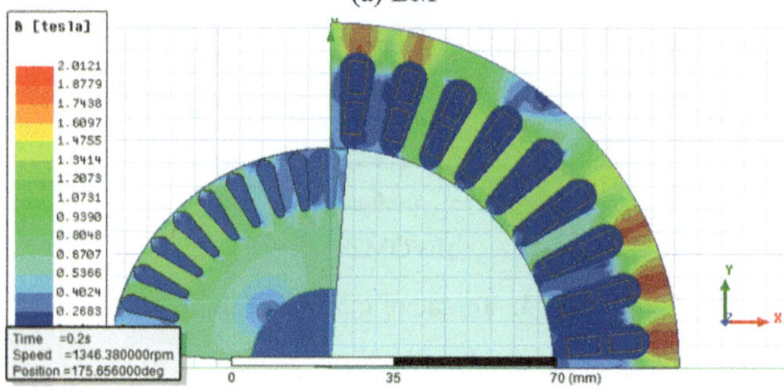

(b) OM1

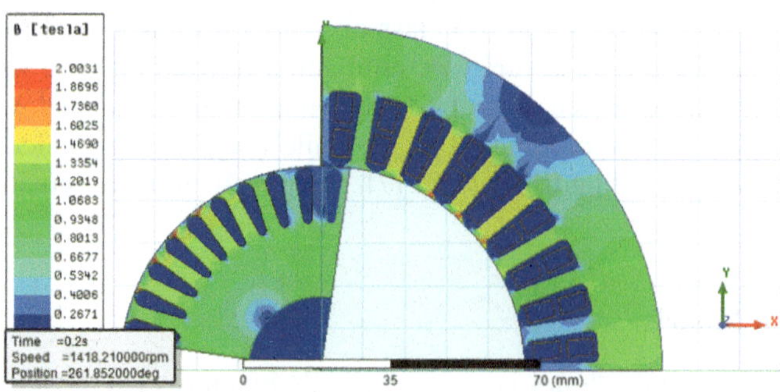

(c) OM2

Fig. 5. Flux density distribution

Additional reason for the modification of stator slots was relatively large areas of stator yoke with high flux density in the BM and OM1. Taking into consideration that slot height was modified and the area of stator yoke was increased thus the flux density in stator yoke is considerably decreased from 1.98 T at OM1 to 1.26 T at OM2. Areas of high flux density are still notable at OM2 in the lower part of stator and rotor teeth. Yet, the areas of magnetic core saturation are greatly reduced at OM2 model.

2.3 Dynamic Model

Motor dynamics is an important part in analysis of motor operation. It gives an overview of motor acceleration with or without load or with a step load coupled to the motor shaft after motor has accelerated. In this paper the transient characteristics of speed, torque and current for two operating regimes: acceleration with a rated load and operation with a step load, equal to the rated load, and coupled to the motor shaft 0.5 s after the motor starting are presented, for the OM2 model. The dynamic models are derived from Simview, a software module of Ansys program. The presented transient characteristics should also verify the accuracy of the analytical model of the optimized motor-OM2. In Fig. 6 are presented transient characteristics of speed for the above mentioned operating regimes.

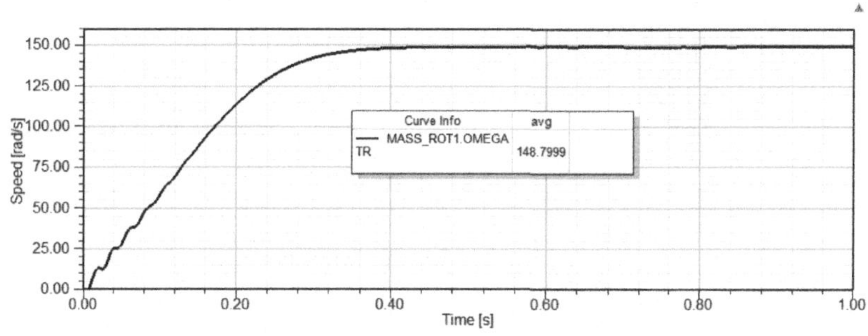

(a) Acceleration with rated load

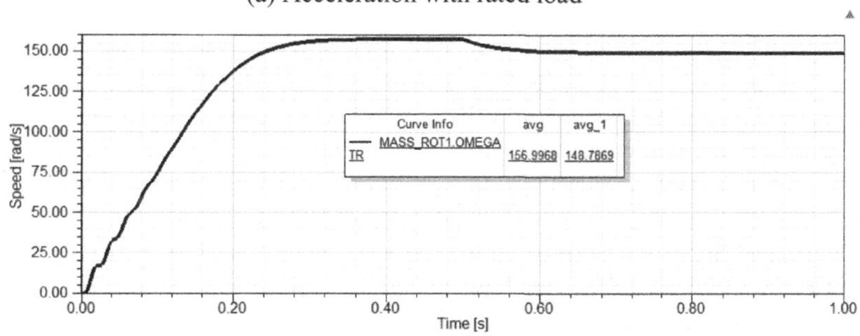

(b) Acceleration with step load

Fig. 6. Transient characteristic of speed-OM2

From the presented characteristics of speed it can be concluded that when motor is accelerated with rated load of 14.8 Nm the acceleration time is longer, approximately 0.4 s, after which the motor reaches the rated speed of 148.8 rad/s or 1421 rpm. The obtained result for the rated speed is in good correspondence to obtained analytical value (Table 1) of 1418 rpm. When the motor accelerates at no load, the acceleration time is shorter, approximately 0.3 s, after which the motor reaches no-load speed of 156.99 rad/s or 1499 rpm. A rated load of 14.8 Nm is coupled to the motor shaft 0.5 s after motor starting. The speed decreases to the rated speed (Fig. 6b).

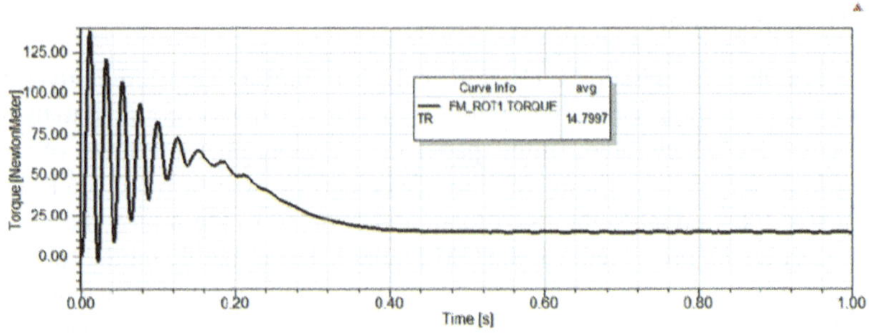

(a) Acceleration with rated load

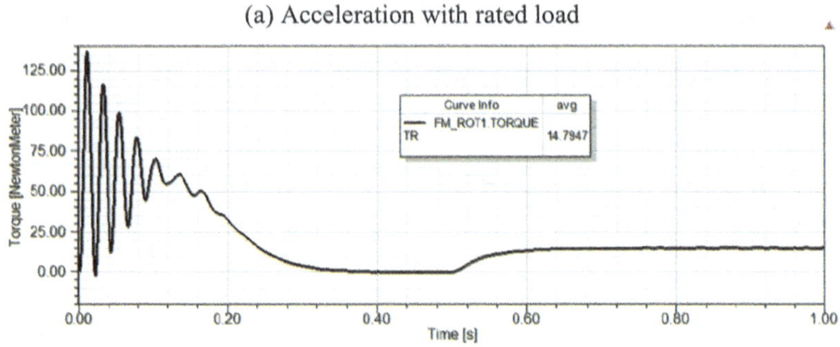

(b) Acceleration with step load

Fig. 7. Transient characteristic of torque-OM2

Similar conclusions can be derived for transient characteristics of torque. After motor acceleration has finished and the transients are suppressed the torque has the value of the rated torque of 14.79 Nm. For the operation with the step load the value of the torque decreases to the no-load torque after the no-load acceleration, and increases again when the step load is applied equal to the rated load (Fig. 7b).

The stator current has high starting value (50 A or rms 35.5 A) which corresponds to the rated current times eight (locked-rotor current ratio of 7.9. in Table 1). After acceleration of the motor is finished, the current decreases to the rms value of the rated current 4.4 A, or in case of the step load it decreases up to the no-load current of rms value of 2.3 A and increases again when the step load is coupled to the motor shaft, up to the rated current (Fig. 8). The comparison between transient and analytical characteristics is presented in Table 4. From the presented transient characteristics it can be concluded that motor accelerates with rated load and maintains the steady-state operation without problems. Similar conclusion can be derived also in case when the rated load is coupled to the motor shaft after the acceleration. The obtained values of torque and current are within expected limits. The similarity of the obtained values of speed, torque and current from the analytical and the dynamic model verifies the accuracy of the models and the obtained results from them.

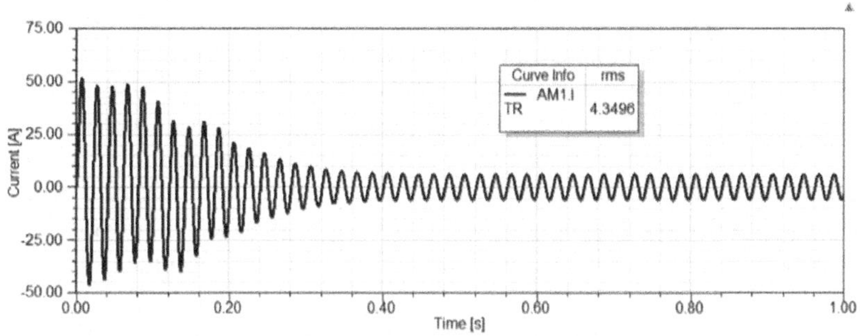

(a) Acceleration with rated load

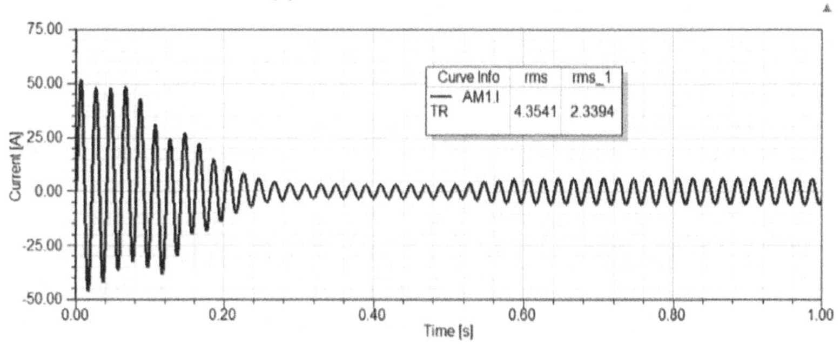

(b) Acceleration with step load

Fig. 8. Transient characteristic of current-OM2

Table 4. Comparison between analytical and dynamic model of OM2.

Parameter	Analytical Model-OM2	Dynamic Model-OM2
Rated speed [rpm]	1418	1421
Rated torque [Nm]	14.8	14.79
Rated current [A]	4.4	4.3
No-load current [A]	2	2.3

3 Conclusion

The need for the more efficient electrical motors has imposed a challenging task to the motor designers. This task becomes even more complex in case of induction squirrel cage motor due to the limited possibilities for its improvement as a result of construction limitations. One of the biggest challenges for the designer is obtaining the optimal design with high efficiency, good power factor, sufficiently big overloading capability while keeping the material consumption and production costs low. Paper proposes the methodology for efficiency improvement of three phase squirrel cage motor by obtaining the best combination of outer stator diameter, motor length, stator slot design, number of conductors per slot and the air gap length in order to achieve high efficiency comparable to IE3 level, good power factor and sufficiently big overloading capability. Moreover, the same parameter of the motor has opposite impact on different operating characteristics of the motor. For example the increase of air gap length increases motor overloading capability and motor cooling but decreases the power factor. Finding the best combination of various parameters that allow obtaining the cost effective model of the motor can be done by optimization techniques i.e. in this paper is used optometric analysis of four parameters: number of conductors per slot, air gap length, outer stator diameter and the motor length, applied in the motor model with modified stator slot. All four parameters are varied in prescribed boundaries, defined on the base of the designer's experience. From the various combinations of these four variables that give numerous motor models, the optimized model OM2 is chosen as the best solution in terms of the motor efficiency, power factor and overloading capability. The weight and the material consumption in obtained optimized model have increased compared to the starting model which certainly increases the cost of the motor. Yet, the efficiency has increased from 78.4 to 86.5% with good power factor of 0.87. Moreover, the proposed optimal model shows very good efficiency at partial load (i.e. 50% or 75% of the rated load) tending to be 86.6 and 87.3, correspondingly. The power factor of OM2 model is high, with value of 0.87 and it is considerably higher than at motor prototype from the producer. The optimized model shows good flux density distribution with very small areas of core saturation. This is mainly due to the modified stator slot and increased outer stator diameter. The drawback of this modification is the increased material consumption and cost of the motor. The optimized model OM2 has satisfactory dynamical behavior at acceleration and steady-state operation with or without load. Moreover, the obtained results of speed, current and torque from the dynamic model of OM2 show satisfactory similarity with the obtained results from analytical model of OM2 thus both models, analytical and dynamical, are verified as sufficiently accurate. From above stated, it can be concluded that motor design requires careful analysis of several major parameters that have biggest impact on motor operating characteristics. Their determination should be the results of optimization processes that are searching for the best combination of those parameters which allow obtaining the model of the motor with high performances. The optimization of the variables should be evaluated in terms of complete specter of operating characteristics, not just regarding one operating characteristic. The proposed model is theoretical, based on computer calculations and simulations. The construction of the motor prototype is affected by many factors such as influence of the air gap length on motor heating, noise and vibration, possibility to obtained high fill factor of the stator slot etc.. The motor prototype should verify the proposed model and its characteristics.

References

1. YazidKaikaa, M., Babaa, F., Khezzar, A., Boucherma, M.: Analytical analysis of rotor slot harmonics in the line current of squirrel cage induction motors. J. Electr. Eng. **37**(1), 10–12 (2006)
2. GökhanYetgin, A., Turan, M.: Efficiency optimization of slitted-core induction motor. J. Electr. Eng. **65**(1), 60–64 (2014). https://doi.org/10.2478/jee-2014-0009

3. Di Tommaso, A.O., Genduso, F., Miceli, R., Nevoloso, C.: Fast procedure for the calculation of maximum slot filling factors in electrical machines. In: 12th International Conference on Ecological Vehicles and Renewable Energies (EVER), Monaco, pp. 1–8 (2017). https://doi.org/10.1109/EVER.2017.7935906
4. Shen, Y., Zhu, C., Wang, H.: Slot optimization design of induction motor for electric vehicle. In: IOP Conference Series: Material Science and Engineering. https://doi.org/10.1088/1757-899X/301/1/012081
5. Kumar, D.S., Manamalli, D.: Development of Induction motor efficiency model and improving at low load condition using controllers. Int. J. Comput. Appl. **75**(16), 1–8 (2013). https://doi.org/10.5120/13192-0521
6. Kim, D.-J., Choi, J.-K., Chun, Y.-D., Koo, D.-H., Han, P.-W.: The study of the stray load loss and mechanical loss of three phase induction motor considering experimental results. J. Electr. Eng. Technol. **9**(1), 121–126 (2014). https://doi.org/10.5370/JEET.2014.9.1.121
7. Kollár, B.: Design considerations by the optimization of PMS motors. ActaTechnica Jaurinensis **8**(1), 77–87 (2015). https://doi.org/10.14513/actatechjaur.v8.n1.355
8. Çunkas, M., Akkaya, R.: Design optimization of induction motor by genetic algorithm and comparison with the existing motor. Math. Comput. Appl. **11**(3), 193–203 (2006). https://doi.org/10.3390/mca11020193
9. Bottauscio, O., Chiampi, M., Donadio, L., Zucca, L.: Measurement of local electric and magnetic quantities inside electrical induction motor. J. Electr. Eng. **55**(10/S), 66–60 (2004)
10. Adamczyk, K.: Rotor resistance estimator based on virtual current sensor algorithm for induction motor drives. Power Electron. Drives **5**(40), 143–158 (2020). https://doi.org/10.2478/pead-2020-0008
11. Končar: Electric motors-catalog. https://koncar-mes.hr/en/archive/catalogues/. Accessed 28 July 2021
12. Boldea, I., Nasar, S.A.: The Induction Machines Design Handbook, 2nd edn. CRC Press, Boca Raton (2010)

Development, Structure and Design of Stamping Tool

Jure Marijić(✉), Marko Vilić, Ivan Grgić, Mirko Karakašić, and Željko Ivandić

Mechanical Engineering Faculty in Slavonski Brod, University of Slavonski Brod, Trg I. B. Mažuranić 2, 35000 Slavonski Brod, Croatia
jmarijic@unisb.hr

Abstract. Market demand for electrical products by standard electrical products defines the demand for stamping tools. The design of the tool is based on the requirements of easy assembly, disassembly and maintenance. The development of such tools requires the designer to meet complex requirements, some of which are only defined at the beginning, so that the most precise specifications possible are necessary right at the start of the design process. In order to meet the required complex product geometry, product tolerance and the number of products required by the customer, the development of a multi-row punching tool is shown. The tool consists of an upper and a lower part, which are appropriately positioned accordingly and fixed to the stamping machine. The force required to stamp and the force required to detach the guide plate from the sheet metal were determined. The functionality of the developed tool and the stamping process is shown in its entirety, from the entry of the sheet into the tool to the exit of the finished product. The products obtained by stepwise stamping are rotors and stators, which are the basic elements of any electrical device.

Keywords: Stamping tool · Tool development · Tool structure · Stamping

1 Introduction

Stamping is the processing of metals or non-metals without separation of particles. As today's trend is towards large-scale and mass production, this technique is an ideal choice because the machines can be automated, very high productivity is achieved, and the price of the products is drastically reduced the higher the production and the degree of automation. The time of production by punching is very short. It is 7 to 10 times shorter than machining by particle separation, and the manufacturing accuracy is in tolerances of the class IT 11 to IT 9 [1].

Press tools are used to fabricate components in large numbers out of sheet metal. The design of press tool involves common steps to be followed to have an efficient and acceptable outcome. Economy, use of unskilled labour, high degree of precision has made press working indispensable [2].

The term punch refers to tools that separate, reshape, or join material on a press. The punch usually consists of a lower part, which is usually fixed, and an upper part, which is

H. Glavaš et al. (Eds.): OTO 2021, LNNS 369, pp. 50–70, 2022.
https://doi.org/10.1007/978-3-030-92851-3_4

movable. It can process metals (steel, aluminum, brass) and non-metals (leather, rubber), usually using strips or individual semi-finished products as preparation depends on the type of processing. They use different motorized presses to achieve the force. Depending on the force needed, the size of the press is chosen. For high volume production, it is common to use the progressive tool shown in Fig. 1 which allows the production of very complex products, complex geometry and dimensions with a small tolerance field. Figure 1 is displaying a progressive tool.

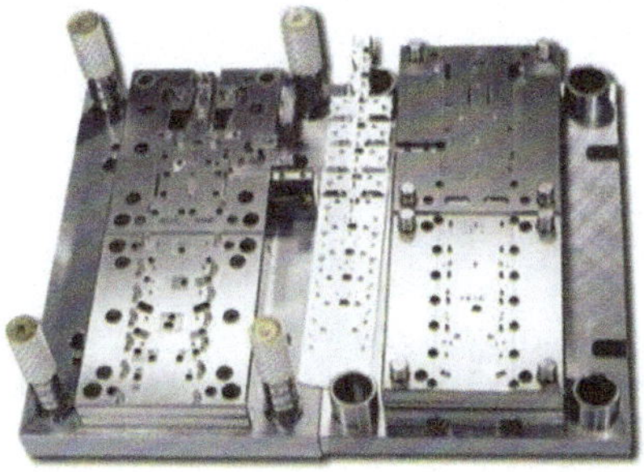

Fig. 1. Progressive tool.

1.1 Progressive Tool

A progressive tool is an alternative tool for the transfer press. With this tool, you can perform a whole series of processing steps with one stroke of the press. Progressive tools are designed for processing strip material. The strip is pushed through the tool with the help of feeder along the tool and a precise strip guide during each stroke.

Progressive tool performs two or more operations at different stages in each stroke. The stock strip is advanced through a series of stations that form one or more distinct press working operations on the strip to get the component [3].

The product is separated from the strip only at the last station. It then usually falls through the opening into the designated tank or the transport system is transporting the product to another control operation.

A progressive press tool has been designed for the purpose. In progressive press tool the final product is obtained by progressing the sheet metal or strip in many stages. In each and every stage the component will get its shape and at the final stage the component is completely ready [4].

Punching waste is also collected and freely disposed of freely in appropriate containers or disposed of via the transport system.

Progressive tools are suitable for continuous production of thin sheets with complex geometry, and the same sheets can be combined into packages. A progressive tool is a highly efficient production tool. It is used to produce very large quantities. However, its reliable operation requires absolutely smooth operation of the production machine, the press.

Stamping tools are mounted and placed on automatic presses. Manual operation is not possible with this type of tool. Common applications of progressive tool are found in the automotive, electronics or household appliance industries. The cost-effective manufacture of complex products in these industries can be partly due to the use of progressive tools (Fig. 2).

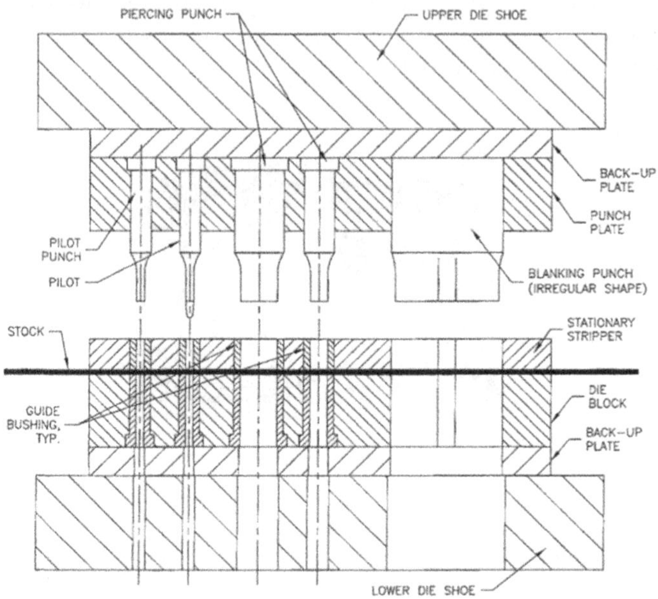

Fig. 2. Standard progressive tool with different elements [5].

2 Development of a Progressive Tool

The first phase of the tool development design process begins with the study of the drawing that design department received from the customer. Detailed analysis is performed to determine the possibility of manufacturing the desired product. The number of tool rows is determined based on the required number of products. If blanks are required due to the complex geometry of the product, the customer is contacted and presented with the stamping product that matches the technological capabilities of the stamping tool. Figure 3 is displaying a product example.

By performing the analysis and confirming the possibility of punching, a layout of strip is developed showing the punching sequences to obtain the desired product. The offset and strip width is chosen so that the usability of the strip is about 90%.

Machining is an operation of forming the desired geometry by performing chip removal over the material. Hence, the appropriate material, tools, and lathes should be used in order to achieve this operation [6].

As every technological process is planned among others also according to the requirements of quality, the quality of machined surface (surface roughness) needs to be also pointed out. The quality of machined surface is directly correlated with the manufacturing process which usually contains various influencing parameters (variables) [7].

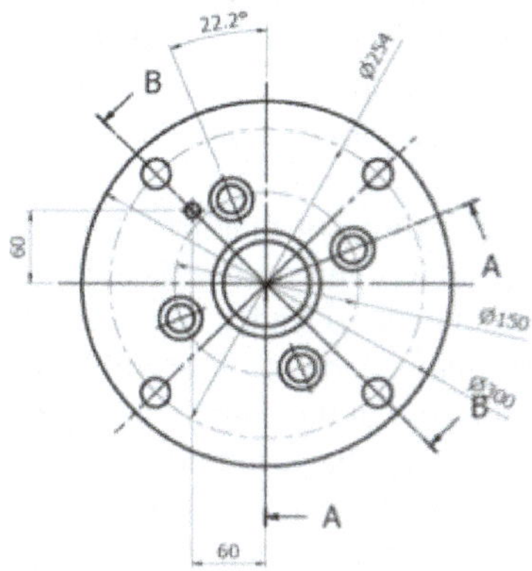

Fig. 3. Product example.

In the process of developing punching stations, the possibilities of the machines (EDM or milling machines) must be taken into account, and experience shows that at least two additional stations are used in the first stages of development. Additional stations are mostly used for additional requirements of the customer for the geometry of the products, which occur in the process of tool development.

When developing stations, the preset offset is taken into account. The offset directly affects the dimensions of the station, as it is necessary to ensure easy assembly, positioning and disassembly of each station.

The Fig. 4 displayed for the two-row toolbar for punching the desired product.

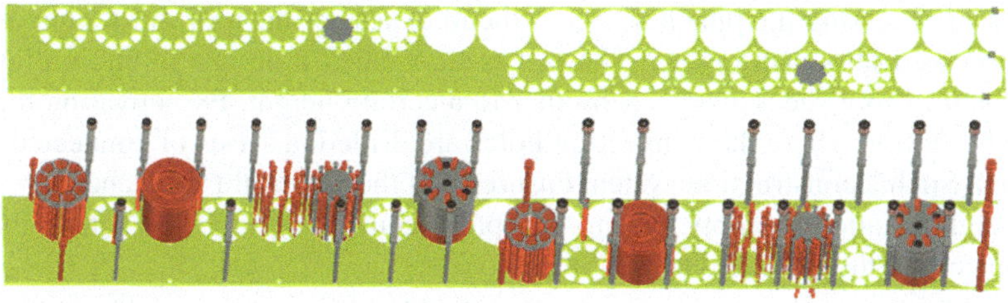

Fig. 4. Layout of strip.

2.1 Layout Strip

According to the defined thickness of the input sheet and the dimensions of the product, the displacement and the width of the input sheet are defined. The rest between the cutting lines of the two stations must be at least two sheet thicknesses.

In the design development of the strip, the first station is drilling holes. The specified hole is later used by the tool to hold the sheet catcher. To ensure the function of the machine operation, it is necessary to provide one of the holes for the catcher with the mechanism. In the case of an unperforated hole, the catcher strikes the solid material and stops the press via a sensor.

The second station is the punching of magnetic holes. The contour is defined by the customer and the puncher and tool matrix inserts are manufactured for the specified station. Matrix inserts are manufactured under the cone and in equal segments. Such design solution ensures easier maintenance and assembly and disassembly as needed.

The third station is punched parts of the outer contour. The showed product contains outer contours with sharp edges. Due to the technological impossibilities of the machines, it is necessary to perform the required work in two stations. One station with pre-punching and the last station with round contour punching. This arrangement ensures a sharp transition from the pre-punching of the parts of the outer contour to the round contour.

The fourth station is the station where the punch core is removed. The process must be predefined by the customer. During the construction process, a uniform contour is taken at the place where the core must be removed. A contour matrix insert should be attached to the lower part of the tool. The cutter presses the contour bar onto the contour stamp via the spring system, thus removing the bar.

The fifth station is empty. Due to the space required for the possibility of positioning and screwing the stations, the fifth station remains empty.

The sixth station is a hole punching station for joining sheets. Depending on the requirements of a certain height, a certain number of sheets connected together, this station performs the function as required. Due to the design of the slider above the sheets, the station can be turned on and off. When the slide mechanism is off, the punching force is not transmitted to the punches and no operation is performed.

The seventh station punches the inner diameter of the product and presses the punches into the strip. Pressing the punches, 2/3 of the thickness of the tool sheet, creates shots that are pressed into each other in the last station. Such a design solution allows the arrangement of tool sheet inside each other to a certain height. By activating the sixth station, the station where the connecting holes are drilled, a series of connected sheets is stopped. Such a construction system ensures that the set height is reached.

The eighth station is empty. Due to the space required for the sheet rotating system during injection, the indicated station is empty.

The ninth station is the station for punching the outer contour of the product, the round part and the system for rotating the tool sheets. The system is realized by an electric motor. The rotation is transmitted through a gearbox, eventually through a belt. The sheets always rotate at the same angle, each in relation to the others. Such a system prevents the influence of uneven flatness of the input sheet, on the verticality of the punched product.

In the case of multi-row tool, a uniform station layout is used for all rows. The last station is the strip cutting station. This completes the stamping process.

2.2 The Structure of the Progressive Tool

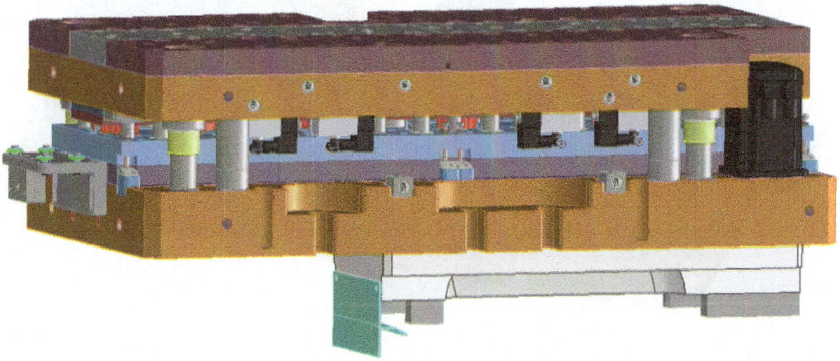

Fig. 5. Progressive tool for stamping.

The tool consists of an upper and a lower part (Figs. 5, 6 and 7).

Tool top:
- cover plate,
- upper plate,
- intermediate plate,
- puncher holding plate,
- guide plate.

Bottom of the tool:
- matrix,
- base plate,
- support plate.

Fig. 6. Tool top.

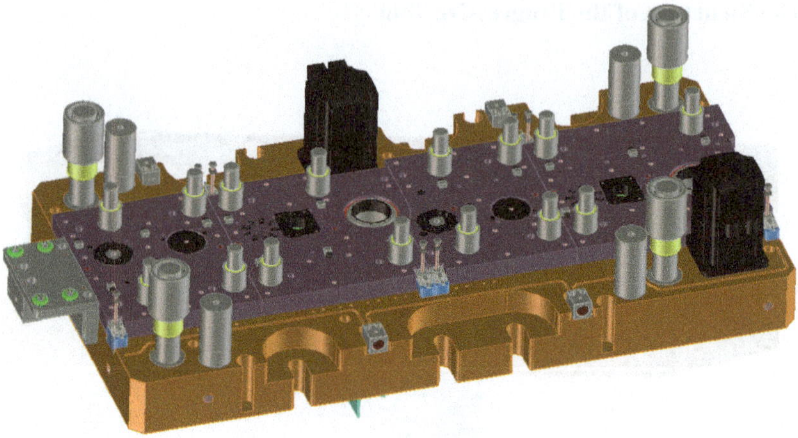

Fig. 7. Bottom of the tool.

Matrix

The matrix or cutting plate is one of the most important parts of the tool, which together with the punch cuts the material. It provides all the contours for the shaping of the work piece and the release for the inserts. The holes of this cutting contour are placed under the cone. The figure shows the possibility of a design cone.

The matrices are cut so that certain stations are built into the same plate for stability. For the need of additional machining, it is necessary to provide two holes Ø10H7 for the possibility of calibration of machining machines. It is obligatory to grind the upper and lower surfaces of the plate to achieve the appropriate parallelism. To facilitate the separation of the matrix from the base plate when all the screws are unscrewed, it is necessary to provide 4 threads M12 that allow to separate the plates through the screw. The screw is screwed into the matrix, thus separating it from the base plate. The M12 thread must be provided in the guide area.

The releases for the inserts on the matrix are made by electro erosion. The choice of this technology ensures the positioning of the insert due to the erosion accuracy. The erosion accuracy is ± 0.005 μm.

To facilitate assembly and disassembly of the tool, the matrix inserts are attached to the base plate from below. Holes are provided on the matrix for guiding elements. The bore tolerances are H5. Such a system ensures that the burning part of the tool is guided downwards.

For multi-row tools, the distribution of stations per matrix should be equal. Such distribution excludes the possibility of differentiation of products by stamping rows.

The figure shows the matrices for a multi-row tool.

The height of the plate on the illustrated matrix is 30 mm. It is determined empirically (Figs. 8 and 9).

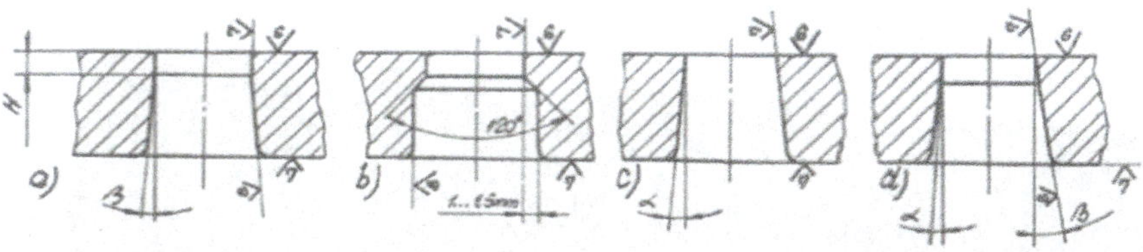

Fig. 8. Possible cone designs.

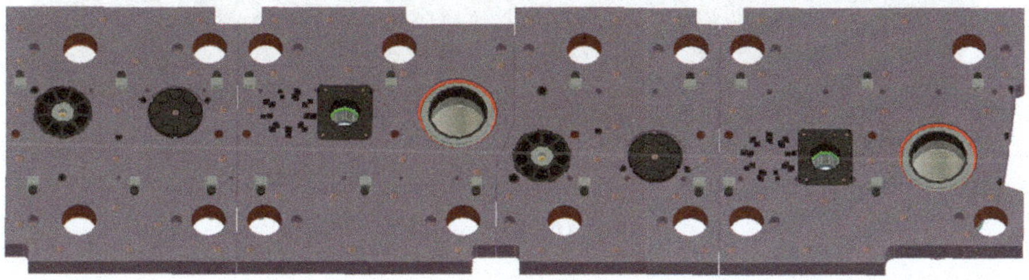

Fig. 9. Matrix.

Base Plate

The base plate has the task of transferring the force from the matrix to the press table and serves as a stable and robust support for the matrix. The openings are slightly larger than those of the matrix due to facilitate the passage of waste. The length of the plate is determined by the lengths of the matrices. Recesses are provided on the base plate for the inserts on which the matrix inserts rest. The inserts are hardened so that they can absorb the cutting force.

It is obligatory to grind the upper and lower surfaces of the plate to achieve the appropriate parallelism. To ensure possible manipulation of the plate, it is necessary to provide at least 4 positions for blind holes with M16 thread for transport during tool assembly.

The height of the base plate is determined empirically depending on the stamping force. It is necessary to know the data on the maximum and minimum height of the inlet for the strip in the press. Each press has defined data for the height of the strip inlet and mounting height.

The base plate has the function of connection and positioning of the tool in the press. According to the calculation of the punching force and based on the center of gravity of the specified force, it is necessary to provide exceptions for the positioning and mounting of tool on the press table.

Holes for large and small guides are provided on the base plate. The bore tolerances are H5. Such a system ensures that the top of the tool is in position with the bottom of the tool (Fig. 10).

Fig. 10. Base plate.

Guide Plate

The function of the guide plate is to guide the punches and remove the tape from the punches. Holes and contours are made according to the corresponding knives with a certain distance. The holes and contours are made by electro erosion. Due to the technological possibilities of this technology, it is possible to achieve the required clearance. The clearance is determined empirically and is up to μm, depending on the punches. In the direction of faster and easier assembly, it is possible to provide recesses for inserts. This selection facilitates maintenance of the punching stations.

Small guide holes are provided on the guide plate. The hole tolerances are H5. Such a system ensures that the burning part of the tool is guided downwards. A 2 mm deep groove is provided on the upper side, which serves as an accumulation of punches lubricant. When the press is lowered, the lubricant is poured between the punches and the hole, ensuring high-quality passage of the punch and reducing friction.

On the lower surface there are recesses for the guide rails. The inlets are designed so that the guide plate can be lowered to the stop. The stops achieve a distance between the guide plate and the 0.2 mm strip. This ensures better punch guidance.

The guide plate is connected to the base plate by springs. The number of springs, the position and the spring force are determined by calculating the force required to separate the plate from the strip. This connection ensures that the sheet is separated from the punches. When the tool, the punches are lifted above the mounting plate while the guide plate remains in the down position until the force required to lift the guide plate is reached. In this way, the punches are continuously guided during the stroke of the press and the strip is separated from the punches.

For the need of additional machining, it is necessary to provide two holes Ø10H7. The height of the plate is 37 mm (Fig. 11).

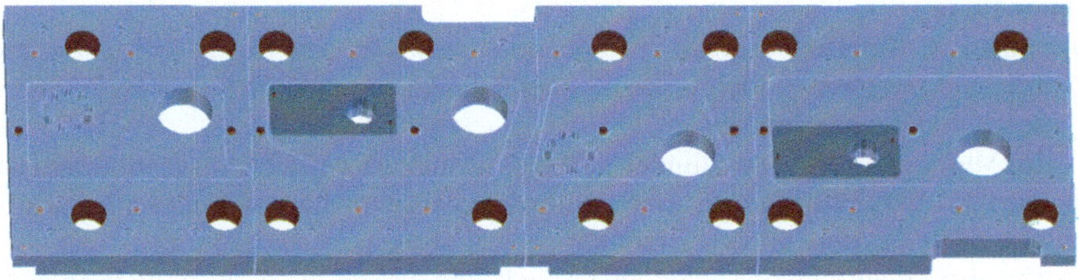

Fig. 11. Guide plate.

Puncher Holding Plate

The function of this plate is to receive, align and pre-position the punchers towards the guide plate. The distance between the plate and the puncher is 0.05 mm per surface. It is fixed to the intermediate plate. The plate is hardened and all holes and recesses for the puncher are made by electro erosion. It is obligatory to grind the upper and lower surfaces to achieve the required parallelism.

It is fixed to the upper plate by M12 screws. For the need of additional machining, it is necessary to provide two holes Ø10H7. The height of the plate is 15 mm (Fig. 12).

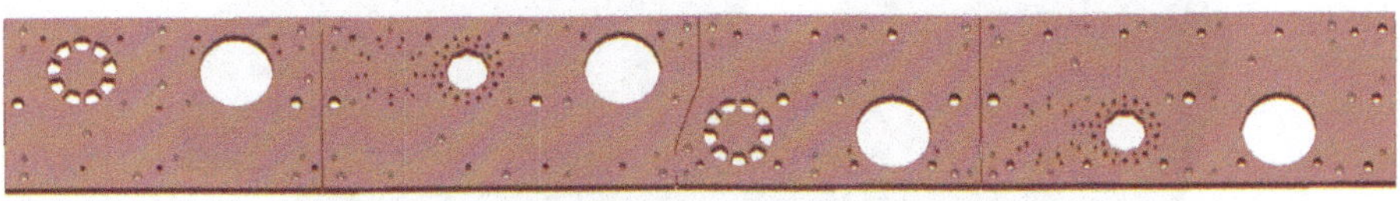

Fig. 12. Puncher holding plate.

Intermediate Plate

The intermediate plate is made to save machining on the belove plate. The punch head rests on the punch holder plate, and there is a release for the punch head in the intermediate plate. The distance between the plate and the punch is 0.5 mm per side.

It is screwed to the holding plate. For the need of additional machining, it is necessary to provide two holes Ø10H7. The intermediate plate is positioned on the mounting plate by means of pins. The height of the plate is 10 mm (Fig. 13).

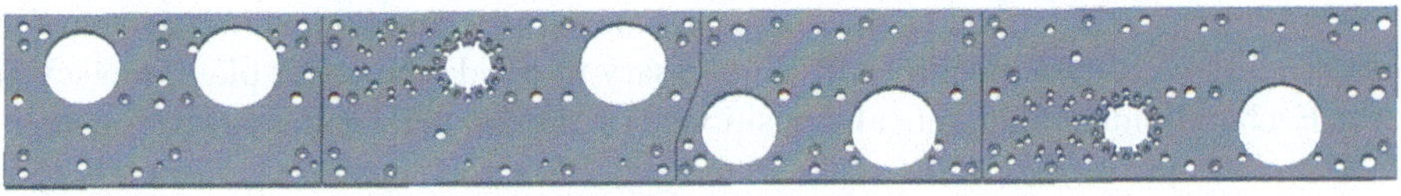

Fig. 13. Intermediate plate.

Upper Plate

The function of the upper plate is to connect the upper part of the tool to the press and transmit the force to the punchers. In punching stations, it is necessary to provide recesses for slide mechanisms.

The slide moves with the cylinder and lands on the surface of the lower slide, transmitting the force to the punchers. Each individual slider should have two capacitive position sensors. Such a system allows maximum control of the puncher under the slider.

The upper plate contains holes and through contours for the supports on the puncher. Such a system facilitates disassembly and assembly of the upper part of the tool. It is possible to pull out the punchers without disassembling the upper part of the tool.

Holes for large and small guides are provided on the upper plate. The hole tolerances are H5. Such a system ensures that the burning part of the tool is guided downwards.

The spring for the guide plate is fixed to the upper plate. It is obligatory to grind the upper and lower surfaces of the plate to achieve the appropriate parallelism. In order to ensure possible manipulation of the plate, at least 4 positions for blind holes with M16 thread for transport must be provided during the tool assembly (Fig. 14).

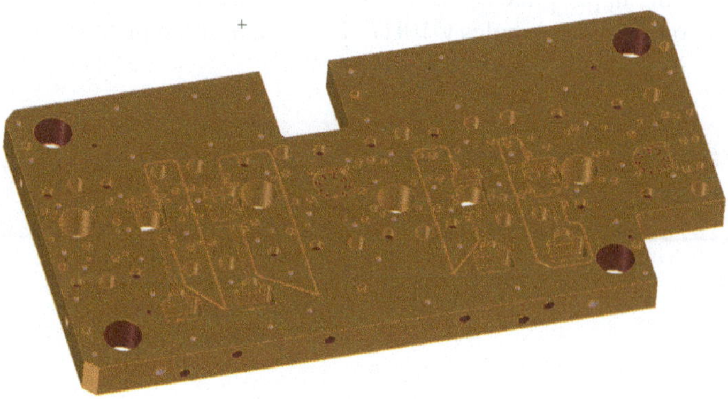

Fig. 14. Upper plate.

Cover Plate

The function of the cover plate is to cover all punchers supports and all sliding mechanisms and cylinders. It prevents the penetration of impurities and thus affects the function of the work.

The dimension is defined by the top plate. Due to the technological limitations of the workbenches of certain machines, it is necessary to divide the cover plate into several parts. Each individual plate must be positioned on the top plate and screwed together.

The plates positioned in the space of the punch support, must be hardened because of the possibility of twisting under the punch shaft during punching. Such distribution of the plates facilitates and speeds up the assembly and disassembly of punchers, and thus the maintenance of the tools. Each plate contains M16 threads with which the plates can be transported (Fig. 15).

Fig. 15. Cover plate.

Support Plate

The function of the support plate is to absorb some of the force generated by the press. The progressive tool is positioned in the press above the specified plate. The positioning is done after calculating the center of gravity of the required cutting force.

The support plate also drains the products with transport systems, it is necessary to provide drainage channels and lead the punching waste out of the evacuation systems. Said plate also contains threads for handling and transport (Fig. 16).

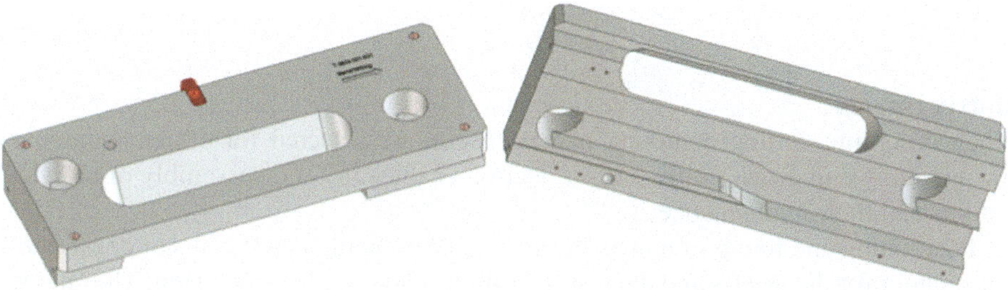

Fig. 16. Support plate.

Punchers

The basic function of the puncher is to punch the input sheet together with the matrix. Depending on the distribution of punching stations, a certain product is formed. We divide punchers into purchased and made on own demand for the design of a given product.

In case of punches made on demand, it is necessary to form the head of the puncher. The puncher head, like the punch, is made by electro erosion. There must be a clearance of about 0.01 mm between the punch and the head. The puncher is pressed into the head and firmly soldered. The puncher head contains a slot of 1mm. The punches are made of CF-H40S.

For punchers with larger diameters, a bracket is made to save material. The bracket is positioned from CF-H40S and screwed to the part. The puncher head should provide a secure hold in the mounting plate (Fig. 17).

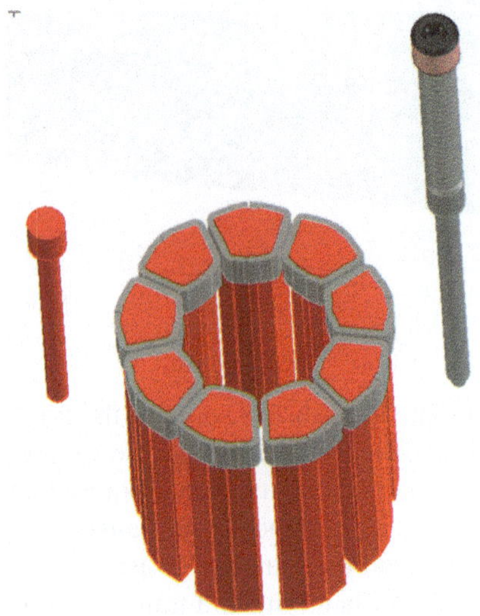

Fig. 17. Punchers.

Matrix Inserts

When designing cutting contours, matrix inserts are selected for easier maintenance and assembly and disassembly. Such selection enables faster disassembly of individual stations and their maintenance.

Positioning and fixing of matrix inserts can be done in several ways of construction. The contour for the inserts and the inserts itself is made by electro erosion. The choice of this technology allows to achieve the necessary tolerances. The insert, which is positioned by the contour, is fixed to the base of the plate with a screw connection. Another design option is the production of a cone on the matrix insert and a contour on the matrix. Selection of this variant ensures positioning by the contour. When selecting a negative cone, the matrix insert is mounted on the bottom side of the matrix and the die is fixed to the base plate.

The cutting contour is determined according to the puncher with a certain distance. The distance depends on the thickness of the sheet to be punched (Fig. 18).

Fig. 18. Matrix inserts.

2.3 Positioning the Progressive Tool in the Press

The positioning of the progressive tool inside the press is done according to the center of gravity of the cut. The calculation of the center of gravity of the cut is processed as the paper progresses. The ideal position of the tool inside the press is when the center of the press coincides with the center of gravity of the cut. After determining the position of the tool inside the press, it is necessary to provide the releases and the through contour for the screws.

The lower part of the tool is screwed to the press table. On the press table are provided grooves for T-segments with thread. Holes for wedges on which the base plate is positioned are also provided on the table. The upper part of the tool is also screwed to the upper table of the press. In this way, the positioning and fastening of the tool to the press is achieved (Figs. 19, 20 and 21).

Fig. 19. Screwing and positioning the lower part of the tool.

Fig. 20. Screwing the upper part of the tool.

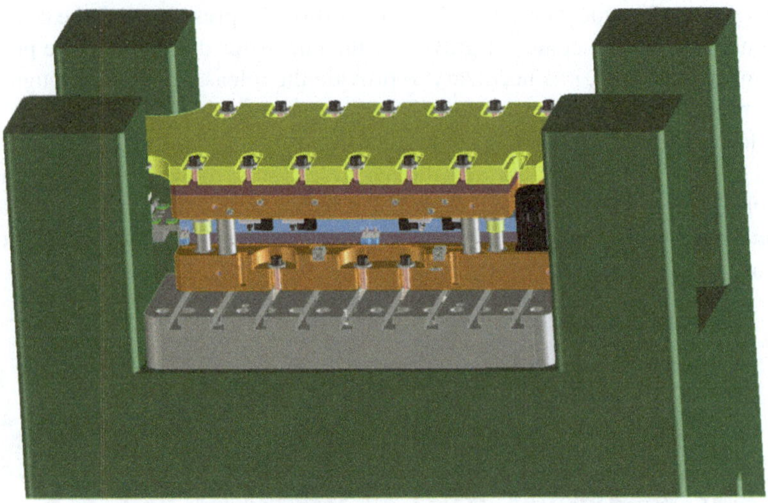

Fig. 21. Progressive tool positioned and attached to the press.

3 Maintain the Progressive Tool

Maintenance is an interdisciplinary activity that brings together experts from the fields of mechanical engineering, electrical engineering, electronics and other professions with the aim of achieving optimal maintenance as well as improving the maintenance of installed equipment [8].

Maintenance is the function of the company entrusted with the constant control of equipment and the performance of certain repairs and inspections, which allows constant functioning and maintenance of production and auxiliary equipment and other devices [8].

Downtime and maintenance costs are interdependent. Reducing maintenance costs, i.e., reducing the number of inspections and maintenance, results in frequent and longer downtimes and thus higher downtime costs. With the other hand, excessive maintenance costs are also not good, as they are an unnecessary expenditure of financial resources [9].

3.1 Preventive Maintenance

Preventive maintenance involves maintenance activities that are performed on a specific schedule before a failure occurs. The condition of the component is assumed to be as good as new after the preventive maintenance is performed [8].

Preventive maintenance can be periodic and preventive, depending on the condition. In periodic preventive maintenance, the maintenance interval can be based on a calendar or on the operating time (service life). In condition-based maintenance, certain variables are measured continuously and action is taken only if the variable is outside certain limits [8].

The advantages of preventive maintenance are the increase in reliability and availability, the ability to plan actions, and the ability to prevent failures. Disadvantages are the impossibility to predict failures, higher costs due to the replacement of parts that could still work [8].

3.2 Corrective Maintenance

Corrective maintenance is the oldest model of machine maintenance and is performed after a failure. No preventive action was taken prior to the failure. Corrective maintenance activities are limited to returning the machine from a state of failure to a state of operation. Today, this approach is used only for auxiliary equipment or in combination with preventive maintenance [8].

The advantages of corrective maintenance are full use of the system's usability and a lower level of technical education. Disadvantages are the possibility of breakdowns and uncontrolled failures [8].

3.3 Maintenance According to Condition

Condition maintenance uses a range of technologies to achieve and maintain optimal operating condition of machinery and individual components. By measuring and determining the trends of physical parameters against known limits or specifications, potential machine problems are identified, analyzed, and corrected before failure occurs. Machine diagnostics is performed to detect faults at the earliest possible stage, to plan machine repair and inspection, and to avoid undesirable consequences for the process [8].

The advantages of maintenance by condition are reduced maintenance costs, increased reliability and availability of the machines. The disadvantages are that it requires good organization and a higher level of knowledge, possession of monitoring equipment and training of personnel [8].

3.4 Maintenance of the Progressive Tool

The progressive tool in theoretical terms can be stamped continuously, 24 h 7 days a week. In a real environment, where the tool is limited in time to one press and depending on the number of products required in a given time, maintenance is as follows.

When the tool does approx. 10000000 strokes, it is disassembled and grinding of the matrix and punches are ground by 0.2 mm. Grinding is performed as long as the cutting elements produce a product with the given dimensions within the tolerance field. The empirical dimension is 10 mm grindability for the matrix and also for the punches. Any grinding above these values will affect the stability of the tool and the product will no longer be within the given tolerances.

Structural elements such as springs are replaced with new elements, bearings are checked and, if necessary, replaced with new ones. The guide plate is cleaned and blown. Punchers are lubricated and the tool is reassembled and set for stamping. An inspection of the cutting elements is performed to determine the consumption condition. Control measurements, height of the matrix and punches are performed and the data is archived in the necessary documentation.

4 Calculating

In this chapter, the calculation for the punching force, the cutting center of gravity, and the usability of the strip are presented. These points should be determined in the initial stage of the design process. The punching force and the center of gravity of the cut must be known at the phases of the design process where the press is selected. The usability of the strip must be determined when defining the tool offset. It is necessary, in terms of punching stability, to ensure a relationship between the usability of the strip and the residue between the punch cuts.

4.1 Punching Force

The stamping force is one of the most important quantities to calculate, since the choice of the type and strength of the press depends on it. The force required is not the same throughout the entire path of penetration of the punch into the material, but changes and consists of phases of elastic and plastic deformation, cutting and fracture of the material.

The punching force depends on the shear strength of the material, the total length of the cut, the thickness of the material, the size of the gap and the condition of the cutting punchers. The size of the gap and the condition of the cutting punchers cannot be expressed by a fixed coefficient. In addition, the cutting punchers gradually become blunt during the operation of the tool, so these coefficients are not taken into account, but the total required force calculated according to the known values increases by 30%.

The required force for punching is calculated according to the expression [1]:

$$F = 1.3 \times L \times s \times \tau_m, \text{N} \tag{1}$$

Where is:

- F – required punching force, N
- 1.3 – coefficient used due to the influence of the clearance between the matrix and the punches and the wear of the cutting edges
- L – total cut length, mm
- s – sheet thickness, mm
- τm – shear strength, N/mm^2.

The total length of the cut is calculated according to the expression

$$L = L_1 + L_2 + \ldots.L_n, \text{mm} \tag{2}$$

Where is:

L_1 – length of the cut at the first station, mm
L_2 – length of the cut at the second station, mm
L_n – length of the cut at each separate station, mm.

Calculation for Tool:

- total length of the cut

$$L = L_1 + L_2 + \ldots L_n$$

$L = 14.14 + (9 \times 42.42) + 14.14 + (2 \times 12.7) + (18 \times 4.1) + (18 \times 4.1) + 119.7 + 188.3 + 20.42 + (9 \times 42.42) + 20.42 + (2 \times 12.7) + (18 \times 4.1) + (18 \times 4.1) + 119.7 + 188.3 = 1820.1$ mm.

- sheet thickness

$$s = 0.5\,\text{mm}$$

- Shear strength (defined on customer's drawing)

$$\tau_m = 400\,\text{N/mm}^2$$

- punching force

$$F = 1.3 \times 1820.1 \times 0.5 \times 400 = 473226\,\text{N}$$

The obtained force is approx. 48 *to* and the press is selected according to this force.

4.2 Center of Gravity of the Cut

The center of gravity of all cutting forces is determined for the purpose of correct positioning of the tool clamp. The tool clamp must be positioned at the center of gravity of all cutting forces, which is especially important for tools with a guide plate. If the center of gravity of such tools is not positioned correctly, incorrect guiding will occur and the cutting edges will be loaded on bending, resulting in shorter tool life and accelerated wear of the sliding surfaces of the press.

The center of gravity of the cut is calculated according to the expression:

$$\mathrm{x} = \frac{\sum \mathrm{L_i} \times \mathrm{a_i}}{\sum \mathrm{L_l}}, \ \mathrm{mm} \tag{3}$$

Where is:

x – position of the center of gravity of the shape along the x axis, mm
L_i – length of an individual cutting line, mm
a_i – distance of the center of gravity of an individual cutting line along the x axis from the origin, mm.

Calculation for Tool:

- Center of gravity of the cut

$$x = \frac{908597.4}{1820.1} = 504.2\,\mathrm{mm}$$

4.3 Strip Usability

The degree of utilization of the strip is an important factor and shows what percentage of the material is utilized, and is calculated according to the expression [1]:

$$\eta = \frac{f \times n}{S} \times 100, \% \tag{4}$$

Where is:

η – degree of strip usability, %
f – surface of one product, mm^2.
n – number of rows.
S – surface of strip, mm^2.

Calculation for Tool:

- Surface of one product

$$f = 189.9\ \text{mm}^2$$

- Number of rows

$$n = 2$$

- Surface of strip

$$S = 404.1\ \text{mm}^2$$

- degree of strip usability

$$\eta = \frac{189.9 \times 2}{404.1} \times 100 = 93.90\%$$

5 Conclusion

Following the worldwide trend of the development of electric cars and the general development of electric products, the aim of the work was to bring the process of manufacturing stators and rotors of electric motors closer to the general public. Stators and rotors are the basic elements of any electrical device and the process of their production is described in more detail. The technological process for manufacturing these elements is stamping.

The paper presents the constructive development of a progressive tool. The technological process of stamping to obtain the required product is presented, shown by the example of product design. The process of developing punching stations is described in detail, and the layout of strip with punching stations is shown. The design of the progressive tool is shown. Design solutions in terms of easier tool assembly, disassembly and maintenance are also presented. The maintenance procedure of the progressive tool is described. The key positions for the progressive tool are shown and described. The mounting and positioning of the progressive tool in the press was performed and the calculated force for performing the punching operation was determined.

In view of today's trend towards large-scale and mass production, and with the emphasis on quality and economy of products, it is necessary to work continuously on the development of tools and equipment.

References

1. Margić, S., Rebec, B.: Stamping Part I. Sveučilišna naklada Liber, Zagreb (1990)
2. Dinesh, K.V., Madesh, R.: Design and stress analysis of washer tool. Int. J. Appl. Theoret. Sci. Technol. **1**(8), 194–200 (2015)

3. Mastanamma, C., Prasada Rao, K., Venkateswara Rao, M.: Design and analysis of progressive tool. Int. J. Eng. Res. Technol. (IJERT) **1**(1) (2012)
4. Jaya Laxmi, B., Vijaya Kumar, Y.: Design of progressive die using FE analysis. Int. J. Res. Modern Eng. Emerg. Technol. **2**(1) (2014)
5. Suchy, I.: Handbook of Die Design, 2nd edn. McGraw Hill, Michigan (2006)
6. Özdemir, M.: Modelling and prediction of effect of machining parameters on surface roughness in turning operations. Technical Gazette **27**(3), 751–760 (2020). https://doi.org/10.17559/TV-20190320104114
7. Šarić, T., et al.: Modelling and prediction of surface roughness in CNC turning process using neural networks. Technical Gazette **27**(6), 1923–1930 (2020). https://doi.org/10.17559/TV-20200818114207
8. Mijović, B.: Maintenance of Machines and Devices. Veleučilište u Karlovcu, Karlovac (2019)
9. Rejac, E.: Terotechnology-Modern Organization of Maintenance of Funds. Informator, Zagreb (1974)

Mass Reduction Methods for Energy Efficiency Improvement of Vehicles

Ivan Palinkas(✉), Jasmina Pekez, Borivoj Novakovic, and Mica Djurdjev

University of Novi Sad, Technical Faculty "Mihajlo Pupin", Djure Djakovica BB, 23000 Zrenjanin, Republic of Serbia
ivan@tfzr.uns.ac.rs

Abstract. Energy efficiency has become imperative in design of vehicles. Energy efficiency can be viewed from multiple aspects like economy to environmental (from cost of transportation to carbon emissions). There are many factors that have influence on energy consumption in any type of transportation (by land, water or air) and one of them is vehicle mass. In this paper the current methods for reducing vehicle mass will be shown.

Keywords: Energy efficiency · Mass reduction · Vehicle

1 Introduction

In the previous century widespread availability of fuel combined with its cheapness provided increase in mobility and great expansion in the transport system. Today, due to changing circumstances, such as future availability of required quantities of fossil fuels and growing concerns about environmental consequences of transport activity, the usage of alternate energy sources and reduction of energy consumption of transport system is becoming top-priority objective of public administrations at all levels [1].

Reduction of mass (weight) represents one of the most important strategies for lowering vehicle emissions and fuel consumption. In vehicles the design of engine, transmission, hybrid, and thermal management technologies have the purpose to reduce energy losses and increase energy efficiency. A part from that, weight reduction lowers the amount of energy needed for vehicle movement, regardless of the propulsion system efficiency [2].

Lightweighting is in especially high demand for products such as [3]:

- Automobiles
- Mass transit systems
- Airplanes
- Helicopters
- Portable tools

H. Glavaš et al. (Eds.): OTO 2021, LNNS 369, pp. 71–80, 2022.
https://doi.org/10.1007/978-3-030-92851-3_5

Development of lightweighting technology brought to significant reductions in weight of the vehicle (as seen on Fig. 1), but even so, the total weight of vehicle and size increased (Fig. 2) due to integration of entertainment, safety features and a system for reduction of exhaust emission. The cause of this are safety regulations and consumer preferences for more spacious and comfortable cars [4].

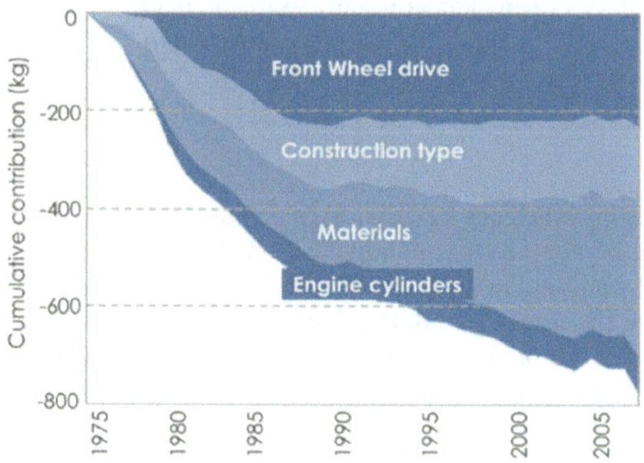

Fig. 1. Cumulative contribution of different approaches to weight reduction in vehicles [5].

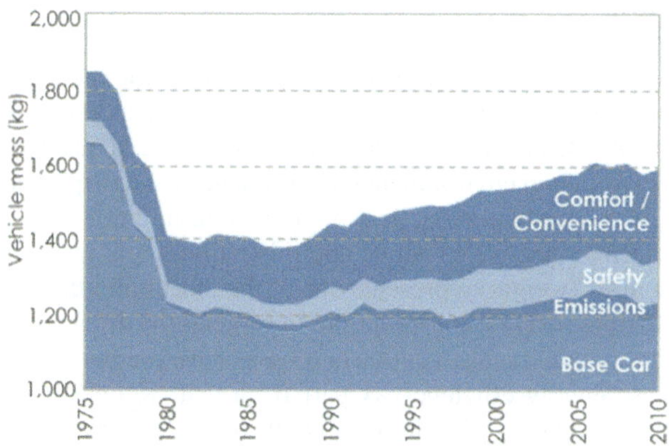

Fig. 2. The evolution of average passenger car mass between 1975 and 2010 [6].

There are several factors that influence on development of lightweighting technologies today (Fig. 3).

Of course, industries that maximize the lightweighting application are automotive and aviation. But due to mentioned drivers of change, it can be seen that lightweighting have its purpose in various applications, from construction (where the purpose of lightweighting is to decrease weight and amount of used materials without the sacrifice of structural integrity) to every vehicle from bicycles to drones.

In rest of the paper the methods for lightweighting will be described.

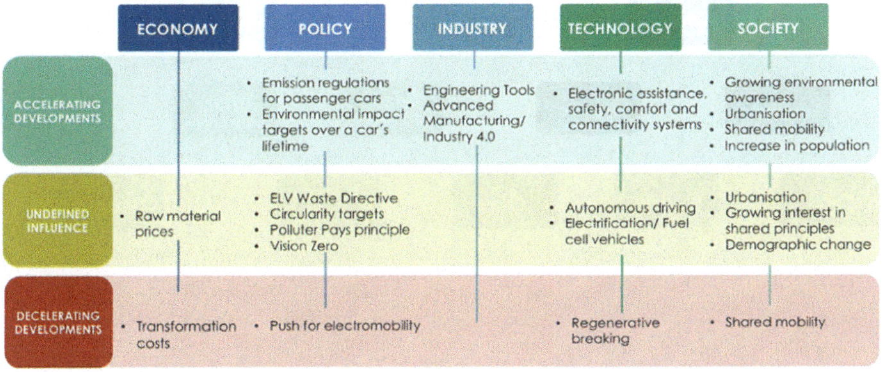

Fig. 3. Drivers of change and their impact on lightweighting development [4].

2 Methods for Lightweighting

2.1 Usage of Different Materials

One way to reduce vehicle mass is to replace traditional manufacturing materials that have high weight with lightweight materials. There are several performance factors that must be taken in consideration when choosing lightweight material: [4].

1. Technical performance – material must be in compliance with mechanical and thermal requirements.
2. Lightweighting potential – characteristics of materials (low density, high strength or their combination through weight-to-strength ratio).
3. Environmental impact – energy demands and emissions from obtaining raw material and manufacturing into application as well as usage of other consumables in these processes (water etc.).
4. Recyclability – possibility to gain materials from manufacturing scraps in order to reduce impact from extraction (economical and environmental impact).
5. Manufacturability – the ease with which material can be manufactured into components. Companies invests in infrastructure and manufacturing capacities, and with development of new materials with different properties there is need for different processing approaches.
6. Material compatibility – ability of material technology to be combined with other materials using existing joining technologies.
7. Cost of lightweighting – represents the additional cost required to reduce weight.

Materials for lightweighting can be viewed from two aspects: the material type and material structure (Fig. 4).

On Fig. 4 are shown the most relevant materials and structures but, depending on industry that uses them, there are more specific materials (for example, in aerospace industry, besides mentioned, materials used for lightweighting are also titanium, ceramics etc.).

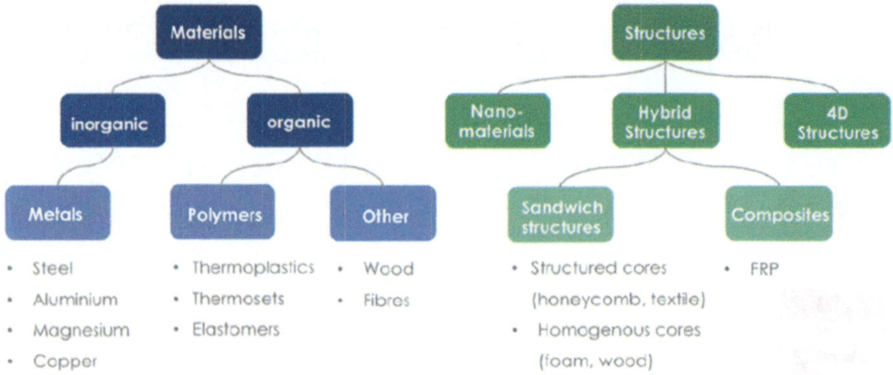

Fig. 4. Overview of the most relevant lightweight materials and structures [4].

On Fig. 5 is diagram that show the usage the materials in automotive industry in year 1995. compared to year 2008. From the figure it can be seen that usage of iron castings is declining. Usage of steel is rising because of development of steel grades (Fig. 6) that provides better strength-to-weight ratio.

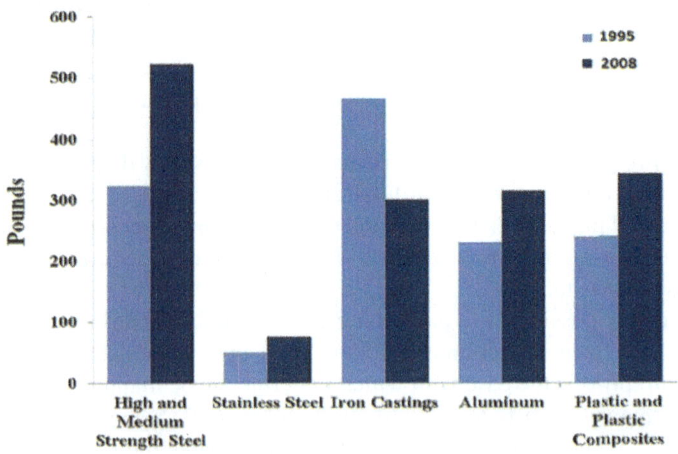

Fig. 5. Growth of advanced materials in automotive for lightweighting and Safety [7].

On Fig. 7 it is shown the percentage usage of lightweight materials in aviation an automotive industry.

Lightweight materials (light metals, aluminum, plastics and composites) already makes around 80% of airplane. The purpose of lightweighting in airplane has two parts: reducing the fuel consumption and increasing the number of passengers and amount of luggage that airplane can carry. Lightweighting in heavy transport have the same purpose.

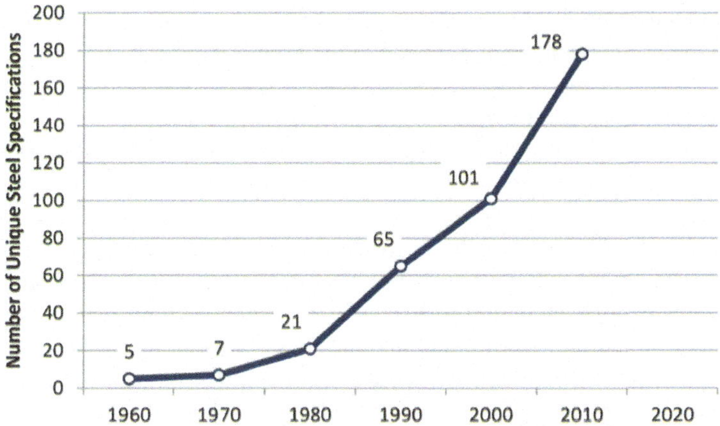

Fig. 6. Growth in steel grades – higher strength steels [8].

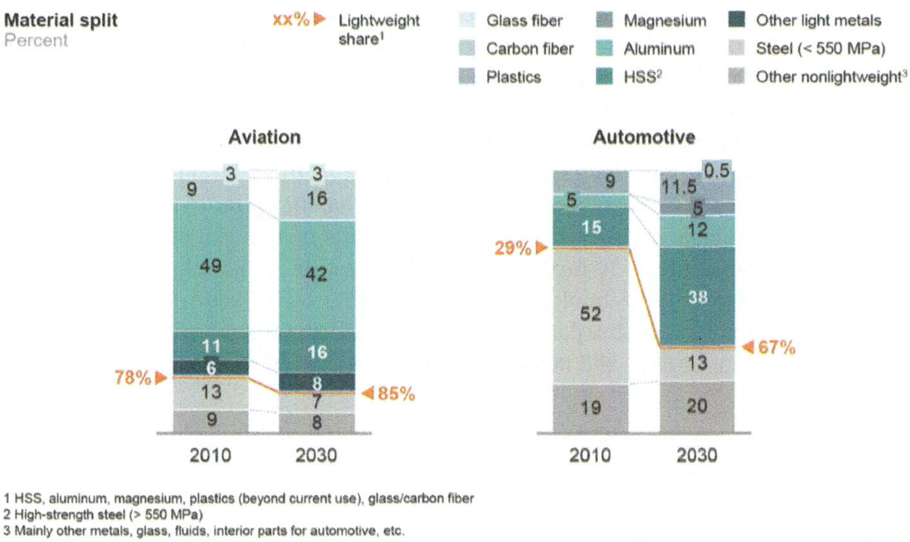

Fig. 7. Usage of lightweight material in aviation and automotive industry [9].

Current research, especially in automotive industry go in the direction of development of composite materials. Composite materials have the greatest strength-to-weight ratio in comparison to other lightweight materials.

2.2 Technology Used in Lightweighting

In the next segment the some of the specific technologies used in lightweighting process will be mentioned. They can be divided in two sections: first represents the technologies in order to manufacture part using lightweight materials and the second represents the ways of joining parts made from lightweight materials [10].

Shaping/Fabricating Technologies

Some specific technologies form manufacturing parts from lightweight materials are [10]:

- Warm forming – is process used in forming parts while sheet is warm in order to improve formability (strain rate sensitivity of the material is much greater allowing enhanced formability including superplastic deformation). This technology provides the means to form part from materials that lacks formability of steels. One of the application is quick plastic forming process for car parts.
- Hot stamping – is similar to warm forming process with the difference that is used to form parts from steel. Using this process parts with exceptional strength level can be produced using specially designed alloys, that influences on vehicle safety.
- Impulse forming – is process where very high forming velocity (50–500 m/s) is used. Achieved improvement using this process, from conventional forming, is improved formability, attained precision shape control, high fidelity embossing and possibility of precision shearing with high quality.
- High-integrity casting of light alloys – is process that offers inexpensive mass production of complex shaped components. In the past, the problem with this process was that parts had poor ductility, toughness and sometimes strenght. But, with the improvement made in evacuating the die-casting chamber prior to casting and using pressure to eliminate porosity, characteristics of parts increase significantly.
- Additive manufacture – is process providing creation of part layer by layer (in contrast to subtractive technologies such as milling etc.). Additive manufacturing enables creation of complex parts with defined infill structures using materials such as thermoplastic, ceramic and metal. It also enables reinforced structures with carbon fiber.
- Beam structures – usage of beam structures (through different materials and shapes) that are joined in order to achieve stiff and strong structural parts of vehicle.

Joining Technologies

Joining technologies in lightweighting are necessary in order to connect special materials or different materials. Some of these technologies are [10]:

- Friction stir welding - is based on the solid state intense mixing of the metals of two components and can join materials in lines or spots. This technology is used in automotive industry.
- Impact welding – consists of metal impact with high velocity and under certain angle generated by explosive or magnetic pulse.
- Self-piercing rivets – technology used for joining two very different materials (for example metal with plastics). Rivets in this technology can be made from aluminium.

- Structural adhesives – adhesives are also used for joining two dissimilar materials, but often used as addition to rivets or similar joining technologies. It can be used as individual joining method but there are concerns with joining parts that needs to be in fixed position until adhesive dries.
- Conformal joining – is technology that uses part design such that features abut one another and eliminate motion. Structures that provide great strength can be made by adjoining of interfering structures

2.3 Design and Optimization in Lightweighting

In design process of vehicle, parts and assemblies are optimized while retaining their functionality and demanded characteristics. The design and optimization of parts, through development of Computer Aided Design/Computer Aided Manufacturing/Computer Aided Engineering (CAD/CAM/CAE) technologies have great impact on lightweighting. Reducing weight and improving mechanical properties can now be done through simulation with usage of Finite Element Analysis and variety of optimization types.

Structural Optimization Types

There are several types of structural optimization methods that is used in design of lightweight parts. One of the classification is by the type of design variables [11]:

- Topology optimization – design variables correspond to the element volume fractions that designs simultaneously the material properties (modulus of elasticity) and density in order to identify what elements to keep and what to discard.
- Sizing optimization – design variables are the element of cross-sectional dimensions, where all elements associated to a property data entry are designed with the same value.
- Topometry optimization – represents a generalization of sizing optimization where each element is designed independently.
- Shape optimization – design variables are scale factors of perturbation vectors that are input directly or by providing basis vectors (containing alternative grid locations) that can be internally converted into perturbation vectors (vectorial difference between basis vector and the original grid locations).
- Topography optimization – special case of shape optimization where perturbation vectors are either perpendicular to the designable region or in a given direction.
- Freeform optimization – special case of shape optimization where program splits any given perturbation into multiple perturbations on a grid by grid basis which increases the variability of design space.

Application of structural optimization can be seen on Fig. 8. There is shown the optimized design of plate. Goal in this example was to improve the stiffness and reduce weight of the component.

With development of additive manufacturing technologies, the design and manufacturing of complex parts has made possible. Optimization provided through additive

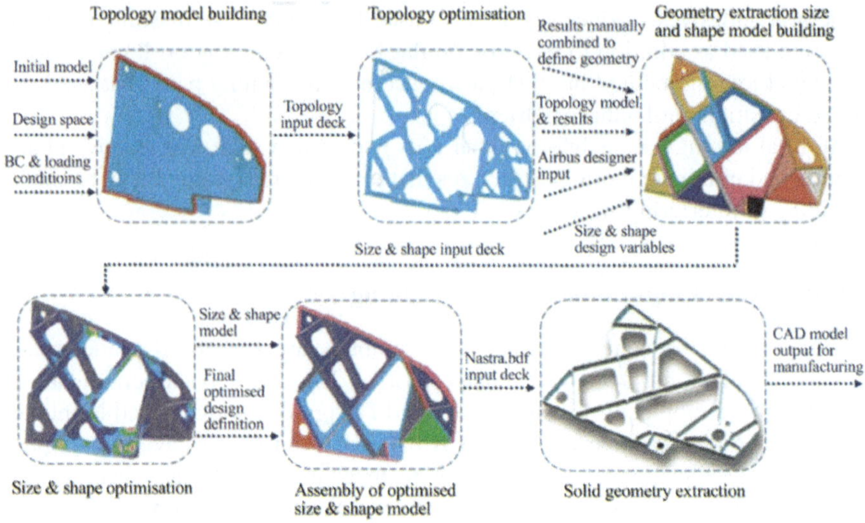

Fig. 8. Example of combination of topology, sizing and shape optimization methods on the design of plate component [12].

manufacturing have one more step and that is creation of infill structures in different configurations in order to decrease the amount of used material without the degradation of part mechanical properties.

From that the multi-scale optimization found its purpose. Additive manufacturing provides realizations of parts with topological optimization in macro-scale and lattice structure in meso-scale, and structural elastic properties can be made with designed gradients by lattice structure optimization [13].

On Fig. 9 can be seen the optimized structures of automotive bracket and lightweight titanium seatbelt buckle used in airplanes made by additive manufacturing.

Further improvement of lattice structures is through usage of bionic and mathematical inspirations seen on Fig. 10.

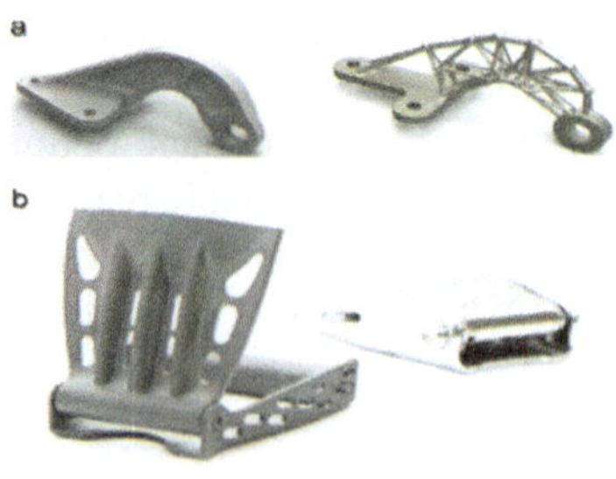

Fig. 9. Examples of optimized structures for manufacturing by additive technologies: a. optimization of automotive bracket; b. 3D printed lightweight titanium seatbelt buckle used in airplanes [13–15]

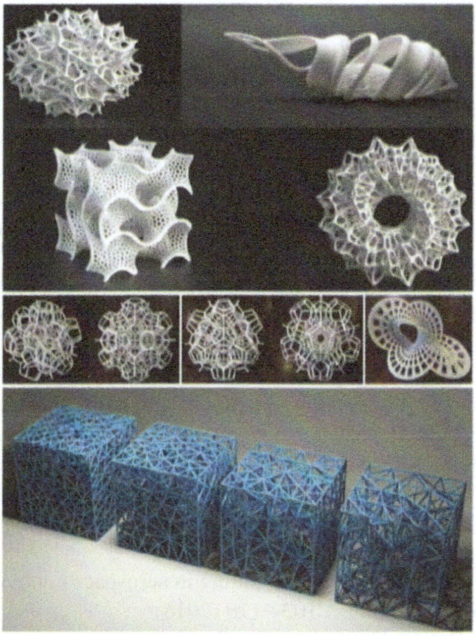

Fig. 10. Examples of AM lattice structures inspired by bionic design and mathematic [13, 16–18].

3 Conclusion

Application of lightweighting technology on transportation have great impact on energy efficiency, environmental protection and also economy. The methods shown in this paper are intertwined. Reduction of mass is achieved through optimized design using lightweight materials. Further development of lightweighting in transportation will reflected through transportation end environmental policies and through development of other technologies for increase of energy efficiency (such as alternative fuels etc.)

References

1. López, E., Monzón, A., Pfaffenbichler, P.C.: Assessment of energy efficiency and sustainability scenarios in the transport system. Eur. Transp. Res. Rev. **4**(1), 47–56 (2011). https://doi.org/10.1007/s12544-011-0063-4
2. Isenstadt, A., et al.: Lightweighting technology development and trends in U.S. passenger vehicles. Int. Council Clean Transp. 1–24 (2016)
3. What are the benefits and challenges of lightweight manufacturing?, https://gesrepair.com/what-are-the-benefits-and-challenges-of-lightweight-manufacturing/. Accessed 27Aug 2021
4. Reiland, J., Bax, L., Ierides, M.: A Vision on the Future of Automotive Lightweighting. Bax & Company, Barcelona (2019)
5. MacKenzie, D., Zoepf, S., Heywood, J.: Determinants of US passenger car weight. Int. J. Veh. Des. **65**(1), 73–93 (2014). https://doi.org/10.1504/IJVD.2014.060066
6. Zoepf, S.M., Automotive Features: Mass Impact and Deployment Characterization, Massachusetts Institute of Technology (MIT) (2010)

7. Office of Energy Efficiency & Renewable Energy, Vehicle Technologies Office, Fact #642: September 27, 2010 Material Content per Light Vehicle, 1995 and 2008, https://www.energy.gov/eere/vehicles/fact-642-september-27-2010-material-content-light-vehicle-1995-and-2008. Accessed 28 Aug 2021
8. Baron, J.: Identifying Real World Barriers to Implementing Lightweighting Technologies and Challenges in Estimating the Increase in Costs, Center for automotive research, (2016)
9. Lightweight, heavy impact. https://www.mckinsey.com/~/media/mckinsey/dotcom/client_service/automotive%20and%20assembly/pdfs/lightweight_heavy_impact.ashx. Accessed 27 Aug 2021
10. Daehn, G.S.: Sustainable design and manufacture of lightweight vehicle structures. In: Alternative Fuels and Advanced Vehicle Technologies for Improved Environmental Performance, Science Direct, pp. 433–461 (2014). https://doi.org/10.1533/9780857097422.2.433
11. Palinkas, I., Desnica, E., Pekez, J., Radovanovic, L.J.: Modern approaches on construction optimization in mechanical engineering. In: Raos, P., Galeta, T., Kozak, D., Raos, M., Stojšić, J., Tonković, Z. (eds.) 10th International Natural Gas, Heat And Water Conference, pp. 260–264 Strojarski fakultet u Slavonskom Brodu, Osijek (2019)
12. Krog, L., Tucker, A., Rollema, G.: Application of topology, sizing and shape optimization methods to optimal design of aircraft components. In: Proceedings of the 3rd AltarirUK HyperWorks Users Conference (2002)
13. Zhu, L., Li, N., Childs, P.R.N.: Light-weighting in aerospace component and system design. Propulsion and Power Research, **7**, 103−119 (2018)
14. Austin-Morgan, T.: Design Optimisation for Additive Manufacturing (2016). https://www.eurekamagazine.co.uk/design-engineering-news/design-optimisation-for-additive-manufacturing/143016/. Accessed 28 Aug 2021
15. T.R. Ltd. (2006). https://www.3trpd.co.uk/portfolio/saving-project-saving-litres-of-aviation-fuel/
16. https://3dprintingforbeginners.com/category/printing-services/. Accessed 28 Aug 2021
17. Schleimer, S., Sagerman, H.: 3D printing mathematics. https://plus.maths.org/content/3d-printing. Accessed 28 Aug 2021
18. 4DID (2013). http://www.4did.net/3d_modeling.html

Development of Compact 12 V Automotive Maintenance Vacuum Pump

Josip Cumin(✉), Daniel Novoselović, and Mirko Karakašić

Mechanical Engineering Faculty, University of Slavonski Brod, Trg. I. B. Mazuranic 2, 35000 Slavonski Brod, Croatia
jcumin@unisb.hr

Abstract. In the automotive industry there is a requirement for quick and clean brake oil maintenance. Since braking oils (DOT 4, DOT 5.1) are hygroscopic, it is recommended that braking oil should be changed every two years. Modern cars are made with intention of making service in specialized workshops which is related to OBD compliant information exchange (date of oil change, factory reset of service interval, special activation of brake servo-motors for oil bleeding etc.). Older cars and especially motorcycles are often serviced in DIY variants and there is a requirement for various kinds of vacuum pumps. Most often used vacuum pumps in automotive service are hand-driven pumps. In this paper, the small electric membrane-operated vacuum pump is utilized for the brake line oil bleeding of cars and motorcycles. The technical data (voltage, running current, operating vacuum) is shown, as well as the technical drawings and bill of materials for the build.

Keywords: Electrical · Vacuum pump · Brake line · Bleeding introduction

1 Introduction

In the automotive industry there is a requirement for quick and clean brake oil maintenance. Since braking oils (DOT 4, DOT 5.1) are hygroscopic, it is recommended that braking oil should be changed every two years.

During the car MOT (TÜV) which is performed every year/every 2 years, depending on country and legislative, the brake oil is tested [1]. This test is performed in order to determine braking oil boiling point [1]. When braking oil absorbs moisture, its boiling point lowers, thus creating conditions for instant creation of high-pressure bubble of evaporated oil. In difficult conditions such as constant braking on downslopes and serpentines, the disc brakes can be overheated, and brakes can be locked-up, or even catastrophic failure can occur.

Modern cars are made with intention of making service in specialized workshops which is related to OBD compliant information exchange (date of oil change, factory reset of service interval, special activation of brake servo-motors for oil bleeding etc.).

Older cars and especially motorcycles are often serviced in DIY variants or small shops and there is a requirement for various kinds of vacuum pumps. Most often used

H. Glavaš et al. (Eds.): OTO 2021, LNNS 369, pp. 81–91, 2022.
https://doi.org/10.1007/978-3-030-92851-3_6

vacuum pumps in automotive service are hand-driven pumps like popular Mityvac oil bleeder [2]. Technical data for Mityvac MV8020 shows that the pump is capable of producing up to 0,85 bar of operating vacuum.

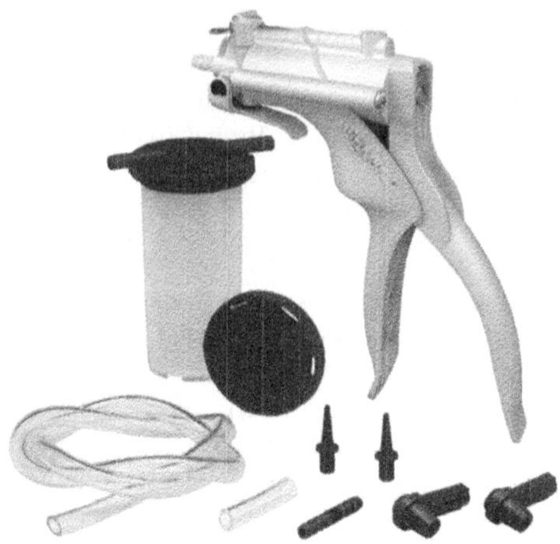

Fig. 1. Mityvac MV8020 brake bleeding kit [2].

For the bleeding operation with the usage of hand vacuum pump from Fig. 1, the operator's hand is busy which can be quite impractical in closed spaces under the car or motorcycle. For this reason, the idea for the 12 V vacuum pump was born.

2 Electrical Vacuum Pump

For this part, the two easily available, electrically operated 12 V vacuum pumps were obtained. Pump with 5 l/min capacity is shown in Fig. 2 (lower pump), while technical data is given in Table 1.

For the larger capacity 40 l/min pump, technical data is given in Table 2, and pump itself is shown in Fig. 1 (upper pump).

Table 1. Technical data for 12 V vacuum pumps.

Data:	5 l/min pump	40 l/min pump
Operating voltage	9 ÷ 14 V (12V)	9 ÷ 14 V (12 V)
Vacuum	≅0.65 bar	≅0.85 bar
Power	6 W	42 W
RPM	3000 min−1	3000 min−1
Coil resistance	6.5 Ω	0.9 Ω

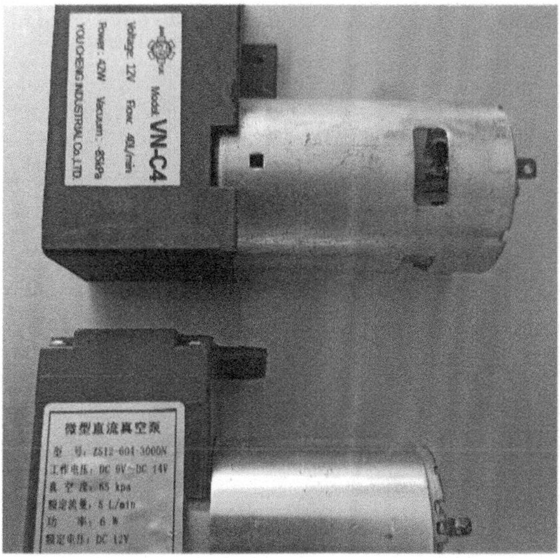

Fig. 2. Electrical 12 V vacuum pumps utilized for brake bleeder (upper – 40 l/min; lower 5 l/min capacity)

On the test bench, a simple fixture was assembled and the vaccum pumps were tested without loads (without oil canister). The data obtained is shown in Table 2 (5 l/min pump), and Table 3 (40 l/min pump).

Table 2. Pump working parameters (5 l/min version)

No:	Voltage/V	Current/A	Vacuum/bar
1	12.6	0.42	0
2	12.6	0.43	0.1
3	12.6	0.46	0.2
4	12.6	0.48	0.3
5	12.6	0.48	0.4
6	12.6	0.47	0.5
7	12.6	0.43	0.6
8	12.6	0.39	0.65
9	12.6	0.3	0.7

Table 3. Pump working parameters (40 l/min version)

No:	Voltage/V	Current/A	Vacuum/bar
1	12.5	3.6	0
2	12.5	3.6	0.1
3	12.5	3.63	0.2
4	12.5	3.96	0.3
5	12.5	3.72	0.4
6	12.5	3.64	0.5
7	12.5	3.55	0.6
8	12.5	3.45	0.65
9	12.5	3.2	0.7
10	12.7	3	0.75
11	12.7	2.68	0.8
12	12.7	2.19	0.85

3 Pump Chassis

For the pump chassis. aluminium AlMg3 (EN AW 5754/3.3535) which has following properties was used [3, 4]:

- hardness 52 ÷ 88 HB
- Young's modulus $E = 68$ GPa
- $\varepsilon_{\text{max}} = 2 \div 19\%$
- yielding strength $R_{\text{p}0.2} = 80 \div 280$ MPa
- ultimate tensile strength $R_{\text{m}} = 200 \div 330$ MPa

Figure 3 shows base of the chassis which consists of L-profile 25 × 25 × 2 mm welded together in the rectangular form with dimensions 210 × 240 mm. A flat cross section 45 × 2 mm was welded to the frame as a support for the vacuum pump carrier.

On top of the frame. the sheet plate of dimensions 200 × 230 × 2 mm was welded as well. In this way high rigidity is obtained.

Figure 4 shows welded base frame with circular cross-section pipe of $\phi 20 \times 3$ mm prepared and welded at angle.

Fig. 3. Base frame of the chassis.

Fig. 4. Handle welded on bottom base frame.

On the bottom section from Fig. 3 the same 25 × 25 × 2 mm profile was welded vertically to accommodate jar. Choice of a simple jar was based on common availability, easy manipulation of waste oil visibility of fluid and good sealing properties. Pneumatic 6 mm hose connector was soldered on top of a jar lid. carefully not to melt or damage the lid seal.

Universal and adjustable pipe carrier was utilized to mount the vacuum pump itself [5]. Electrical system is very simple. Pump gets power form 12 V battery. The electrical circuit is switched on by switch located in the handle of chassis for easy access.

Figure 5 shows finished prototype of electrical vacuum pump.

Fig. 5. Electrical vacuum pump prototype.

4 Pump Operating Characteristics

Vacuum pump working parameters for both pumps are shown in Tables 1, 2 and 3. Now, the assembly of the prototype changes behavior of the pumps since the jar has it's own volume and the whole system operates on the basis of open pneumatic system where intake pressure and flow can be monitored through connecting pipe then into jar. pump itself and through exhaust port of the pump.

Efficiency of the system depends on the losses (resistance in the pipes. connectors. viscosity of the fluid etc.). Vacuum pump assembly has proven itself very versatile in the shop. The prototype was often used for sucking-out used brake oil (DOT4). over-filled motorcycle oil from the engine or oil tank for the extraction of ATF oil from servo steering system for the extraction of cooling fluid for the extraction of excess oil from front forks of several motorcycles since the oil level is set by repair manual and heavily affects motorcycle forks compression and damping characteristics. In the Fig. 6 the front fork oil level for Yamaha Bulldog BT 1100 is shown [6]. The same motorcycle had been repaired and the oil level vas easily obtained on site by prototype vacuum pump just seconds after pump was started.

For shop-oriented usage of the prototype some technical parameters were measured and are shown in Tables 4 and 5 (based on the suction of 350 ml of fluid as shown in the Fig. 7). The choice of these types of fluids is based on the variety of oils used in the automotive industry from common 10 W–40 multi-grade oils for older types of ICE engines up to newer type of multi-grade low friction oils (which have additives and are purposed for newer low friction engine types) like 5 W–30 [7].

Data shown in Tables 4 and 5 is in correlation with data from Tables 2 and 3. Here is shown that high viscosity fluids such as heavy motorcycle oils tend to create high vacuum

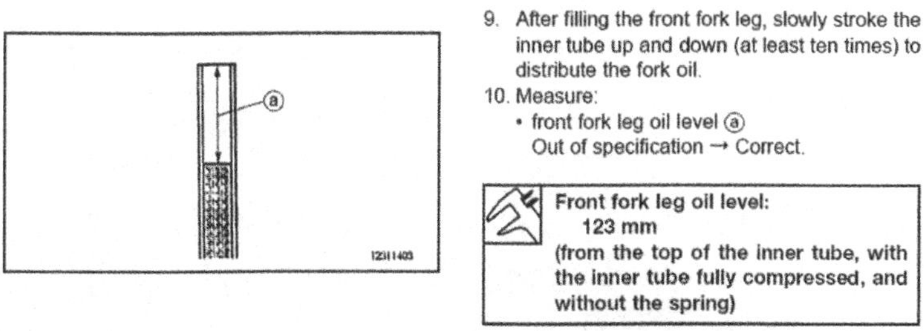

Fig. 6. Yamaha Bulldog front fork oil level [6].

Table 4. Operating technical data for the suction of various fluids for 40 l/min pump.

Fluid:	Voltage. V	Current. A	Suction time. s
Braking oil DOT4	12.8	3.2	7.2
Motul 300 V 10 W–40	12.6	2.2	33.9
Motul Motocool coolant (−37 °C)	12.6	3.6	5.9
Total Quartz Ineo ECS 5W–30	12.3	2.2	19.5
INA Hipenol GTL SAE 90	12.6	2.2	1:32.6 min

Table 5. Operating technical data for the suction of various fluids for 5 l/min pump.

Fluid:	Voltage. V	Current. A	Suction time. s
Braking oil DOT4	12.6	0.53	13.4
Motul 300V 10W–40	12.6	0.45	40.3
Motul Motocool coolant (−37 °C)	12.5	0.41	10.8
Total Quartz Ineo ECS 5W–30	12.6	0.45	29.2
INA Hipenol GTL SAE 90	12.6	0.32	1:43.5 min

on the suction side of the pump and the current of the pump drops. Similar, when pump is used to suck-in low viscosity fluid such as water, coolant or brake oil – there is high fluid flow and pump rotates at higher rpm. demanding more current.

In the Fig. 7 maintenance of braking oil is shown. Old brake oil removal from the brake oil container is shown (motorcycle KTM 990 Supermoto 2007) (Fig. 8).

Figure 9 shows front brake bleeding on KTM 950 supermoto motorcycle. As opposed to conventional bleeding procedure this approach is effortless (Conventional brake bleeding process: press brake lever. while lever is being pressed open bleed screw with 8 mm wrench for 90–180°, then oil is bled from the caliper. turn back bleeding screw, release brake lever. pump brake lever few times and repeat again).

On electric pump the operator just needs to connect bleeding hose to the bleed screw nipple, turn on the pump and pour fresh oil in the container/maintain oil level.

Fig. 7. Bench setup for measuring vacuum pump operating conditions in sucking 350 ml of various fluids (data shown in Tables 4 and 5).

Fig. 8. Old brake oil removal from the brake oil container. KTM 950 supermoto.

Fig. 9. Front brake bleeding. KTM 950 supermoto.

5 Bill of Materials

Here. the bill of materials (BOM) for the chassis is given.

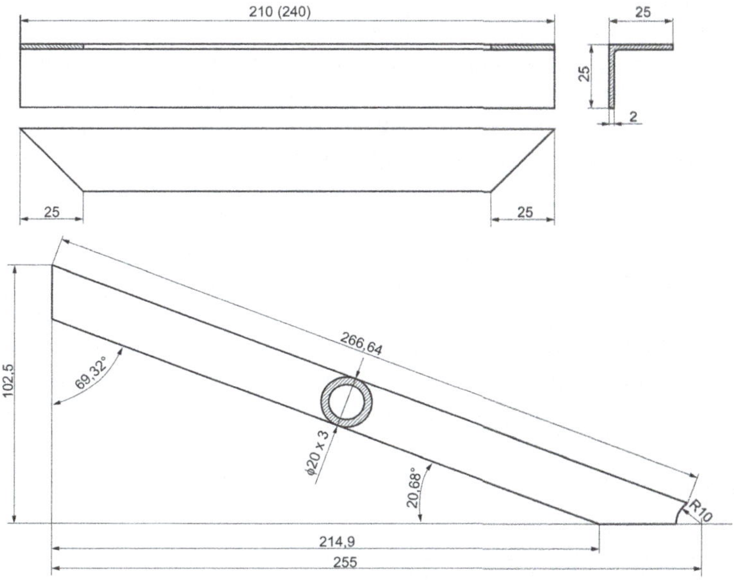

Fig. 10. Aluminium profiles for the chassis of prototype.

Bill of materials:

- 2 pieces of AlMg3 L-profile of dimensions 25 × 25 × 2 mm. length 210 mm
- 2 pieces of AlMg3 L-profile of dimensions 25 × 25 × 2 mm. length 240 mm
- 2 pieces of AlMg3 L-profile of dimensions 25 × 25 × 2 mm. length 100 mm
- 2 pieces od AlMg3 pipe. ϕ20 (outside diameter) × ϕ14 (inside diameter). cut as shown in Fig. 10
- 1 piece of AlMg3 pipe. ϕ20 × 3 mm. length 180 mm
- 1 piece of AlMg3 plate with dimensions 200 × 230 × 2 mm

All the welding was not prescribed by WPS it was done with empirical welding parameters in home shop with TIG welder and AlMg3 (5754) welding wire in the shielding of technically pure argon gas. The only difference to the standard TIG welding machine was usage of TIG gas lens for WP26 torch with No. 7 nozzle and 2.4 mm tungsten rod.

For the future work, there is plan for new chassis which will incorporate high power, light weight Li-Po battery. Li-Po batteries are less prone to discharging. have high energy at demand and are largely used in the electrical automotive industry [8]. Downside is that they require special smart chargers and battery management service (BMS) [9].

6 Conclusion

In this article the prototype of vacuum 12 V. electrically driven vacuum pump is shown. Technical data for the 5 l/min and 40 l/min pumps were measured and are given in the respective tables of this paper.

The bill of materials (BOM) for the prototype chassis is listed. as well as technical parameters for the AlMg3 material from which chassis was made.

For the mechanical shops technical data for the suction of various automotive fluids is given. Pump itself was proven in the field mostly for removal of old braking fluid from the braking fluid reservoir and for the brake line oil bleeding procedure where the operator has both hands free as opposed to the standard hand operated Mityvac-like vacuum pumps.

For the future work, there is plan for new chassis which will incorporate high power, light weight Li-Po battery. Li-Po batteries are less prone to discharging. have high energy at demand and are largely used in the electrical automotive industry. Downside is that they require special smart chargers and battery management service (BMS). This new chassis will be smaller since this battery is much lighter and smaller than lead-acid one. Also. plan is to install voltmeter/ampere meter in the chassis to monitor state of the Li-Po battery since they must not be overly discharged.

References

1. Centar za vozila Hrvatskke (CVH): Nova oprema u stanicama za tehnički pregled vozila. Stručni bilten, vol. 165, pp. 28–54. https://www.cvh.hr/media/2926/cvh_bilten_165_nova-oprema_2018_web.pdf. Accessed 29 July 2021

2. SKF: Brake bleeding kit MV8020. https://www.skf.com/mityvac/products/hand-vacuum-pressure-pumps/hand-pump-kits/mv8020. Accessed 29 July2021
3. The steel.com: Aluminij lim AlMg3 - H22. https://www.thesteel.com/hr/Aluminij-lim-AlMg3-H22-T440120004. Accessed 29 July2021
4. Makeitfrom.com: 5754 (AlMg3. 3.3535. A95754) Aluminum. https://www.makeitfrom.com/material-properties/5754-AlMg3-3.3535-A95754-Aluminum. Accessed 29 July2021
5. Fero Term: Cijevna obujmica. https://www.fero-term.hr/obujmica-za-cijev-50-55-s-gum-6-4-bez-vijka-i-tiple/prid/8889. Accessed 29 July2021
6. Yamaha. BT 1100 5JN1-AE1 Service Manual. Belgarda S.p.a, Italy (2001)
7. Zoldy, M.: Engine oil test method development. Tehnički vjesnik. **28**(3), 1012–1016 (2021)
8. Barcellona, S., Brenna, M., Foiadelli, F., Longo, M., Piegari, L.: Analysis of ageing effect on li-polymer batteries. Sci. World J. **2015**, 1–8 (2015)
9. Rimpas, D., Kiatipis, A.: Charging strategy effect on lithium polymer battery capacity: a case study. Int. J. Energy Environ. **11**(2), 107–118 (2020)

Techno-Economic Analysis of the Justification of the Machine Part Reparation Procedure on the Excavator

Eleonora Desnica[1(✉)], Vladimir Jakovljević[2], Imre Kiss[3], Ljiljana Radovanović[1], and Luka Đorđević[1]

[1] Technical Faculty "Mihajlo Pupin", University of Novi Sad, Zrenjanin, Republic of Serbia
desnica@tfzr.uns.ac.rs

[2] EPS, RB Kolubara, Lazarevac, Republic of Serbia

[3] Faculty of Engineering, University Politehnica Timisoara, Hunedoara, Romania

Abstract. Dragline excavator EŠ 5/45 is used in the exploitation of coal in open-pit mines RB Kolubara. It is a complex technical system consisting of a larger number of subsystems, assemblies, and elements. Therefore, the maintenance of this machine implies a set of complex procedures, methods, and various repair activities that need to be undertaken to keep the excavator in good (working) condition and to work with the allowed level of reliability and the optimal level of productivity and economy during its life cycle. The paper presents an algorithm and description of activities undertaken during the repair of the machine part - the jaws of the jaw clutch coupling of the electric motor on the excavator EŠ 5/45, including the techno-economic analysis of the justification of this reparation procedure.

Keywords: Excavator · Jaw clutch coupling · Reparation · Techno-economic analysis

1 Introduction

During the exploitation of open-pit mines, unexpected breakdowns on machines inevitably occur, even though the planned measures have been taken as part of regular maintenance. By maintenance of mining machines, we mean the overhaul of all assemblies that are an integral part of these machines. This activity takes place according to the established maintenance plan for the current year or as needed.

For simpler failures, preparation is not particularly important, as they can often be fixed by general maintenance personnel without the use of spare parts [1, 2].

For more complex works, preparation is important and often takes a significant part of the time of the overall repair. That is why the handy spare parts warehouses had been established as one of the measures for faster intervention.

Also, the assemblies are regenerated in the workshop and, as such, are placed in the warehouse so that they can be ready to mount in case of a sudden failure. The principle of aggregate replacement of assemblies is quite present, and there the complete aggregate

H. Glavaš et al. (Eds.): OTO 2021, LNNS 369, pp. 92–103, 2022.
https://doi.org/10.1007/978-3-030-92851-3_7

(motor, clutch, brake drum coupling, gear unit) is removed from the subject and sent for repair, and the complete one, previously regenerated, is set instead [3].

The size of machine failures determines the place of repair either on the work site, in a open-pit mine's workshop, or in a specialized workshop. The role of the maintenance team, in addition to the appropriate quality and experience of the staff, requires a good supply of appropriate tools, spare parts, as well as good off workshop mobility [3, 4].

Repair is a precisely prescribed sequence of operations that should be consistently performed on damaged and wrecked machine parts or constructions in order to restore the original condition and regain working capacity. The most important result of the repair is economically in the form of cost reduction since this procedure avoids the production of a new part that would replace the damaged one [5–7].

In this paper, in order to make an adequate decision based on the techno-economic analysis of the reparation of the machine part - the jaws of the jaw clutch coupling of the electric motor on the excavator EŠ 5/45, a repair algorithm consisting of a series of specific procedures is set.

2 Dragline Excavator EŠ 5/45

The machine type EŠ 5/45, used in the exploitation of coal in the open-pit mines of RB Kolubara, belongs to the category of excavators with a pulling bucket, the so-called dragline excavator. A dragline excavator is used on surface mines for excavation and loading or excavation and direct transfer of excavated masses, both soft and loose masses, as well as solid but previously crushed (mined) masses. The dragline excavator is structurally destined for excavating the material below the ground level (zero line) on which it is located, i.e., for in-depth work. It, therefore, digs up the material "towards itself". The appearance of this excavator is presented in Fig. 1 [1, 2].

Fig. 1. The dragline excavator EŠ 5/45 [1, 2]

3 Reparation of the Jaws on the Jaw Clutch Coupling of the Electric Motor on the Excavator

3.1 Jaw Clutch Coupling

Couplings connect two coaxial or approximately coaxial rotating elements in order to transmit torque and rotational motion. Couplings are most often used to connect two shafts or parts of the same shaft. They are also used to connect shafts with elements for transmitting circular motion (e.g., gears, sprockets, etc.). They are placed at the ends of the shafts as close to the bearings as possible. The following are often connected: the shaft of the propulsion machine with the shaft of the working machine, the transmission shaft, etc. [8, 9].

The jaw clutch coupling of the dragline excavator's EŠ 5/45 electric motor belongs to the flexible jaw clutch couplings group. These couplings consist of two rims. Each rim has a certain number of jaws on the front surface and the same number of interspaces by which a large force is transmitted from one shaft to another. In terms of construction, in relation to rigid jaw clutch coupling, in these couplings, the front grooves are larger than the jaws so that a flexible muffled element (inlay) can be placed between the metal claws (Fig. 2) [1, 8, 9].

Fig. 2. Flexible jaw clutch couplings with elastic inserts

3.2 Coupling Failure Diagnostics

Before starting the repair process, it is necessary to determine why there is partial or complete damage, whether due to the service life or due to the operating conditions. A more common case is the operating conditions that are not appropriate, i.e. the elements in the system are outside the standard limits of operation and functioning. The authors of the paper [10], in their research, clearly state that one of the main problems of couplings is the increased level of vibration on machines, electric motors. Given this statement, it is necessary to approach a modern method of vibrodiagnostics and analysis of the spectrum of the state of machines in which vibration excitation occurs. The method of vibration measurement is probably the best-known method within the scope of this concept, as the most widespread concept in this segment [11].

Mechanical vibration measurements are performed using sensors in the form of anaccelerometer. Depending on the analysis, two types of accelerometers can be used, one for measuring the vibration velocity mm/s and the other for analyzing the condition of bearings, gears (Pk Accelaration G's).

Also, whether the operating conditions affect the coupling can be found out through the method of diagnostics of the alignment of the input shaft and the output shaft from the coupling. The alignment of the shafts is checked by the laser method, and the causes of the coupling defect can be concluded on the basis of the report [12].

The coupling of shaft misalignment and crack in rotor systems will produce more complex non-linear dynamic characteristics in the vibration response and increase the difficulty of the fault identification greatly [13].

3.3 Algorithm and Description of Activities During Jaw Clutch Coupling Reparation [14]

To make an adequate decision based on the techno-economic analysis of the machine part's reparation, it is necessary to set the reparation algorithm. The algorithm for repairing the jaw clutch coupling of an electric motor consists of a series of specific procedures (Fig. 3):

1. disassembly;
2. washing, cleaning and degreasing;
3. defect plan;
4. techno-economic analysis and decision;
5. preparation of technical documentation;
6. elaboration of the technological procedure of reparation;
7. reparation procedure;
8. final control;
9. assembly, trial operation and running-in.

During the coupling **disassembling**, it must be taken care not to cause any additional damage due to carelessness. When disassembling, it is necessary to take into account:

- not to damage parts of the assembly;
- not to cause additional damage to the already damaged machine part;
- preserve the original damage.

While removing the coupling from the shaft, take care not to damage the inner surface of the coupling that is in contact with the shaft. Also, pay attention not to damage the existing shaft keyway.

After disassembling the coupling, it is **washed, cleaned and degreased**. This operation aims to see the type and amount of damage to the coupling. Cleaning means removing dirt and deposits from all surfaces. For this operation, it is necessary to provide a mobile washing machine with hot water, as well as a chemical degreaser for manual removal of grease.

The next step is **defect planning**. The defect plan defines all the parameters that will be focused attention on during the defect procedure. *The Defect Leaflet* and *Defect Work Order* forms are supporting documentation used when planning a defect.

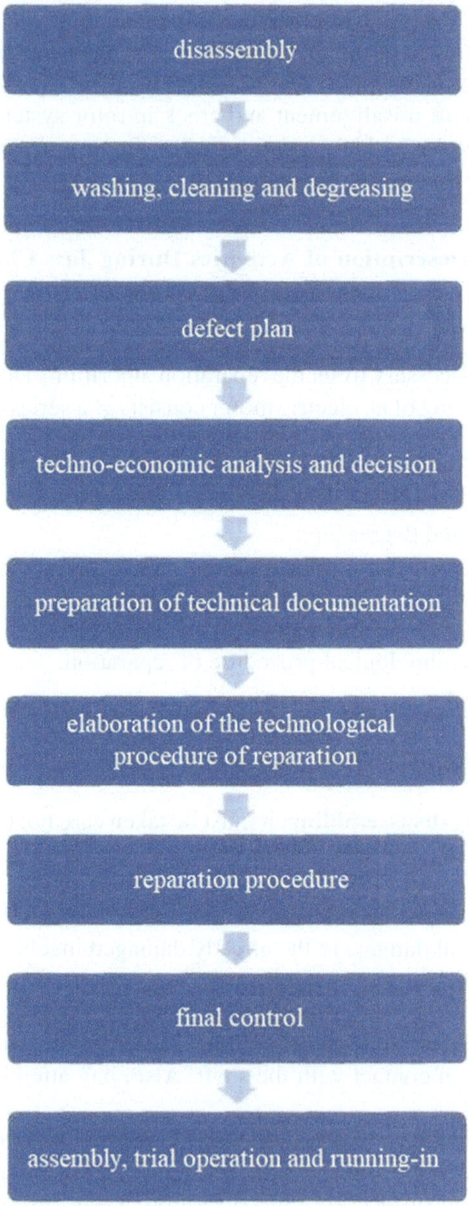

Fig. 3. Algorithm for performing jaw clutch coupling repair [14]

After the detailed defect plan has been made, the coupling is defective. The purpose of defecting is to provide us with an insight into the current state, based on which a decision is made on future activities aimed at bringing this machine element to the required state.

Defecting is carried out according to the defect plan and measurement control sheets attached to the working order form for defecting. The most important thing is to check all tolerated surfaces and the accuracy of individual surface's shapes. The coupling must be tested with ultrasound and magnetic flux testing to check for internal damage and surface cracks.

The degree of repair success depends on the reliability of the collected data. That why it is important that the damages are analyzed in detail to make the right decision on how to conduct the repair in the best way. The defect results are presented through the forms *Defect Report/Request for Works* and *Measurement and Control Sheet.*

The next step that is being taken is the **techno-economic analysis** of the justification of reparation, as well as making a decision based on the processed *Defect Reports* and *Measurement and Control Sheets* on the scope and type of work to be undertaken during reparation, in order to eliminate the shortcomings. The following reparation procedure is carried out only if the techno-economic criteria show the justification of the application of the adopted technology. Otherwise, a new part will be made.

If, based on the image obtained by the magnetic flux method, surface cracks have been ascertained and, if the existence of deep cracks or other significant physical damages has been determined by the ultrasonic method, a new coupling is made.

If the shaft hub is worn, an impermissible clearance or the circularity of the hub opening is found, the keyway is damaged, some of the jaws are broken, it is proceeded towards the jaw clutch coupling repair.

Specific measures and activities are defined through the *Defect List* and the *Defect Report - request for works.*

Repairs on the machine parts of dragline excavator EŠ 5/45 was performed in the spring of 2018, when a detailed defect plan was carried out. In the case of the jaw clutch coupling of the electric motor of the dragline excavator EŠ 5/45, damage to the jaws of lower intensity in the form of worn out sides of the jaws was determined by defecting, so that it is possible to repair them successfully.

The technical documentation is an integral part of the reparation procedure. If the damaged machine part does not have the accompanying technical documentation, it must be made based on values of the required geometric, kinematic, and mechanical characteristics. The technical documentation includes workshops and other drawings, accompanying sketches, etc.

Based on the analysis of the damage, type of base material, function, and application of the machine part, the repair method is selected. Repair, depending on the method of execution, can be mechanical (cold) and thermal. Mechanical repair is performed without heat input. The thermal repair is accompanied by the input of a large amount of heat, which is reflected in the appearance of deformations and residual stresses, so additional thermochemical treatments are necessary. Despite these problems, thermal repair is dominant in the repair of machine parts. After that, the **elaboration of the technological procedure** follows, which includes:

- selection of tools and accessories (grinding and hand grinder; gas welding machine and industrial thermometer; electrode dryer; electric chamber furnace for heat treatment; universal lathe PA 1000M/4000; vertical planer-cutter, type 7410);
- selection of additional material (elektrode);
- preheating temperature selection (temperature of 150 °C);
- protection of non-treated surfaces (special thermo-resistant material).

After a detailed defect has been performed, the **reparation** procedure is started. Activities for repair of the coupling are performed according previously determined dynamics, and based on the *Technological procedure*, prescribed by the reparation technologist.

In this paper the emphasis is on the techno-economic analysis of reparations (and not on the detailed technological procedure). The following figures (Fig. 4 5, 6 and 7) show some of the technological procedures for repairing the jaws of the jaw clutch coupling of an electric motor of an dragline excavator EŠ 5/45 (machine, accessories, tools or accessories).

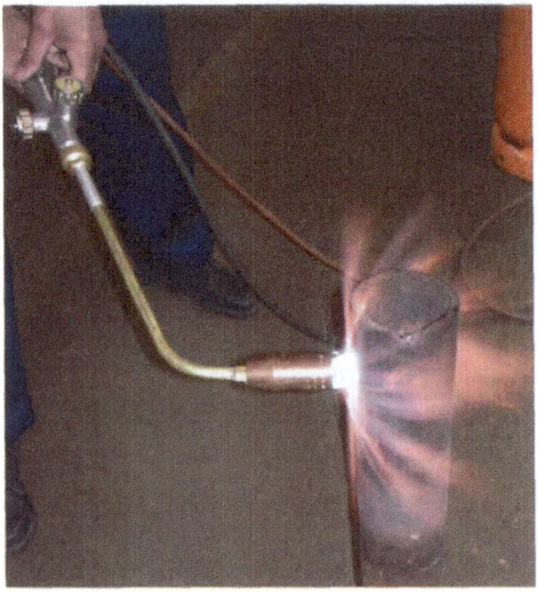

Fig. 4. Intensive preheating procedure with a gas welding machine [14]

After the reparation works are completed, the **final control** of the coupling is performed, based on which the efficiency of the reparation is determined, i.e., whether the treated element has been restored to working capacity. The controller determines whether the measurements and tolerances on the coupling are in accordance with the one specified in the accompanying workshop drawing. A template made of sheet metal with the geometry of the jaw interspace is used as a controller.

Fig. 5. Manual metal arc (MMA) welding [14]

Fig. 6. Processing jaw clutch coupling on the universal lathe PA 1000M/4000 [14]

Fig. 7. Processing of welded sides of jaws of the jaw clutch coupling on vertical planer-cutter, type 7410 [14]

The next step in this process is **assembly**. The repaired coupling is mounted in the machine with the electric motor and reducer, after which the machine is put into **trial operation and running-in**. After the successful completion of the test operation, the machine with the built-in coupling is mounted on the excavator, after which the device's functional tests are started, and then the commissioning.

3.4 Applied Measures on Safety and Health at Work and Fire Protection

In order to provide safety at work during the performance of works, it is necessary to permanently undertake and provide all measures of safety at work following the applicable legal regulations, normative acts, and instructions developed for that purpose. All supervisory, technical staff, and each worker in his/her work field are responsible for the implementation of the appointed protection measures during the execution of works in terms of tasks and obligations provided by the company's acts.

4 Techno-Economic Analysis of the Justification of Reparation Procedure

The selected technology of repairing the jaws of the jaw clutch coupling of the electric motor of the dragline excavator EŠ 5/45 enables the implementation of the techno-economic analysis on the justification of its application. By simply comparing the price of the repaired part and the new part's price, we come to an assessment of the justification of the procedure.

Based on the simulated process of making a new coupling, we come to the cost of making it. The following parameters are included in the price of the new jaw clutch coupling (based on the workshop's internal price list), Table 1:

Table 1. The price of the new jaw clutch coupling [14]

Parameters	Cost
Procurement of castings (steel cast, 120kg)	400 €
Preparation and elaboration of technical documentation	60 €
Making of technological documentation	20 €
Machining of castings	1200 €
Product quality control	15 €
In Total	**1695 €**

The price of repaired jaw clutch coupling includes the following parameters (based on the workshop's internal price list), Table 2:

By analyzing the costs of reparations and comparing them with the costs of making a new jaw clutch coupling for an electric motor, we conclude that the reparation procedure is a more economically acceptable solution than making a new one. The costs of the

Table 2. The price of repaired jaw clutch coupling [14]

Parameters	Cost
Defecting	40 €
Preparation and elaboration of technical documentation	30 €
Making of technological documentation	20 €
Reparation	460 €
Material (electrode)	8 €
Product quality control	15 €
In Total	**573 €**

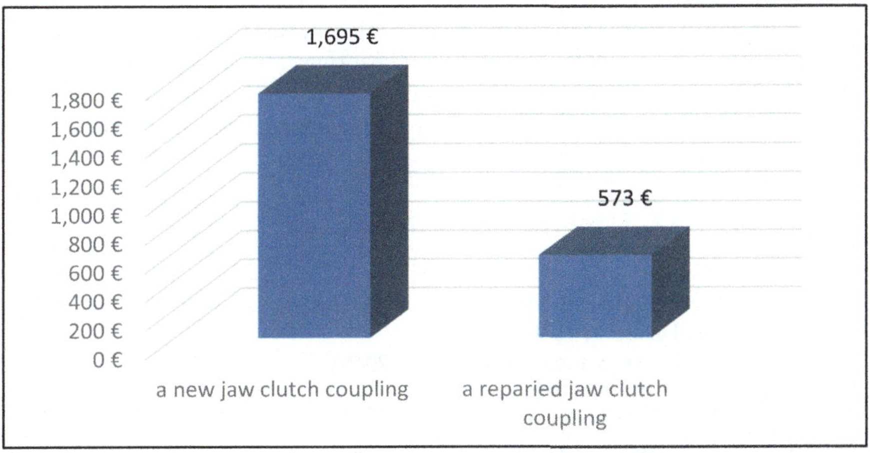

Fig. 8. Comparative costs of making a new jaw clutch coupling and it's reparation

reparation procedure participate with one third in the total costs of making a new coupling (Fig. 8).

Also, the repaired jaw clutch coupling meets all technological parameters in terms of required geometry and mechanical properties, determined by product quality control performed after the intervention that restored working capacity and can re-perform its purpose in an acceptable number of work cycles.

5 Conclusion

If we consider the importance of the repair from the point of view of the factory, the result is:

- disposal of damaged and worn parts is reduced;
- the number of spare parts in the warehouse is minimized;

- costs are reduced due to production interruptions due to sudden failures (accidents) by applying a quick solution to the problem;
- costs are additionally reduced due to the application of an economically more favorable solution in relation to the production of a new part.
- Observed globally, the application of reparation can significantly reduce:
- consumption of non-renewable energy sources;
- consumption of mineral wealth;
- pollution of the planet.

The paper analyzes the possibility and justification of repair and presents the details of the technology of reparation of the jaws of the jaw clutch coupling, where the justification and economy of the chosen solution are confirmed and in which the costs of the repair procedure represent one-third of the total costs of a newly made coupling.

The thermal repair was applied to this coupling, one of the basic procedures for repairing elements of an assembly in the process industry due to the relatively short time required for repair, costs, and the possibility of ensuring satisfactory quality.

The above suggests that the reparation procedure for each maintenance participant should be the first choice for resolving emergency situations in the maintenance process of technical systems.

References

1. Pantelić, M.: Technologies for Maintenance of Mining Machines on Surface Coal Mines Janko Stajčić a.d., Lazarevac (2011)
2. Internal technical documentation. RB Kolubara (2018)
3. Jakovljević, V.: Revitalization of rotary excavators at surface mines. Master thesis, Technical faculty "Mihajlo Pupin", Zrenjanin (2012)
4. Adamović, Ž., Radovanović, Lj.: Reliability of technical systems, Technical faculty "Mihajlo Pupin", Zrenjanin (2008)
5. Adamović, Ž.: Maintenance technology, Technical faculty "Mihajlo Pupin", Zrenjanin (2005)
6. Ašonja, A.: Economic justification of bearing assembly repair. Agric. Tech. **33**(1), 67–73 (2008)
7. Jakovljević, V., Desnica, E., Prvulović, S.: Repair procedure of belt wagon BRs 1200 crawler unit. IMK-14 Res. Dev. Heavy Mach. **26**(2), 29–34 (2020). https://doi.org/10.5937/IMK2002029J
8. Ognjanović, M.: Machine Elements. University of Belgrade, Faculty of Mechanical Engineering, Belgrade (2009)
9. Gligorić, R.: Machine Elements. University of Novi Sad, Faculty of Agriculture, Novi Sad (2015)
10. Grega, R., Homišin, J., Krajňák, J., Urbanský, M.: Analysis of the impact of flexible couplings on gearbox vibrations. Sci. J. Silesian Univ. Technol. Ser. Transp. **91**, 43–50 (2016). https://doi.org/10.20858/sjsutst.2016.91.4
11. Amari, S., McLaughlin, L., Pham, H.: Cost-effective condition-based maintenance using markov decision processes. In: RAMS '06 Annual Reliability and Maintainability Symposium, Newport Beach, California, pp. 464–469 (2006). https://doi.org/10.1109/RAMS.2006.1677417

12. Novaković, B., Desnica, E., Radovanović, Lj., Ivetić, R., Đorđević, L., Labović Vukić, D.: Optimization of industrial fan system using methods laser alignment. Appl. Eng. Lett. **6**(2), 62–68 (2021). https://doi.org/10.18485/aeletters.2021.6.2.3
13. Zhao, W., Hua, C., Wang, D., Dong, D.: Fault diagnosis of shaft misalignment and crack in rotor system based on MI-CNN. In: Wahab, M.A. (ed.) Proceedings of the 13th International Conference on Damage Assessment of Structures. LNME, pp. 529–540. Springer, Singapore (2020). https://doi.org/10.1007/978-981-13-8331-1_39
14. Internal documentation and the Machine Workshop Reports of OC Kolubara Metal, RB Kolubara (2018)

Optimal Maintenance of the Electric Vehicle Battery Storage Level in Prosumer Power Network

Nemanja Mišljenović[1(✉)], Mia Stanić[1], Goran Knežević[1], and Josip Jakab[2]

[1] Fakultet elektrotehnike, računarstva i informacijskih tehnologija, Kneza Trpimira 2B, 31000 Osijek, Croatia
nemanja.misljenovic@ferit.hr

[2] Fakultet za dentalnu medicinu i zdravstvo, Crkvena 21, 31000 Osijek, Croatia

Abstract. Previous passive electricity consumers by the integration of the renewable energy sources in their possession are becoming consumers with their own generation, or so-called prosumers. Current market conditions stimulate the prosumer to use the produced electricity from his own power plant mainly for his own needs because the surplus electricity that he delivers to the power network is paid at a much lower price than the one at which the customer pays for the consumed electricity. Owning an electric vehicle gives the prosumer an additional opportunity to optimally use the produced electricity from their own sources. This paper presents two models for maintaining the battery charge level of an electric vehicle in combination with expected electricity generation from photovoltaic system. The first model enables optimal maintenance of the battery storage level of the electric vehicle by applying only charging process, while the second model enables the charge and discharge of the electric vehicle battery while being connected to household power network. The models have the possibility of single-level and two-level charging power. The model with two levels of charging power gives better results in terms of optimal maintenance of the charge level of the battery storage. The charging and discharging model shows the advantages of maintaining the charge level in terms of savings. The simulations are conducted for one summer and one winter day.

Keywords: Renewable energy sources · Prosumer · Electric vehicle · Optimization

1 Introduction

The global emissions of CO_2 in transport increased by less than 0.5% in 2019, due to improved efficiency, electrification and use of biofuels. Nevertheless, transportation still contributes with 24% to direct CO_2 emissions, according to [1]. Consequently, the transition to electric vehicles is necessary for the future. Furthermore, the European Commission aims to reduce greenhouse gas emissions by 40% by 2030, compared to 1990, according to [2]. Throughout all sectors, from electricity generation, industry,

H. Glavaš et al. (Eds.): OTO 2021, LNNS 369, pp. 104–118, 2022.
https://doi.org/10.1007/978-3-030-92851-3_8

office buildings to households and transportation, there must be a reduction in greenhouse gas emissions in order to achieve this. The European Commission aims to achieve zero greenhouse gas emissions by 2050, according to [3]. Investment in renewable energy sources and the purchase of electric cars have been encouraged in order to achieve these goals. In order to find the optimal solution for the electricity network operation, projects have also been launched. As one of the EU members, Croatia follows the directive and adopts a rulebook in which the "nZEB" term (building with almost zero energy) is mentioned for the first time in 2015, and from December 31st of 2019, becomes mandatory for all energy dependent facilities. It states that a minimum share of 20% of the used energy must be obtained from renewable systems [4]. Furthermore, the conditions of produced electricity sale in Croatia, where the selling price is significantly lower than the purchase price of electricity, encourage prosumers to consume the generated electricity for their own needs, according to [5].

Planning for charging and discharging intervals of electric vehicles in the tenant's common parking lot is a solution proposed in the literature [6]. Planning aims to ensure the maximum profit for all users of electric vehicles, providing ancillary services to the distribution network, as well as meeting the electricity requirements of all consumers. A comparison of the stationary battery storage and an electric vehicle battery storage use in combination with a photovoltaic power plant is presented in the literature [7]. Four different tariff structures of electricity were taken into account, and in the case of using the battery storage of an electric vehicle and a photovoltaic power plant, the results showed greater savings.

Managing of charging and discharging process of the electric vehicle battery connected to a home installation and in combination with a photovoltaic power plant was investigated in literature [8]. The optimization method is based on historical data and forecasting of all variables used by the system. Reduction in the exchange of electricity with the network, improvement of own electricity needs and self-sustainability are the results of the optimization application, based on the forecast. A combination of a stationary system for energy storage, electric vehicles and production from a photovoltaic power plant in response to electric vehicles charging was investigated in the literature [9]. The results showed that the stored electricity from the photovoltaic power plant could be used for charging electric vehicles in later hours. A central model for energy management in a household is presented in the paper [10]. The energy management model optimizes the use of household appliances, electricity generation, electricity storage and electric vehicles and is based on mixed integer linear programming (MILP). Optimized controlled charging of electric vehicles in order to avoid peak loads is presented in the literature [11]. The method is based on real-time measurement to minimize charging costs. The importance of consumption and production management has been explored in literature [12]. By applying consumption, production, battery storage and electric vehicle management, the greatest savings were achieved.

With the integration of renewable energy power plants into their ownership, former passive consumers of electricity become consumers with their own production, i.e. prosumers. Therefore, they need to use the produced electricity for their own needs in order to increase profit.

An optimizing model for the maintenance of the electric vehicle battery storage charge level to minimize the costs of the prosumer is proposed in this paper. In the proposed model, it is possible to maintain the charge level of the battery storage with one or two levels of charging power, which corresponds to the current situation.

This paper consists of five chapters. The second chapter describes a mathematical model for optimal use of an electric vehicle battery storage under market conditions. The objective function and constraints are described in detail. In the third chapter, the proposed optimization model is applied to a prosumer who owns a photovoltaic power plant and an electric vehicle. The results of the proposed model are presented in the fourth chapter. A conclusion and a proposal for future work are given in the last chapter.

2 The Mathematical Model for Optimal Maintenance of the Electric Vehicle Battery Storage Level in Market Conditions

2.1 Mathematical Formulation of the Optimization Problem

The proposed model aims to minimize the household costs of the prosumer who owns the production and electricity storage. The objective function is to minimize household costs by optimally maintaining the desired electric vehicle battery charge level in combination with a photovoltaic power plant.

The optimization problem can be written as:
Minimize:

- *purchased electricity - sold electricity*

Constrains:

- *electricity balance*
- *state of charge of the battery storage*
- *limitation of the maximum current power from the network*
- *limitation of the maximum current power into the network*

The minimization of costs that would be achieved in the household is defined by the following relation:

$$min \sum_{t=1}^{n_t} (Pk_t \cdot ck_t - Pp_t \cdot cp_t) \tag{1}$$

where:

Pk_t – electricity purchased from the network in the period t
ck_t – the price of purchased electricity from the network in the period t
Pp_t – electricity sold to the network in the period t
cp_t – the price of electricity sold to the network in the period t
n_t – simulation period

The first part of relation (1), i.e. multiplication of purchased electricity from the network and the price, represents the cost of purchasing electricity from the network, which is then reduced by the generated income of selling electricity to the network, which is the second part of relation (1), for each period t. The simulation period t refers to hourly values.

The electricity balance equation is given by the following relation:

$$L_t + P_{ch1_t} \cdot b_{1_t} + P_{ch2_t} \cdot b_{2_t} - P_{PV_t} - Pk_t + Pp_t - P_{dch1_t} \cdot d_{1_t} - P_{dch2_t} \cdot d_{2_t} = 0 \quad (2)$$

where:

L_t – household electricity demand in the period t
P_{ch1_t} – electricity required to charge the electric vehicle battery storage in period t, higher power level
b_{1_t} – binary decision variable for higher charging power in the period t
P_{ch2_t} – electricity required to charge the electric vehicle battery storage in period t, lower power level
b_{2_t} – binary decision variable for lower charging power in the period t
P_{PVt} – electricity produced by a photovoltaic power plant in a period t
P_{dch1_t} – electricity obtained by discharging of the electric vehicle battery storage in period t, higher power level
d_{1_t} – binary decision variable for higher discharge power in the period t
P_{dch2_t} – electricity obtained by discharging of the electric vehicle battery storage in period t, lower power level
d_{2_t} – binary decision variable for lower discharge power in the period t

The state of charge (SOC) of the electric vehicle battery storage is given by the relation, if $t = 1$, i.e. the first period:

$$SOC_t = SOC_{start} + \eta_{ch} \cdot P_{ch1_t} \cdot b_{1_t} - \frac{P_{dch1_t}}{\eta_{dch}} \cdot d_{1_t} + \eta_{ch} \cdot P_{ch2_t} \cdot b_{2_t} - \frac{P_{dch2_t}}{\eta_{dch}} \cdot d_{2_t} \quad (3)$$

For all other periods, if $t \neq 1$, the relation will hold:

$$SOC_t = SOC_{t-1} + \eta_{ch} \cdot P_{ch1_t} \cdot b_{1_t} - \frac{P_{dch1_t}}{\eta_{dch}} \cdot d_{1_t} + \eta_{ch} \cdot P_{ch2_t} \cdot b_{2_t} - \frac{P_{dch2_t}}{\eta_{dch}} \cdot d_{2_t} \quad (4)$$

where:

SOC_t – the state of charge of the electric vehicle battery storage at the end of the period t
SOC_{start} – the state of charge of the electric vehicle battery storage at the start of the simulation period (t_1)
SOC_{t-1} – the state of charge of the electric vehicle battery storage for the past period ($t - 1$)
η_{ch} – coefficient of charging efficiency of battery storage
η_{dch} – coefficient of discharging efficiency of battery storage

The binary decision variables of the charge and discharge are defined as integers by the relation:

$$b_{1_t}, d_{1_t}, b_{2_t}, d_{2_t} \in \mathbb{Z} \tag{5}$$

The limitations on the binary decision variables of the charge and discharge values are given by the relations (6–9):

$$0 \leq b_{1_t} \leq 1 \tag{6}$$

$$0 \leq b_{2_t} \leq 1 \tag{7}$$

$$0 \leq d_{1_t} \leq 1 \tag{8}$$

$$0 \leq d_{2_t} \leq 1 \tag{9}$$

The constrain of simultaneous charging and discharging is given by the relation:

$$b_{1_t} + d_{1_t} + b_{2_t} + d_{2_t} \leq 1 \tag{10}$$

The limit of the maximum current power purchased from the network in period t is given by the relation:

$$0 \leq Pk_t \leq Pk_{max} \tag{11}$$

The limit of the maximum current power sold to the network in period t is given by the relation:

$$0 \leq Pp_t \leq Pp_{max} \tag{12}$$

The state of battery storage charge is set by the relation:

$$SOC_{min} \leq SOC_t \leq SOC_{max} \tag{13}$$

2.2 Optimization Technique Applied in the Proposed Model

Mathematical optimization methods can find solutions to various problems that require a minimum or maximum value of the objective function that meets the given constraints.

The proposed model is presented as an optimization problem of finding the minimum cost in the household, which is solved with mixed integer linear programming (MILP). The mean value of the electricity produced is used in order to take into account the uncertainty of the expected electricity production. The model is written in the GAMS programming language, where the MILP problem is solved by using a CPLEX solver.

3 Case Study

3.1 Model Description

The data that will be used in the model for the produced electricity of the photovoltaic power plant was obtained from the photovoltaic geographic information system (PVGIS). The location of Osijek was chosen, and the installed power of the photovoltaic power plant was 10 kW. Two random dates were selected in order to better see the difference in electricity production from the photovoltaic power plant between the two extreme seasons, i.e. summer and winter. It will be taken on July 14 for the summer and on January 5 for the winter season. An average of the produced electricity value was made for three years, i.e. from 2014 to 2016 for selected days. Household consumption is modelled based on the assumption of average electricity requirements during summer and winter. The authors assumed electricity requirements during summer and winter. The selected vehicle is the Nissan Leaf, as one of the best-selling battery electric vehicles. The vehicle specifications for the battery and its charging are listed in Table 1. The state of charge of the electric vehicle battery storage is enabled to be arbitrarily maintained within the desired limits. Other input data for the model, apart from consumption and production, are the price of purchased electricity and the price of sold electricity. Costs related to electricity supply will be based on the supply model of "Hrvatska elektroprivreda" (HEP) as the main electricity supplier in the Republic of Croatia. According to HEP, i.e. to the White tariff model, the price of the higher tariff is 1.07 HRK/kWh, while the price of the lower electricity tariff is 0.58 HRK/kWh. The average price of electricity sold in the network is 0.33 HRK/kWh, for the case when the purchased electricity is higher than the electricity sold to the network in a monthly period, and this case is observed. The model also sets the maximum and minimum current power that can be taken from and delivered to the network. For power purchased from the network, it is the limit of 13.8 kW, and for power that can be sold to the network, the limit is 10 kW. Two cases will be considered for each scenario and each annual period. The first case will show charging with only one power $P_{ch1_t} = 7$ kW, which will also be the discharge power in cases where it is possible to transmit electricity back to the network. In the second case, two levels of charging power are used, which amount to, for the first level $P_{ch1_t} = 7$ kW, and for the second level $P_{ch2_t} = 2$ kW. The same two-tier discharge power approach is used in the case of discharge.

3.2 Simulation Scenarios

Simulations of optimal maintenance of the electric vehicle battery storage level will be conducted for summer and winter days through four scenarios. The expected electricity production for the summer season, i.e. 14 July, is shown in Fig. 1, while Fig. 2 shows the expected electricity production for the winter season, i.e. 2 January.

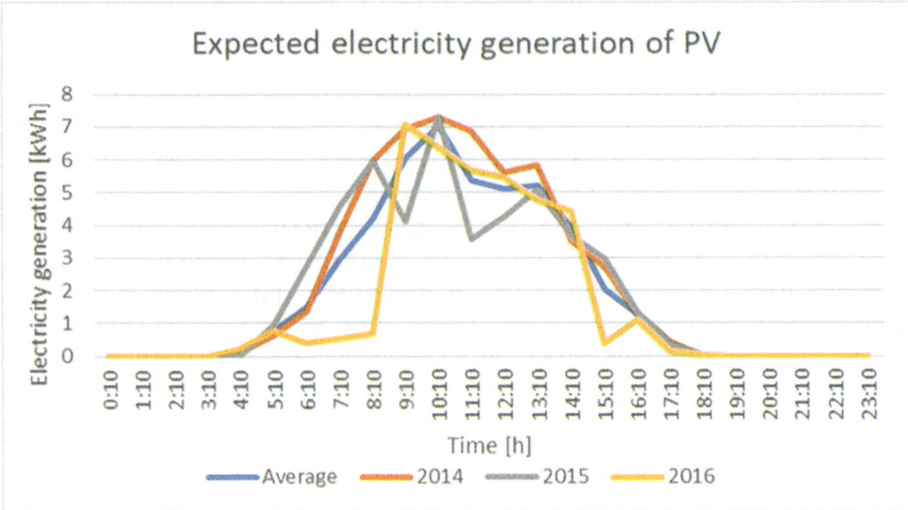

Fig. 1. Expected electricity production for a summer day

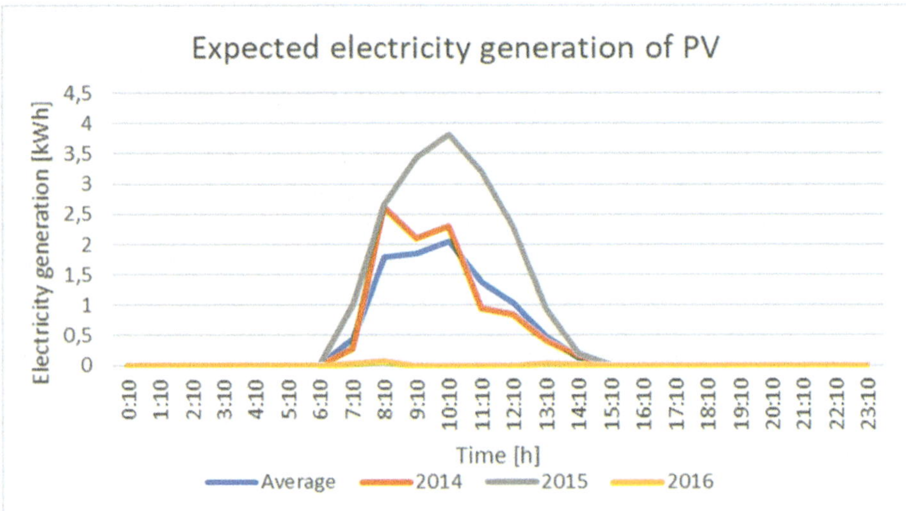

Fig. 2. Expected electricity production for a winter day

Figure 3 shows the expected daily consumption of one household for a summer day, with the average expected production of electricity from a photovoltaic power plant. Figure 4 shows the expected daily consumption of one household for a winter day, with the average expected production of electricity from a photovoltaic power plant.

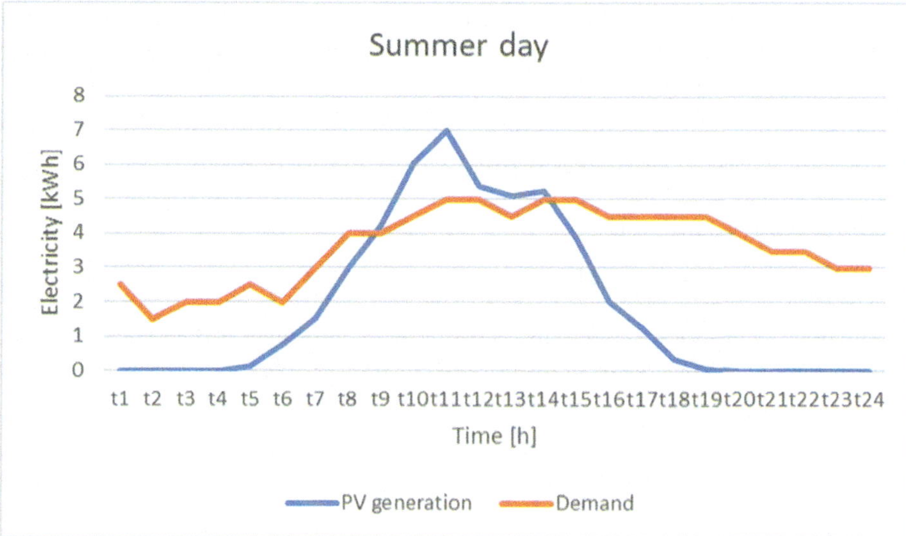

Fig. 3. Daily expected household consumption for a summer day with average expected PV production

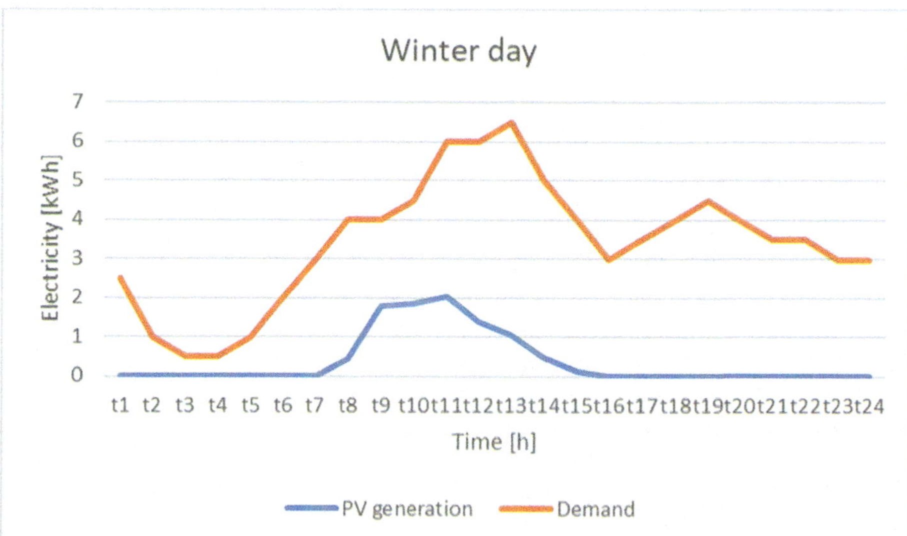

Fig. 4. Daily expected household consumption for a winter day with average expected PV production

According to the White tariff model, the total price of energy and network usage fees is 0.84 HRK/kWh during the day and 0.41 HRK/kWh during the night, without a fee for renewable sources and high-efficiency cogeneration and VAT. For the winter period, the high tariff lasts from 7 a.m. to 9 p.m., the low tariff from 9 p.m. to 7 a.m., and for the summer period, the high tariff is from 8 a.m. to 10 p.m., and the low one from 10 p.m. to 8 a.m. A special fee for renewable sources and high-efficiency cogeneration is added to these values, which amounts to 0.105 HRK/kWh.

After the added VAT of 13%, the price of electricity for a high tariff is:

$$C_{K,D} = (0.84 + 0.105) \cdot 1.13 = 1.07 \text{ HRK/kWh} \quad (14)$$

while for a low tariff, it is:

$$C_{K,N} = (0.41 + 0.105) \cdot 1.13 = 0.58 \text{ HRK/kWh} \quad (15)$$

The price of electricity sold to the network by the prosumer is 0.33 HRK/kWh and is obtained as the average value of the price of electricity itself without transmission and distribution costs with a certain coefficient, at low and high tariffs:

$$C_p = 0.9 \cdot \frac{0.49 + 0.24}{2} = 0.33 \text{ HRK/kWh} \quad (16)$$

The vehicle to be used in the examples is the Nissan Leaf, as one of the best-selling battery electric vehicles. Vehicle specifications regarding battery and charging are listed in Table 1.

Table 1. Battery and charging specifications for the Nissan Leaf electric vehicle

Nissan leaf acenta			
Technology	Lithium-ion	Vehicle charger	Type 2 – 6.6 kW connection
Total voltage (V)	350	Quick charge connector	ChaDeMo 50 kW
Capacity (kWh)	40		
Battery charging - 50 kW quick charger 40 min–60 min (charging time depends on charging conditions, including the type of fast charger, battery condition and capacity, and ambient temperature)			
Battery charging - external chargers/wallbox of 7.4 kW 7.5 h (charging time depends on charging conditions, including the type of fast charger, battery condition and capacity, and ambient temperature)			
Battery charging - 220 V/10 A or 2.2 kW home socket 21 h (charging time depends on charging conditions, including the type of fast charger, battery condition and capacity, and ambient temperature)			
Charging cable (220 V/10 A or 2.2 kW home socket) 6 m EVSE cable			
Charging cable (external chargers or wall box of 7.4 kW) Cable mode 3/T2 - T			

The proposed model uses battery charging via an external charger, i.e. wallbox, with a charging power of 7 kW or 2 kW. A value of $\eta = 95\%$ efficiency of charging and discharging for the electric vehicle battery storage was taken for this charging method.

4 Results

4.1 Scenario 1: Charging and Discharging with SOC ~100% Requirement in t = 24 h

In this scenario, the vehicle is all day on the charger, i.e. it will not be used for driving, and therefore serves as a electricity storage. The optimal case of vehicle battery charging and discharging is observed in order to achieve savings when buying or selling electricity, considering the low and high tariff for electricity in summer and winter. The condition given to the model is that the vehicle is fully charged by the end of the day, i.e. that it is SOC ~100% in period t = 24 h.

Figure 5(a) shows the first case when the vehicle battery is charged and discharged with only one power, and Fig. 5(b) shows the case when two different levels of charging and discharging power are used.

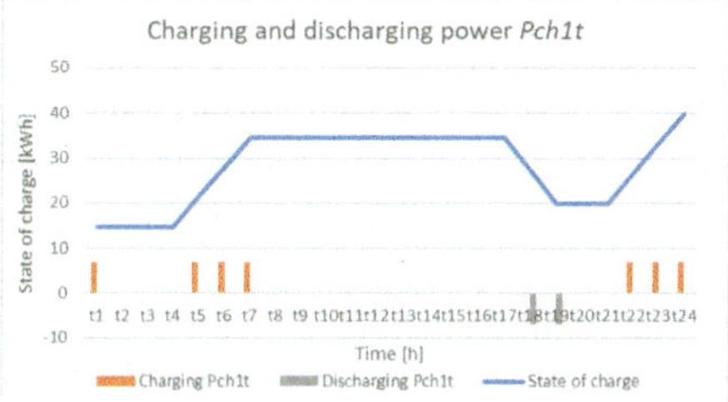

(a)

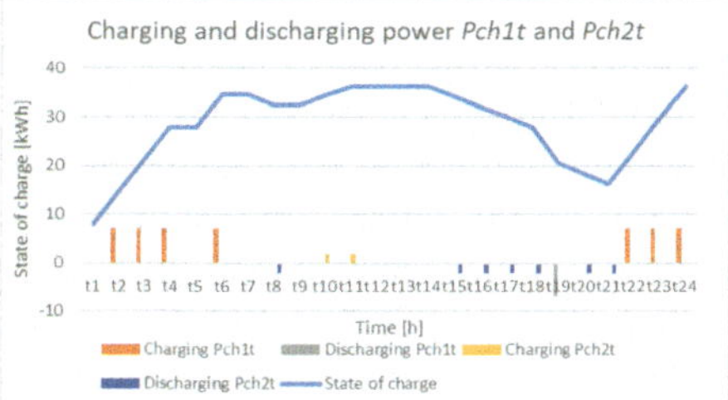

(b)

Fig. 5. Comparison of optimized vehicle battery charging and discharging with power P_{ch1_t} (a) and combination of power P_{ch1_t} and P_{ch2_t} (b) for a summer day

Regarding the savings achieved for a summer day in this scenario, the case with two levels of charging and discharging power is more favorable, with a difference of approximately 15% or 6.54 HRK per day. The reason for this is the possibility of better use of own electricity because it is possible to exchange electricity in finer steps.

For a winter day, in the case of battery charging and discharging with only one power, it can be observed that the battery is discharged only when maximum consumption of household occurs, Fig. 6(a). The reason for this is the low production of the photovoltaic power plant in respect to household load. In Fig. 6(b), the battery is discharged with power P_{ch2_t} almost throughout the day in fine steps to ensure the overall lower price of electricity for the prosumer and increase in profit. Thus, the required electricity is recharged from the battery, not purchased from the grid. It can be observed that the battery charging takes place only at a low tariff.

Since the high potential of solar radiation cannot be used during the winter day, the combination of charging with two power levels proved to be the main factor for considerable savings, of approximately 30% or 19.28 HRK per day.

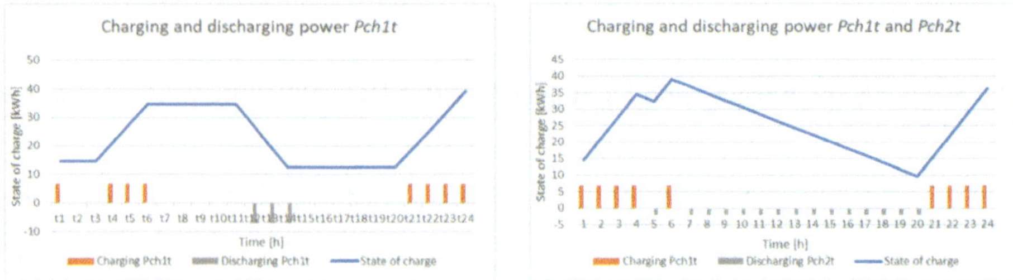

(a) (b)

Fig. 6. Comparison of optimized vehicle battery charging and discharging with power P_{ch1_t} (a) and combination of power P_{ch1_t} and P_{ch2_t} (b) for a winter day

4.2 Scenario 2: Charging and Discharging Without Set Vehicle Usage Requirements

In this scenario, the battery has the possibility to charge and discharge as in the previous scenario. The vehicle (battery) is available on the charger all day, and it is assumed that the vehicle will not be used, so the battery only serves as electricity storage. Since it is desirable for the battery life that the SOC does not fall below 20% of the charge, limits are set to a minimum charge of 20%.

According to Fig. 7, a difference in the optimal maintenance of the battery storage level can be seen. The electric vehicle battery is charged at a low tariff and discharged at a high tariff in order to minimize costs, as shown in Fig. 7(a). In Fig. 7(b), the results show fine regulation of maintaining the battery storage level compared to Fig. 7(a). The difference is the appearance of P_{ch2_t}, which alternately charges and discharges the battery resulting with savings of approximately 20% or 6.28 HRK per day.

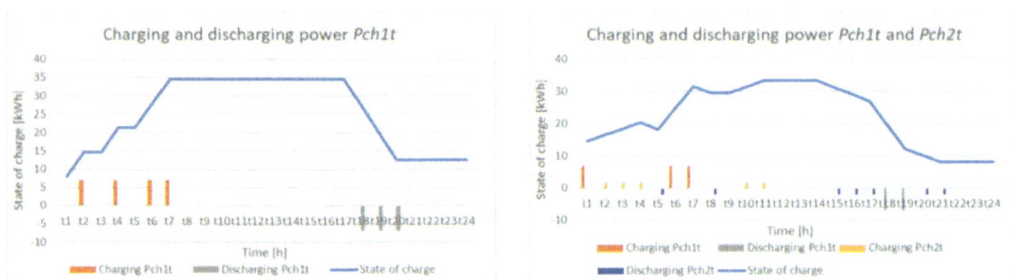

(a) (b)

Fig. 7. Comparison of optimized vehicle battery charging and discharging with power P_{ch1_t} (a) and combination of power P_{ch1_t} and P_{ch2_t} (b) for summer day

Maintaining the optimal storage level of the electric vehicle battery is more favorable for case 7. (b) because the system is allowed to use a lower power level to compensate for the necessary differences in electricity consumption and production.

In the case of a winter day from Fig. 8(b), the two-level charge uses the power P_{ch2_t} to gradually discharge the battery throughout the day, after the battery is first charged to almost maximum capacity in the morning at a low tariff. According to the results, this ultimately showed savings of approximately 16% or 9.94 HRK per day, compared to the first case, Fig. 8(a).

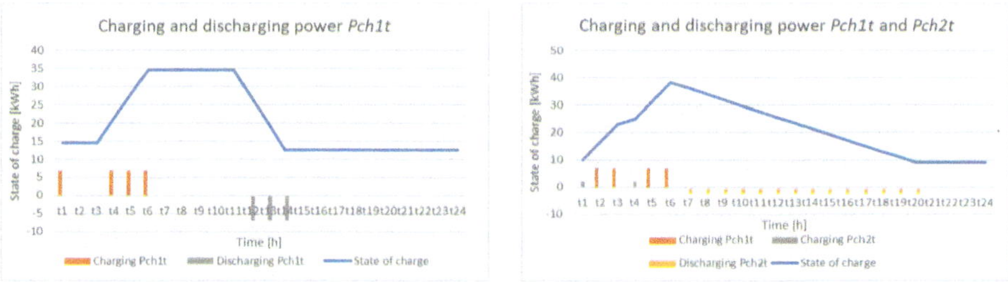

(a) (b)

Fig. 8. Comparison of optimized vehicle battery charging and discharging with power P_{ch1_t} (a) and combination of power P_{ch1_t} and P_{ch2_t} (b) for winter day

4.3 Scenario 3: Using the Vehicle During the Day with Ability of Battery Charging Only

Under this scenario, it is assumed that the vehicle owner uses the vehicle between 8 a.m. and 4 p.m., for example, to go to work. This means that the vehicle, i.e. the battery, is not present on the charger all day as it was in previous scenarios. The requirement is that the vehicle is fully charged at 7 a.m., and it is assumed that the vehicle is returned to the charger with a 25% charge level at 5 p.m. Also, since this is a real case, only charging the battery is possible (not discharging).

According to Fig. 9, it is noted that the differences between (a) and (b) are minimal, i.e. there was no use of the power P_{ch2_t} in the second case, Fig. 9(b). Therefore, in both cases, the battery is charged only with the power P_{ch1_t} in the first hours of the day at a low tariff.

Starting from the fact that in this scenario, it is not possible to discharge the battery as in previous scenarios and that the vehicle is not available all day on the charger, it is clear that this scenario will be the most unfavorable so far in terms of savings and final costs. The final costs are the same for both cases, according to Fig. 9.

According to Fig. 10, charging takes place similarly to a summer day, in the first hours of the day at a low tariff.

From this scenario, it can be concluded that it is not possible to achieve savings when combining a photovoltaic power plant and an electric vehicle that serves as storage because when production is available, an electric vehicle is unavailable for charging. Furthermore, the downsides of this scenario are the inability to empty the electricity storage and the equal price of electricity during the low tariff. All these factors influenced the simulation result. The final costs are the same for both cases, according to Fig. 10. It

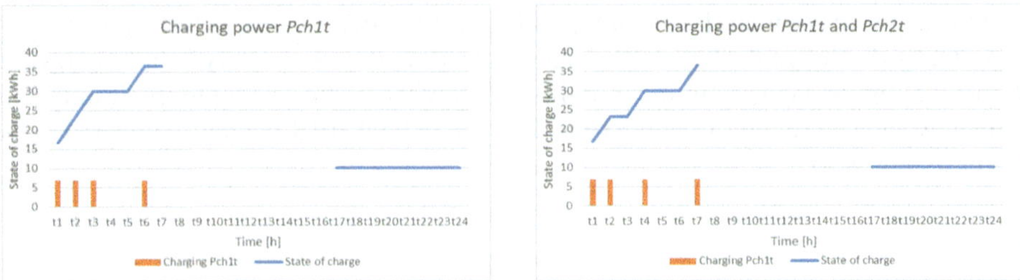

(a) (b)

Fig. 9. Comparison of optimized vehicle battery charging with power P_{ch1_t} (a) and combination of power P_{ch1_t} and P_{ch2_t} (b) for summer day

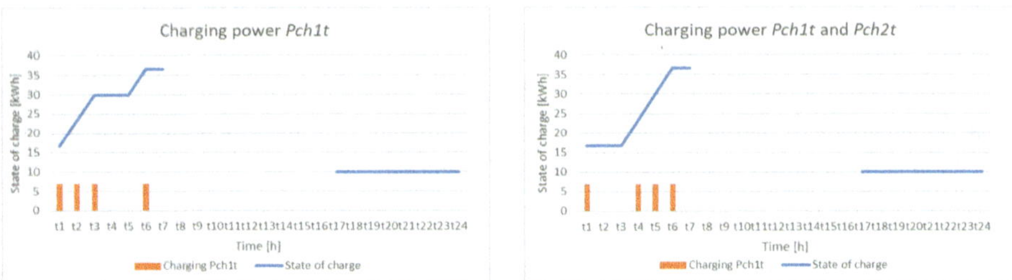

(a) (b)

Fig. 10. Comparison of optimized vehicle battery charging with power P_{ch1_t} (a) and combination of power P_{ch1_t} and P_{ch2_t} (b) for winter day

is necessary to apply the possibility of discharge, and in the case of a photovoltaic power plant, stationary electricity storage needs to be installed in order to achieve savings.

4.4 Scenario 4: Using the Vehicle Overnight with Ability of Battery Charging Only

The assumption of this scenario is that the electric vehicle is needed for the user at night, so it will have to be charged to 100% capacity only during the day. The vehicle is not present on the charger from 9 p.m. to 7 a.m., and the request for a full charge is determined for 8 p.m. Also, in this scenario, the possibility of discharge is not available, only charging the battery storage.

Charging takes place at a low tariff and when the production of a photovoltaic power plant is higher than the demand, Fig. 11(a). From Fig. 11(b), it can be observed that in the case of two-level charging power, the achievement of the desired charge level takes place in finer steps, so the difference in household consumption and production power can be used, compared to the results in Fig. 11(a). Although the savings are negligible, the two-level way of maintaining the charge level of the electric vehicle battery storage has advantages such as reduced battery degradation and household power limiter. The savings are approximately 1% or 0.55 HRK per day.

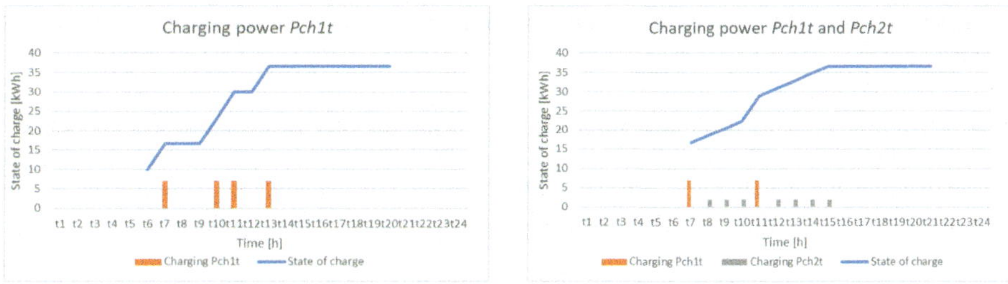

(a) (b)

Fig. 11. Comparison of optimized vehicle battery charging with power P_{ch1_t} (a) and combination of power P_{ch1_t} and P_{ch2_t} (b) for summer day

In the case of a winter day and significantly lower production, no savings were achieved, i.e. the total cost of electricity is the same in both cases, Fig. 12.

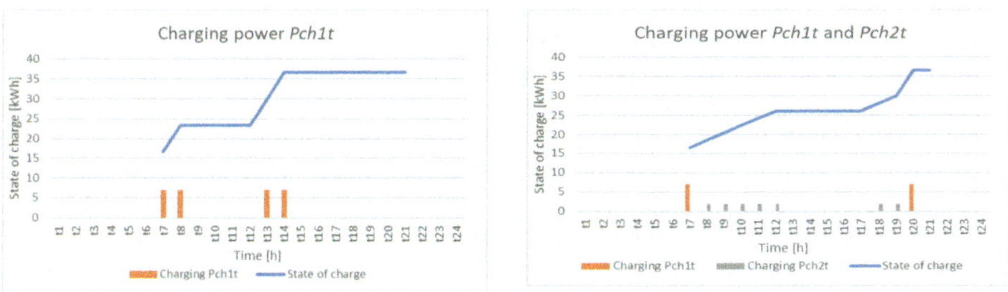

(a) (b)

Fig. 12. Comparison of optimized vehicle battery charging with power P_{ch1_t} (a) and combination of power P_{ch1_t} and P_{ch2_t} (b) for winter day

From Fig. 12(a), it can be seen that the battery is charged when the electricity production is still present, and demand is reduced. Utilization of own electricity can be noticed in the second case, Fig. 12(b). When there is electricity production from a photovoltaic power plant, the battery is charged instead of selling electricity to the grid.

5 Conclusion and Future Work

This paper presents an optimization model for maintaining the battery storage level of an electric vehicle in market conditions for the prosumer. Prosumer is a consumer with his own production. In order to demonstrate the possibility of developing an effective short-term plan for optimal maintenance of the battery storage level of an electric vehicle, the model is applied to several cases. The results show a significant benefit in the case of two-level charging and discharging power compared to single charging power level. In future work, the existence of additional battery storage for the electricity storage of the prosumer in combination with an electric vehicle and a photovoltaic power plant can be investigated.

Acknowledgments. This work was supported by Faculty of Electrical Engineering, Computer Science and Information Technology Osijek under projects IZIP 2020 "Optimalna veličina izgradnje proizvodnog sustava aktivnog kupca" and by Croatian Science Foundation under the project "Prosumer-rich distribution power network" (project number: UIP-2020-02-5796).

References

1. International Energy Agency (IEA) – Tracking transport. https://www.iea.org/reports/tracking-transport-2020. Accessed 20 Aug 2020
2. European Commission: Climate action - 2030 climate & energy framework (2014)
3. European Commission: COM(2011) 112 - A roadmap for moving to a competitive low carbon economy in 2050. European Environment Agency (2011)
4. Ministarstvo graditeljstva i prostornoga uređenja RH: Tehnički propis o racionalnoj uporabi energije i toplinskoj zaštiti u zgradama (2015)
5. HEP – Tarifni model. https://www.hep.hr. Accessed 20 Aug 2020
6. Mehrabi, A., Kim, K.: Low-complexity charging/discharging scheduling for electric vehicles at home and common lots for smart households prosumers. IEEE Trans. Consum. Electron. **65**(3), 348–355 (2018). https://doi.org/10.1109/TCE.2018.2864548
7. Bjarghov, S., Korpås, M., Zaferanlouei, S.: Value comparison of EV and house batteries at end-user level under different grid tariffs. In: IEEE International Energy Conference, Limassol (2018). https://doi.org/10.1109/ENERGYCON.2018.8398742
8. Giordano, F., et al.: Vehicle-to-home usage scenarios for self-consumption improvement of a residential prosumer with photovoltaic roof. IEEE Trans. Ind. Appl. **56**(3), 2945–2956 (2020). https://doi.org/10.1109/TIA.2020.2978047
9. Gong, H., Ionel, D.M.: Optimization of aggregated EV power in residential communities with smart homes. In: IEEE Transportation Electrification Conference & Expo, Chicago (2020). https://doi.org/10.1109/ITEC48692.2020.9161532
10. Hou, X., Wang, J., Huang, T., Wang, T., Wang, P.: Smart home energy management optimization method considering energy storage and electric vehicle. IEEE Access **7**, 144010–144020 (2019). https://doi.org/10.1109/ACCESS.2019.2944878
11. Simolin, T., Rauma, K., Järventausta, P., Rautiainen, A.: Optimised controlled charging of electric vehicles under peak power-based electricity pricing. IET Smart Grid **6**(3), 751–759 (2020). https://doi.org/10.1049/iet-stg.2020.0100
12. Aoun, A., Ibrahim, H., Ghandour, M., Ilinca, A.: Supply side management vs. demand side management of a residential microgrid equipped with an electric vehicle in a dual tariff scheme. Energies **12**(22), 4351 (2019). https://doi.org/10.3390/en12224351

Application of ISO 80000-2:2009 in the Description of Technical Systems

Damjan Cvenić, Ivanka Ferčec, and Hrvoje Glavaš(✉)

Faculty of Electrical Engineering, Computer Science and Information Technology Osijek, University of Osijek, Kneza Trpimira 2b, 31000 Osijek, Croatia
{ivanka.fercec,hrvoje.glavas}@ferit.hr

Abstract. Technical systems are inconceivable without documentation describing their implementation, use and maintenance. Documentation is usually based on three languages - the manufacturer's domicile language, English and mathematics. Mathematics is not only the first foreign language that we learn, but it is also the only foreign language that everyone learns. By mastering mathematics, we often focus on the result and over the years we may forget about proper notation. This can be overcome by means of the ISO 80000 standard. A standard is a document issued by one or more organizations whose goal is to bring order into a certain field. Standards are transferred to lower instances through a hierarchical structure. The ISO 80000 standard consists of fourteen documents under the general title Quantities and units. This paper provides an analysis of the requirements of Part 2 of the standard called "Mathematical signs and symbols to be used in the natural sciences and technology", and exemplifies its application in scientific journals and Master's theses at engineering faculties. The presence of the standard was assessed by analyzing the elements described by the standard in papers published in 2018 in the following five scientific journals: Tehnički vjesnik/Technical Gazette, Engineering Review, Acta Polytechnica Hungarica, the American Journal of Electrical Power and Energy Systems and the International Journal of Renewable Energy Research. In addition to these journals, the presence of the standard was also analyzed in Master's theses at two engineering faculties, i.e. the Faculty of Electrical Engineering, Computer Science and Information Technology in Osijek and the Mechanical Engineering Faculty in Slavonski Brod, and the results are presented in a clear and concise graphical format. The conclusions of the performed analyses indicate that the standard is generally only partly implemented, with the exception of Tehnički vjesnik/Technical Gazette, where the standard is fully implemented, and the importance of adhering to normative documents is stressed in its Instructions for Authors.

Keywords: Standard · ISO 80000 · Technical systems · Maintenance · Mathematics

1 Introduction

Almost all systems are inconceivable without documentation containing their description. Technical systems are specific because numbers and mathematical expressions are

H. Glavaš et al. (Eds.): OTO 2021, LNNS 369, pp. 119–130, 2022.
https://doi.org/10.1007/978-3-030-92851-3_9

used in their description. The basic rule stating that if something cannot be represented by means of numbers, then we do not know anything about it, imposes the need to write mathematical texts correctly. The importance of proper notation is reflected in the fact that part of the international ISO 80000 standard is entirely dedicated to mathematical writing. The aim of the paper is to explain the concept of standard, standardization and its application by using a practical example. The first part of the paper delineates a hierarchical structure of standardization through examples of international and local organizations and indicators of the number of the normative documents published. Standards at the level of the Republic of Croatia are available through the Croatian Standards Institute, so is the standard that regulates mathematical writing. The ISO 80000 standard consists of 14 normative parts that periodically undergo revision and change. The rules prescribed by the second part of *ISO 80000-2:2009 Quantities and units—Part 2: Mathematical signs and symbols to be used in the natural sciences and technology* are presented partially in textual and tabular form. In order to show the application of the standard, papers and Master's theses were analyzed that were published during 2018 in five scientific journals and at two engineering faculties, respectively. The results of the analysis are presented graphically, and based on them, conclusions are drawn about the application of the standard in everyday practice.

2 Standardization and Standards

2.1 Standardization

Standardization is the activity of establishing an optimum degree of order for common and repeated use in a given context with regard to potential problems or actual problems, [1]. Standard refers to a document approved by a recognized body that provides guidelines whose compliance is not mandatory and may apply to a process or a production method, [1]. The function of a standard is to efficiently connect systems, provide quality of service and reduce a variety of different systems, as well as to describe processes and products. The application of the standard is mainly related to technical systems and system organization, but in the past decade, it has also related to the field of energy, where the advantage of standardization has been recognized and applied when defining the guidelines of energy policy, [2].

Standardization is based on the principles of consensus, involvement of all stakeholders, operational transparency, state-of-the-art technology and the coherence of the set of standards, [3]:

- Consensus is a "general agreement characterized by the absence of sustained opposition to substantial issues by any important part of the concerned interests and by a process that involves seeking to take into account the views of all parties concerned and to reconcile any conflicting arguments".
- All stakeholders must be involved in the preparatory stage of the standard and in the drafting stage of the standard.
- Operational transparency during the entire standard development process is important because the public must have access to documents, information about the body

preparing the standard, and information about the stages of standard preparation and development. The public must be informed in a timely and appropriate manner.
- State-of-the-art technology at a particular time is based on proven scientific, technical and experiential knowledge defining current standards.
- The coherence of the set of standards implies that two standards must not be contradictory, i.e. when adopting a new standard for something for which there already exists a standard, the old standard must be withdrawn in order for the new standard to be published and enter into force.

2.2 Standardization Organizations

Standardization organizations can be divided into local and international. At the European level (i.e. local in relation to the authors), these organizations are the European Committee for Standardization (CEN), the European Committee for Electrotechnical Standardization (CENELEC) and the European Telecommunications Standards Institute (ETSI).

European Committee for Standardization (CEN) [4] brings together 34 European countries. With regard to its level of activity, CEN is below the ISO organization. There were 1,198 documents produced during 2018 (technical specifications, European standards, technical reports, CEN workshop agreements, and CEN guides), of which 1,076 were European standards. At the end of 2018, there were 3,498 documents in preparation. At the same time, there were 16,979 active documents, of which 15,305 were European standards. The Committee consists of 2,134 technical bodies, which are partly joint committees with the European Committee for Electrotechnical Standardization (CENELEC), [5].

European Committee for Electrotechnical Standardization (CENELEC) [6] is a European committee responsible for standardization in the field of electrical engineering, which prepares voluntary standards for trade between countries and supports the unity of the European Common Market. This Committee actively participates in 34 European countries and consists of 383 different technical bodies. In 2018, the Committee developed 383 standards. At the end of 2018, there were 7,085 active standards, while 97 standards are awaiting publication, [5].

European Telecommunications Standards Institute (ETSI) [7] is an institute in Europe responsible for standards in systems, applications and services in information and communication technologies, which later began to be applied all over the world. It has 64 member countries and 872 members. The Institute has published 46,765 standards since its establishment, and more than 17 million standards have been downloaded. It is currently a partner in a global project aimed at developing 4G and 5G mobile communications, [5].

Local organizations act not only in Europe but also internationally, and work on standards with organizations around the world.

International standards and standardization organizations are organizations that operate throughout the world. They are headquartered in one country with multiple branches around the world. Standards are adopted independently or in collaboration with other organizations operating either globally or only in a small geographical area. The standards adopted in most cases solve problems that concern the whole world, so the final solution is a compromise made up of several possible solutions. The largest organizations are the International Organization for Standardization - ISO [8], Deutsches Institut für Normung - DIN [9], and the British Standards Institution - BSI, [10].

International Organization for Standardization (ISO) is an organization that started to operate in 1946, when 25 delegates met at the Institute of Civil Engineers in London to facilitate the international coordination and unification of industrial standards. The International Organization for Standardization (ISO), which currently has 164 member countries and 3,573 technical bodies in charge of developing standards, officially began operating in February 1947. The organization is headquartered in Geneva, Switzerland. Since it was founded, 23,163 international standards have been published covering almost all aspects of technology and manufacturing. During 2018, as many as 1,637 standards were published.

Deutsches Institut für Normung (DIN), i.e. the German Institute for Standardization, was founded in 1917. This Institute is an independent platform for standardization in Germany and worldwide that supports innovative solutions related to business digitization and research projects, which has 3,575 working committees. It has published 40,352 standards.

British Standards Institution (BSI) is a company that takes care of other organizations achieving excellence around the world. It has 193 member countries and 84,000 customer clients. BSI has published a total of 50,529 British standards.

The increase of knowledge and activities inevitably leads to a greater number of necessary compromises in solving the identified challenges and consequently the number of new standards. Figure 1 shows a trend referring to the number of ISO and DIN standards in the last couple of decades.

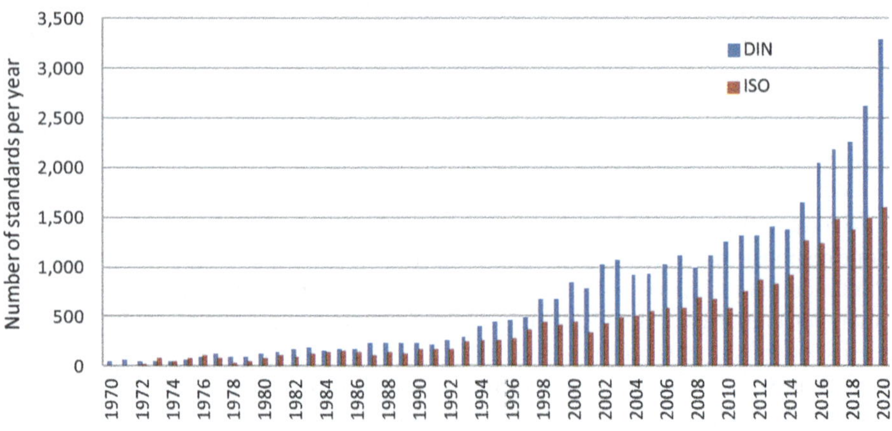

Fig. 1. Number of ISO and DIN standards published by year

Croatian Standards Institute (Hrvatski zavod za norme – HZN). The standards are accessed through the national standards organization of the Republic of Croatia. The Croatian Standards Institute is an independent, public and non-profit institution whose goals are as follows: to increase the safety level of products and processes, to protect human health and the environment, to promote the quality of products, services and processes, as well as appropriate use of materials and energy, to improve product efficiency, to control variety, to ensure compatibility and interchangeability, and finally, to remove technical barriers to international trade. The Croatian Standards Institute is a member of the International Organization for Standardization (ISO), the International Electrotechnical Commission (IEC), the European Committee for Standardization (CEN), the European Committee for Electrotechnical Standardization (CENELEC), and the European Telecommunications Standards Institute (ETSI), [3].

2.3 Standards

ISO 80000 Standard. The ISO 80000 standard was completed in 2009 and it replaced the IS 31 standard, which provided guidelines in scientific and educational documents on how to use the units of measurement, physical quantities, and formulas containing the aforementioned. This standard is also known as ISO/IEC 80000, because ISO is not the only standards developing organization. The International Electrotechnical Commission (IEC) played a great role in adopting Part 6 - Electromagnetism, Part 13 - Information science and technology, and Part 14 - Telebiometrics related to human physiology.

Under the general title Quantities and units, the ISO 80000 standard consists of the following parts:

- Part 1: General
- Part 2: Mathematical signs and symbols to be used in the natural sciences and technology
- Part 3: Space and time
- Part 4: Mechanics
- Part 5: Thermodynamics
- Part 6: Electromagnetism
- Part 7: Light
- Part 8: Acoustics
- Part 9: Physical chemistry and molecular physics
- Part 10: Atomic and nuclear physics
- Part 11: Characteristic numbers
- Part 12: Solid state physics
- Part 13: Information science and technology
- Part 14: Telebiometrics related to human physiology.

In 2017, Part 14 was withdrawn from the standard in line with the votes cast by international organization board members, and the standard currently has 13 parts. Parts 2, 3, 4, 5, 7, 9, 10, 11 and 12 were revised in 2019. Part 8 of the standard was changed in 2020. Part 1 of the standard is awaiting revision and will be replaced by ISO/CD 80000-1.2.

ISO 80000-2:2009 Standard - Mathematical Signs and Symbols to be Used in the Natural Sciences and Technology. A standard which gives general information about mathematical signs and symbols, their meanings, verbal equivalents and applications is known as ISO 80000-2:2009 Quantities and units—Part 2: Mathematical signs and symbols to be used in the natural sciences and technology. The recommendations in ISO 80000-2:2009 are intended mainly for use in the natural sciences and technology, but they also apply to other areas where mathematics is used, [11]. The research described in this paper is based on the given standard.

Variables, Functions and Operators. Variables such as x, y, etc., and running numbers, such as i in $\Sigma_i x_i$ are printed in italic (sloping) type. Mathematical constants, whose values never change, are printed in Roman (upright) type, e.g., $\mathrm{e} = 2{,}718\,218\,8\ldots$; $\pi = 3{,}141\,592\ldots$; $\mathrm{i}^2 = -1$. A comma, semicolon or some other appropriate symbol can be used as a separator between numbers or expressions. The comma is generally preferred, with the exception of numbers with the comma as a decimal separator. In individual tables, for each individual field of mathematical notation, the standard provides a handful of examples by linking symbols, the meaning of the symbols, verbal equivalents, and remarks and examples. Table 1 shows examples from four fields.

Table 1. Examples of mathematical notation

Field	Sign, symbol, expression	Meaning, verbal equivalent	Remarks and examples
Mathematical logic	$p \Rightarrow q$	p implies q, if p, then q	$q \Leftarrow p$ has the same meaning as $p \Rightarrow q$ $\Rightarrow$ is the implication symbol
Sets	$B \subseteq A$	B is included in A, B is a subset of A	Every element of B belongs to A $A \supseteq B$ has the same meaning as $B \subseteq A$
Standard number sets and intervals	$(a, b]$	Left half-open interval from a excluded to b included	$(a, b] = \{x \in \mathrm{R} \mid a < x \leq b\}$ The notation $]a, b]$ is also used
Miscellaneous signs and symbols	$a \triangleq b$	a corresponds to b	When $E = kT$, then 1 eV $\triangleq$ 11 604,5 K When 1 cm on a map corresponds to a length of 10 km, one may write 1 cm $\triangleq$ 10 km

In addition to examples given in Table 1, the standard also provides examples in the fields of elementary geometry (e.g. AB||CD, whose verbal equivalent is: the straight line AB is parallel to the straight line CD), operations (e.g. $a \cdot b$, $a \times b$, $a\ b$, ab, whose verbal equivalents are: a multiplied by b, a times b), combinatorics (e.g. $n!$, whose verbal equivalent is: factorial), functions (e.g. f, g, h, etc.), exponential and logarithmic functions (e.g. a^x, whose verbal equivalents are: a to the power of x, exponential function to the base a of argument x), circular and hyperbolic functions (e.g. arccos x, whose verbal equivalent is: arc cosine of x), complex numbers (e.g. $|z|$, whose verbal equivalent is: modulus of z), matrices, coordinate systems, scalars, vectors and tensors, transforms and special functions.

3 Application of ISO 80000-2:2009 in Scientific Journals and Master's Theses

By adopting the way of writing mathematics related expressions proposed by the standard, it is possible to unambiguously analyze all types of mathematical notation and identify deviations from normative recommendations. In order to determine the application of the standard in everyday practice, we selected five scientific journals and two engineering faculties, where we conducted an analysis of Master's theses published in 2018.

3.1 Scientific Journals

The application of the standard in everyday practice was analyzed in the following five different journals: Tehnički vjesnik/Technical Gazette, Engineering Review, Acta Polytechnica Hungarica, the American Journal of Electrical Power and Energy Systems and the International Journal of Renewable Energy Research.

Tehnički vjesnik/Technical Gazette (ISSN 1330-3651) is a journal that has been published since 1994 at Josip Juraj Strossmayer University of Osijek. The publishers are the Mechanical Engineering Faculty in Slavonski Brod, the Faculty of Electrical Engineering, Computer Science and Information Technology in Osijek and the Faculty of Civil Engineering and Architecture in Osijek. The editorial board of the journal requires all authors to comply with the ISO 80000 standard. The impact factor of the journal is 0.644. The journal publishes 6 issues per year, [12]. All manuscripts must undergo an external, mostly international, double-blind review process and, when accepted for publication, they are made freely available for access for all users to read and download.

Engineering Review (ISSN 1330-9587) is an international journal that publishes original research papers related to the analysis of structures, materials and new technologies in the field of mechanical engineering, shipbuilding, basic engineering sciences, electrical engineering, computer science and construction. The journal has been published since 1970 by the University of Rijeka and it is the successor of the Proceedings of Engineering Faculties in Rijeka. As of 2012, the journal publishes 3 issues per year; before that, it published 2 issues per year, [13]. The impact factor of the journal is 0.627. Only scientific papers are published. Before publication, all manuscripts must undergo external, mostly international, peer review.

Acta Polytechnica Hungarica - Journal of Applied Sciences (ISSN 1785-8860) is an international peer-reviewed scientific journal of Óbuda University, the Hungarian Academy of Engineering and the IEEE Hungary Section. The journal has been published since 2004. Depending on the number of papers, it publishes 7 to 8 issues per year, and in the last two years it published 10 issues each year, [14]. The impact factor of the journal is 1.806.

American Journal of Electrical Power and Energy Systems (ISSN 2326-912X) is an American journal of energy and energy systems but the topics of the papers published in the journal are not exclusively related to these areas. The first volume was published in 2012, and the journal publishes 2 issues per year (with the exception of the first year), [15].

International Journal of Renewable Energy Research (ISSN 1309-012) is an international journal dedicated to renewable energy research. The journal has 9 volumes (i.e. the first volume was published in 2011) and it publishes 4 issues per year. The impact factor of the journal is 5.127, [16].

In order to analyze the application of the standard, the year 2018 was selected for two reasons. The first and most important reason is that it is the last year for which there exist complete records required for the preparation of the final paper [5], and the second reason is that we have a local maximum in the publication of papers at the national level. The second reason is not relevant because the journals covered in our analysis publish papers by foreign authors whose number does not correspond to the national trend. According to the Croatian scientific bibliography (CROSBI), 2018 is the year of the local maximum after 2011, when we recorded the largest number of papers published in journals by Croatian scientists, Fig. 2.

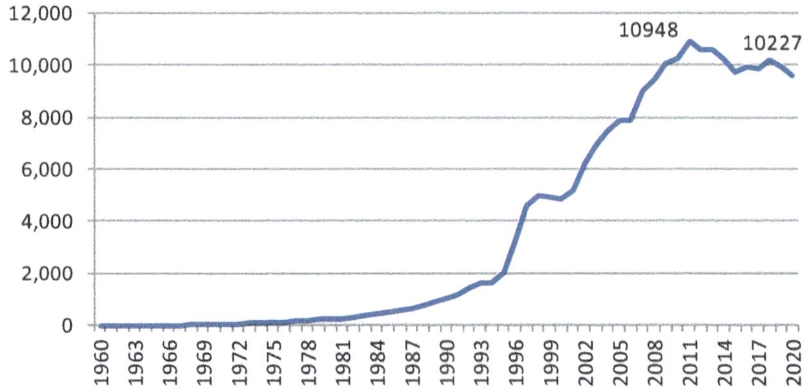

Fig. 2. Number of papers published in journals by Croatian scientists (CROSBI data)

The number of papers published in selected journals during 2018 can be seen in Fig. 3.

All papers published in 2018 were taken into consideration such that they were classified into five categories in accordance with the content in which the elements were observed that are regulated by ISO 80000-2:2009. In doing so, the first category contains

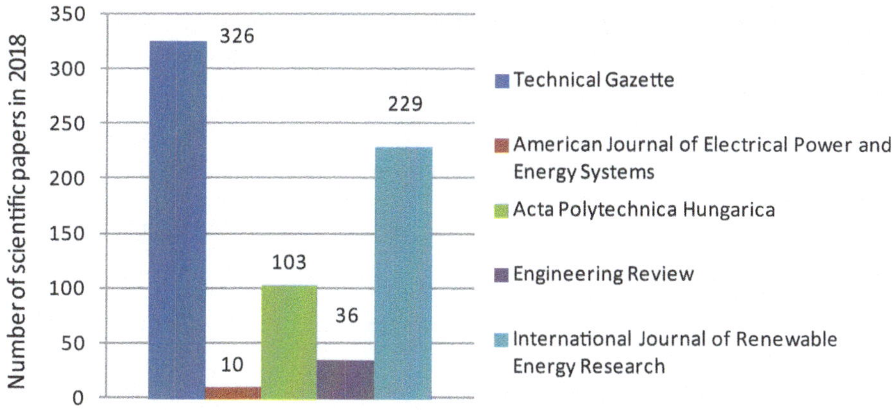

Fig. 3. Number of papers published in analyzed journals during 2018

papers whose 80% to 100% content can be regulated by the standard. Each subsequent category contains 20% less mathematical text. The second category encompasses papers in which there is from 60% to 80% of mathematical text, the third category presents papers with mathematical text in the papers ranging between 40% and 60%, while in the fourth and the fifth category there is from 20% to 40% and from 0% to 20% of mathematical text in the papers, respectively. The analysis was conducted by the first author, it is a subjective review and there is a possibility of deviation from the actual situation. Figure 4 shows the ratio of the number of pages by issues for 2018, i.e. by papers published in issues depending on the field in which the standard is applied. The number of pages containing elements of the standard is expressed as a percentage for each of the five categories with mathematical text.

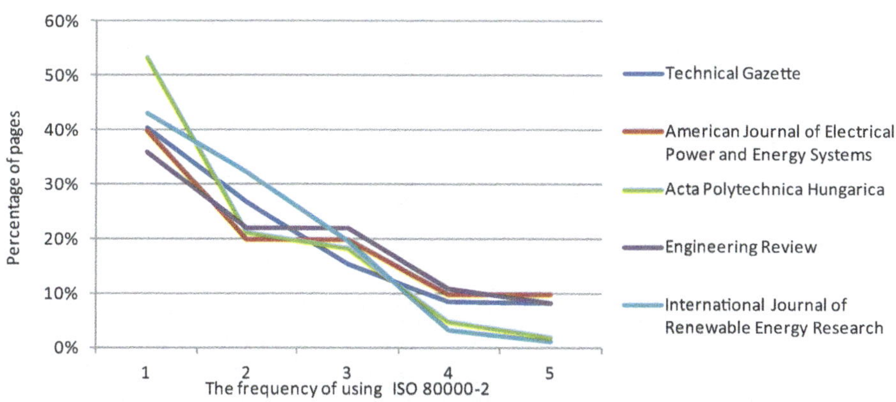

Fig. 4. Ratio of the number of pages to the frequency of using ISO 80000-2:2009

Correct application of ISO 80000-2:2009 in the aforementioned journals was also evaluated in the analysis (i.e. a degree to which the rules were observed) within a range of grades between 1 (worst) and 5 (best) (Table 2).

Table 2. Evaluation of the application of ISO 80000-2:2009

Journal	Grade
Tehnički vjesnik/Technical Gazette	5
Engineering Review	4
Acta Polytechnica Hungarica	4
American Journal of Electrical Power and Energy Systems	5
International Journal of Renewable Energy Research	3

3.2 Master's Theses

The analysis of Master's theses included theses defended during 2018 at two engineering faculties in Croatia, i.e. the Faculty of Electrical Engineering, Computer Science and Information Technology (FERIT) in Osijek and the Mechanical Engineering Faculty in Slavonski Brod. The scope of application of ISO 80000-2:2009 and its presence expressed as a percentage was analyzed, as well as a degree to which standard-related guidelines were followed. Figure 5 shows the number of papers classified into five categories. The first category consists of papers containing 80% to 100% of mathematical text, the second category encompasses papers in which there is from 60% to 80% of mathematical text, and in the third category there is from 40% to 60% of mathematical text in the papers, while in the fourth and the fifth category, there is from 20% to 40% and from 0% to 20% of mathematical text in the papers, respectively.

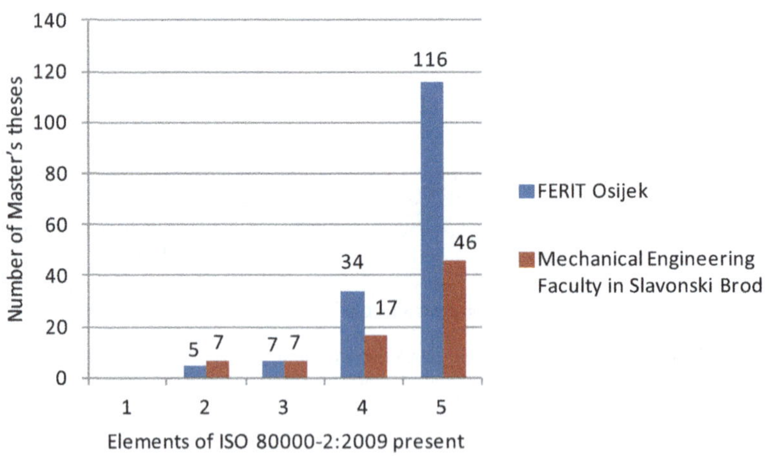

Fig. 5. Dependence of the presence of ISO 80000-2:2009 elements on the number of Master's theses

As can be seen in Fig. 5, 60% of Master's theses defended at FERIT contain up to 20% of mathematical text, while this number amounts to 48% at the Mechanical Engineering Faculty. There are no theses composed of 80% to 100% of mathematical text. Correct applications of the standard in individual Master's theses were analyzed and evaluated. The graph in Fig. 6 compares the total number of Master's theses and the number of Master's theses that are fully compliant with the standard.

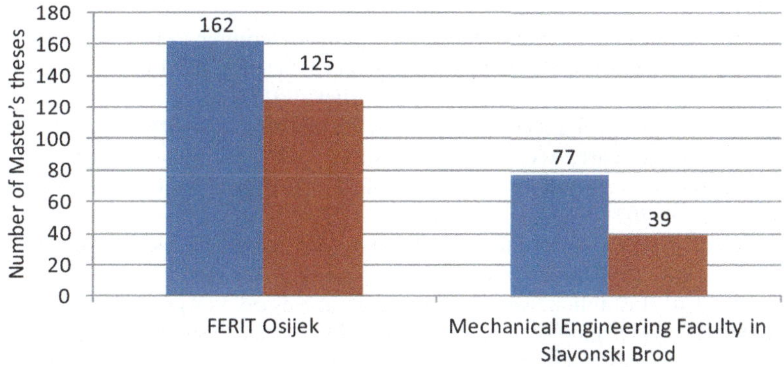

Fig. 6. Number of Master's theses marked in blue and Master's theses that achieved full compliance with ISO 80000-2:2009 marked in red

4 Conclusion

Technical systems are inconceivable without documentation describing their implementation, use and maintenance. Mathematics is a language used in the description of technical systems. In order to write mathematical text correctly, we must respect the rules, i.e. the standards. A standard is a document issued by one or more organizations whose goal is to bring order into a certain field. The ISO 80000 standard brings order into the field of mathematical writing. The ISO 80000 standard consists of fourteen documents under the general title Quantities and units. ISO 80000-2:2009 Quantities and units - Part 2: Mathematical signs and symbols to be used in the natural sciences and technology provides recommendations for writing mathematical texts. Scientific journals in the field of technical sciences publish papers that mostly contain mathematical signs and symbols. The share of papers based on mathematics as a whole is up to a maximum of 10%. The application of the ISO 80000-2:2009 standard is at a high level. Master's theses analyzed as part of our research follow the guidelines stipulated by ISO 80000-2:2009 in the range of 20 to 70%. The results of this research also indicate that the standard is significantly more frequently present in Master's theses in the field of mechanical engineering.

References

1. ISO/IEC Guide 2: 2004 Standardization and related activities - General vocabulary

2. Glavaš, H., Zovko-Ribić, F., Dorić, D., Talapko, D.: Development of energy management standards. In: Žagar, D., Martinović, G., Rimac Drlje, S., Miličević, K. (eds.) 2018 International Conference on Smart Systems and Technologies (SST), pp. 185–189. Faculty of Electrical Engineering, Computer Science and Information Technology Osijek, Josip Juraj Strossmayer University of Osijek, Osijek (2018). https://doi.org/10.1109/SST.2018.8564633
3. HZN Croatian Standard Institute: Objectives and principles of standardization. https://www.hzn.hr/default.aspx?id=89. Accessed 30 Apr 2021
4. European Committee for Standardization. www.cen.eu/about/CENinFigures. Accessed 30 Apr 2021
5. Cvenić, D.: Application of ISO 80000-2:2009 in the description of technical systems. Final paper. Faculty of Electrical Engineering, Computer Science and Information Technology Osijek, Josip Juraj Strossmayer University of Osijek (2019)
6. European Committee for Electrotechnical Standardization. www.cenelec.eu/aboutcenelec/whatwedo/factsandfigure. Accessed 30 Apr 2021
7. European Telecommunications Standards Institute. https://www.etsi.org/media-library. Accessed 30 Apr 2021
8. International Organization for Standardization. www.iso.org. Accessed 30 Apr 2021
9. Deutsches Institut für Normung. www.din.de. Accessed 30 Apr 2021
10. British Standards Institution. www.bsigroup.com. Accessed 30 Apr 2021
11. ISO 80000-2:2009 Quantities and units—Part 2: Mathematical signs and symbols to be used in the natural sciences and technology. https://www.iso.org/standard/31887.html
12. Technical Gazette. https://hrcak.srce.hr/tehnicki-vjesnik. Accessed 30 Apr 2021
13. Engineering Review. https://hrcak.srce.hr/engineering-review. Accessed 30 Apr 2021
14. Acta Polytechnica Hungarica. http://acta.uni-obuda.hu/. Accessed 30 Apr 2021
15. American Journal of Electrical Power and Energy Systems. http://www.ajepes.org/index. Accessed 30 Apr 2021
16. International Journal of Renewable Energy Research. https://www.ijrer.org/ijrer/index.php/ijrer. Accessed 30 Apr 2021

Testing the Procedure for Optimization of Cascaded Hydropower Plants and Wind Power Plants Using the IEEE 14 Bus System

Robert Noskov[1], Krešimir Fekete[2](✉), Ružica Kljajić[2], and Zvonimir Klaić[2]

[1] HOPS (Croatian Transmission System Operator), Osijek, Croatia
robert.noskov@hops.hr
[2] Faculty of Electrical Engineering, Computer Science and Information Technology Osijek, University of Osijek, Kneza Trpimira 2b, 31000 Osijek, Croatia
{kresimir.fekete,ruzica.kljajic,zvonimir.klaic}@ferit.hr

Abstract. In real-life power system, optimization methods are used on a daily basis to obtain optimal dispatch for various types of power plants. In this paper, the existing optimization method is used to maximize profit of three cascaded Hydro Power Plants (HPP) and to co-ordinate HPPs dispatch according to Wind Power Plants (WPP) production forecast. A two-stage optimization method was used with first stage design to obtain optimal HPPs dispatch plan and to maximize profit by selling electricity on a day-ahead market. First optimization stage is defined as mixed-integer liner programming (MILP) problem. Obtained dispatch plan together with the WPPs production forecast is used to check power flows in the transmission network and if line congestion occurs, second optimization stage is needed, and it is defined as quadratic programming (QP) problem. The objective of second optimization stage is to find re-dispatch plan that would be minimal deviation from an initial dispatch but it would eliminate line congestions. Proposed optimization strategy is tested using IEEE 14 Bus Test System and the results are presented in this paper.

Keywords: Hydro power plant · Wind power plant · Optimization · Dispatch

1 Introduction

The problem of finding optimal short-term operation schedule (dispatch) of cascaded hydropower plants (HPPs) is well studied in the research literature. For example, good optimization models can be found in [1]. Due to electricity market reform, researchers developed optimization models that take into consideration variable electricity market prices [2, 3]. The main challenge regarding cascading HPPs is how to properly model nonlinear discharge characteristics of the HPP as well as take into consideration hydrological couplings between the HPPs in the cascade [4]. The situation becomes more complex in the modern power system with connected wind power plants (WPPs). Uncertainty of WPP production can be captured with stochastic programming and good examples are [5, 6]. Paper [7] presents the two-stage optimization model for optimization

H. Glavaš et al. (Eds.): OTO 2021, LNNS 369, pp. 131–145, 2022.
https://doi.org/10.1007/978-3-030-92851-3_10

of cascaded HPPs in the presence of the WPPs. In the first stage, an optimal day-ahead dispatch plan of the HPPs is obtained and it maximizes profit from selling energy to the day-ahead electricity market [7]. The second optimization stage is started only when the optimal dispatch plan of the HPPs together with the forecasted WPPs production causes congestion in the transmission network. As it is explained in [7]: "The objective of the second optimization stage is the minimization of the re-dispatching of cascaded hydropower plants in order to avoid possible congestion."

In this paper, IEEE 14-bus Test system is used to evaluate the optimization procedure developed in [7]. The original test system [8] consists of 5 generators, 11 loads, 16 lines, and 3 transformers. It is used as a reference system for many case studies in its original or modified configuration. Transient stability after three-phase faults with different locations and fault clearing time is analyzed in paper [9] using IEEE 14 Bus System. In [10] IEEE 14 bus system was used to test Static Var Compensator (SVC) integrated into the system at the weakest node. The load was changed in the range of 5–45% in order to test functionality and justification of SVC installation. Due to its complexity 14 bus system is used often in transient stability studies, for example [11, 12]. Beside above-mentioned application, 14 bus system can also be used for system optimization and minimizing production costs while respecting system constraints and managing congestions [13, 14], minimizing power losses [15] and other possibilities regarding electricity market (for example calculating locational marginal price (LMP) of electricity market [16]).

Paper has a following structure: in Sect. 2 optimization procedure is described and presented with block diagram. Model and modifications made on a IEEE 14-bus Test System along with data regarding WPPs and HPPs are given in Sect. 3. Optimization results, given graphicly and in tables are given in Sect. 4. At the end of the paper short conclusion and comments are given.

2 Used Optimization Procedure

The used optimization procedure is initially introduced in [7] and it is developed for minimization of hydropower plants re-dispatching due to wind power plants output uncertainty. The stochastic nature of wind power can lead to re-dispatching of cascaded HPP's, in case of a surplus in power production, in order to alleviate transmission network congestion. The object of the proposed optimization is to minimize the given effect with two optimization stages. The optimization is conducted day-ahead (d − 1) on a daily basis and split into hourly segments. The necessary input parameters, given in the block diagram in Fig. 1, are the HPP's technical data, data from the previous optimization period, hydrological data, and local water inflow forecast to the river and reservoirs supplying the HPP's as well as the forecast for the electricity market, both the day-ahead and future market prices. The data from the previous optimization period provides the starting reservoir levels as well as the last known water discharge which can be necessary due to the water delay. Hydrological data provides the water inflow from upstream to downstream reservoirs as well as the reservoir levels when the optimization period ends.

The first optimization stage uses mixed-integer linear programming (MILP) and calculates the day-ahead HPP dispatch plans (hourly discharge or production plan) and the day-ahead reservoir management plans (hourly reservoir water levels). Then, the transmission system operator (TSO) uses the calculated HPP dispatch plans and the hourly WPP forecast in order to determine the day-ahead hourly transmission network power flow. If the results do not show any congestions in the transmission network, the optimization ends as there is no need for further optimization in the second stage. However, if the hourly power flow shows congestions, the second optimization stage is needed.

The second stage inputs are all outputs from the first stage, hourly wind forecast with the given forecast error, congested transmission line capacities, and power transfer distribution factors (PTDF) for the congested lines. The second optimization stage uses quadratic programming (QP), in combination with chance-constrained programming (CC), and gives the changed hourly dispatch planes of the HPPs and the changed hourly reservoir management plans so that no congestions occur in the transmission network. The TSO once again calculates the hourly power flow with the newly optimized data and if there still are congestions in the network, the second stage optimization is once again started. Once there are no congestions, the optimization ends. The second stage optimization ensures that re-dispatching is minimized so that HPPs water management plan changes minimally.

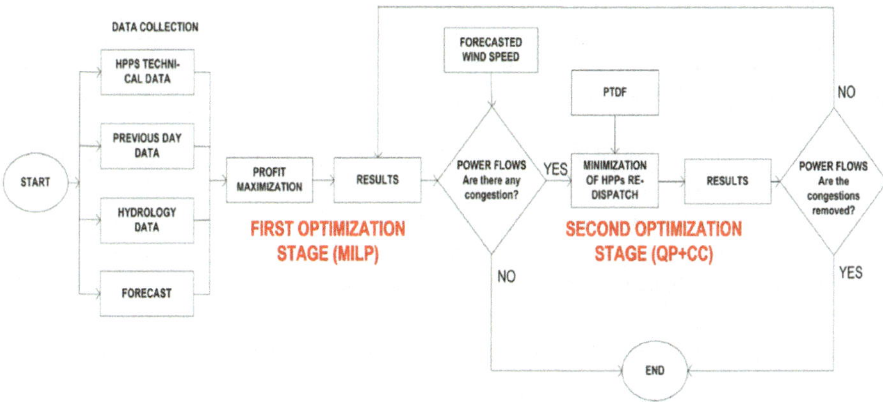

Fig. 1. Block diagram of the used optimization procedure

The objective function is the first optimization stage is defined as maximization of the expression (1):

$$\sum_{t=1}^{24}\left[\rho(t)\cdot\sum_{i=1}^{n_i}\sum_{a=1}^{n_{ai}}\left[z_a(i,a,t)\cdot\mu_{amin}(i,a)\cdot Q_{amin}(i,a)+\sum_{j=1}^{n_{jai}}\mu_{as}(i,a,j)\cdot Q_{as}(i,a,j,t)\right]\right]$$
$$+\rho_f\cdot\sum_{i}^{n_i}\left[V(i,t=24)\cdot\sum_{i\in\Gamma_i}\mu_f(i)\right]-C\cdot S(i,t) \tag{1}$$

Part I of the objective function represents the income from selling energy to the day-ahead electricity market where:

t - Time interval in the optimization period (discrete variable) [hour]
i - HPP index where $i = 1, \ldots, ni$; ni – total number of HPPs
a - Production unit index where $a = i, \ldots, n_{\text{ai}}$; n_{ai} – total number of production units in the HPP i
j - Index of the linear segment in the discharge characteristic of the production unit, $j = 1, \ldots, n_{\text{jai}}$; n_{jai} - total number of linear segments in the discharge characteristic of the production unit a in the HPP i
$\rho(t)$ - Forecasted day-ahead electricity market price [€/MWh]
$z_a(i, a, t)$ - An integer variable that is used to model forbidden zones of the discharge characteristic of production unit$a$ in HPP i
μ_{amin} - Production equivalent that describes a minimum discharge segment of production unit a in HPP i [MWh/Heh].
Q_{amin} - Minimal discharge of production unit a in HPP i [HE]
$\mu_{as}(i, a, j)$ - Production equivalent of the linear segment j in the discharge characteristic of production unit a in HPP i [MWh/Heh]
$Q_{as}(i, a, j, t)$ - Discharge of the linear segment j in the discharge characteristic of production unit a in HPP i during the hour t [HE]

Part II is the expected income from the usage of water that remained in reservoirs at the end of the optimization period, with following parameters:

ρ_f - Forecasted average future electricity price (the average price for the next optimization period - next day) [€/MWh]
$V(i, t)$ - Volume of water in the reservoir of HPP i at the end of hour t [HE]
$\mu_f(i)$ - Production equivalent of HPP i that is used for calculation of HPP production in the future optimization period [MWh/HEh]

Part III represents the penalty cost for spillage of water where:

C - Artificial penalty cost associated with water spillage [€/MWh]
$S(i, t)$ - The total spillage of water by HPP i during hour t [HE]

Optimization constraints are defined as:

$$0 \leq Q_{as}(i, a, j, t) \leq Q_{as,\max}(i, a, j) \tag{2}$$

$$z_a(i, a, t) \in \{0, 1\} \tag{3}$$

Where

$Q_{as,max}$ - Maximal discharge of the linear segment$j$ in the discharge characteristic of production unit a in HPP i. The unit is HE

Discharge constraints can be divided on constraints regarding discharge of the generation units and constraints which take in regard output power. Total discharge of the production units and the discharge through the individual discharge are described with Eqs. (4) and (5). Output electrical power and discharge are described with Eqs. (6) and (7):

$$Q_a(i,a,t) = z_a(i,a,t) \cdot Q_{a\min}(i,a) + \sum_{j=1}^{n_{jai}} Q_{as}(j) \tag{4}$$

$$Q(i,t) = \sum_{a=1}^{n_{ai}} Q_a(i,a,t) \tag{5}$$

$$P_a(i,a,t) = z_a(i,a,t) \cdot \mu_{amin}(i,a) \cdot Q_{amin}(i,a) + \sum_{j=1}^{n_{jai}} \mu_{as}(i,a,j) \cdot Q_{as}(i,a,j,t) \tag{6}$$

$$P(i,t) = \sum_{a=1}^{n_a} P_a(i,a,t) \tag{7}$$

Where parameters:

$Q_a(i,a,t)$ - Total discharge through production unit a in HPPi during hour t [HE]
$Q_{a\min}(i,a)$ - Minimal discharge of production unit a in HPP i [HE]
$\overline{Q}_{as}(j)$ - Maximal discharge through linear segment j of production unit a [HE]
$Q(i,t)$ - Total discharge of HPP i during hour t [HE]
$P_a(i,a,t)$ - Output electric power of the unit$a$ in the HPP i during hour t [MW]
$Q_{as}(i,a,j,t)$ - Discharge of the linear segment j in the discharge characteristic of production unit a in HPP i during the hour t [HE]
$P(i,t)$ - Total output power of HPP i during hour t [MW]

Hydrological constraints of the HPP reservoirs are defined as follows:

$$V_{\min}(i) \leq V(i,t) \leq V_{\max}(i); \tag{8}$$

$$Q(i,t) = \sum_{a=1}^{n_{ai}} Q_a(i,a,t); \tag{9}$$

$$0 \leq S(i,t). \tag{10}$$

Where:

$V(i,t)$ - Volume of water in the reservoir of HPP i at the end of hour t [HE]
$V_{\min}(i)$ - Minimal volume of the reservoir HPP i [HE]
$V_{\max}(i)$ - Maximal volume of the reservoir HPP i [HE]

As stated, water reservoirs located at the same river are coupled and coupling is expressed with equation:

$$V(i,t) = V(i,t-1) - Q(i,t) - S(i,t) + Q(i-1,t-\tau) + S(i-1,t-\tau) + w(i,t) \tag{11}$$

τ - Time delay
$w(i,t)$ - Local inflow in the reservoir i during hour t [HE]

Constraint expressed with Eq. (12) needs to be fullfiled if the owner of the HPPs contracted a long-term bilateral contract:

$$P(i,t) \geq P_f(i,t). \tag{12}$$

$P_f(i,t)$ - The amount of electric energy in every hour$t$ for every HPP i that is contracted by bilateral contract, expressed as constant power per hour [MW]

After first optimization stage is carried out, if there are congestions, second optimization stage needs to be carried out. Objective function of the second optimization stage is the minimization of the equation:

$$\sum_{t=1}^{24}\sum_{i=1}^{n_i}\left[D(i,t) - \sum_{a=1}^{n_{ai}} \mu_a \cdot Q_a(i,a,t)\right]^2 + C \cdot S(i,t) \tag{13}$$

where first part of the equation is minimization of the square deviation of old dispatch plan and the new dispatch plan and second part of the equation is to minimize the possible spillage. In this equation following parameters are used:

$D(i,t)$ - Output active power of HPP i during hour t, obtained in the first optimization stage [MWh/h]
$\mu_a(i,a)$ - Production equivalent of the discharge characteristic around the operating point for production unit a in the HPP i, obtained in the first optimization stage [MWh/HEh]

Second optimization stage uses constraints (2) and (8–12). It is also assumed that result regarding water volume at the end of the last hour is optimal but if re-dispatch is needed, this value can be altered as long as these alterations remain minimal. Minimal and maximum value (Vsetmin and Vsetmax) are defined by user:

$$V_{set\,\min}(i) \leq V(i,t=24) \leq V_{set\,\max}(i) \tag{14}$$

In second optimization stage, it is necessary to obtain power flows. If there is congestion in transmission line, re-dispatch is necessary. Available transfer capacity (ATC) of the lines need to be calculated and included in optimization process because active power flow beyond ATC results in HPP output power reduction. The connection between ATC and HPP output power is defined as

$$\sum_{v=1}^{n_v} P_v(v,t) \cdot \mathrm{PTDF}_{WPP\,v,k} \;\geq\; ATC_k(t) \tag{15}$$

$P_v(v, t)$ - Forecasted production of WPP v in the hour t [MW]
v - WPP index, $v = 1, \ldots, n_v$, where n_v is the total number of WPPs
$\mathrm{PTDF}_{\mathrm{WPPv,k}}$ - PTDF which determines the change of active power flow of transmission line k due to active power change of the WPP i
ATCk(t) - The available transfer capacity of the transmission line k in the hour t, [MWh/h]

Additional constraint regarding the HPP re-dispatch is defined as

$$\sum_{v=1}^{n_v} P_v(v,t) \cdot \mathrm{PTDF}_{WPP\,v,k} \;-\; \sum_{i=1}^{n_i} \left[(D(i,t) - P(i,t)) \cdot \mathrm{PTDF}_{HPP\,i,k}\right] \;\leq\; ATC_k(t) \tag{16}$$

where

$\mathrm{PTDF}_{\mathrm{HPPv,k}}$ - PTDF which determines the change of active power flow of transmission line k due to active power change of the HPP i

Re-dispatch and spillage need to be minimized and this is obtained by objective function (13) in case when WPP production causes congestion. WPP production value is deterministic which means that also constraint (16) is deterministic. On the other hand, wind speed is stochastic so Eq. (16) needs to be modified as

$$-ATC_k(t) \;-\; \sum_{i=1}^{n_i} \left[(D(i,t) - P(i,t)) \cdot \mathrm{PTDF}_{HPP\,i,k}\right] \;\leq\; \sum_{v=1}^{n_v} P_{v,cr}(v,t) \cdot \mathrm{PTDF}_{WPP\,v,k} \tag{17}$$

where

$P_{v,cr}(v, t)$ - Critical value of WPP power plant [MW].

3 Modifications of IEEE 14 Bus Test System

For the purpose of the optimization procedure testing, the original IEEE 14 Bus Test System [8] is modified and modeled in Power World Simulator [17]. The aforementioned model was modified for the sake of this paper and the following changes were made:

- Synchronous compensators connected to buses 3, 6, and 8 were replaced with generators;

- Generators connected to buses 2, 3, and 6 represent HPP generators;
- A generator connected to bus 8 represents a wind farm;
- A generator connected to bus 1 replaces the neighboring power system in which energy is exported through interconnected lines 1–2 and 1–5;
- Assumed peak load in a 24-h period are shown in Table 1 and assumed 24-h wind speed forecast together with forecasting error and forecasted WPP output power are shown in Table 2;
- Loads of the remaining hours in the observed day were recalculated according to the daily load diagram of the Croatian power system taken on 27.11.2012. [18] and they are shown in Fig. 2;

The day-ahead electricity price was taken from the EEX for 27.11.2012. and is shown in Fig. 2.

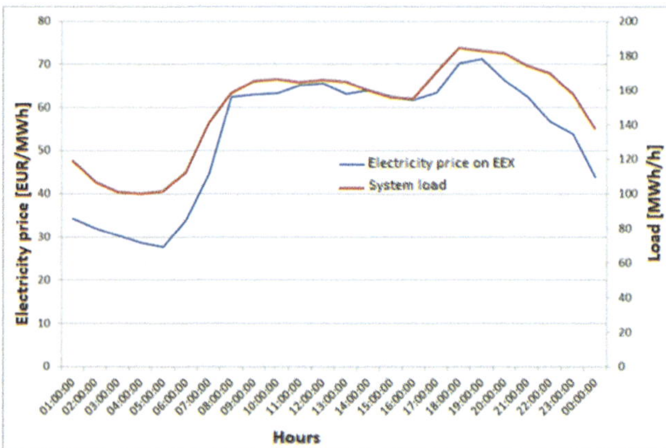

Fig. 2. Electricity prices on the day-ahead market and load diagram of the test system

The assumed electricity price, at which it will be possible to sell electricity in the future, is 50 €/MWh. The hydrological system is modeled containing 3 cascaded accumulation HPPs with its generators connected to buses 2, 3, and 6. The first one in line (HPP 1) is located at bus 6 and it has a large accumulation of seasonal character. The next in line (HPP 2) is located at bus 3 and has a smaller capacity accumulation (compensational pool for daily accumulation). The last one (HPP 3) is located at bus 2 and is the largest one of the 3 with its accumulation being also daily. Each HPP consists of 2 generators. The technical data of the HPPS are given in Table 3. The unit HE (hour equivalent) is used for water discharge as well as water volume. It corresponds to the water flow of 1 m^3/s during one hour.

The assumed wind park consists of 42 wind turbines. The rated power of each turbine is 1.5 MW so the WPPs total rated power is 63 MW. The IEEE.14 Bus Test System with the aforementioned modifications is shown in Fig. 3.

Table 1. Active and reactive bus load

Bus	1	1	2	3	4	5	6	7	8	9	10	11	12	13	14
Active power [MW]	0	0	21.7	20	47.8	7.6	11.2	0	0	29.5	9	3.5	6.1	13.5	14.9
Reactive power [MVAr]	0	0	12.7	5	−3.9	1.6	7.5	0	0	16.6	5.8	1.8	1.6	5.8	5

Table 2. Wind speed forecast and WPP output power for observed planning period

Hour	1	2	3	4	5	6	7	8	9	10	11	12
Forecasted wind speed [m/s]	4.5	4.6	5	7	10	12	12	11	9.5	9	9.4	9
Forecasting error [%]	0.1	0.6	1	1.4	1.9	2.2	2.6	3.1	3.2	3.3	3.35	3.4
WPP power [MW]	3.15	3.53	5.25	18.9	55.1	63	63	63	47.8	41	46.4	41
Hour	13	14	15	16	17	18	19	20	21	22	23	24
Forecasted wind speed [m/s]	8	10	8	7.5	7	8	9	9.2	10	9.5	11	8
Forecasting error [%]	3.45	3.5	3.55	3.6	3.65	3.7	3.75	3.8	3.85	3.9	3.95	4
WPP power [MW]	28.9	55.1	28.9	23.6	18.9	28.9	41	43.6	55.1	47.8	63	28.9

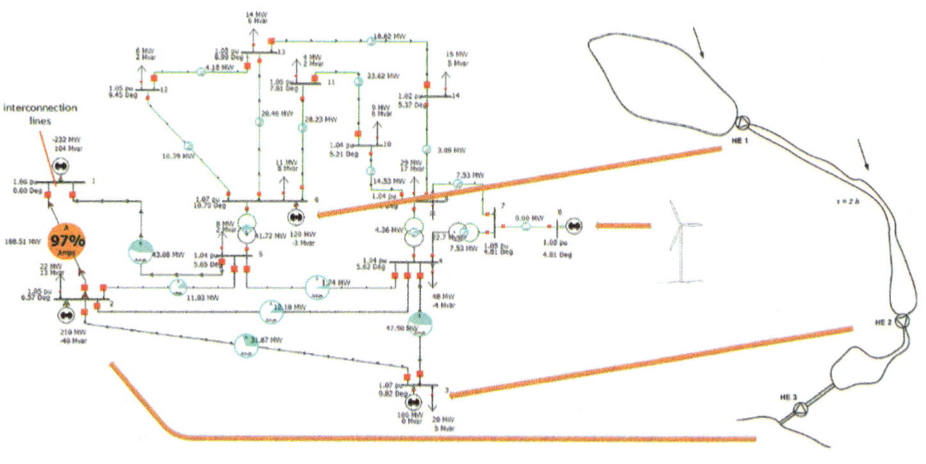

Fig. 3. Modified IEEE 14 bus test system

Table 3. HPPs technical data

		HPP1		HPP2		HPP3	
Rated power [MW]		2 × 60		2 × 50		2 × 108	
Forecasting error [%]		0.1		0.6		1	
Linearized discharge characteristic		Qas, max [HE]	μas [MWh/HEh]	Qas, max [HE]	μas [MWh/HEh]	Qas, max [HE]	μas [MWh/HEh]
	Technical minimum	30	0.67	40	0.47	40	1.05
	1. segment	30	0.67	40	0.47	40	1.05
	2. segment	40	0.5	30	0.42	30	0.8
The maximal volume of water in accumulation		540 000 000 m^3 150 000 HE		3 600 000 m^3 1000 HE		5 040 000 m^3 1400 HE	
Minimal volume of water in accumulation		180 000 000 m^3 50 000 HE		720 000 m^3 200 HE		1 080 000 m^3 300 HE	
Accumulation volume at the start of the planning period		360 000 000 m^3 100 000 HE		1 800 000 m^3 500 HE		2 520 000 m3 700 HE	
Local inflow [HE]		25		5		–	
The time delay of water from the upstream accumulation [h]		–		2		–	
Long term bilateral contracts [MWh/h]		50		40		80	
Allowable daily discharge of water from the reservoir		12600000 m^3 3500 HE		no limit		no limit	

4 Results of the Optimization Procedure

4.1 First Optimization Stage

The first optimization stage gives optimal dispatch for HPPs in order to achieve maximal profit by selling electricity on the day-ahead market and from future selling of hydro energy saved in reservoirs at the end of the planning period. Key parameters which define the HPP dispatch plan are forecasted electricity price on the day-ahead market and forecasted future electricity price at the end of the planning period.

Obtained HPPs production plan is correlated with electricity prices and is given in Fig. 4. During nighttime planned production is equal to planned bilateral contracts and HPPs do not participate in the day-ahead market; instead, they operate with their minimum capacity. During the day, when electricity prices are higher, planned production is equal to rated power.

Water volume in reservoirs also changes as is shown in Fig. 5. HPP1 reservoir dictates discharge for all downstream reservoirs, since the incoming inflow is low, the volume

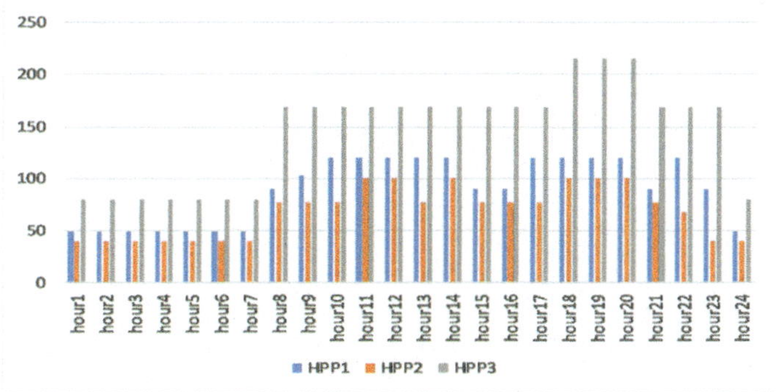

Fig. 4. Calculated day-ahead dispatch plan of the HPPs during the planning period

of HPP1 reservoir constantly decreases at a rate of 3500 HE (12600000 m^3). In order to maximize profit, all of the water needs to be used. HPP2 and HPP3 volumes are in margins between the minimum and maximum volume with HPP2's tendency to increase volume in the last 4 h of the planning period. This increase is due to electricity prices which are lower than the presumed future price (50 €/MWh). Water stored in HPP2 reservoir can be used in both HPP2 and HPP3 which leads to bigger profit from saving water than selling it on the day-ahead market during the last 4 h.

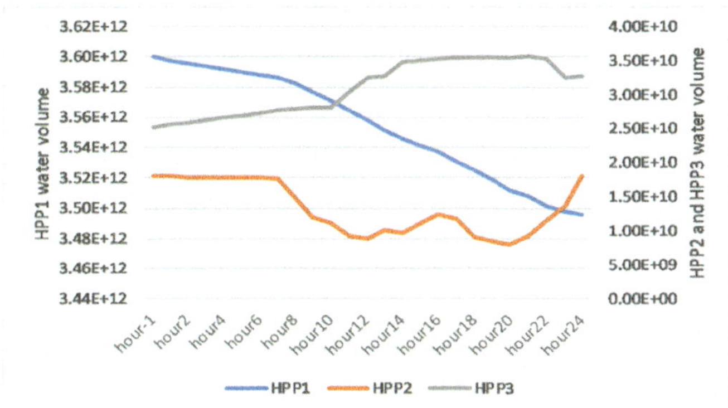

Fig. 5. Calculated water volume in reservoirs at the end of each hour

4.2 Second Optimization Stage

In the first optimization stage optimal HPPs dispatch plan is obtained and these data, along with forecasted production from WPPs, are used as input data for second optimization stage. Using the aforementioned data, power flows are calculated in order to check if any congested (critical) lines exist. Power flow results show that line 1–2 could be congested and it is considered to be critical.

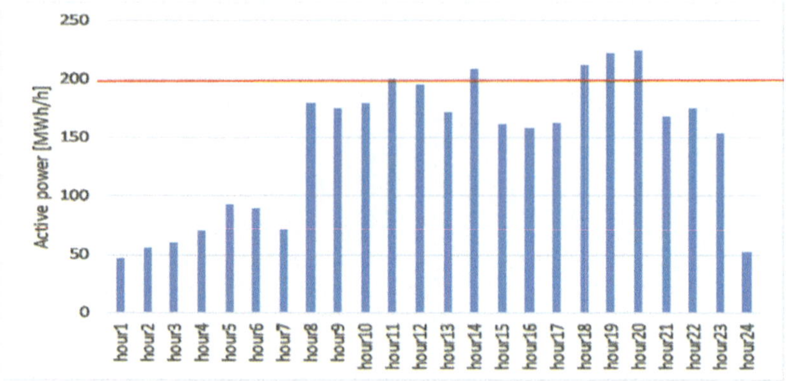

Fig. 6. Power flows on the critical line

Power flows on a critical line during the planning period are shown in Fig. 6. Transfer capacity of line 1–2 is 200 MWh/h with the red line representing this transfer capacity. All values which are above the transfer capacity are considered to be congestions.

Since congestions are possible, the second optimization stage is needed to re-dispatch HPPs output power and remove congestion. Figure 7 shows the re-dispatching of HPPs power during the planning period. The first hour of congestion is hour 11 but due to the time delay of water, HPP1 is the first one to reduce its output power one hour before (hour 10). The biggest change is in the output power of HPP 3. This HPP has the biggest contribution in power flow changes of the critical line since it is nearest to it.

Fig. 7. HPPs output power re-dispatching

As in the previous optimization stage, the reservoir volumes are also calculated, and they are shown in Fig. 8. In this case, volumes are similar to the first optimization stage. The goal of the second optimization stage is to find optimal dispatch to ensure minimal offset from the reservoir volume values obtained during the first optimization stage.

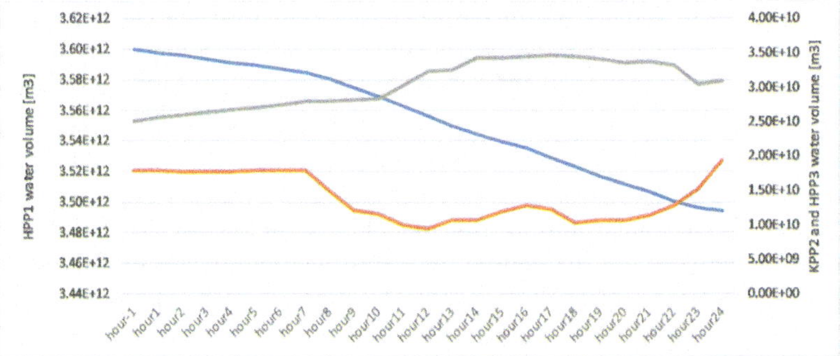

Fig. 8. Water volume in reservoirs at the end of each hour obtained in the second optimization stage

Results obtained in the second optimization stage and re-dispatched HPPs are used for new power flow calculations in order to eliminate line congestions. Figure 9 shows a comparison of line active power flow before and after re-dispatching was made. Congestions are eliminated and critical line 1–2 is uncongested during the whole planning period and optimization process can end.

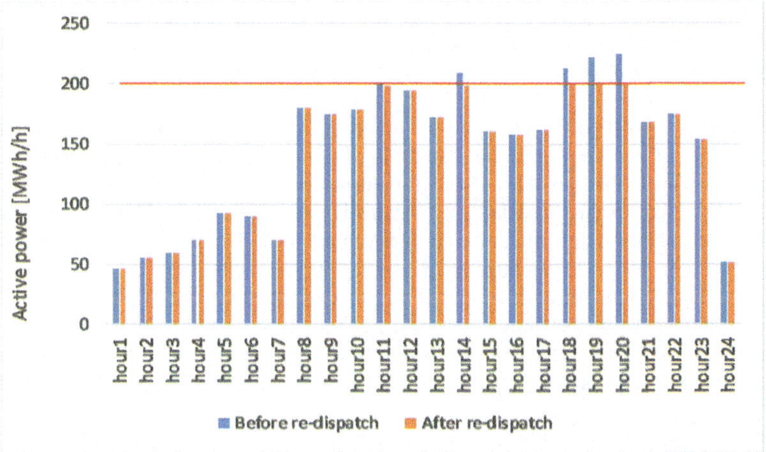

Fig. 9. Active power flow comparison after the first and second optimization stage

5 Conclusion

Optimization of cascaded HPPs dispatch plan is quite challenging problem especially in modern power systems where WPPs are connected. Two-stage optimization procedure proposed in [7] is tested using IEEE 14 Bus Test Case. In first optimization stage optimal dispatch and accumulation plan was obtained and since congestions occurred it was necessary to perform second optimization stage. As the results showed, if there are congestions in the system, optimal plan needs to be modified but these modifications need to be minimal. Parallel to HPPs dispatch plan, accumulation management plan needs to be made. Accumulated water can be used when electricity prices are higher, so it is necessary to make a decision when the right time is to use it. Since WPP production is stochastic, HPPs production needs to be dispatched in a way to eliminate congestions. Results shows that two-stage optimization model is useful tool for alleviating congestion in transmission network that are caused by WPPs uncertainty and second optimization stage ensures that deviations from original HPPs dispatch plan are minimal.

References

1. Kerola, M.: Calibration of an optimisation model for short-term hydropower production planning. Helsinki University of Technology (2006)
2. Fosso, O.B., Belsnes, M.M.: Short-term hydro scheduling in a liberalized power system. In: International Conference on Power System Technology, PowerCon 2004, vol. 2, pp. 1321–1326 (2004). https://doi.org/10.1109/ICPST.2004.1460206
3. Conejo, A.J., Arroyo, J.M., Contreras, J., Villamor, F.A.: Self-scheduling of a hydro producer in a pool-based electricity market. IEEE Trans. Power Syst. **17**(4), 1265–1272 (2002). https://doi.org/10.1109/TPWRS.2002.804951
4. Chang, G.W., et al.: Experiences with mixed integer linear programming based approaches on short-term hydro scheduling. IEEE Trans. Power Syst. **16**(4), 743–749 (2001). https://doi.org/10.1109/59.962421
5. Fleten, S.E., Kristoffersen, T.K.: Short-term hydropower production planning by stochastic programming. Comput. Oper. Res. **35**(8), 2656–2671 (2008). https://doi.org/10.1016/j.cor.2006.12.022
6. Vardanyan, Y., Amelin, M.: The state-of-the-art of the short term hydro power planning with large amount of wind power in the system. In: 8th International Conference on the European Energy Market (EEM), pp. 448–454 (2011).https://doi.org/10.1109/EEM.2011.5953053
7. Fekete, K., Nikolovski, S., Klaić, Z., Androjić, A.: Optimal re-dispatching of cascaded hydropower plants using quadratic programming and chance-constrained programming. Energies **12**(9), 1604 (2019). https://doi.org/10.3390/en12091604
8. IEEE 14-Bus System - Illinois Center for a Smarter Electric Grid (ICSEG). https://icseg.iti.illinois.edu/ieee-14-bus-system/. Accessed 10 May 2021
9. Iyambo, P.K., Tzoneva, R.: Transient stability analysis of the IEEE 14-bus electric power system. In: AFRICON 2007, pp. 1–9 (2007). https://doi.org/10.1109/AFRCON.2007.4401510
10. Mishra, P., Holkar, V., Choubey, A.: Enhancement of voltage profile for IEEE-14 bus system by using STATIC-VAR compensation (SVC) when subjected to various changes in load. J. Res. Stud. Sci. Eng. Technol. **1**(2), 27–33 (2014)
11. Anuar, A., Wahab, M.A.A., Arshad, S.N.M., Romli, M.I.F., Bakar, A.H.A., Bakar, M.A.A.: Transient stability for IEEE 14 bus power system using power world simulator. In: Journal of Physics: Conference Series, vol. 1432, p. 012009, January 2020
12. Pattanaik, P.P., Panigrahi, C.K.: Stability and fault analysis in a power network considering IEEE 14 bus system. In: 2nd International Conference on Inventive Systems and Control (ICISC), pp. 1134–1138 (2018). https://doi.org/10.1109/ICISC.2018.8398981
13. Asija, D., Soni, K.M., Sinha, S.K., Yadav, V.K.: Assessment of congestion condition in transmission line for IEEE 14 bus system using D.C. optimal power flow. In: 7th India International Conference on Power Electronics (IICPE), pp. 1–6 (2016). https://doi.org/10.1109/IICPE.2016.8079332

14. Dashtdar, M., Najafi, M., Esmaeilbeig, M.: Calculating the locational marginal price and solving optimal power flow problem based on congestion management using GA-GSF algorithm. Electr. Eng. **102**(3), 1549–1566 (2020). https://doi.org/10.1007/s00202-020-00974-z
15. El-Fergany, A.A., El-Arini, M.: Minimization of energy loss using integrated evolutionary approaches. In: IEEE Electrical Power and Energy Conference, pp. 323–329 (2012). https://doi.org/10.1109/EPEC.2012.6474973
16. Bouddou, R., Benhamida, F., Belgacem, M.: Profit analysis and optimization in a deregulated electricity market using DCOPF based LMP. In: International Conference on Applied Automation and Industrial Diagnostics (ICAAID), pp. 1–6 (2019). https://doi.org/10.1109/ICAAID.2019.8934977
17. PowerWorld. https://www.powerworld.com/solutions/students. Accessed 20 Mar 2021
18. Croatian Transmission System Operator System scheme. https://www.hops.hr/wps/portal/en/web/hees/data/present. Accessed 19 Mar 2021

Procedures for Testing and Maintenance of Electric Motors for the Purpose of Determining the Correctness and Reliability at Operating Conditions

Venco Ćorluka(✉)

Faculty of Electrical Engineering, Computer Science and Information Technology Osijek, University of Osijek, Kneza Trpimira 2b, 31000 Osijek, Croatia
venco.corluka@ferit.hr

Abstract. Electric motors today are represented in all branches of industry and their supervision is a very complex task to research. Improper maintenance and proper checking of electric motors can cause great damage to the drive. In order to determine the condition of electric motors, it is necessary to perform diagnostics of basic parameters in order to determine the reliability and thus the service life of the motor. This is why fault diagnosis is important for the maintenance of electrical machines. The assessment of the condition of the correctness and reliability of the electrical machine is carried out by the method of measuring the basic parameters. The measured data is compared with the nominal data of the correct motor provided by the manufacturer. Diagnostic measurements are performed on the measurement of winding operating resistance, winding impedance, temperature, current, voltage, power, torque, speed of noise and vibration. The measurement is performed for all three operating conditions: no load, nominal and maximum load. Spectral analysis of these quantities gives a picture of the electromechanical state of the motor. For example, the mechanical state of a motor can be determined from the frequency spectrum of noise and vibration, and the condition of each mechanical part can be estimated on the basis of individual harmonics. The analysis of the measured data provides an assessment of reliability and suggests the method and time of servicing. By applying the appropriate sensors and placing them in the right places, the required sizes can be measured quickly and efficiently, and thus the condition of the motor can be assessed. Non-contact measurement of parameters such as temperature, noise, vibration, current, voltage, speed creates a basis for the development of an automated measuring system, which significantly speeds up and facilitates the assessment of reliability and correctness, especially when it comes to a large number of motors. Proper and periodic measurement as well as proper maintenance can extend motor life and result in increased productivity and financial savings. This paper presents the procedures for measuring the required quantities and making an assessment of reliability and correctness on the same type of motor with a different number of operating hours.

Keywords: Reliability · Maintenance · Spectrum · Processing

H. Glavaš et al. (Eds.): OTO 2021, LNNS 369, pp. 146–159, 2022.
https://doi.org/10.1007/978-3-030-92851-3_11

1 Introduction

Three-phase asynchronous motors today surpass all other types of electric motors in terms of the number of installed units. Thanks to the intensive development and application of frequency converters, the asynchronous machine, even in terms of control characteristics, has approached DC machines, and it can be used in very demanding drives. Unexpected failure and failure of a machine from operation can cause high material costs and process downtime of various kinds. By applying adequate diagnostic methods, failures can be detected in time, ie at an early stage of their occurrence. In that case, we make engine changes at the most technologically convenient moment, and the planning of overhaul and purchase of spare parts is facilitated. All this information is useful for both the user and the equipment manufacturer. Recently, modern diagnostic methods are being introduced in the production lines of machines, as an additional segment of thc final control. Diagnostics of electrical machines is an area that is developing very intensively in world technical practice. Various methods are mentioned in the literature by which it is possible to establish a whole range of failures of electrical machines. Many external indicators, such as oscillation of ammeter hands, oscillations of rotational speed or increased vibrations, can indicate irregularities in the operation of electrical machines. For reliable fault diagnosis, it is necessary to develop accurate signal processing and observation algorithms, so that the cause can be determined from the obtained data. Therefore, it is very important to know the behavior of the correct machine in operation. Their stator winding is always derived as three-phase whether its phases are connected in a triangle or in a star. The design of the rotor winding depends on the size of the machine and the method of starting. Previously, large asynchronous machines were operated with sliding rings. In these machines, the rotor winding is three-phase, as in the stator, the winding is insulated to ground, and its terminals are connected to sliding coils [1]. This rotor design was used in motors driven by resistor actuators. Testing of electrical machines, including asynchronous ones, is carried out primarily to determine its mechanical and electrical characteristics, but also to check the compliance of the manufactured asynchronous motor with the design data. Tests can be divided into three categories: serial, type and special tests. Batch tests are a set of tests aimed at confirming key features on the machine and detecting any gross manufacturing errors. They are carried out on each machine produced in a particular series. The scope of serial testing is determined by the manufacturer, depending on production experience and market expectations. Type tests are performed on only one machine of a certain series. They determine all the features of the asynchronous machine that are determined by the design documentation. The last group of tests consists of special tests, which are carried out at the customer's request, regardless of the series and type tests. Their content and eligibility criteria are defined by a contract or special rules (e.g. for shipping, transport, explosion-proof installations, etc.). Here, the customer also defines the standards according to which the test in question should be carried out [2]. The most common case where such tests are applied are insulation systems of high voltage machines, special requirements when measuring machine noise, explosion tests, etc. For all tests it is necessary to have predefined acceptance criteria to make it clear whether the test result is acceptable or not. Eligibility criteria are determined by the design documentation or relevant international or recognized national standards. Figure 1 shows a system for measuring and diagnosing failures on induction motor bearings.

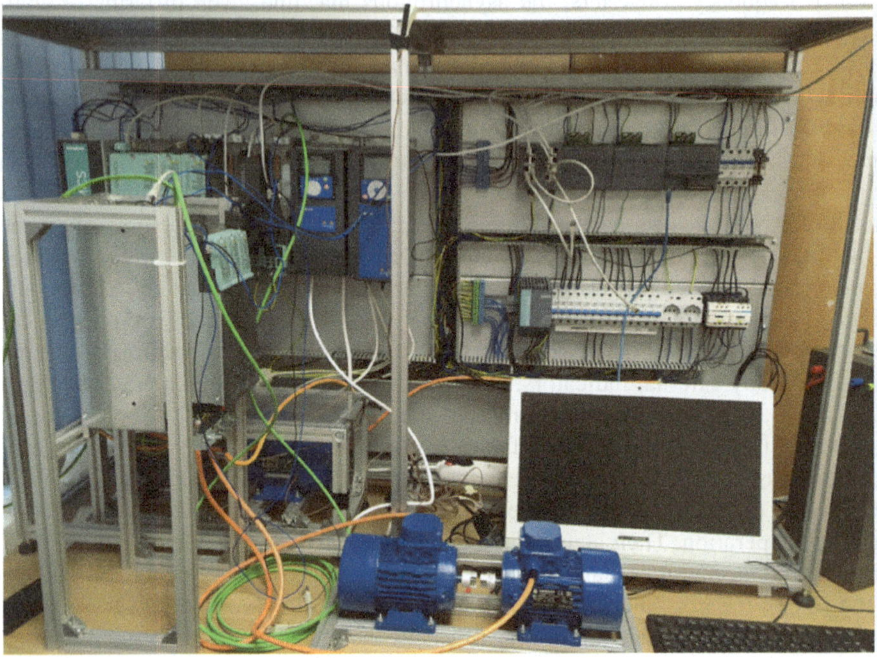

Fig. 1. Experimental measuring system for determining faults in induction motor bearings.

2 Measurement of Ohmic Resistance of Stator and Rotor Windings

The stator windings, in the case of pulley motors and rotors, are three-phase alternating windings. In most companies, the calculated values for the accepted ohmic resistances are calculated values, which define the allowable lower and upper limit of the deviation of the measured resistance as a percentage of the ohmic resistance from the calculated data. In the calculation data, the machine designer gives the ohmic resistances per phase, calculated at 20 °C, regardless of whether the winding is connected in a triangle or a star [2]. Due to the performance of the motor, as well as the specificity of the conditions and the required operating speed in the test station, it is sometimes not possible to measure the ohmic resistance per phase. This mainly applies to motors whose stator (or rotor) winding is connected in a triangle, and the motor is for example made in a closed version. The tester has only winding leads available and is able to measure only the ohmic resistance between the winding terminals. Such an approach to measuring the ohmic resistance also facilitates the calculation of losses in the motor winding because a unique formula is used for all winding designs:

$$P_{Cu} = 1.5 \cdot I^2 \cdot R \tag{1}$$

Where P_{Cu} are the losses in the winding, I is the current flowing through the winding, and R is the ohmic resistance of the winding measured between the terminals. This resistance is measured on all three pairs of terminals U1-V1, U1-W1 and V1-W1.

Calculate the measured resistance between the terminals to the phase resistance, which is done for the winding connected in a triangle according to the expression:

$$R_f = \frac{3}{2} \cdot R \tag{2}$$

and for a winding connected to a star according to the expression:

$$R_f = \frac{1}{2} \cdot R \tag{3}$$

where R_f is the mean value of the phase and R is the mean value of the resistance measured between the terminals. The mean phase resistance can now be converted to a temperature of 20 °C. The value thus calculated is compared with the calculated data. It is very important to measure the winding temperature as accurately as possible during the measurement. The most accurate value is given by built-in thermometers, if the engine has them. If, on the other hand, the engine does not have a built-in thermal probe, the temperature of the housing or terminal is measured.

3 Measurement of Insulation Resistance of Stator and Rotor Windings

Insulation is one of the most sensitive parts of an electrical machine and, especially in the past, failures due to damage and deterioration of insulation were very common. Modern insulation methods, as well as the latest insulation materials, have improved the mechanical and insulating properties. Measuring the insulation resistance is in some ways similar to measuring the ohmic resistance, only in this case the insulation resistance is measured, not the conductor. One end of the DC voltage source is connected to the terminal of the phase to which the insulation resistance is measured, while the other end is connected to ground. In practice, the DC voltage connected to high-voltage motors is 1000 V. While the voltage is connected, the current flowing through the insulation is measured. Insulation consists mainly of insulators with a specific conductivity of less than 10^{-6} S/m. The device that measures the insulation resistance is called a megohmmeter or inductor. It has a custom scale for reading the insulation resistance which, due to the height of the applied voltage and very low current, is measured in megaomes. As the current through the insulator changes its character depending on the start of the measurement, the insulation resistance is read at precisely defined moments. U.S. IEEE regulations define the insulation resistance reading after the first and tenth minutes from onset measurements. European regulations, based on, define that the insulation resistance is read after 15 and 60 s from the start of the measurement. However, failures of large machines caused solely by aging insulation are relatively rare. Much more frequent failures are caused by the penetration of foreign materials (oils, metals…) into the insulation. Damage to the insulation of the phase leads from the machine, which is caused by vibrations, sometimes occurs during operation. For these reasons, it is necessary to include isolation tests in regular diagnostic tests [3]. The basic methods of testing the state of insulation are: low dc voltage test (insulation resistance and polarization factor measurement), high dc voltage test, leakage current measurement, dielectric loss factor (tgδ) measurement, capacity measurement, partial discharge measurement and high voltage dielectric strength test, pulse test. Except for the housing and package, all machine parts and windings that are not connected to high voltage must be earthed. If it is a final test, it needs to be conducted after a warm-up attempt, which is the worst possible case for the engine insulation system. If the winding "withstands" the test voltage for all 60 s without a ground fault, it is considered correct. The final high voltage test is the last test performed on the machine. After all the characteristics are recorded, and all sizes are measured, there are only 60 s left, which the motor separates from delivery or reassembly and re-insulation of the windings. To avoid unpleasant surprises, it is necessary to check the correctness of the windings at all stages.

3.1 Rotor Winding Failures

Rotor cage failures occur especially with larger machines. The most common reasons are the high temperatures that develop in the cage and the high forces to which the rotor is exposed, especially during take-off. The causes of failure in cast rotors can be in the poor construction of the cage, while in rotors with welded rings there are sometimes poor joints between the rod and the ring [14]. In these cases, the fault site heats up more, which further weakens it, and eventually cracks. The places of the most common cracking are precisely the joints of the ring and the rods, as well as the parts of the rods outside the rotor package that are free. Similar failures can also occur as a result of small cage displacements within the package, which occur due to alternating heating and cooling of the rotor. In cases of intermittent operation, when the machine is exposed to large changes in speed or frequent starts, there is a risk of failure due to material fatigue. Early indicators of these failures are pulsations in rotational speed, stator current, and machine leakage currents. An example of a total rotor cage failure of a high voltage motor is shown in Fig. 2.

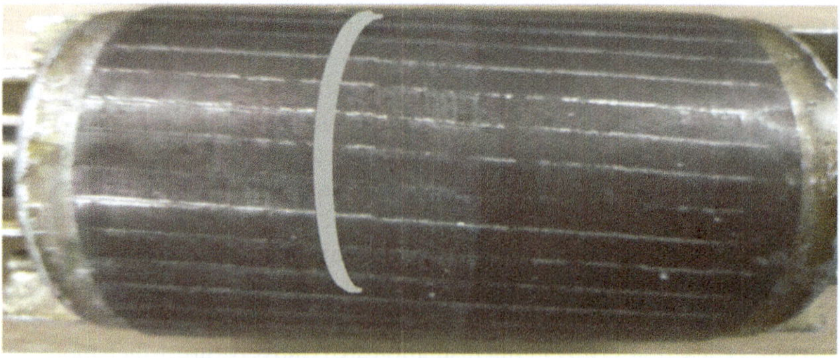

Fig. 2. Example of rotor cage failure

4 Temperature Measurement

Temperature measurement is a common method of monitoring the condition of electrical machines and drives from overheating [4]. Continuous temperature measurement can prevent damage to stator windings and packages, rotor windings and packages, as well as bearing damage (usually combined with vibration measurement). To measure the temperature continuously or intermittently, sensors are used in the design: thermistor, thermocouple, resistant thermometers.

Thermocouples - the work is based on the fact that when two different metals are joined at one end, and the joint is at a different temperature from the free ends, a voltage is induced at the ends of those two metals. The amount of induced voltage is proportional to the difference between these two temperatures. When measuring, it is important that the measured temperature safely reaches the joint of both wires, and therefore they are very often installed on the plates that are inserted into the machine in, for example, ventilation channels. Thermocouples measure temperatures from −250 °C to 3000 °C, depending on the materials used. The most commonly used:

- copper/constant −250 to 400 °C
- iron/constant −200 to 850 °C
- chromel/alumel −200 to 1100 °C
- platinum/platinum rhodium 0 to 1400 °C

Resistance thermometers - elements that determine the temperature from a change in resistance. They have a positive temperature coefficient of resistance and good stability [13]. These are specially made wire resistors that are installed in a characteristic place in the machine with the purpose that the temperature can be measured at that place. Alloys with a high temperature coefficient are used for production, e.g. platinum, nickel. This is most often a Pt 100, a resistor that has a resistance of 100 Ω at 0 °C. It is very important that the measuring current be small because otherwise the meter develops excessive Joule losses, which heat it up and distort the measurement results. Resistance thermometers are better than thermocouples for diagnostic purposes due to the small voltage values induced on the thermocouple. This voltage for the measuring devices needs to be increased which, sometimes, introduces an inaccuracy.

4.1 Bearing Temperature Measurement

The bearing temperature is measured at idle. This test determines the correct installation of the bearing and its lubrication. The test lasts until the bearing temperature stagnates. With larger engines, this can take up to three hours, while with secondary examination almost after an hour and a half. In practice, there are cases that the bearing temperature will not stagnate. This happens when the bearing is not mounted correctly, when it is pinched. The bearing temperature is measured using built-in resistance thermometers. If they are not present, thermocouples (e.g. copper-constant thermocouple) are placed in tubes used to supply grease for bearing lubrication. In this way, the temperature of the outer ring of the bearing is measured.

4.2 Thermal Imaging Diagnostics

The method is based on recording parts of the drive with special thermal imaging cameras. The thermal imaging camera records in the infrared range, and displays the temperatures of the recorded surfaces in different colors. Older versions with liquid nitrogen cooled are bulky, so they are recordings limited to external installations. Newer digital versions work at room temperature, are small in size, and allow recording inside the machine housing. In diagnostic terms, the method can be assessed as moderately effective due to its limitation to visible parts. Gives good results in detecting overheating at junctions, fuses, busbars, conductive insulators (poor contact) and housings (overheating due to stray currents, magnetic flux concentration or specific problems related to high current leads). Measurements are best performed in poor visibility (night or day with cloudy weather). During the measurement, the tested parts of the system should be loaded with at least 50% of the rated currents, in order to reduce possible errors due to the conversion of currents to rated loads.

5 Noise and Vibration Measurement

The noise of electric machines of asynchronous motors is divided into three types. These are aerodynamic noise, mechanical noise and magnetic noise. Aerodynamic noise is caused by the flow of coolant as a result of engine ventilation. Mechanical noise is caused by the rotation of the rotating parts of the engine (bearings, sliding wheels). The last type of noise is caused by magnetic conditions in the machine, which is why it is called magnetic noise [3]. Aerodynamic and mechanical noise are present at and below the supply voltage, while the saturation of the magnetic circuit has not yet occurred. Many failures of electrical machines are accompanied by an increase in vibration of the machine or its individual parts. In the case of many irregularities, vibrations can be reliable indicators of the same. Vibration measurements can detect failures of rotor bearings, rotor cages and a number of other failures. Due to the fact that most failures are also manifested through an increase in vibration, the experience of the person or the quality of the expert system is important for interpreting the results of vibration measurements and making judgments about the possible cause. Nowadays, vibration measurement is one of the most important diagnostic methods in plants. The most common causes of increased vibration of rotary electrical machines are:

- imbalance of the rotor,
- bearing damage,
- incorrect installation and foundation,
- damage to the foundation,
- local resonance of structural parts,
- accidental touching of rotating and stationary parts
- electrical and magnetic asymmetries.

Vibrations can be felt well by touch. Three basic quantities are measured:

vibration amplitude (from 0 to 10 kHz), vibration speed (10 Hz to 10 kHz) and vibration acceleration (1 kHz to >100 kHz).

The choice of size for measurement and analysis depends on the machine being tested as well as the frequencies of interest for the tests. Vibration measurements are performed with vibrometers, proximeters (induced voltage depends on the distance of the object from the sensor), accelerometers (piezoelectric sensors, acceleration measures but are also used to measure speed and displacement; the desired value is obtained by integration). Permissible vibration amounts are determined by regulations.

6 Current Measurement

For current purposes, the drive current signals must be adapted to the measuring devices. Current transformers and shunts are used for this purpose. Shunts are not suitable for measuring large currents as large currents on shunts develop large losses which can create difficulties in shunt construction. In addition, they give relatively small signals. In high-voltage devices, current transformers are also used to isolate measuring equipment from high voltages [7]. Current measuring transformers, as well as voltage ones, are required to have a constant transmission ratio and a small phase shift. For control measurements, as well as in cases when it is not possible to use built-in transformers, current clamps are used. Current clamps are a special design of current measuring transformers that enable connection without the need for prior interruption of the circuit. They are built so that the iron core can be disassembled and covered by a current-carrying conductor. This conductor represents the primary winding of the current transformer, while the secondary is wrapped around the core itself. It is possible to quickly perform a series of measurements with current clamps, as they are easily switched from one conductor to another. This makes them very suitable for diagnostic measurements in operating conditions. "DC transformers" based on the Hall probe principle are used to measure DC and low frequency currents.

7 Shaft Voltage Measurement

Irregularities in the magnetic circuit of an electric machine can cause the existence of a small resultant current in the shaft, which will induce an electromotive force between its ends. This electromotive force is called the shaft voltage, which, if large enough, can cause current to flow through the motor bearings. According to tests conducted in [10], voltages whose effective value exceeds 300 mV can cause current to flow through the motor bearings. This current is dangerous because it causes electrocorrosion on the bearings, which destroys the bearings, ie reduces their service life. In order to avoid the flow of bearing currents, the insulation of one, if not both bearings, according to the mass is carried out. This procedure terminates bearing current circuit. Shaft voltage occurs on all major AC electric rotary machines, and in asynchronous motors the most common causes are:

- the resulting comprehensive flows around the shaft
- higher harmonics of the spatial distribution of flow
- stray currents of foreign sources.

There are two ways to measure shaft voltage. The first measures the voltage drop at the ends of the shaft, so that the stator yoke is included. In the second case, the shaft voltage is measured between the grounded housing and the end of the shaft on which the bearing is insulated. If no bearing currents flow, the shaft voltages measured in both ways do not differ. If, on the other hand, the bearing currents flow, a significantly higher value of the shaft voltage would be measured in the first way than in the second case. Namely, in the first case the induced shaft voltage is reduced only by the voltage drop in the shaft, while in the second case it is also reduced by the voltage drops in the bearing, base plate and motor housing.

8 Data Processing

Data collection is the recording of measured data, the conditions in which the recording was performed (machine load, temperatures, voltages, etc.). The method of data recording is primarily determined by further data processing, ie operations that will be performed recorded data to execute. The data collected must be accurate, of good quality and with as little noise and interference as possible. Many diagnostic systems record data over a long period of time. This method of recording enables quality monitoring of the condition of the machine as well as the development of possible failures and malfunctions. In this case, it is very important that the data is recorded systematically and permanently. In some plants, the tested object is physically distant from the place where data processing and diagnostics are performed, and when recording the size, one should keep in mind the way the data will be transmitted (current signals, optics…), and sometimes on the spot. perform partial processing. In order to be able to diagnose the condition of the machine on the basis of the collected data, it is often necessary to perform a series of operations on the collected data. This is precisely the part in which computers are used the most and for the longest time, and in which a high level of process automation is possible. Data processing can be performed during operation (online) or after the data is recorded and stored (off-line). The most widespread are data processing methods, for diagnostic purposes spectral analysis, correlation and time averaging. The development of measurement and computer technology has enabled the wider application of these methods. Spectral analysis - is the basis of the method of converting signals recorded in the time domain to the frequency domain. By analyzing the signals in the frequency domain (currents, vibrations…) it is possible to detect various faults. Switching to a frequency domain is a display of the frequency spectrum of the recorded signal and the corresponding amplitudes. The basis of the method is the Fourier transform [5]. The idea of the transformation is to show the periodic signal of period T, given by the expression, with uniformly spaced frequency components, which are obtained as:

$$G(f_k) = \frac{1}{T} \int_{-T/2}^{T/2} g(t) \cdot e^{-j2\pi f_k k'} dt \tag{4}$$

where k represents the k-th harmonic of the fundamental frequency f. Conversely, from the frequency spectrum the time function is obtained according to the expression:

$$g(t) = \sum_{k=-\infty}^{\infty} G(f_k) \cdot e^{j2\pi f_k t} \quad (5)$$

In order to obtain a satisfactory sampling frequency transformation accuracy with FFT, it must meet the Nyquist criterion, which means that the recording must be performed with at least twice the frequency of the highest frequency we want to analyze. In stator current analysis, side harmonics occur at a relatively small distance from the fundamental frequency. Therefore, the Fourier transform needs to be performed at a sufficiently high resolution. In order to increase the transformation resolution N times, it is necessary to collect N times more samples, since it is the number of samples that determines the transformation resolution. Classic FFT gives a spectrum covering frequencies from 0 Hz to the maximum recorded frequency fmax. Higher resolution can be obtained in several ways. The first way is to increase the degree of transformation thus increasing the time required to carry out the transformation. Another possibility is to reduce the maximum frequency. This solution is not always advantageous since it loses more frequencies needed to determine the state of eccentricity. The next solution is to increase the recording time, which automatically gives a better resolution, provided that the consistency of the measured size is ensured throughout the recording.

9 Testing of an Induction Motor for Various Faults of the Rotor Winding by Spectral Analysis of the Stator Current

The circuit diagram used to record the stator current is shown in Fig. 4. The AM asynchronous motor is connected trough a shaft to a DC motor with independent excitation which serves as a load on the induction motor. The magnitude of the load is determined by the regulation of the direct current of the independent excitation, in this case it is 5 A. The stator current AM is measured using current measuring transformers CMT and current clamps CP trough an AD card connected to a computer where the measurement signals are processed (Fig. 3).

Appearance of the current spectrum, at three different states at a load of 5 A: for a machine in good condition, Fig. 4, a machine with one damaged rod, Fig. 5 and a machine with two damaged bars, Fig. 6.

In normal operation, the stator current is sinusoidal and contains only the fundamental harmonic and noise harmonics.

In case there is one damaged rod on the rotor, the fundamental harmonics and harmonics at 75 Hz, 120 Hz and 250 Hz appear in the stator current.

In case there are two damaged rods on the rotor, in the stator current we have the same harmonics at 75 Hz, 120 Hz and 250 Hz but a larger amount.

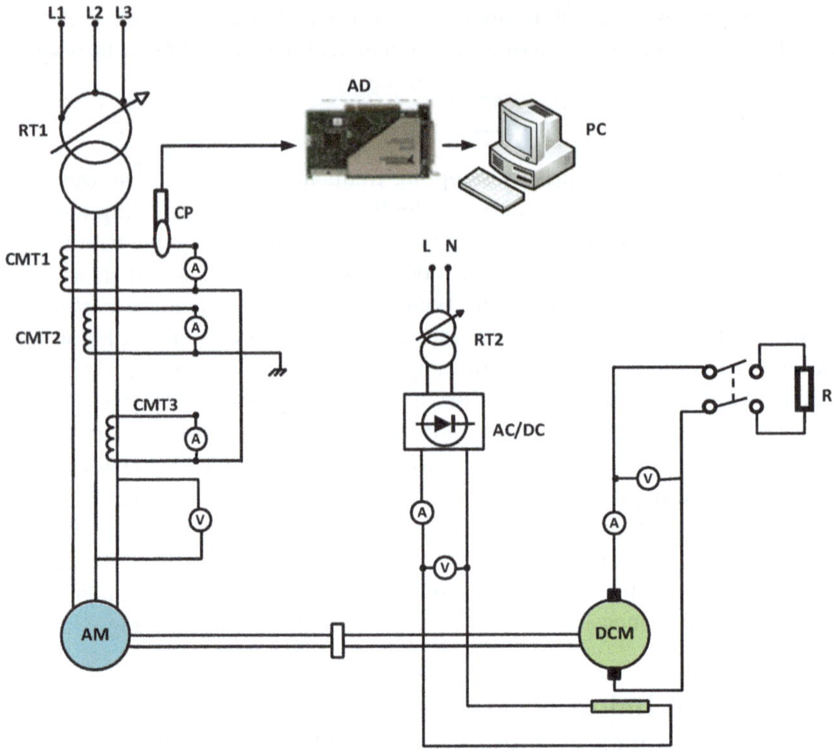

Fig. 3. Laboratory system for recording the stator current spectrum of an induction motor

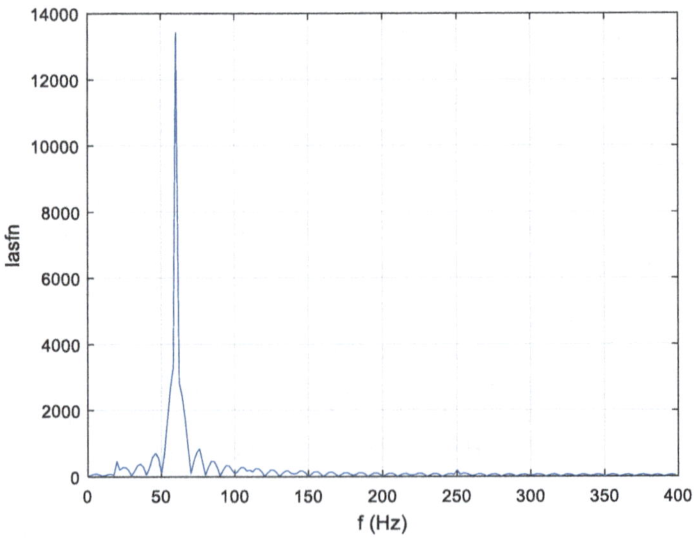

Fig. 4. Appearance of the current spectrum at a load of 5 A for the machine in normal operation when working on the network

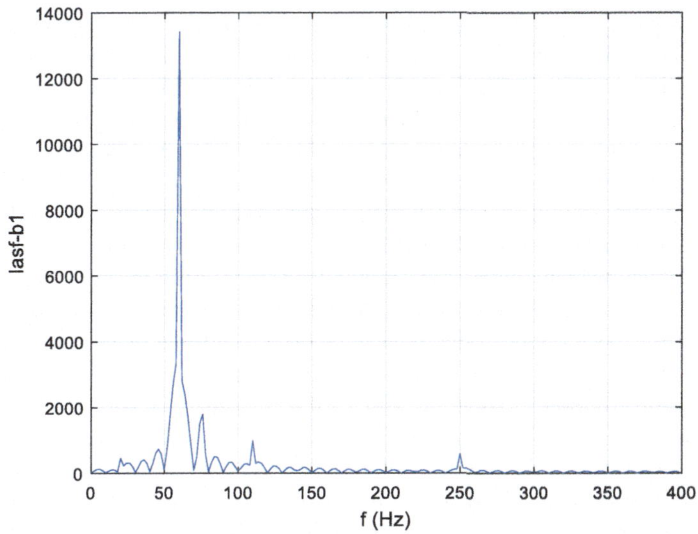

Fig. 5. Appearance of the current spectrum at a load of 5 A for a machine with one cut rods when working on the network

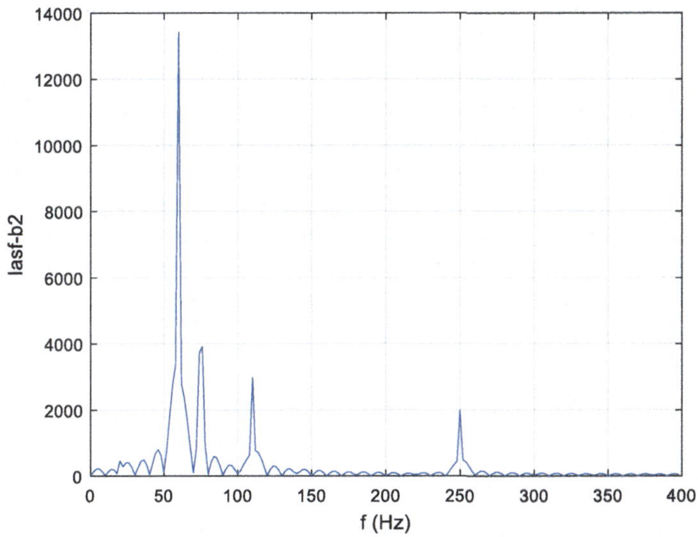

Fig. 6. Appearance of the current spectrum at a load of 5 A for a machine with two cut rods when working on the network

10 Conclusion

Diagnosis is the timely or periodic determination of the condition of a machine in order to assess the reliability of further operation and suggest ways and scope of servicing [6]. The role of diagnostics is to detect malfunctions of the machine or its individual parts at the earliest possible stage. Monitoring the condition of the machines facilitates the planning of maintenance and repairs, thus reducing the downtime due to the replacement and repair of machines to a minimum. In cases where quality mon toring is carried out, many unnecessary plant shutdowns are avoided. The basis of diagnostics is a comparison of actual and desired behaviors or machine parameters. In addition to theoretical knowledge, it is very useful and experiential knowledge of the behavior of the machine in certain fault modes. Quality monitoring and diagnostics requires the observation and processing of a range of electrical quantities. Many of these tasks today are solved using computers. The process of fault diagnosis is very demanding since different faults can be equally manifested, ie reflected on the operation of the drive. The diagnosed fault is assessed and based on the assessment, a decision is made to continue or stop the operation of the plant. In cases when the deviation is small and when there is no danger of major damage, the plant continues to operate, and the data obtained are used to plan repairs and purchase spare parts. In case of larger deviations, overhaul is planned as soon as possible. In the event of very large deviations, ie faults that are dangerous for further operation, the drive is stopped and the fault is rectified. The method of spectral analysis of stator current proved to be one of the most interesting methods. This method is applicable to asynchronous machines of a wide range of powers and applications. In addition, it does not require the installation of additional measuring members, nor the shutdown of the drive. For these reasons, much of the work is devoted to this method. The method is based on the fact that each occurrence of asymmetry (damage) of the rotor cage, in the spectrum of the stator current causes the appearance of side harmonics, distant from the mains frequency by the amount of 12sf±. The condition of the motor cages can be determined from the values of the side harmonics. From the recording it can be concluded that the values of the side harmonics change significantly with the load of the machine. For these reasons, it is important to record at the highest possible load on the machine. The changes in the values of the harmonics are shown in the table and graphically. The power supply from the frequency converter reduces the values of the side harmonics in relation to the values when working on a rigid network for the same load. Such values may lead to the conclusion that the machine is in better condition than the actual one. In addition, changes in harmonics, by changing the load are less pronounced.

References

1. Tavner, P.J., Penman, J.: Condition Monitoring of Electrical Machines. Wiley, New York (1987)
2. Vas, P.: Parameter Estimation, Condition Monitoring, and Diagnosis of Electrical Machines. Clarendon Press, Oxford (1993)
3. Stone, G.C., Boulter, E.A., Culbert, I., Dhirani, H.: Electrical Insulation for Rotating Machines: Design, Evaluation, Aging, Testing, and Repair, vol. 21. Wiley, Hoboken (2004)
4. Bonnett, A.H., Soukup, G.C.: Cause and analysis of stator and rotor failures in three-phase squirrel-cage induction motors. IEEE Trans. Ind. Appl. **28**(4), 921–937 (1992)
5. Benbouzid, M.E.H., Nejjari, H., Viera, M.: Induction motor faults detection using advanced spectral analysis technique. In: ICEM 1998, Istanbul, Turkey, 2–4 September 1998, pp. 1849–1854 (1998)
6. Wolf, R.: Osnove električnih strojeva. Školska knjiga, Zagreb (1989)

7. Ban, D., Wolf, R., Cettolo, M.: Metode detekcije kvara kaveza asinkronog motora. Elektrotehnika **34**(6), 241–250 (1991)
8. Stegić, M.: Vibracijska dijagnostika električnih strojeva. I simpozij EEDEEQ, Rovinj (1996)
9. Thomson, W.T.: A review of on-line condition monitoring techniques for three - phase squirrel - cage induction motors - past present and future. In: IEE SDEMPED 1999, Spain, September 1999
10. Bartelmus, W., Zimroz, R.: A new feature for monitoring the condition of gearboxes in non-stationary operating conditions. Mech. Syst. Sig. Process. **23**, 1528–1534 (2009)
11. Cruz, S.M.A., Marques Carsodo, A.J.: Rotor cage fault diagnosis in three-phase induction motors, by extended Park's vector approach. In: ICEM 1998, vol. 3, pp. 1844–1848 (1998)
12. Haji, M., Toliyat, H.A.: Pattern recognition – a technique for induction machines rotor broken bar detection. IEEE Trans. Energy Convers. **6**(4), 312–317 (2001)
13. Parlos, A.G., Kim, K., Bharadwaj, R.M.: Sensorless detection of induction motor faults. In: IEEE SDEMPED 2001, Grado, Italy, 1–3 September 2001, pp. 663–670 (2001)
14. Penman, J., Sedding, H.G., Lloyd, B.A., Fink, W.T.: Detection and location of interturn short circuits in the stator windings of operating motors. IEEE Trans. Energy Convers. **9**(4), 652–658 (1994)

Electro-Thermal Induction Motor Model

Adam Skeledžija and Vedrana Jerković Štil(✉)

Faculty of Electrical Engineering, Computer Science and Information Technology Osijek, Kneza Trpimira 2b, 31000 Osijek, Croatia
vedrana.jerkovic@ferit.hr

Abstract. Even though electrical and thermal conditions of any electrical machine are interconnected, in the most cases these conditions are separately analyzed. On the other hand, electrical machines are designed to achieve desired operating point with minimal material cost. Due to increase of load or reduced cooling possibilities, in industrial surrounding electrical machines thus can easily be overheated, shortening the machine life cycle and increasing maintenance costs. With the development of sophisticated control principles, induction motors became the most widely used machines in industry. For all these reasons, in this paper the simple electro-thermal model of induction motor for prediction of motor temperature in different operating conditions is presented. Further on, dynamical behavior of presented model is analyzed through simulations and experimental measurements.

Keywords: Induction motor · Electrical model · Thermal model · Simulation · Measurement

1 Introduction

The main goal of induction motor design is to achieve desired operating point with minimal material cost. In [1] a method for optimization of motor design based on genetic algorithm is proposed. The objective functions are torque, efficiency and cost. The optimization is carried out considering motor characteristics, magnetic stress and thermal limits. As pointed out in [2], to achieve smaller and more efficient design of electrical motor, it is necessary to have an accurate prediction of motor thermal performances. As stated in [3] a large amount of heat is transferred through a reduced number of motor elements. Further on, modern electrical drives with induction motors are driven with frequency converters, ensuring variable speeds and operating points. It is well known that self-cooling induction motors at low speeds are easily overheated due to reduced air flow. Also, different motor torques at the same speed will give different motor thermal conditions.

To take into consideration all these impacts on thermal condition of a specific motor, electrical and thermal models should be coupled in one electro-thermal dynamical model. This approach can be found in only few scientific papers. In [4] an electro-thermal analysis of permanent magnet synchronous motor with outer rotor embedded into rear rim of a hybrid drive vehicle for public transport is given. In order to characterize effects

H. Glavaš et al. (Eds.): OTO 2021, LNNS 369, pp. 160–171, 2022.
https://doi.org/10.1007/978-3-030-92851-3_12

of shaft torque and speed on the overall electro-magnetic and thermal performance of a permanent magnet synchronous motor authors in [5] present a two-way iteratively coupled electro-thermal modeling framework for permanent magnet synchronous motors. Two finite-element based software have been used for coupled electro-magnetic performance calculation and simulation of temperature distribution of the motor. In [6] authors propose high thermal superconducting motor design method based on electro-thermal model.

In [7] authors introduce coupling of electrical and thermal model of induction motor with two-mass mechanical model used for motor mechanical part modeling. The two-mass mechanical model is introduced to monitor impact of motor start-up at motor thermal condition. Presented simulation results show that model introduced in [7] gives promising results in predicting certain motor part temperatures, but without experimental proof. Model presented in [7] is the basis of electro-thermal model used in this paper. The main goal of this paper is to build a simple electro-thermal model of induction motor which would provide sufficiently accurate prediction of induction motor thermal conditions in case of continuously loaded motor and intermitted motor load. Well predicted motor thermal conditions ensure motor operation below temperature limits, decrease the possibility of failure and simplify maintenance of the whole industrial drive.

2 Thermal and Electrical Modeling of Induction Motor

In this section thermal and electrical models of induction motor will be explained. For motor thermal conditions modeling a rather simple induction motor thermal model consisting of three heat sources is chosen. The electrical motor model is 5-th order induction motor model in rotor flux oriented dq – frame.

2.1 Induction Motor Thermal Model

By comparing two laws, Ohm's law for electrical quantities and Fourier's law for thermal quantities, an analogy can be made between basic thermal and electrical quantities. According to Ohm's law, the current I flowing through an element is equal to the ratio of the difference between the potential ΔU and the electrical resistance of that element R_e [8]:

$$I = \frac{\Delta U}{R_e} \tag{1}$$

Similar to Ohm's law, according to Fourier's law, the heat flux q is equal to the ratio of the temperature difference at the beginning and end of the object $\Delta\vartheta$ and the thermal resistance R_t, according to the following relation [8]:

$$q = \frac{\Delta\vartheta}{R_t} \tag{2}$$

Considering these two laws in Table 1. An analogy is made between electrical and thermal quantities.

Table 1. Analogy between thermal and electrical quantities.

Electrical quantities	Thermal quantities
Electrical current I [A]	Heat flow q [W]
Voltage difference $\boldsymbol{\Delta U}$ [V]	Temperature difference $\Delta\vartheta$ [K]
Electrical resistance $\boldsymbol{R_e}$ [$\boldsymbol{\Omega}$]	Thermal resistance R_t [$\mathrm{K} \cdot \mathrm{W}^{-1}$]
Electrical capacity $\boldsymbol{C_e}$ [F]	Heat capacity C_t [$\mathrm{J} \cdot \mathrm{K}^{-1}$]

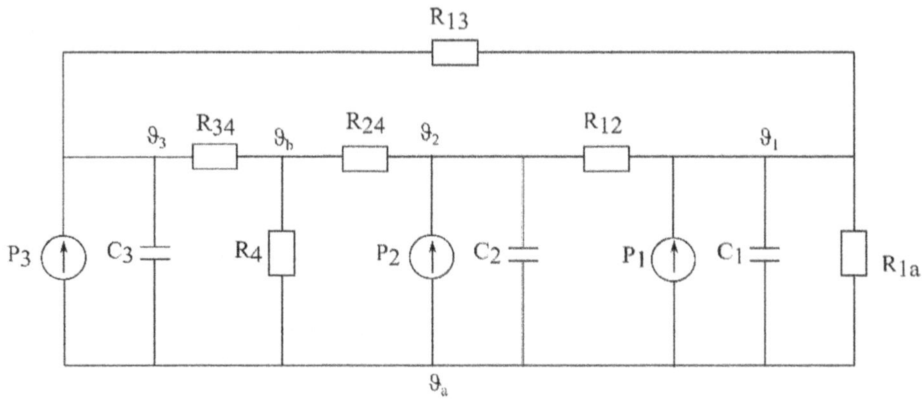

Fig. 1. Induction motor thermal model.

The thermal model of induction motor presented in [7] consists of four power sources and four heat capacities. In this paper a simplification is made and three heat capacities are used: C_1 - frame stator and iron, C_2 - stator winding and C_3 - rotor. The rotor consists of a rotor winding, a rotor iron and a shaft. Also, this model consists of six resistances $R_{1a}, R_{12}, R_{13}, R_{24}, R_{34}$ and R_4 which represent the thermal resistances between the individual motor parts, as shown in Fig. 1. In simplified thermal model there are three power sources that represent the electrical losses in particular motor parts: stator iron losses P_1, stator winding losses P_2 and rotor winding losses P_3.

The simplified thermal model is described by Pseudo-Kirchhoff's law, the main equations can be obtained as shown in [7]:

$$\frac{d\vartheta_1}{dt} = \frac{1}{C_1}\left[P_1 - \frac{1}{R_{1a}}(\vartheta_1 - \vartheta_a) - \frac{1}{R_{12}}(\vartheta_1 - \vartheta_2) - \frac{1}{R_{13}}(\vartheta_1 - \vartheta_3)\right] \tag{3}$$

$$\frac{d\vartheta_2}{dt} = \frac{1}{C_2} \cdot \left[[P_2 - \frac{1}{R_{24}}(\vartheta_2 - \vartheta_4) - \frac{1}{R_{12}}(\vartheta_2 - \vartheta_1)]\right] \tag{4}$$

$$\frac{d\vartheta_3}{dt} = \frac{1}{C_3}\left[P_3 - \frac{1}{R_{13}}(\vartheta_3 - \vartheta_1) - \frac{1}{R_{34}}(\vartheta_3 - \vartheta_4)\right] \tag{5}$$

where $\vartheta_1, \vartheta_2, \vartheta_3, \vartheta_a$ and ϑ_b represent network node temperatures, as shown in Fig. 1.

2.2 Induction Motor Electrical Model

The equations that describe dynamical behavior of induction motor electrical values can be found in [9]. The model is presented in rotor flux oriented *dq*-system.

$$\boldsymbol{u}_{sdq} = \boldsymbol{R}_s \boldsymbol{i}_{sdq} + \omega_s \boldsymbol{\Psi}_{sdq} + \frac{d}{dt}\boldsymbol{\Psi}_{sdq} \tag{6}$$

$$\boldsymbol{u}_{rdq} = \boldsymbol{R}_r \boldsymbol{i}_{rdq} + (\omega_s - \omega_{el})\boldsymbol{\Psi}_{rdq} + \frac{d}{dt}\boldsymbol{\Psi}_{rdq} \tag{7}$$

Where $\boldsymbol{u}_{sdq}$ represents stator voltage vector, $\boldsymbol{R}_s$ stator resistance matrix, $\boldsymbol{i}_{sdq}$ stator current vector, ω_s synchronous speed, $\boldsymbol{\Psi}_{sdq}$ – stator flux vector, $\boldsymbol{u}_{rdq}$ rotor voltage vector, $\boldsymbol{R}_r$ rotor resistance matrix, $\boldsymbol{i}_{rdq}$ rotor current vector, ω_{el} electrical speed of the motor, and $\boldsymbol{\Psi}_{rdq}$ rotor flux vector.

The equations for the stator and rotor flux are given as follows [9].

$$\boldsymbol{\Psi}_{sdq} = \boldsymbol{L}_{sdq}\boldsymbol{i}_{sdq} + \boldsymbol{L}_{mdq}\boldsymbol{i}_{rdq} \tag{8}$$

$$\boldsymbol{\Psi}_{rdq} = \boldsymbol{L}_{rdq}\boldsymbol{i}_{rdq} + \boldsymbol{L}_{mdq}\boldsymbol{i}_{sdq} \tag{9}$$

Where $\boldsymbol{L}_{sdq}$ represents stator inductance matrix, $\boldsymbol{L}_{rdq}$ rotor inductance matrix, and $\boldsymbol{L}_{mdq}$ mutual inductance matrix of stator and rotor.

After calculating motor fluxes and currents, it is necessary to determine the electrical torque of the motor [10]:

$$m_e = \frac{3}{2}p\left(\Psi_{sd} i_{sq} - \Psi_{sq} i_{sd}\right) \tag{10}$$

The dynamical behavior of induction motor speed is described with the following equation [10]:

$$J\frac{d\omega}{dt} = M_e - M_t \tag{11}$$

Where J represents moment of inertia, ω motor speed, M_e motor electrical torque and M_t load torque. In this paper one-mass mechanical machine model is used, because motor warm-up tests have much slower thermal dynamics than motor start-up.

2.3 Induction Motor Losses

Induction motor losses can be divided into stator copper losses, rotor copper losses, iron losses, mechanical losses due to ventilation and friction in bearings, and additional losses [11]. Of the total value of losses, the largest amount are losses in the stator copper, followed by losses in the copper of the rotor and iron.

Losses in a winding are generated due to the flow of electric current through the winding. They are proportional to the product of a square of the current and the resistance of a winding. In order to reduce the losses in the winding, it is necessary to use conductors of larger cross-section. In the *dq* - system, losses in the winding are modeled according to following expressions [8]:

$$P_{scu} = 1{,}5 \cdot R_s \cdot \left(i_{sq}^2 + i_{sd}^2\right) \tag{12}$$

$$P_{rcu} = 1{,}5 \cdot R_r \cdot \left(i_{rq}^2 + i_{rd}^2\right) \tag{13}$$

Losses in iron occur due to the induction of eddy currents and the hysteresis loop. In the *dq* system losses in iron are modeled according to the expression [8]:

$$P_{sfe} = 1{,}5 \cdot \frac{\left(U_{qs}^2 + U_{ds}^2\right)}{R_s} \tag{14}$$

$$P_{rfe} = 1{,}5 \cdot \frac{\left(U_{qr}^2 + U_{dr}^2\right)}{R_r} \tag{15}$$

Additional losses are modeled according to the electrical power, according to the expression [8]:

$$P_{ost} = 0{,}02 \cdot P_{el} \tag{16}$$

2.4 The Dependence of Electrical Resistance on Motor Temperature

The change of temperature has a significant effect on the electrical resistance of the motor windings and should be taken into account. When electrons move through a conductor, they heat up a conductor and that phenomenon causes an increased resistance of the conductor, according to the expression [12]:

$$R = R_0(1 + \alpha\Delta\vartheta) \tag{17}$$

Where R represents electrical resistance of the conductor changed by temperature, R_0 – electrical resistance of conductor at room temperature, α temperature resistance coefficient and $\Delta\vartheta$ is relative temperature in respect to the room temperature.

3 Electro-Thermal Model of Induction Motor

The electro-thermal model of induction motor is built in Simulink and will be presented in this section. As its name suggests, the electro-thermal model shown in Fig. 2 consists of two parts, the electrical part and the thermal part.

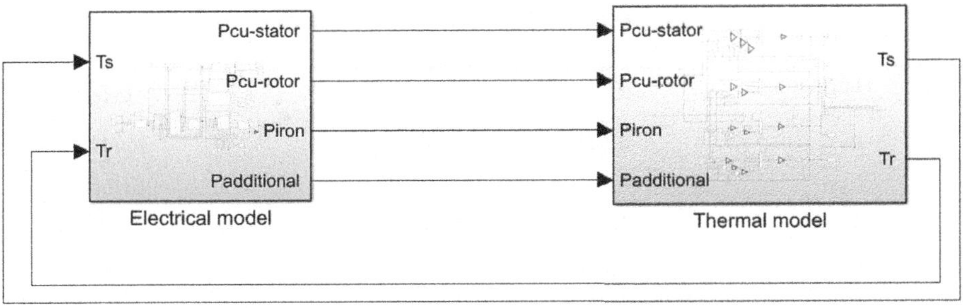

Fig. 2. Electro-thermal induction motor model.

Figure 2 shows that the input of the thermal model are the losses of the induction motor that affect the motor thermal conditions, and the output of the thermal model are the temperatures that affect the operation of the motor. The losses that represent output from the electrical model are: Joule losses in the stator and rotor windings, losses in iron and additional losses. In the model, the additional losses are equally distributed to stator winding losses and iron losses, i.e. they are added to P_1 and P_2 in Eqs. (3) and (4). The temperatures that enter the motor model are stator winding temperature and rotor temperature.

The parameters of tested electrical motor are given in Table 2, and the parameters used in simulation model are given in Table 3. In Table 2 R_s represents stator resistance, R_r rotor resistance, L_s stator inductance, L_r rotor inductance, L_m mutual inductance and J moment of inertia. Parameters in Table 3 are calculated according to known dimensions and materials of tested motor, as can be found in [13]. Two simulations were performed: continuously loaded motor warm-up and motor warm-up in intermittent mode.

Table 2. Tested induction motor parameters

Motor parameter	Value	Measurement unit
R_s	1,3	Ω
R_r	1,04	Ω
L_s	0,15999	H
L_r	0,15999	H
L_m	0,1539	H
J	0,13	kgm^2

The motor is loaded with rated load of 4 kW and the simulation time is set to 7200 s due to a small temperature change over time. Figure 3 shows simulated motor losses in time.

Table 3. Thermal resistances and capacities

Thermal resistance	R $[\frac{°C}{W}]$
R_{12}	0,111
R_{13}	0,220
R_{24}	1,896
R_{34}	1,276
R_{1a}	0,042
R_4	0,015

Thermal capacitance	C $[\frac{J}{°C}]$
C_1	30042,96
C_2	819,589
C_3	4588,94
C_4	1006

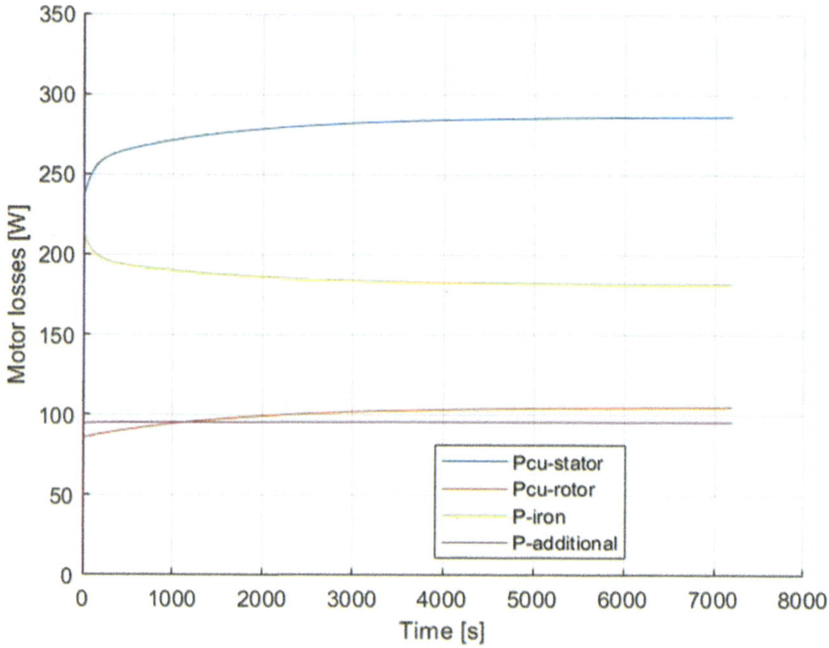

Fig. 3. Induction motor losses in case of continuously loaded induction motor.

In the second simulation, the operation of a motor with dynamical load change is shown. This dynamic of motor loading is in compliance with continuous operation intermitted loading periodic duty S6, according to IEC 60034-1 standard [14]. As the motor is loaded with a rated power of 4 kW for five minutes, and is idle for next five minutes, the intermitted mode is S6 50%. The simulation time is set to 7200 s. Figure 4 shows the simulated motor losses in time.

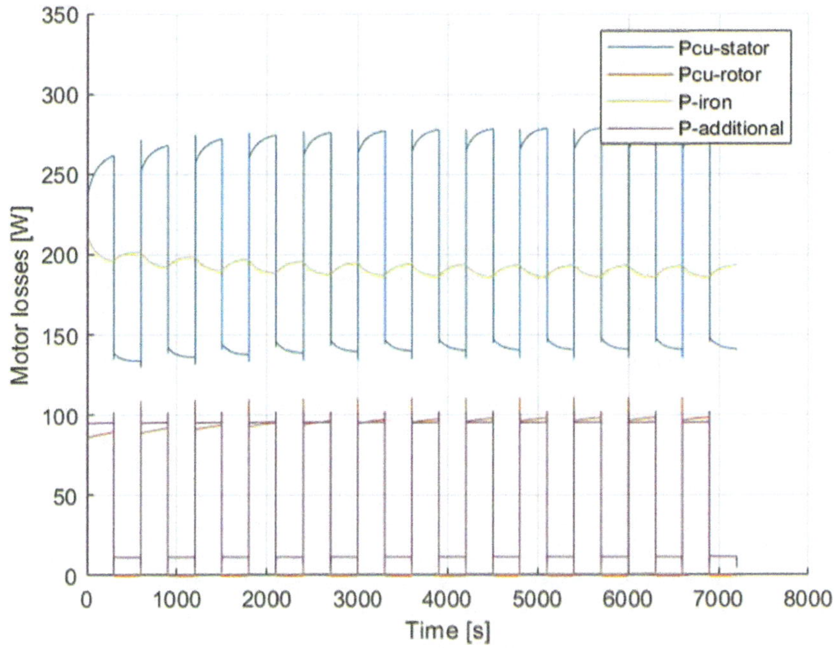

Fig. 4. Induction motor losses in case of intermitted motor load.

4 Experimental Results

The laboratory setup scheme is shown in Fig. 5. The laboratory setup consists of Siemens LOGO! Power supply - 24 V power supply, Telemecanique (RMPT50BD) - measuring bridge for switching ports of Pt - probes, NI myRIO and Pt 100 probes built into Končar induction motor E5AZ 112M-4T.

To adjust the voltage level of a measured signal, the Pt100 probe is connected to the measuring bridge, which is supplied from LOGO! 24 V Power supply. The output from the measuring bridge is connected to NI myRIO where the data are processed. Communication with NI myRIO on a PC and data access is established via an USB cable. Measured signals are recorded in NI LabView software.

As in simulations, two experiments were conducted. In the first warm-up experiment the tested motor was continuously loaded with rated power, 4 kW for 5310 s. At the beginning the motor was at the room temperature. During the experiment two temperatures were recorded, the stator winding temperature, via PT100 probe, and the motor frame temperature, via Fluke 568 infrared thermometer. The experimental results are compared with the simulation results in Fig. 6.

Temperature diagrams presented in Fig. 6 show that simulated and measured temperatures have almost the same dynamical behavior over time. The differences in the dynamical behavior are maximal at the beginning of the warm-up test and are present because of thermal model simplifications. The differences between measured and simulated dynamical responses arise because the temperature is measured at one point, whereas in simulation the average temperature of a certain motor part is given. As

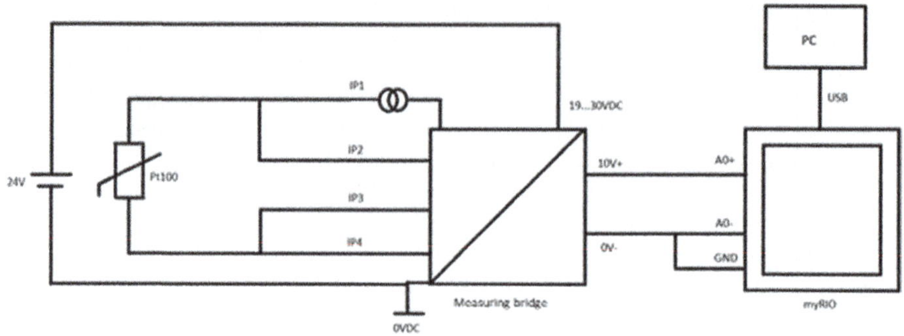

Fig. 5. Induction motor temperature measurement laboratory setup.

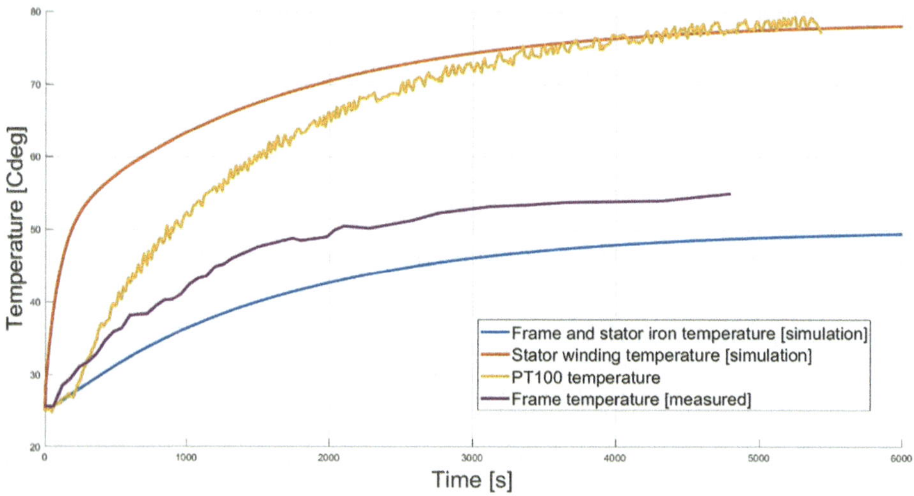

Fig. 6. Experimental and simulation results for motor warm-up at continuous rated load.

expected, when measuring the frame temperature, it was noticed that the frame temperature near the windings and iron is higher than the frame temperature near the fan. The temperature oscillations were ±10 °C. The simulation shows the average temperature of a certain part of the motor and calculates it as a homogeneous whole. The simulated and measured stator winding temperatures in the steady state are approximately the same values, and the simulation response of the frame and stator iron temperature and the measurements of the stationary frame temperatures differ by approximately 5 °C.

Figure 7 shows the experimental and simulation results of motor warm-up in the case of intermittent operation S6 50% lasting 3000 s. In this case the motor was pre-heated to operating temperature of idle operation.

The dynamical behavior of the measured and simulated temperature is approximately the same, but in Fig. 7 a larger difference between the measured and simulated stator windings temperature can be observed. The Pt100 sensor is located near the stator windings and it takes more time to equalize the sensor temperatures in the intermittent motor mode and the windings thus the oscillations of the amplitude of the measured temperature are less pronounced compared to the simulated one. The simulated and measured frame temperatures have approximately the dynamical behavior with minor deviations in the steady state explained for the previous case.

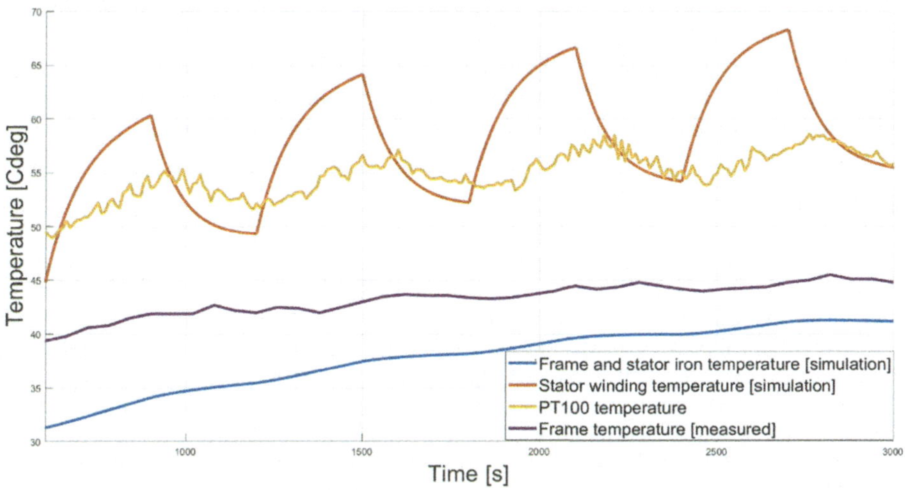

Fig. 7. Experimental and simulation results for motor warm-up in intermittent mode.

5 Conclusion

In this paper an electro-thermal model of induction motor is introduced. The model consists of induction motor electrical model in rotor flux oriented *dq*-frame and simplified thermal model with three heat sources, rotor electrical losses, stator electrical losses and stator iron losses. The connections between electrical and thermal models are motor power losses and winding temperatures. The motor power losses are calculated in motor electrical model and are used as inputs in motor thermal model. The winding temperatures of stator and rotor are calculated in thermal model and serve as input in electrical model for estimation of winding resistances.

Two motor warm-up tests are simulated and conducted at induction motor laboratory set-up, the warm-up test of continuously loaded motor with rated load and the warm-up test in case of intermittent operation S6 50%.

Even though the simulated temperature responses slightly differ from measured temperatures, the dynamical behavior of simulated and measured temperatures are almost the same for both cases, continuously loaded motor and intermittent operation. The differences in temperatures dynamical behavior can be minimized by introducing more complex motor thermal model, e.g. model presented in [15]. Further on, the simulated and measured stator winding temperatures are equal in steady state. As the stator winding temperature is more critical value than the temperature of motor frame and iron

[2] the presented results prove that introduced electro-thermal model can be used for prediction of maximal temperatures for a specific operation of induction motor in certain industrial drive. The use of proposed electro-thermal model thus can simplify the selection of appropriate induction motor in project planning phase of an industrial drive. The selection of a proper induction motor in planning phase of industrial drive assures maximal efficiency, minimal energy losses and avoidance of motor overheating and thus minimizes further maintenance time and costs.

In the further research the proposed model will be improved to minimize differences between simulated and measured temperatures. The improved model could be used to predict motor thermal conditions in design phase of new induction motors.

References

1. Çunkaş, M., Akkaya, R.: Design optimization of induction motor by genetic algorithm and comparison with existing motor. Math. Comput. Appl. **11**(3), 193–203 (2006)
2. Boglietti, A., Cavagnino, A., Staton, D.A.: Thermal analysis of TEFC induction motors. In: 2003 38th IAS Annual Meeting on Conference Record of the Industry Applications Conference, vol. 2, pp. 849–856 (2003). https://doi.org/10.1109/IAS.2003.1257630
3. Boglietti, A., Cavagnino, A., Lazzari, M., Pastorelli, M.: A simplified thermal model of variable speed self cooled industrial induction motor. In: Conference Record of the 2002 IEEE Industry Applications Conference. 37th IAS Annual Meeting (Cat. No. 02CH37344), pp. 723–730 (2002). https://doi.org/10.1109/ias.2002.1042640
4. Hruška, K., Kindl, V., Pechánek, R.: Concept, design and coupled electro-thermal analysis of new hybrid drive vehicle for public transport. In: Proceedings of 14th International Power Electronics and Motion Control Conference, EPE-PEMC 2010, pp. S4-5–S4-8 (2010). https://doi.org/10.1109/EPEPEMC.2010.5606895
5. Tikadar, A., Kumar, N., Joshi, Y., Kumar, S.: Coupled electro-thermal analysis of permanent magnet synchronous motor for electric vehicles. In: 2020 19th IEEE Intersociety Conference on Thermal and Thermomechanical Phenomena in Electronic Systems (ITherm), pp. 249–256 (2020). https://doi.org/10.1109/ITherm45881.2020.9190562
6. Masson, P.J., Tixador, P., Ordonez, J.C., Morega, A.M., Luongo, C.A.: Electro-thermal model for HTS motor design. IEEE Trans. Appl. Supercond. **17**(2), 1529–1532 (2007). https://doi.org/10.1109/TASC.2007.898115
7. Irokwe, N.V., Okoro, O.I.: Electrical-thermal coupling of induction machine for improved thermal performance. Niger. J. Technol. **35**(4), 912–918 (2016). https://doi.org/10.4314/njt.v35i4.28
8. Oti, S.E.: Thermal modelling of induction machine using the lumped parameter model. Doctorate degree thesis, University of Nigeria, Nsukka (2014)
9. Josipović, F.: Mathematical model of induction machine in MATLAB. Diploma degree thesis, Faculty of Electrical Engineering, Computer Sciences and Information Technology Osijek, Croatia (2016)
10. Leonhard, W.: Control of Electrical Drives. Springer, Heidelberg (2001). https://doi.org/10.1007/978-3-642-56649-3
11. Bašić, M.: Vector control system of self-excited induction generator including iron. Doctorate degree thesis, Faculty of Electrical Engineering, Mechanical Engineering and Naval Architecture, Split, Croatia (2013)
12. Edutorij, Fizika 2. https://edutorij.e-skole.hr/share/proxy/alfresco-noauth/edutorij/api/proxy-guest/452e1469-e362-4711-abcb-6f535c3b5254/html/7571_Elektricni_otpor.html. Accessed 17 June 2021

13. Skeledžija, A.: Induction motor thermal model. Diploma degree thesis, Faculty of Electrical Engineering, Computer Sciences and Information Technologies Osijek, Croatia (2021)
14. Electrical engineering portal. https://electrical-engineering-portal.com/10-duty-types-three-phase-asynchronous-motors#intermittent-periodic-duty. Accessed 25 Aug 2021
15. Huai, Y., Melnik, R.V.N., Thorgesen, P.B.: Computational analysis of temperature rise phenomena in electric induction motors. Appl. Therm. Eng. **23**, 779–795 (2003)

Fault Diagnostic and Protection of Power Transformers Grounded with a Common Grounding Resistor

Marko Šarić[1], Predrag Marić[2], Dino Masle[3], and Krešimir Fekete[2](✉)

[1] HEP DSO Elektroslavonija Osijek, Franje Šepera 1a, Osijek, Croatia
marko.saric@hep.hr

[2] Faculty of Electrical Engineering, Computer Science and Information Technology Osijek, University of Osijek, Kneza Trpimira 2B, Osijek, Croatia
{predrag.maric,kresimir.fekete}@ferit.hr

[3] Transmission Area Osijek, HOPS TSO, Vukovarska cesta 217, Osijek, Croatia
dino.masle@hops.hr

Abstract. The power transformer is the main and the most expensive component of the power substation, so successful fault diagnostic and protection is an essential part of its maintenance. Protection engineers doing the scheduled maintenance and protection testing need to be familiarized with transient phenomena that can occur in normal operation. Substations with two parallel transformers sharing a common grounding resistor are very frequent in high-voltage networks since a common grounding resistor limits the single-phase to ground-fault current. However, the unfavorable aspect of this type of transformer grounding system is the increased probability of unwanted trips of the transformer protection during the transient operational states, which has the consequence of a supplied low-voltage network outage. Under transformer energization, an appearance of the inrush current can trigger the transformer differential protection as well as circulating current between transformers in parallel operation caused by sympathetic inrush. A single-phase to ground fault in a high-voltage network can cause circulating currents between neutrals of transformers and trip the differential protection too. The paper presents a simulation model of two transformers grounded with a common grounding resistor adequate for transformer transient operational states analysis. Performed simulations of the inrush and sympathetic currents emphasized the diagnosing of the second harmonic distortion dominance in the current spectrum as an efficient tool in the prevention of unwanted trips of differential protection. The analysis of the simulated circulating currents between transformer neutrals in the paper has examined and discussed the proposed solutions for avoidance of unnecessary transformer outage caused by the single-phase to ground fault in a high-voltage network.

Keywords: Power transformer protection · Inrush current · Circulating currents · Common grounding resistor · Single-phase to ground fault

1 Introduction

Most common faults in power systems are line-to-ground faults and therefore the transformer grounding plays a significant role for safe and efficient system operation [1, 2].

H. Glavaš et al. (Eds.): OTO 2021, LNNS 369, pp. 172–187, 2022.
https://doi.org/10.1007/978-3-030-92851-3_13

As a compromise of technical and economic demands, the grounding of the two parallel transformers is usually performed with a common grounding impedance as processed in [2–4]. While choosing the type of grounding impedance, it is preferable to obtain as small a current as possible during the ground fault to limit the touch potentials, while on the other hand, the higher current results in lower internal overvoltages that can strain the insulation of the transformer [3, 4]. Grounding impedance should be dimensioned to make the ground fault current higher than steady-state current to ensure fast tripping of the grounding protection device and to keep the voltage of the healthy phase within the limited values [5]. Differential protection is an essential transformer protection function, hence incorrect operation of this protection leads to loss of load. Transformer inrush current can cause unwanted differential protection trips. Under the transformer energization on the primary side, an inrush current can reach ten times the rated primary current and magnetic flux can reach two times nominal value [6, 7]. These currents can last from a few cycles to many seconds and can occur when a transformer is energized, when another transformer in parallel operation is energized or the system recovers from a fault outside of the protection zone as explained in [8]. Application of external voltage can cause the saturation of the transformer core - that is inevitable, and operation of differential protection is unfounded [1, 8–10]. That same current can also cause saturation of the current transformers, hence, internal faults can be misinterpreted as inrush current. The presence of the second harmonic content in the differential current indicates the inrush current and this can be used to prevent incorrect trips of the transformer differential protection [1, 8–10]. Sympathetic inrush occurs during parallel transformers energization [11–13]. In this case, a single transformer is operating while the second transformer is energizing [12] and thus causes the circulating current between the two transformers which remains due to small damping i.e. large time constant – R/X [12, 14]. Any mismatch in impedance or voltage will cause a current to circulate between transformers (to equalize set voltage) [15, 16]. Disturbances such as transformer tap changer switching, three-phase reclosing at the high voltage side can cause circulating current [17] too. Under a single-phase to ground fault in the HV network, circulating current is present between transformer neutrals on the MV- side and can cause protection trips of both transformers [18].

The paper constitutes five chapters. The Introduction is followed by the simulation model of the transformer used for the protection function testing and verifying. The protection performance has been examined simulating the energization process of the transformer and faults on the different locations in Chapters 3 to 5. The results of the simulations are supported by the Conclusion.

2 Simulation Model

In the simulation mode, two three-phase transformers were used. Transformer details are listed in Table 1. Tap changer with 21 position (1,5% additional voltage per tap) was also included in the model. Figure 1 shows magnetization characteristic of the transformer core used in simulation.

Equations (1), (2) and (3) show impedance between individual windings at 35 kV [19, 20]. Neglecting the active resistance, the reactances of individual windings mapped to base voltage are calculated according to (4), (5) and (6).

$$x_{d1-2,35} = \frac{u_{k1-2,d}}{100} \times \frac{U_b^2}{S_n}[\Omega] \tag{1}$$

Table 1. Transformer nameplate details of TR1 and TR2

Voltage ratio (W1/W2): 110/35 kV
Vector group: YNyn0(d5)
Rated power: 40 MVA
Short-circuit voltage: uk = 11,3%
Copper losses: 155 kW
No-load losses: 28 kW

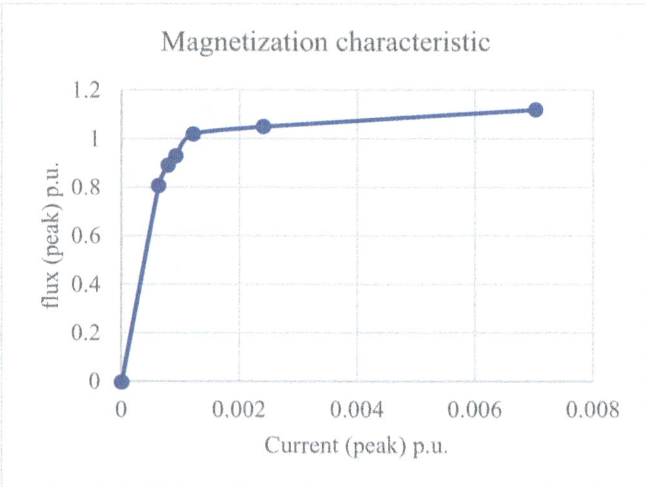

Fig. 1. Magnetization characteristic of the 110/35 kV transformer core

$$x_{d2-3,35} = \frac{u_{k2-3,d}}{100} \times \frac{U_b^2}{S_n}[\Omega] \tag{2}$$

$$x_{d3-1,35} = \frac{u_{k3-1,d}}{100} \times \frac{U_b^2}{S_n}[\Omega] \tag{3}$$

$$X_{d1} = \frac{1}{2}(x_{d1-2,35} + x_{d3-1,35} + x_{d2-3,35}[\Omega] \tag{4}$$

$$X_{d2} = \frac{1}{2}(x_{d1-2,35} + x_{d2-3,35} + x_{d1-3,35}[\Omega] \tag{5}$$

$$X_{d3} = \frac{1}{2}(x_{d1-3,35} + x_{d2-3,35} + x_{d1-2,35}[\Omega] \tag{6}$$

Based on the obtained values, it is possible to display the direct, inverse and zero components scheme of a transformer as it is shown in Fig. 2. As the transformer is a non-rotating element, direct and inverse components are equal [10, 15, 19, 20].

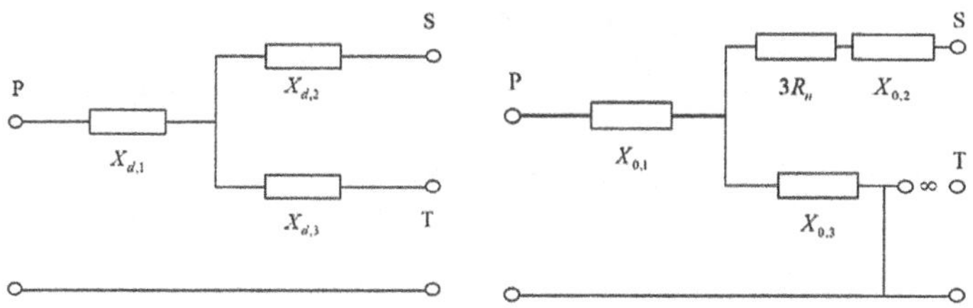

Fig. 2. Direct, inverse and zero component scheme

Figure 3 shows zero component scheme of two parallel transformers. It is shown that a triple value of resistance of resistor is added to the reactance of secondary winding. If a zero component current is injected at point P (primary) i.e. if a single-phase to ground fault occurs at HV side, current will flow between the secondary S-S.

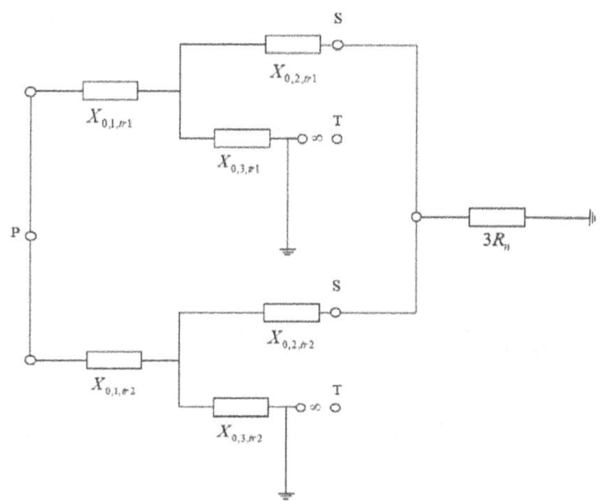

Fig. 3. Zero component scheme of parallel transformers

In the simulation model, a multifunctional protection relay with differential, restricted earth fault (REF) and overcurrent protection was implemented. Emphasis was given to differential and REF protection as two main protection functions of simulated relay. Differential protection provides basic, fast, sensitive and selective protection in the event of short circuit faults within the protection zone (transformer), and is stable during faults outside the protection zone. Protection zone of this protection is the zone between CT of primary winding and the CT of the secondary winding. Differential protection is based on current comparison (Kirchhoff's current law) as it is shown in Fig. 4. [21] Measuring element constantly measures I1 and I2, and if a difference of current marks occurs the relay trips. When switching transformers, high inrush current may occur. Inrush currents can reach ten times rated current, and are characterized by a 2nd harmonic content – this harmonic is not present in the case of a fault so this content is used as a blocking logic for unwanted trips of protection by inrush (Harmonic restraint) [21].

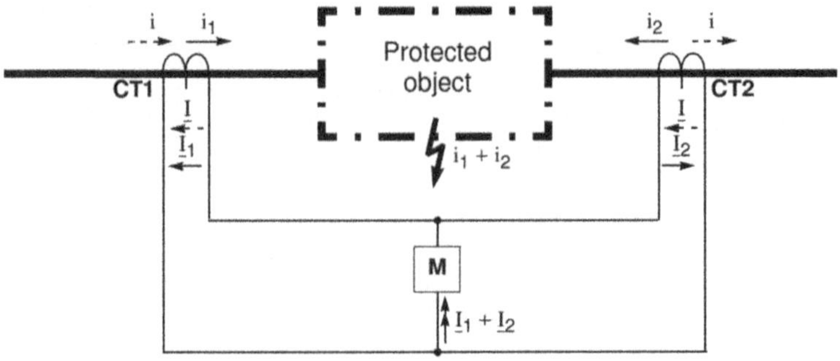

Fig. 4. Principle of differential protection

Restricted earth fault protection (REF) detects earth faults in power transformers, neutral grounding transformers and transformer starpoint (neutral). It is a type of differential protection, as it measures the fundamental wave of the current flowing in the starpoint ($\underline{I}_{SP}$) and the fundamental wave of the sum of the phase currents ($3\underline{I}_0$), and compares them. At Fig. 5 [21], principle of REF protection on a transformer winding was shown. During earth faults in the protection zone, starpoint current $\underline{I}_{SP}$ occurs and flows to the fault location, but also the residual current from the system flows to the fault location – currents are in phase and there is a condition for trip [21].

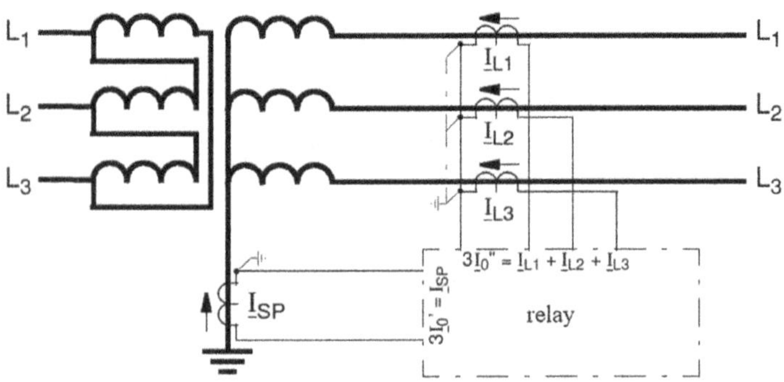

Fig. 5. Principle of REF protection

According to transformers and grounding resistor ratings, following protection settings have been acquired:

Table 2. Protection function settings

Differential	IdMin: 0,30 p.u	IdUnre: 10 p.u	I2/I1 Ratio:15%	I5/I1 Ratio:25%
REF	IdMin:0,10 p.u	time setting: 0,01 s		

Table 2 shows settings of differential and REF protection. *IdMin* represents Sect. 1 sensitivity, which is set by a multiplication of base current (winding 1 current). Section 1 represents the most sensitive part of differential characteristic i.e. minimum operating diff. current. This setting is set in order to avoid unwanted operation for the minimum value of the operating differential current, which includes maximum error of CTs, maximum error due to voltage regulation and additional security factor. *IdUnre* represents unrestricted protection limit, multiplication of base current (W1 current).This part is "instantaneous" part of differential protection, used when the fault is beyond any doubt internal. The reference for this setting is maximum three-phase short circuit current at MV busbars. *I2/I1* represents the maximal ratio of 2nd current harmonic to fundamental current harmonic, *I5/I1* represents the maximal ratio of 5th harmonic to fundamental.

3 Common Grounding Resistor Protection Performance

Performance of the protection of the common grounding resistor has been examined on the simulation model of two parallel 110/35 kV transformers with common low-impedance grounding resistor of 70 Ω to limit the single-phase to ground fault current to maximal value of 300 A on the 35 kV voltage side - Fig. 6. Modeled protection functions are transformer differential and restricted earth fault protection with included auxiliary protection equipment - current transformers (CTs) −150/1 A on the 110 kV side and 600/5 A on 35 kV side and voltage transformers (VTs). Both 110/35 kV transformers are represented with the YN0yn0d vector group.

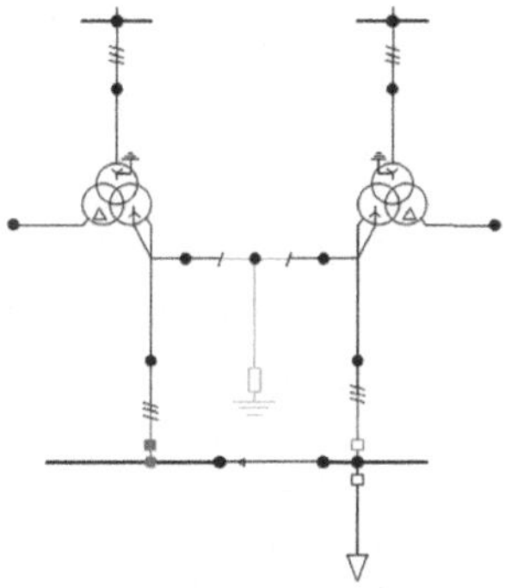

Fig. 6. Simplified schematic presentation of the network model with protection equipment

Differential protection performance has been examined for single-phase to ground fault in case where a single transformer is operating - Fig. 7 – left side or both transformers are operating – Fig. 7 – right side. Protection zone of differential protection includes the zone between two CTs (primary and secondary side) - Fig. 7 - left side, but excludes the substation busbars - Fig. 7- right side.

As shown on Fig. 7, the relay tripped opening the circuit-breakers (labeled as red circles) on high-voltage side and the low-voltage side of the transformer when a fault occurred inside the protection zone and did not trip when the fault occurred outside the protection zone, thus fulfilling the selectivity criterion.

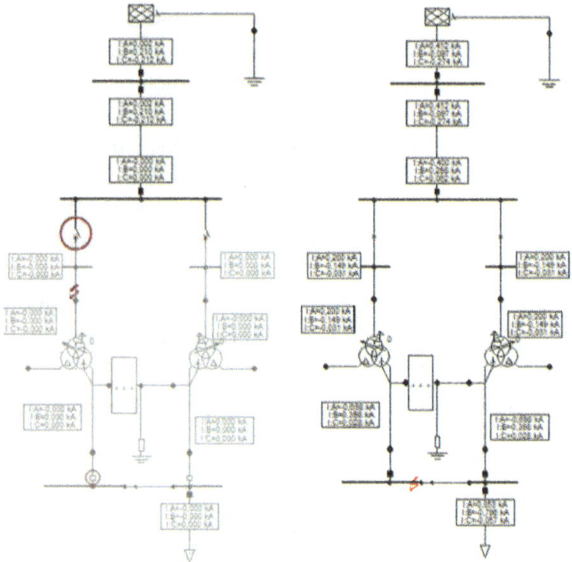

Fig. 7. Single-phase to ground fault inside the protection zone (left side of the figure) and outside the protection zone (right side of the figure)

During internal earth-faults, the sum of phase currents (MV side) and residual current (current at transformer neutral) is different from zero [12], and should trip REF protection. The performance of the REF protection has been examined during the single phase-to ground fault within the protection zone when a single transformer is operating -Fig. 8 - left side and outside the protection zone when both transformers are in parallel operation -Fig. 8 -right side. The simulation diagrams on Fig. 9 indicate on REF protection trip caused by the differential current between the residual current of the secondary side of the transformer and the resistor current during the single-phase to ground fault within the protection zone when a single transformer operates. The trip signal is shown on the upper diagram, the differential current is presented on the second diagram (14 s A (2,5 p.u.) is above the 0,1 p.u. threshold value - Table 2), while the third diagram presents the currents within the protection zone. The last diagram presents the vector sum of the phase and neutral currents at the MV side of the transformer.

The currents during the single-phase to ground fault outside of the protection zone when two transformers are in parallel operation are presented on Fig. 10. Since the differential current of 0,08 p.u. (the second diagram) did not reach the threshold value of the 0,1 p.u. during the simulated fault, the REF protection did not trip. The differential current is presented on the second diagram, the currents within the protection zone are presented on the third diagram, while the last diagram presents the vector sum of the phase and neutral currents at the MV side of the transformer.

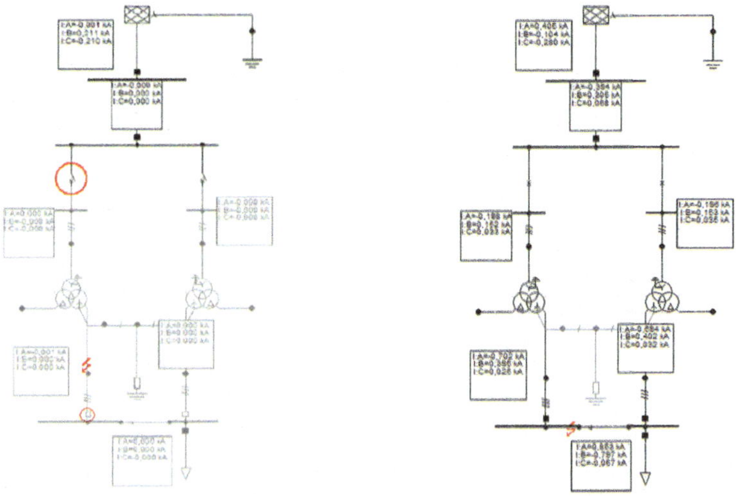

Fig. 8. Single-phase to ground fault inside the protection zone for testing REF protection when a single transformer operates - left side of the figure, single-phase to ground fault outside the protection zone for testing REF protection when both transformers operate - right side of the figure.

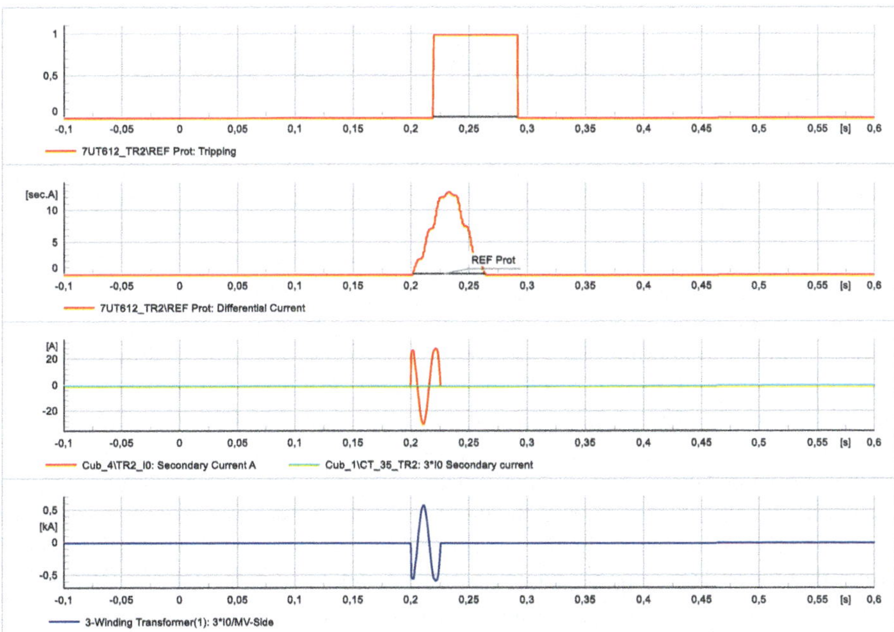

Fig. 9. Diagrams associated with REF protection testing during the single phase-to ground fault inside the protection zone when a single transformer operates

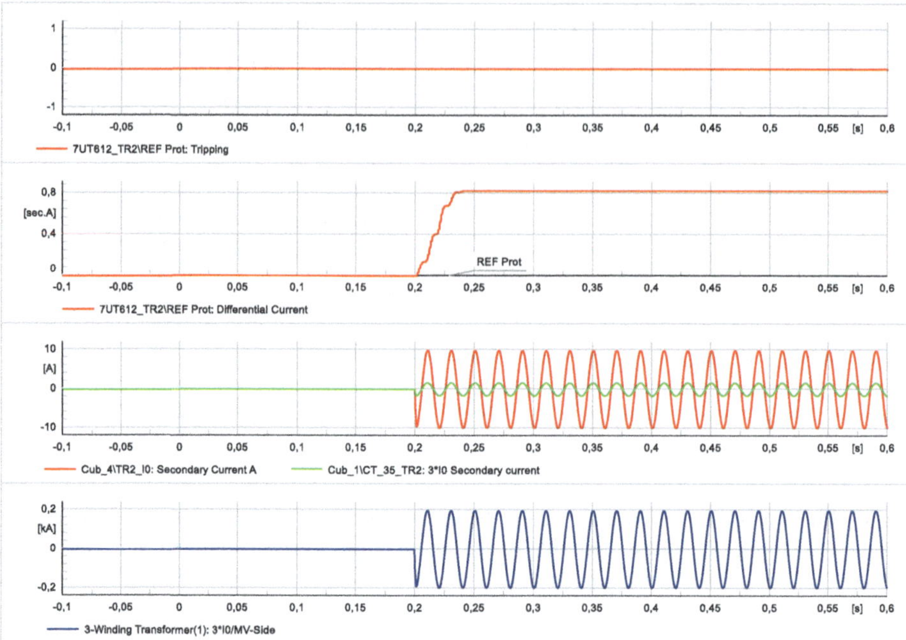

Fig. 10. Diagrams associated with REF protection testing during the single phase -to ground fault outside the protection zone when both transformers operate

4 Inrush Current Influence on Transformer Protection Performance

Simulation of a single transformer energization has been performed at time-instant the worst in terms of possibility of the inrush current occurrence - when the magnetic flux can double the initial value [6–10]. The energization process began at the time instant t = 0, 1 s when the HV-side phase "C" voltage obtained zero value which has resulted in corresponding magnetic flux increase (magnetic flux of phase C) for more than two-times from the initial value Fig. 11 – the third diagram. Phase voltages of all three phases remained within the nominal values during the energization process while the highest value of magnetic flux of phase "C" resulted in the occurrence of the highest in-rush current in phase "C" – Fig. 11. The differential current, because of the transformer in-rush occurrence, obtained the highest value at phase "C" – Fig. 12 second diagram and could lead to an unwanted differential protection trip – dotted line at second diagram of Fig. 12 presents the threshold. However, simulated values do show significant changes during the transformer energization, hence the differential protection did not trip because of settings that block 2nd harmonic shown in Table 2 and the transformer was successfully energized – as it is shown at Fig. 11 and Fig. 12.

The unwanted differential protection trip during the transformer energization process is prevented by analyzing the harmonic spectrum of all three-phase at HV-side of the transformer. The second harmonic presence in phase currents indicates the transformer energization and thus is used for blocking the differential protection tripping as it shows

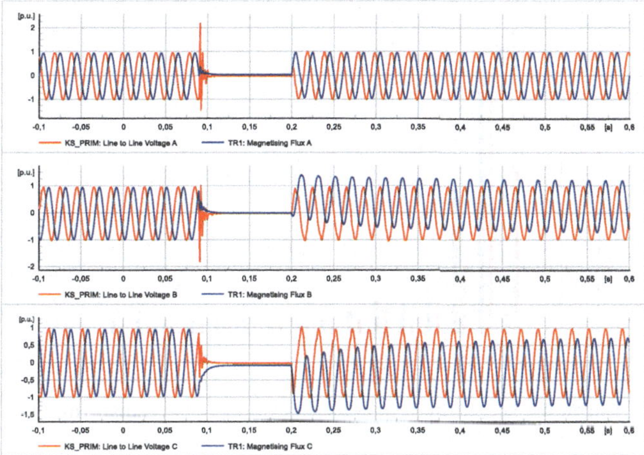

Fig. 11. Phase-voltages and magnetic-fluxes at HV transformer side; phase A -the first diagram, phase B- the second diagram, phase C-the third diagram

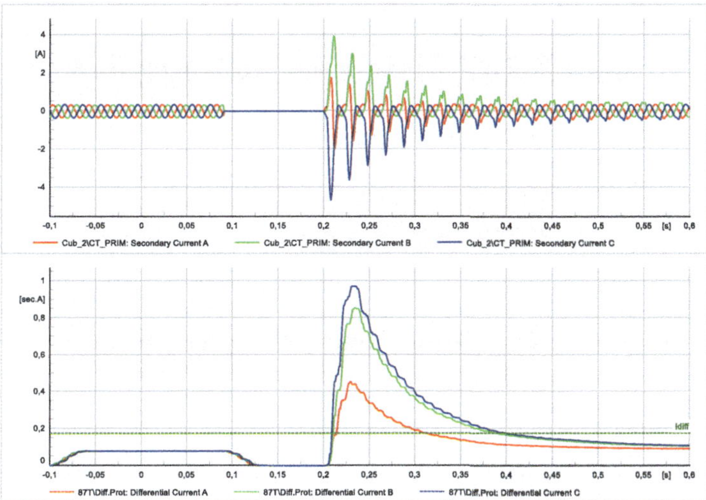

Fig. 12. Inrush currents at the HV - transformer side – the first diagram (values at differential protection CT); differential currents at the HV - transformer side – the second diagram (values at differential protection CT)

on Fig. 13 - 55% of 2nd harmonic presence. This setting is over the threshold setting at Table 2. The manufacturer recommended I2/I1 ratio.

The same differential protection agenda is applied as well when two transformers are in parallel operation. The appearance of the inrush current during the energization of a single transformer causes the inrush current i.e. sympathetic inrush current in the already energized transformer. Figure 14 shows phase currents of TR1 and TR2 during

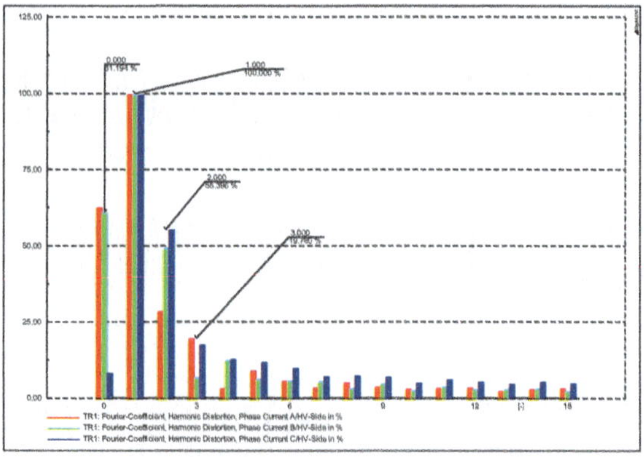

Fig. 13. Harmonic spectrum of the phase currents at the transformer HV - side

TR2 switch-on at the most unfavorable moment because the magnetic flux almost doubles the value. There is an increase in the amplitudes of currents at the switch-on.

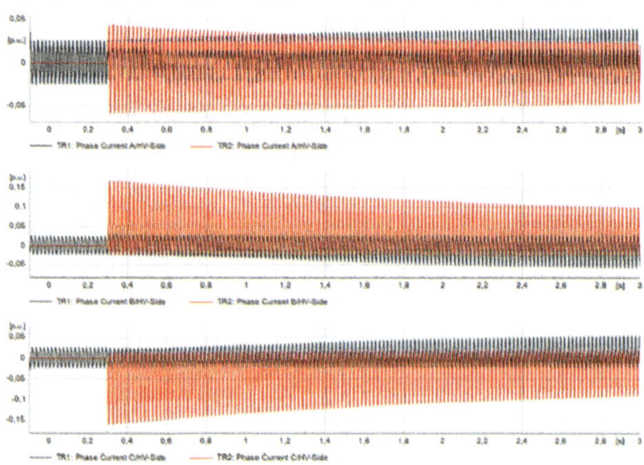

Fig. 14. Waveforms of phase currents at HV transformer side for both transformers; phase A – the first diagram, phase B – the second diagram, phase C – the third diagram

Harmonic spectrum of phase currents at the HV side of both transformers is shown at Fig. 15. The second harmonic presence occurred here as well, but the settings from the Table 2 blocked differential protection trips because the percentage rose above the set threshold. Figure 16 shows differential currents at both transformers. First diagram of Fig. 16 is associated with the first transformer, where the differential current is below the dotted line – setting threshold was not crossed. Second diagram on Figure 16 shows differential currents for the second transformer, where differential current crossed the threshold, but the differential protection trip was blocked by 2nd harmonic I2/I1 ratio shown in Table 2. Second harmonic percentage of the second transformer can be seen at Fig. 15 in the second diagram.

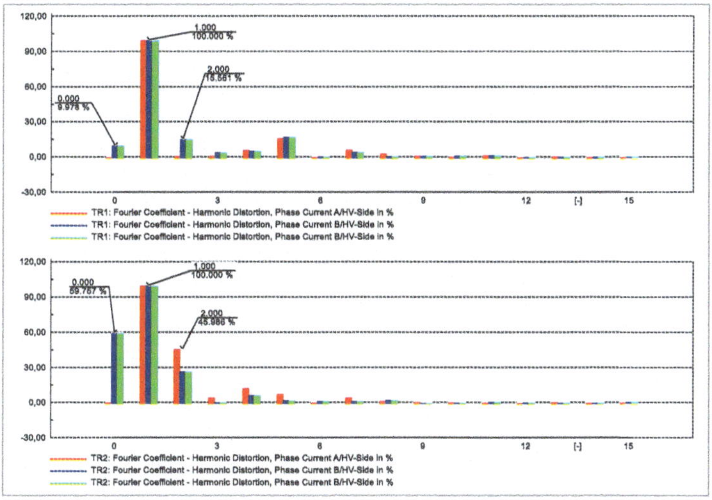

Fig. 15. Harmonic spectrum of the phase currents at the transformer HV side; first transformer - first diagram, second transformer – second diagram

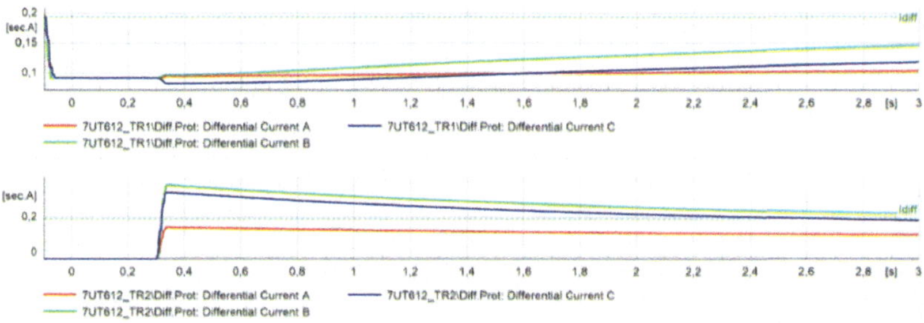

Fig. 16. Differential currents at the HV transformer side; first transformer – first diagram; second transformer – second diagram (values at differential protection CT)

5 Unwanted Circulating Currents Between TR1 and TR2

Single-phase to ground fault was simulated on the HV power line. It can be seen that there is no current through the grounding resistor but there is a circulating current between two transformers - Fig. 17. Current waveforms at HV/MV side during the simulated fault are shown at Fig. 18. First three diagrams of Fig. 18 show phase current and their increase in amplitude when fault occurred, Fourth diagram shows residual current at HV and the fifth one shows there is residual current at MV side – i.e. circulating current.

Figure 19 shows voltage (first diagram) and current waveforms at transformer neutral during fault. As it shows on the first diagram, voltage is minimal and that indicates there is no operating condition. Second diagram shows there is no current through the resistor (CT RN_Io – CT located between resistor and grounding at Fig. 6). Second diagram also shows a circulating current flowing between secondary windings through common grounding recorded by CTs TR1_Io and TR2_Io (located between transformer neutral

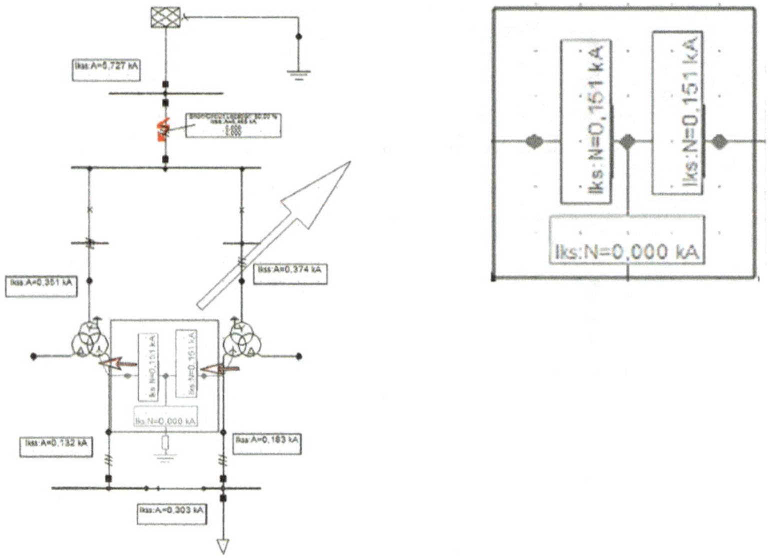

Fig. 17. Single-phase to ground fault at HV network with circulating currents

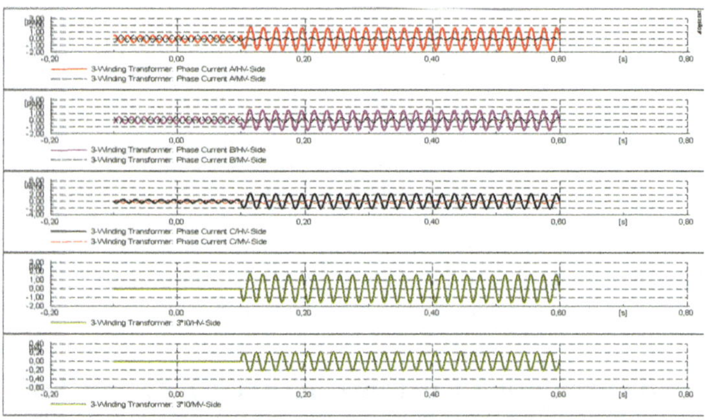

Fig. 18. Current waveforms during single-phase to ground fault

and resistor). That could indicate there is a fault in the 35 kV network that needs to be cleared.

It can be concluded that the occurrence of circulating current between the neutral and secondary windings of two parallel-connected transformers is a very dangerous phenomenon because it can cause unwanted, non-selective trips of short-circuit protection of transformer neutral (resistor protection). This phenomenon was discovered in practice, by unwanted protection trips. Such faults, in 400, 220 or 110 kV - Croatian transmission network, can cause power outages in large areas, even in the whole country depending on voltage level unless the phenomenon is detected and removed in a timely manner.

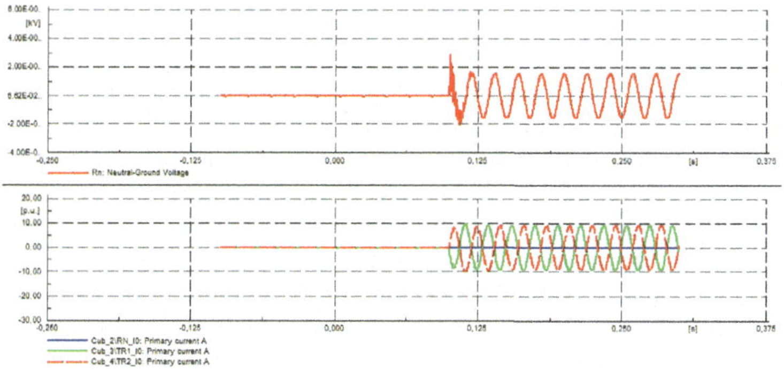

Fig. 19. Voltage and current waveforms at transformer neutral

Resistor protection is used as a backup protection for single-phase to ground faults for 35 kV feeders (distribution network) and coordination between DSO and TSO is necessary for keeping protection settings updated and graded. The most important parameter for quantifying the amplitude of the circulating currents and its occurrence is the zero sequence impedance between the windings [17, 18]. The proposal for solving this problem provides a technical solution with numerical relays, which recommends that the condition for the operation of protection against bridging of resistors is the existence or absence of current flowing through the resistor [17, 18]. A CT built into resistor housing would measure the current through the resistor. With this proposal, it is possible that the resistor can be bridged i.e. no current flows through the CT but the fault is present and it is possible that CT becomes saturated under high magnitudes of current and there will be no conditions for protection to operate [17, 18]. Second solution is to have certain types of protection devices, which have a sufficient number of analog current inputs and filters with the possibility of summing these two analog values (phasors) from different sources [18] i.e. the concept of "differential protection of the resistor". If circulating currents at transformer neutral occur, the sum is equal to zero, but during single-phase to ground fault in a 35 kV network, the summation amount is equal to the current through the resistor and there is a condition for trip. Figure 20 shows the basic concept of this derived protection.

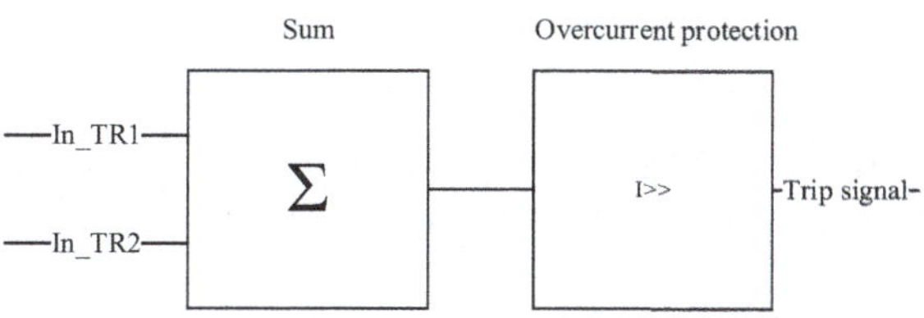

Fig. 20. Basic principle of proposed solution for circulating currents at neutral

6 Conclusion

The transformer can be seen as one of the most important elements of the power system, so it is vital to define and design its protection in order to minimize consequences and to extend its service life. Differential and restricted earth fault protection are the backbone of transformer protection. Nowadays modern protection relays can be programmed to withstand and recognize disturbances, to react quickly and properly. Substations of 110/35 kV connect transmission and distribution networks therefore protection of transformers in these substations is vital for maintaining the stability of the entire distribution network and end customers. Substations with two transformers that share a common grounding resistor are a common practice in Croatia. The choice for a common grounding resistor is the requirement to limit single-phase to ground fault current to a certain value and this construction is economically acceptable compared to other solutions. Transformer inrush current, which occurs under switch on of transformer, can cause an unselective trip of differential protection but solution was given by the detection of second harmonic content in the spectral current separation. In addition, the sympathetic inrush current can occur. Occurrence of circulating currents between neutrals of transformers in parallel operation due to a single-phase to ground fault in the HV network is an unwanted phenomenon. To define the applicable technical solution, it is necessary to analyze each specific and possible situation, taking into account the amount and nominal data of power transformers, available type of protection device, location and characteristics of current transformers and the need and possibility of replacing or upgrading them. The simulation network model is not an actual system but can serve as a model for testing various hazardous operating conditions and faults that may occur within substations that connect transmission and distribution networks and create a power grid.

References

1. Hewitson, L.G., Brown, M., Balakrishnan, R.: Pratical Power System Protection. Elsevier, Oxford (2004)
2. Yu, L., Henriks, R.L.: Selection of system neutral grounding resistor and ground fault protection for industrial power systems. In: Industry Applications Society 38th Annual Petroleum and Chemical Industry Conference, pp. 147–153 (1991)
3. IEEE Recommended Practice for System Grounding of Industrial and Commercial Power Systems. IEEE SA 3003. 1-2019, pp. 1–79 (2019)
4. Ćućić, R., Komen, V., Živić Đurović, M.: Neutral point concept in distribution networks. Eng. Rev. **28**, 77–89 (2008)
5. Fu, Z., Wang, N., Huang, L., Zhang, R.: Study on neutral point grounding modes in medium-voltage distribution network. In: 2014 International Symposium on Computer, Consumer and Control, Taichung, (2014)
6. Chen, N., Wang, C., Zhang, J., Yang, F., Chen, Y., Hong, W.: The inrush current analysis and restraining method of energizing no-load transformer. In: 2016 3rd International Conference on Systems and Informatics (ICSAI), pp. 204–207 (2016)
7. Alibašić, E., Marić, P., Nikolovski, S.: Transient phenomena during the three-phase 300 MVA Transformer energization on the transmission network. Int. J. Electr. Comput. Eng. **6**(6), 2499–2505 (2016)

8. IEEE Guide for Protecting Power Transformers, IEEE SA C37.91-2008 (Revision of IEEE SA C37.91-2000), pp.1–139 (2008)
9. Alstom: Network Protection & Automation Guide (2011)
10. Heathcote, M.: The J & P Transformer Book, 13th edn., Elsevier, Oxford (2011)
11. Bronzeado, H., Yacamini, R.: Phenomenon of sympathetic interaction between transformers caused by inrush transients. Sci. Measur. Technol. IEE Proc. **142**, 323–329 (1995)
12. Ziegler, G.: Numerical Differential Protection. Siemens, Nuremberg (2012)
13. Masle, D., Benović, J., Tolić, I., Marić, P.: The effect of transformer operating conditions on occurrence of inrush current with respect to differential protection settings. In: 12th HRO CIGRE Session, Šibenik (2015)
14. Tsourakis, G., Milis, I., Prousalidis, J.: Transformer sympathetic inrush: a case study. In: 8th Mediterranean Conference on Power Generation, Transmission, Distribution and Energy Conversion (MEDPOWER 2012), Cagliari (2012)
15. Hong-Ming, S., Tao, Z., Shao-Feng, H., Ou, L.: Study on a Mal-operation case of differential protection due to the interaction between magnetizing inrush and sympathetic inrush. In: 2014 IEEE PES General Meeting | Conference & Exposition, National Harbor, MD (2014)
16. Qi, X., Yin, X., Zhang, Z.: Sympathetic inrush current in a transformer and a method for its identification. IIEEJ Trans. Electr. Electr. Eng. **11**(4), 442–450 (2016)
17. Calzolari, G., Saldana, C.: Power transformers in parallel which share a neutral resistance: how to manage the circulating current between the secondary windings. In: IPST International Conference on Power System Transients, Canada (2005)
18. Masle, D., Rošić, M., Benović, J., Marić, P.: On protection of medium voltage neutral resistor during single phase to ground fault in high voltage grid. In: 12th Symposium on Power System Management, Split (2016)
19. Dommel, H.W.: EMTP Theory Book. Microtran Power System Analysis Corporation, Vancouver (1995)
20. Watson, N., Arrillaga, N.: Power Systems Electromagnetic Transients Simulation. The Institution of Engineering and Technology, London (2003)
21. Siemens. SIPROTEC Differential Protection 7UT612, V4.0 Manual (2002)

Simulation of the Grid Connected Battery Storage with Ability of Active and Reactive Power Control

Zvonimir Šimić(✉), Josip Bilandžić, Danijel Topić, and Antun Damjanović

Faculty of Electrical Engineering, Computer Science and Information Technologies Osijek, J. J. Strossmayer University of Osijek, Kneza Trpimira 2b, 31000 Osijek, Croatia
zvonimir.simic@ferit.hr

Abstract. In this paper the simulation of the battery storage system connected to the grid with the possibility of active and reactive power control is presented. Considered battery storage system has the ability to inject the active power when there is a lack of active power and the reactive power when there is a lack of reactive power in the grid. Amount of the active and the reactive power injected to the grid are defined according to the active and reactive power demand. Battery storage system is connected to the grid over the two-way DC-AC converter with the IGBT switches. Model of the battery storage system is based on Typhoon HIL grid connected battery storage, but some improvements have been made on the model. Model is consisted of two parts. The first part is the battery storage model in the schematic editor and the second part is the monitoring and the control interface in SCADA. Model in the schematic editor is consisted of electric and signal parts and all used components are described in the paper. Simulation of the battery storage system with the possibility of active and reactive power control in Typhoon HIL is done and the impact on the grid voltage and the frequency is visible according to the simulation results.

Keywords: Battery storage · Active power · Reactive power · Battery storage modeling · Battery storage simulation

1 Introduction

In the power system, there is a strong impact of active power to the grid frequency and reactive power to the grid voltage. In the conventional power system, energy flow is in the one way, from electricity production in the large power plants, over the transmission and distribution grid, to the end-customers. The production output power needs to be equal to sum of the consumption power and the system losses to keep power system in the balance. The nominal grid frequency is 50 Hz in the Europe and 60 Hz in the USA, if the grid frequency is lower or higher than the nominal frequency, the power system is unbalanced.

The grid frequency is lower than the nominal frequency if the production output power is lower than the sum of the consumption power and the system losses. To increase

H. Glavaš et al. (Eds.): OTO 2021, LNNS 369, pp. 188–204, 2022.
https://doi.org/10.1007/978-3-030-92851-3_14

the grid frequency, it is required from generators in the power plants to increase the output power. The grid frequency is higher than the nominal frequency if the production output power is higher than the sum of the consumption power and the system losses. To decrease the grid frequency, it is required from generators in the power plants to decrease the output power.

While the active power balance is related to the frequency, the reactive power balance is related to the voltage. If there is a lack of the inductive reactive power, the grid voltage is lower than the nominal voltage. It is required to inject additional inductive reactive power to increase voltages. In conventional power systems, condenser batteries and synchronous machines are usually used for reactive power compensation. If there is a exceed of the inductive reactive power, the grid voltage is higher than the nominal voltage. It is required to take the exceed of inductive reactive power from the grid to decrease voltages. Highest voltages are occurred during periods of low consumption which were leading to needs for regulation coil installations.

In the power grids with the high amount of electricity produced in renewable energy sources (RES) is hard to keep the balance between production and consumption whole time because of the variability of RES, especially wind and solar energy. There is a possibility to install a battery energy storage system (BESS) which have ability of active and reactive power control which than have impact to the grid frequency and the voltage. Working principle of the BESS can be observed in the four cases:

- If the consumption of active power is higher than the active power available in the grid, BESS injects active power to the grid.
- If the consumption of active power is lower than the active power available in the grid, BESS receives active power from the grid.
- If the consumption of reactive power is higher than the reactive power available in the grid, BESS injects reactive power to the grid.
- If the consumption of reactive power is lower than the reactive power available in the grid, BESS receives reactive power from the grid.

Different cases are possible thanks to using of controllable bidirectional AC-DC converter so the energy way, the current phase angle, which is directly related to the active and reactive power and the amount of power can be easily controlled.

There are many researches related to the using of battery storage systems with ability of active and reactive power control in the last decade. Main focus is on the using of BESS for that applications.

A new control strategy for PV system with BESS in microgrids, with possibility of voltage and frequency control is presented in [1]. The control strategy, proposed in [2], presents effective coordination between inverter V-f and P-Q control, MPPT control and energy storage charging and discharging control. A cascade control method based on external controller that regulates the active and reactive power to generate reference currents and internal current controller that produces signals for driving inverter switches is presented in [3]. In [4], a set of PI controller for active and reactive power in the system control is presented.

The ability of reactive power compensation of PV system in combination with battery storage is effective for improving voltage according to [5]. In [6], an energy management algorithm for a grid connected PV battery system is presented. In [7], detailed mathematical model and a control algorithm of a three-phase grid-connected intelligent hybrid system is presented. Battery energy storage system can be used to store energy produced in PV system for later use or to store energy from the grid when the price of electricity is low according to [8].

The possibility of active and reactive power control of battery storage is very important in weak distribution networks where change in the amount of load power leads to significant change of voltage [9]. In [10], battery storage system is used to compensate fluctuations of PV output power caused by solar irradiation and temperature changes. Use of battery storage systems in power grids can help to keep the balance between production and consumption of electricity in power system [11]. In [12], a design of an voltage source converter with ability of bidirectional active and reactive power control in a grid connected battery system is presented.

A method for improving the voltage stability, by injecting correct amount of active and reactive power from BESS, is presented in [13]. Batteries connected to the grid using controllable inverter have ability to provide reactive power to the grid, without using of any other compensation device, which is presented in [14]. In [15], Voltage Source Inverter (VSI) control method with ability to control active and reactive power exchange between the single-phase grid and the load, is presented. Active and reactive power control method, between a battery bank and single-phase local grid, is proposed in [16].

The main contribution of this paper is to show the impact of BESS to grid frequency and voltage through the simulation in Typhoon HIL software package. Decision on the amount of active/reactive power injected to/received from the grid is made according to load active and reactive power. Model in Typhoon HIL is based on the grid connected battery storage model with some modifications and improvements. Controllable three phase load is connected in the grid side of converter, frequency signal is enabled in SCADA interface, signal part is modeled in two separate parts to make it simpler and clearer. Frequency and voltage monitoring on trace graphs, load control unit, battery state of charge on trace graph, separate trace graphs for active and reactive power and digital displays are added in SCADA interface. Amount of active/reactive power injected to/received from the grid cannot be selected manually because it is determined automatically according to load active and reactive power.

This paper is divided into four chapters. First chapter gives introduction to the needs for frequency and voltage regulation in the power grid. The model of the BESS is presented in Chapter 2. In Chapter 3, the simulation of the BESS is done and results are presented. Results discussion and conclusion are given in Chapter 4.

2 Battery Storage Model

Battery storage system model is consisted of two main parts, model in the Typhoon HIL program part called schematic editor and SCADA control and monitoring interface. Model in the schematic editor is consisted of two main parts, first is the electric part and second is the signal part. Block diagram of battery energy storage system model is presented in Fig. 1.

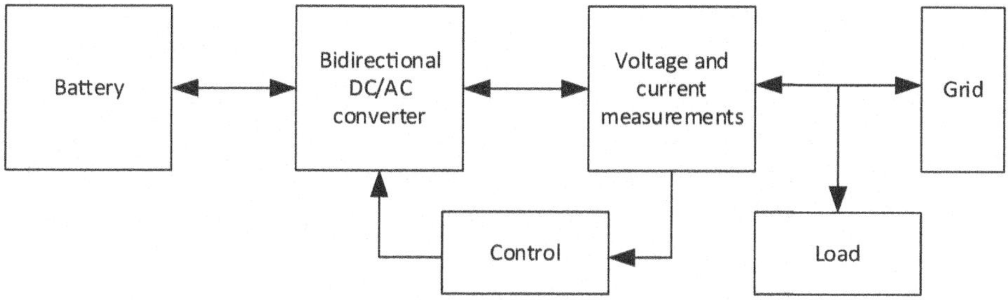

Fig. 1. Block diagram of battery storage system model.

2.1 Electric Part of the Battery Storage Model

Electric part of the BESS model is consisted of battery, bidirectional AC-DC IGBT converter, filter, current and voltage measurements, grid and the controllable load. Schematic diagram of electric part of battery storage model is presented in Fig. 2.

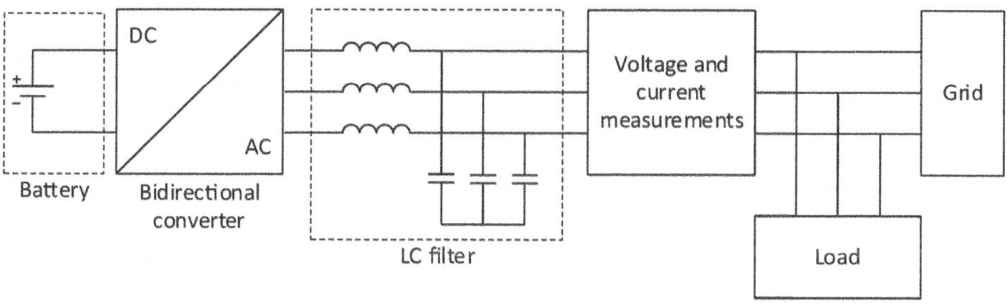

Fig. 2. Schematic diagram of battery storage model.

Battery. Battery in Typhoon HIL is modelled as a controlled voltage source with a series resistance. There are four standard predefined battery technologies available to choose: Lead-acid (Pb-acid), Lithium-ion (Li-ion), Nickel-Cadmium (NiCd) and Nickel-Metal Hydride (NiMH). It is possible that user defines own battery type if he uses some other battery technology which is not predefined in software. For all battery technologies, the nominal voltage, battery capacity in Ah and the initial state of charge (SOC) can be manually defined. In this paper, Li-ion battery technology is used. Figure 3 presents battery component connected in the battery storage model.

Converter. As a bidirectional converter is used three-phase two-level inverter. This component is modeled with using of the ideal IGBT switches and diodes. Inverter can

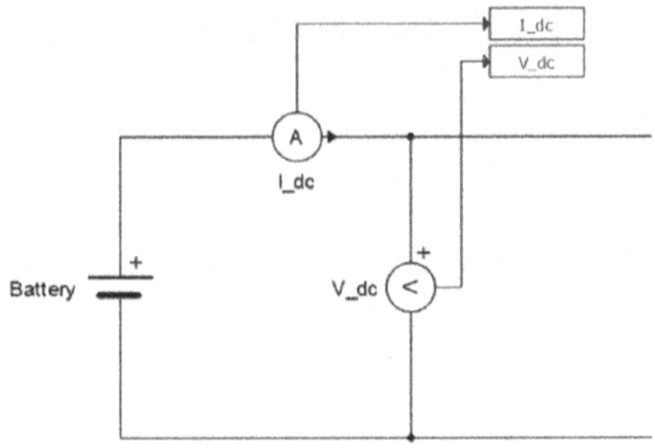

Fig. 3. Battery component used in battery storage model.

be controlled with the using of internal modulator, with the model control signals and with the digital inputs per switch or per leg. In this case, inverter is controlled with the using of internal modulator which means that internal PWM modulator is used to driving switches. Four signals are required for this control mode, first is signal for enabling the internal modulator and other are reference input signals for all three phases. Inverter can work at the fixed and the variable carrier frequency. In this case, operation mode is at fixed carrier frequency of 10 kHz. Inverter component, with reference signals inputs and enable signal input, is presented in Fig. 4.

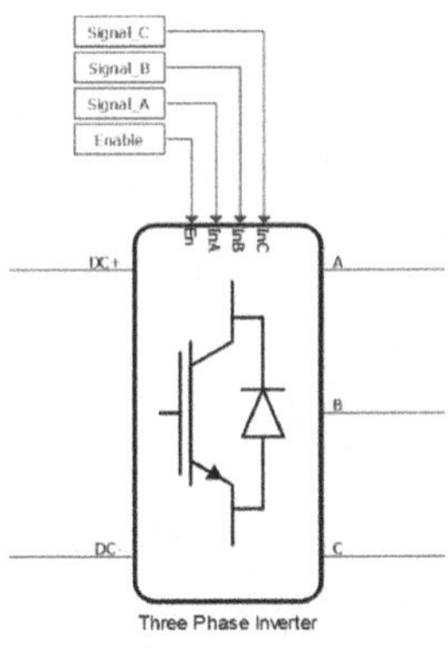

Fig. 4. Inverter component used in battery storage model.

Filter. Electronic components produce harmonics which are unwanted in the power grid. Passive LC filter is used to reduce harmonics injected in the power grid from the inverter. Passive LC filter, used in this model, is presented in Fig. 5.

Current and Voltage Measurements. Current measurements are used for measuring battery current, phase currents and phase RMS currents. Voltage measurements are used for measuring battery voltage, phase voltages, line voltages, phase and line RMS voltages. Measurements components have output signals that can be used for model signal

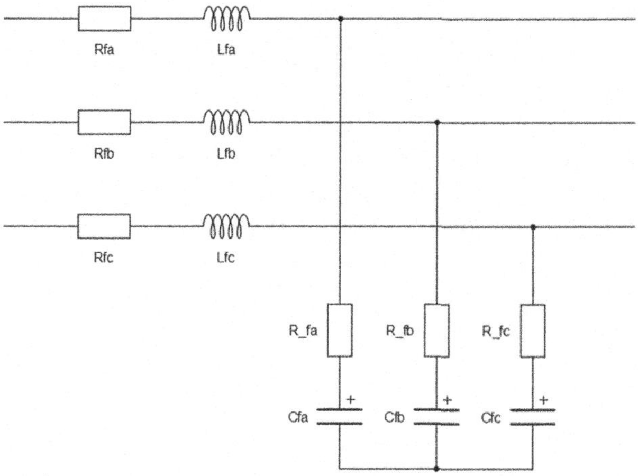

Fig. 5. LC filter used in battery storage model.

part and in the SCADA interface. Figure 6 represents current and voltage measurements. On the left, there are current and RMS currents measurements, in the middle, there are phase voltages and phase RMS voltages measurements, on the right, there are line voltages and line RMS voltages measurements.

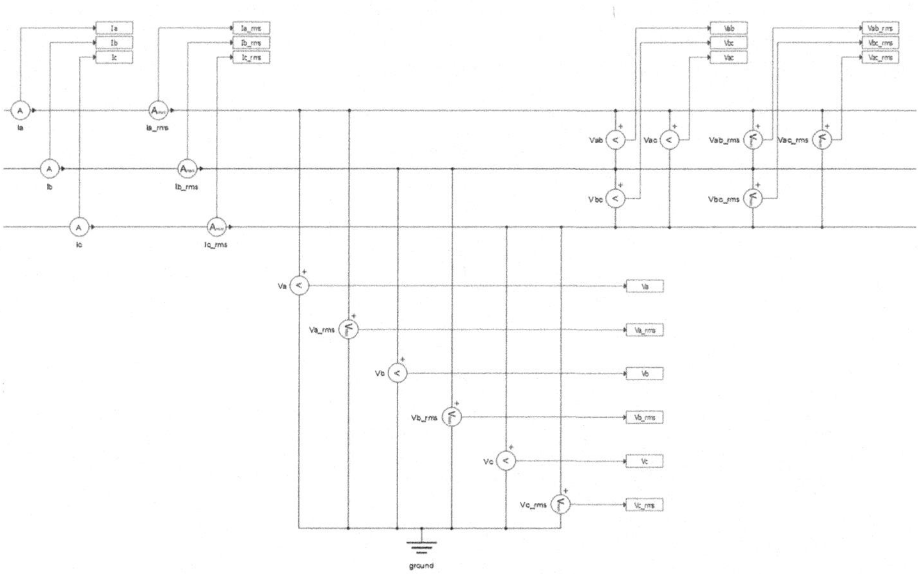

Fig. 6. Current and voltage measurements.

Grid. Grid in Typhoon HIL is modelled as three single phase voltage sources with the series resistances and inductances, presented in Fig. 7. This is used for simulation of symmetric three phase grid. There is possibility to measure phase currents and phase and line voltages. Nominal line voltage and grid frequency can be manually defined. Grid can be defined in three different ways, with resistance and inductance per phase which is used in this case, then with short circuit capacity and power factor and with short circuit capacity and *X*/*R*. In this case nominal line voltage is 400 V. nominal frequency is 50 Hz, resistance per phase is 0.0001 ohm and 20 μH. This results with low grid impedance and high *X*/*R* ratio which means that grid is very strong.

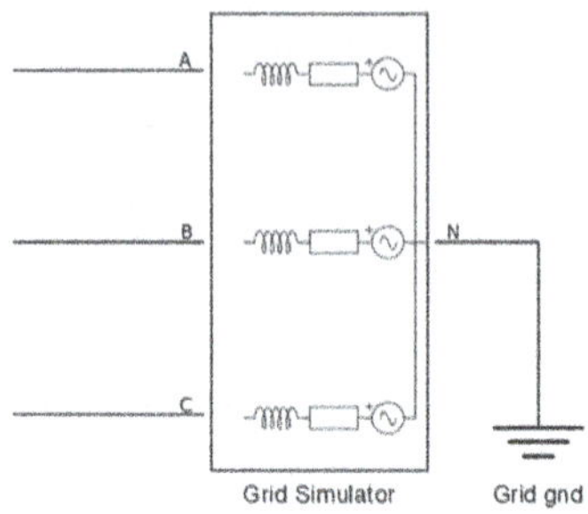

Fig. 7. Grid component used I battery storage model.

Load. As a controllable load is used three-phase constant power load component, presented in Fig. 8. This component is modeled as the sine current sources which are controlled to achieve defined power flow according to the active and reactive power references from SCADA or through the signal inputs.

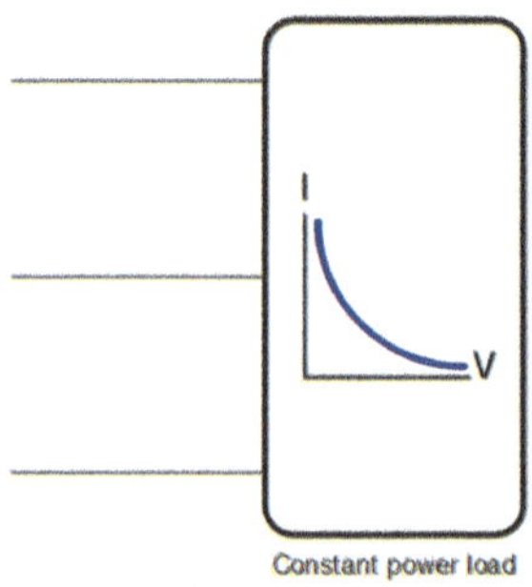

Fig. 8. Load component used in battery storage model.

2.2 Signal Part of the Battery Storage Model

Signal part of the BESS model is consisted of power calculation and reference signals calculation. Electric connections and components are in black color while signal connections, components and blocks are in blue color so these two parts can be easily differed from each other.

Power Calculation. According to the measured phase voltages and phase currents, power meter component calculates instantaneous active power, instantaneous reactive power, average active power, average reactive power, apparent power and power factor. Calculated power values are required in SCADA for monitoring. Block model for calculating power values is presented in Fig. 9.

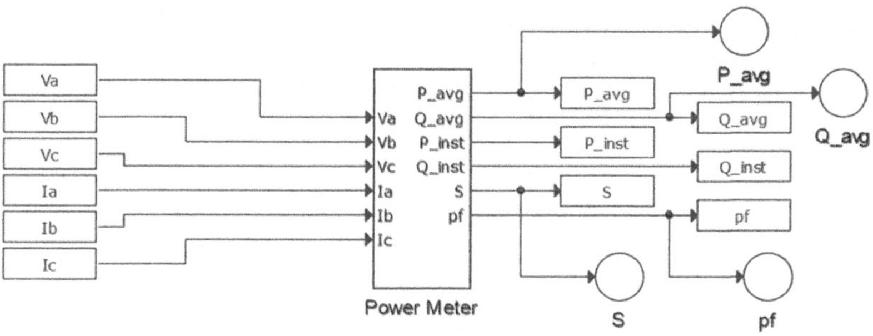

Fig. 9. Block model for calculating power values.

Reference Signals Calculation. Measured phase current are transformed from three phase abc system to dq rotating reference. Direct and inverse components are then subtracted from direct and inverse reference currents. Reference currents are achieved through SCADA interface. Difference between reference and measured currents is then regulated with PI regulator. Achieved signals are transformed from dq rotating reference to three phase abc system. These three signals are used for control of inverter with internal PWM modulator. Block model for calculating reference signals for driving three phase two-way converter is presented in Fig. 10.

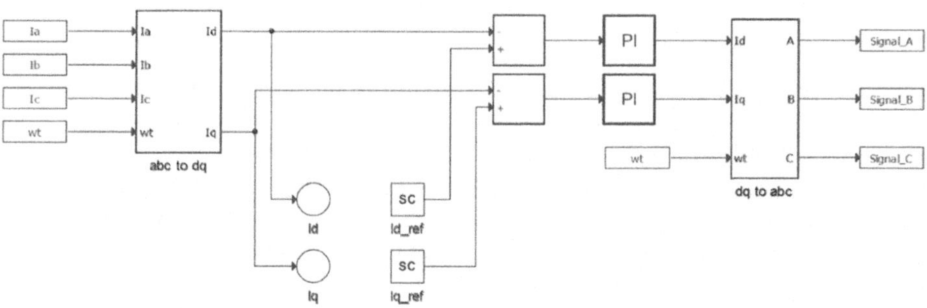

Fig. 10. Block model for calculating reference signals.

2.3 SCADA Interface

The purpose of the SCADA interface in this battery storage model is monitoring of phase currents, phase and line voltages, frequency, battery state of charge (SOC), battery voltage and current, active and reactive output power of the battery storage during simulation. There is possibility to control amount of load power.

Voltage and Currents Measurements. SCADA windows presented in Fig. 11 shows values of phase and line RMS voltages as well as values of phase RMS currents. Digital displays get values from analog signals received from measuring instruments connected in electric model in the real time during simulation.

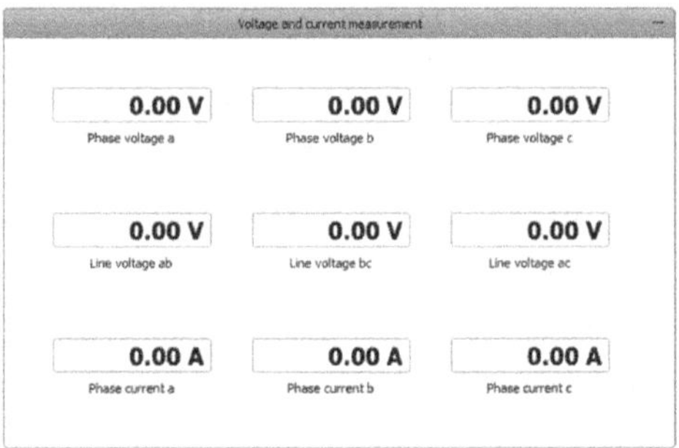

Fig. 11. SCADA voltage and current measurements.

Frequency and Voltage Monitoring. It is required to monitor frequency and voltage in the real time during simulation. SCADA windows for frequency and voltage monitoring are presented in Fig. 12. Upper window is consisted of trace graph and digital display for monitoring frequency value while window below is consisted of trace graph and digital display for monitoring line voltage RMS value.

Load Control. SCADA window for load control is presented in Fig. 13. There are two trace graphs, first graph shows amount of load active power while second graph shows amount of load reactive power. On the right side, there are two digital displays that show load active and reactive power. There are four buttons for different cases and reset button. By pressing a button, the amount of active and reactive power is changed according to observed case. By pressing a reset button, amount of active and reactive power is changed to default values

Battery Monitoring. Battery monitoring SCADA window is presented in Fig. 14. First graph shows battery SOC and it is very important to monitor battery SOC in the real time. It is recommended to keep battery SOC between 20% and 80%. Second graph shows amount of active power and third graph shows amount of reactive power. If amount of

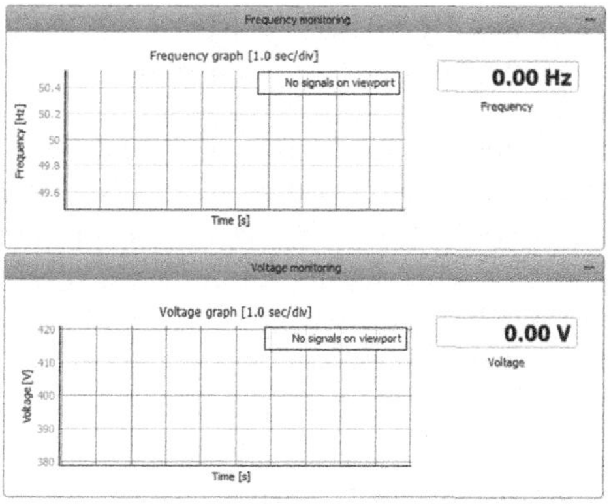

Fig. 12. SCADA frequency and voltage monitoring.

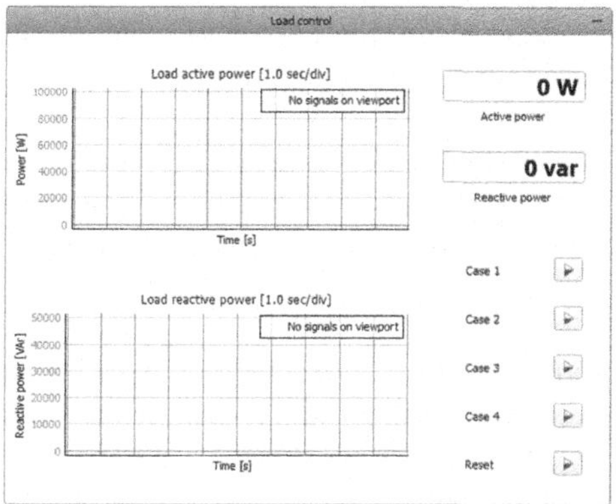

Fig. 13. SCADA load control.

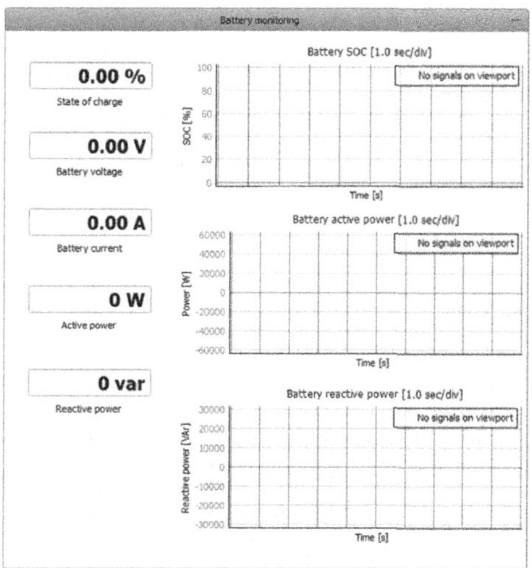

Fig. 14. SCADA battery monitoring.

power is positive, it means that BESS inject active or reactive power to the grid and battery is discharging. If amount of power is negative, it means that BESS takes active or reactive power from the grid and battery is charging.

On the left, there are five digital displays. First display shows battery SOC in percent, second display shows battery voltage and third display shows battery current. If the battery current is positive, flow of energy is from battery to the grid, it means that battery is discharging. If the battery current is negative, flow of energy is from grid to the battery, it means that battery is charging. Last two displays show amount of active and reactive power delivered to and received from the grid.

3 Battery Storage Simulation Results

In this chapter, results of simulations will be presented. Simulations are done with and without battery storage system to see the impact of charging and discharging battery storage. Four different cases are simulated as following:

- Case 1: Load active power is increased, reactive power is remained the same.
- Case 2: Load active power is decreased, reactive power is remained the same.
- Case 3: Load reactive power is increased, active power is remained the same.
- Case 4: Load reactive power is decreased, active power is remained the same.

Default values of controllable load connected to the grid are 40 kW of active power and 20 kvar of reactive power. Figure 15 presents comparison of frequency when values of controllable load are set to default. Left SCADA window shows frequency when battery storage is not available while right SCADA window shows frequency when battery storage is connected. It can be seen that battery storage causes decrease of frequency and frequency with battery is between 49.998 and 49.999 Hz.

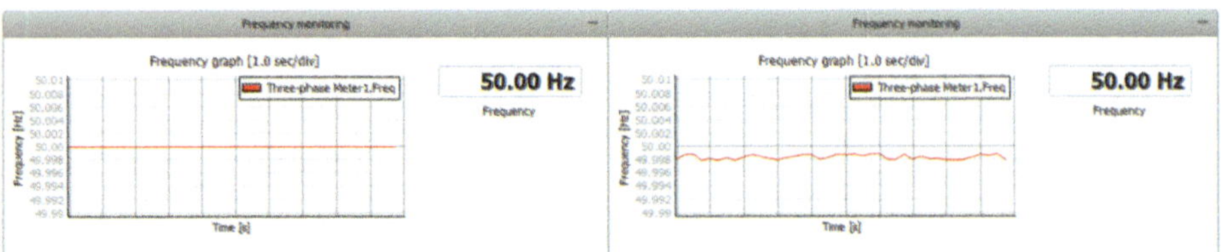

Fig. 15. Comparison of frequency for default load values.

Figure 16 presents comparison of voltage when values of controllable load are set to default. Left SCADA window shows voltage when battery storage is not available while right SCADA window shows voltage when battery storage is connected. It can be seen that line voltage without battery is 399.67 V and with battery is 399.86 V.

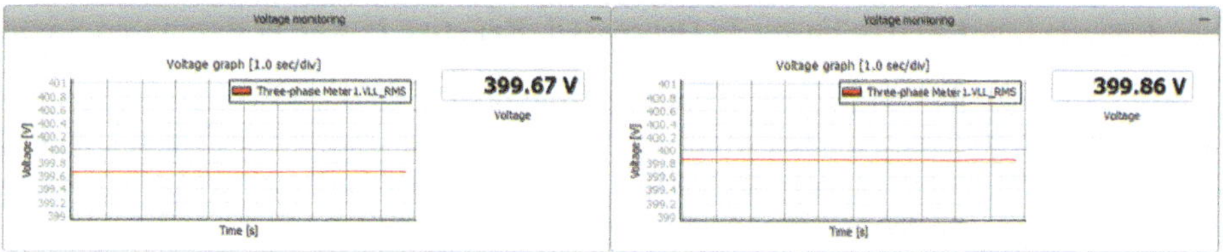

Fig. 16. Comparison of voltage for default load values.

3.1 Case 1: Load Active Power Is Increased, Reactive Power Is Remained Same

In this case, load active power is increased from 40 kW to 80 kW while load reactive power remains 20 kvar. Amount of active power injected from the battery storage to the grid is 40 kW, battery storage is discharging in this case to cover increased consumption of active power.

Figure 17 presents comparison of frequency for the case 1. Left SCADA window shows frequency when battery storage is not available while right SCADA window shows frequency when battery storage is connected. It can be seen that with battery storage connected frequency is between 49.997 and 49.998 Hz.

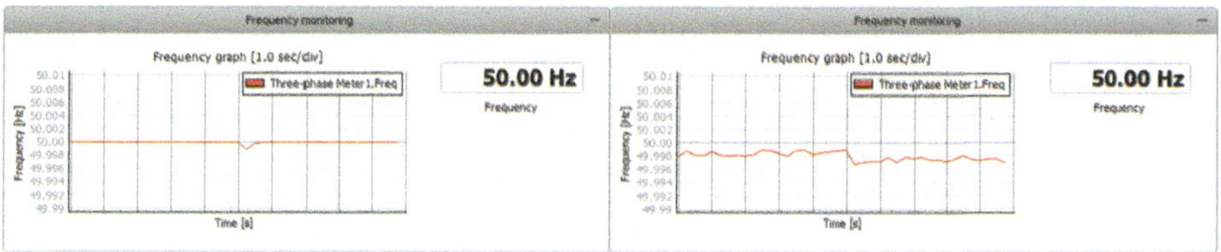

Fig. 17. Comparison of frequency for case 1.

Figure 18 presents comparison of voltage for the case 1. Left SCADA window shows voltage when battery storage is not available while right SCADA window shows voltage when battery storage is connected. It can be seen that line RMS voltage without battery is 399.65 V and with battery is 399.76 V.

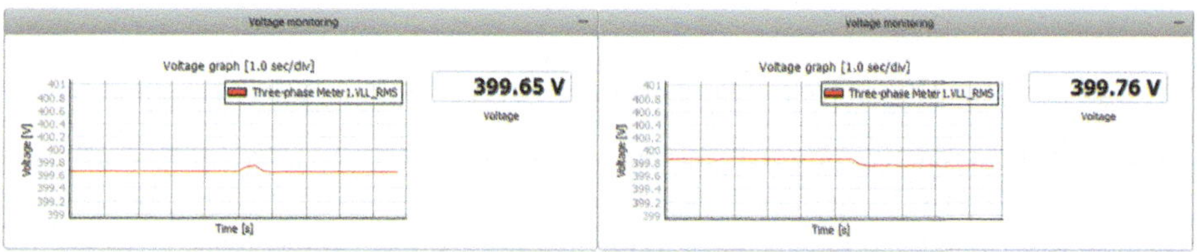

Fig. 18. Comparison of voltage for case 1.

3.2 Case 2: Load Active Power Is Decreased, Reactive Power Is Remained Same

In this case, load active power is decreased from 40 kW to 0 kW while load reactive power remains 20 kvar. Amount of active power received from the grid to the battery storage is 40 kW, battery storage is charging in this case to absorb reactive power excess in the grid caused by decreased consumption of active power.

Figure 19 presents comparison of frequency for the case 2. Left SCADA window shows frequency when battery storage is not available while right SCADA window shows frequency when battery storage is connected. It can be seen that with battery storage connected frequency is between 49.997 and 49.998 Hz.

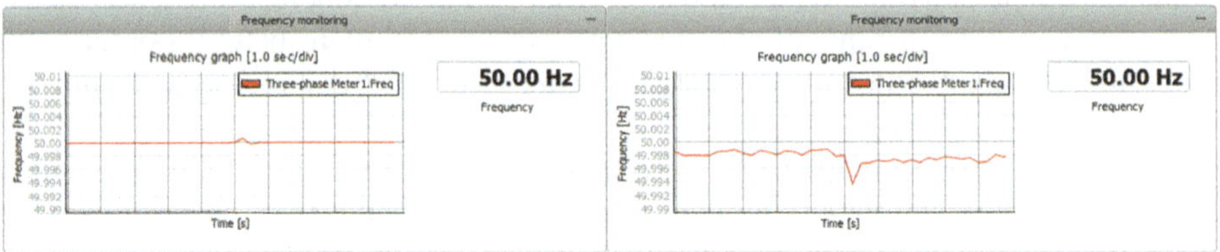

Fig. 19. Comparison of frequency for case 2.

Figure 20 presents comparison of voltage for the case 2. Left SCADA window shows voltage when battery storage is not available while right SCADA window shows voltage when battery storage is connected. It can be seen that line RMS voltage without battery is 399.68 V and with battery is 399.93 V.

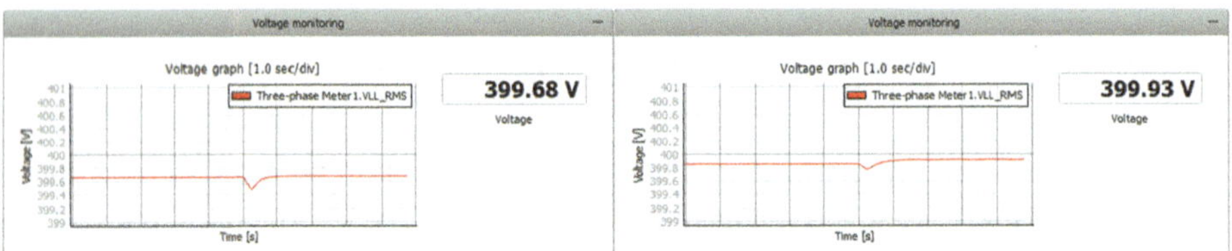

Fig. 20. Comparison of voltage for case 2.

3.3 Case 3: Load Reactive Power Is Increased, Active Power Is Remained Same

In this case, load reactive power is increased from 20 kvar to 40 kvar while load active power remains 40 kW. Amount of reactive power injected from the battery storage to the grid is 20 kvar, battery storage is discharging in this case to cover increased consumption of reactive power.

Figure 21 presents comparison of frequency for the case 3. Left SCADA window shows frequency when battery storage is not available while right SCADA window shows frequency when battery storage is connected. It can be seen that with battery storage connected frequency is between 49.998 and 49.999 Hz.

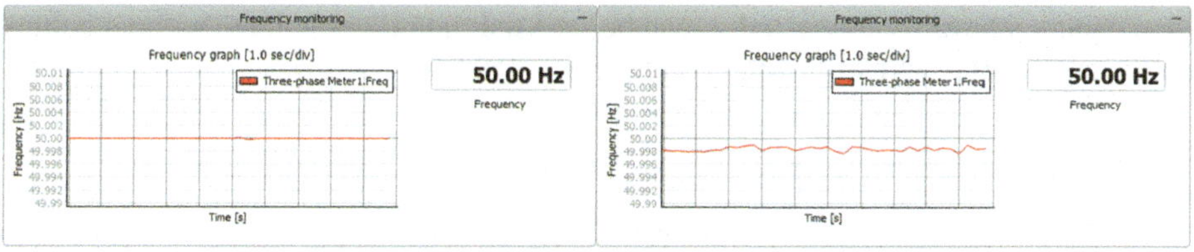

Fig. 21. Comparison of frequency for case 3.

Figure 22 presents comparison of voltage for the case 3. Left SCADA window shows voltage when battery storage is not available while right SCADA window shows voltage when battery storage is connected. It can be seen that line RMS voltage without battery is 399.33 V and with battery is 399.73 V.

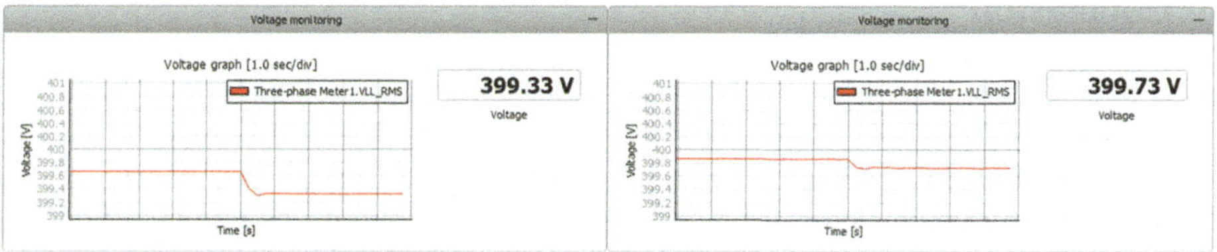

Fig. 22. Comparison of voltage for case 3.

3.4 Case 4: Load Reactive Power Is Decreased, Active Power Is Remained Same

In this case, load reactive power is decreased from 20 kvar to 0 kvar while load active power remains 40 kW. Amount of reactive power received from the grid to the battery storage is 20 kvar, battery storage is charging in this case to absorb reactive power excess in the grid caused by decreased consumption of reactive power.

Figure 23 presents comparison of frequency for the case 4. Left SCADA window shows frequency when battery storage is not available while right SCADA window shows frequency when battery storage is connected. It can be seen that with battery storage connected frequency is between 49.998 and 50 Hz.

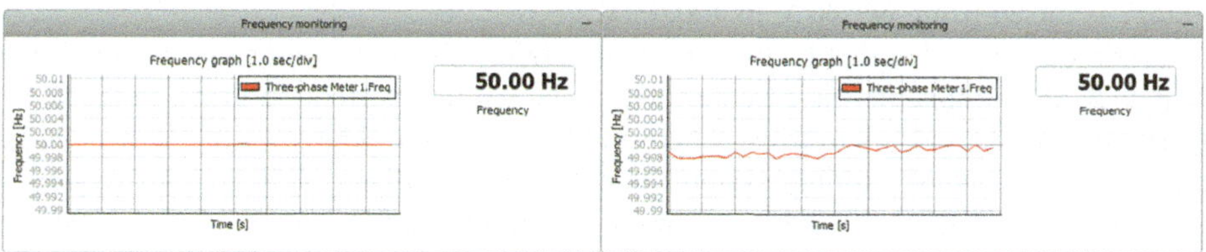

Fig. 23. Comparison of frequency for case 4.

Figure 24 presents comparison of voltage for the case 4. Left SCADA window shows voltage when battery storage is not available while right SCADA window shows voltage when battery storage is connected. It can be seen that line RMS voltage without battery storage is 400.00 V and with battery storage is 400.03 V.

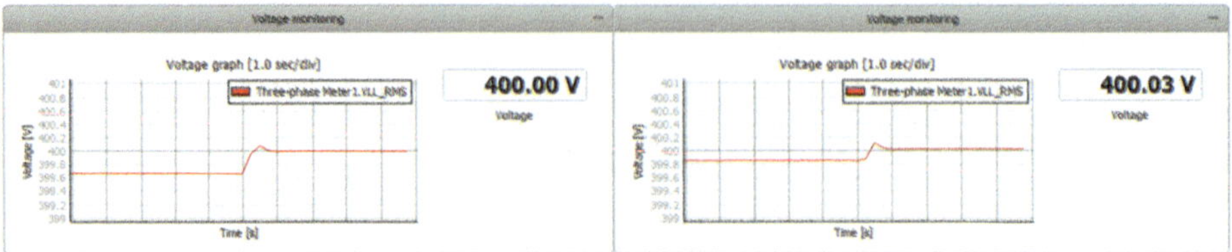

Fig. 24. Comparison of voltage for case 4.

4 Conclusion

In the future, active distribution network and microgrids with high amount of RES generation will not be possible without storage systems. Most often used storage technology on the low and medium voltage is battery storage. Battery storage is connected to the grid with the bidirectional converter with ability of control flow and amount of energy between battery storage and the grid. In this paper, BESS with ability of active and reactive power control is simulated using Typhoon HIL software.

According to the simulation results, impact of the connecting battery storage on the grid frequency and the voltage can be seen. Connecting battery storage causes slightly lower frequency than without battery storage and leads to frequency oscillating for 0.001 to 0.002 Hz. Reason for that is the output power of battery storage which is not constant because battery current is calculated for every simulation step. Changing of battery current in every step lead to output power changes whole time of simulation.

Voltage is slightly higher with battery storage than without battery storage in all four scenarios. Impact of the battery storage on the grid voltage is not significant and the highest voltage increase is visible in case 3 where the voltage with connected battery storage is for 0.3 V higher than without battery storage. In case 3, without battery storage, the voltage drop is from 399.67 V to 399.33 V and with the connected battery storage, the voltage drop is from 399.86 V to 399.73 V. In case 4, without battery storage, the voltage is raised from 399.67 V to 400.00 V and with the connected battery storage, the voltage is raised from 399.86 V to 400.03 V.

It can be concluded that battery storage systems have not significant impact on the frequency and voltage in this case because of the strong distribution grids. In the weak distribution grids and microgrids, where the load amount changes affect grid voltage and frequency, battery storage systems can lead to significant voltage and frequency improvements. In the microgrids and weak distribution grids, battery storage systems have ability to regulate grid voltage with possibility of reactive power regulation. In the strong distribution grids, transformer voltage tap and capacitor banks have more significant impact on grid voltage than BES. Installation and maintenance costs of BES is much higher than costs of capacitor banks and transformer voltage tap which is one additional reason for not using BES for voltage regulation.

Acknowledgement. This work was funded by the European Union through the European Regional Development Fund's Operational Programme Competitiveness and Cohesion under project KK.01.1.1.04.0034 Connected Stationary Battery Energy Storage.

References

1. Indu, V., Bindumol, I.K.: A hybrid photovoltaic and battery energy storage system with P-Q and V-f control strategies in microgrid. In: Proceedings of 2015 International Conference on Power, Instrumentation, Control and Computing (PICC). pp. 1–6. IEEE, Thrissur, India (2015). https://doi.org/10.1109/PICC.2015.7455747
2. Adhikari, S., Li, F.: Coordinated V-f and P-Q control of solar photovoltaic generators with MPPT and battery storage in microgrids. IEEE Trans. Smart Grid **5**(3), 1270–1281 (2014). https://doi.org/10.1109/TSG.2014.2301157
3. Giannoutsos, S.V., Manias, S.N.: A cascade control scheme for a grid connected Battery Energy Storage System (BESS). In: Proceedings of 2012 IEEE International Energy Conference and Exhibition (ENERGYCON), pp. 469–474. IEEE, Florence, Italy (2012). https://doi.org/10.1109/EnergyCon.2012.6348200
4. Nayak, P., Patel, T., Munda, N., Mallick, R.K.: Optimal P-Q control of solar photovoltaic based microgrids with battery storage. In: Proceedings of 2019 Innovations in Power and Advanced Computing Technologies (i-PACT), pp. 1–6. IEEE, Vellore, India (2019). https://doi.org/10.1109/i-PACT44901.2019.8959518
5. Gira, N., Dahiya, A.K.: Solar PV-BES in distribution system with novel technique for DC voltage regulation. Eng. Sci. Technol. Int. J. **23**(5), 1058–1067 (2020). https://doi.org/10.1016/j.jestch.2020.01.004
6. Chakir, A., et al.: Optimal energy management for a grid connected PV-battery system. Energy Rep. **6**(3), 218–231 (2019). https://doi.org/10.1016/j.egyr.2019.10.040
7. Rezvani, A., Khalili, A., Mazareie, A., Gandomkar, M.: Modeling, control, and simulation of grid connected intelligent hybrid battery/photovoltaic system using new hybrid fuzzy-neural method. ISA Trans. **63**, 448–460 (2016). https://doi.org/10.1016/j.isatra.2016.02.013
8. Nath, S., Wu, J.: Online battery scheduling for grid-connected photo-voltaic systems. J. Energy Storage **31** (2020). https://doi.org/10.1016/j.est.2020.101713
9. Rouco, L., Sigrist, L.: Active and reactive power control of battery energy storage systems in weak grids. In: Proceedings of 2013 IREP Symposium Bulk Power System Dynamics and Control - IX Optimization, Security and Control of the Emerging Power Grid, pp. 1–7. IEEE, Rethymno, Greece (2013). https://doi.org/10.1109/IREP.2013.6629422
10. Zhang, W., Xu, D., Lou, X., Yan, W., Yang, W.: Power management of battery energy storage system using model free adaptive control. In: Proceedings of 2018 IEEE 7th Data Driven Control and Learning Systems Conference (DDCLS), pp. 798–801. IEEE, Enshi, China (2018). https://doi.org/10.1109/DDCLS.2018.8516079
11. Lawder, B.M.T., et al.: System (BESS) and Battery Management System (BMS) for grid-scale applications. Proc. IEEE **102**(6), 1014–1030 (2014)
12. Gundogdu, B., Gladwin, D.T.: Bi-directional power control of grid-tied battery energy storage system operating in frequency regulation. In: Proceedings of 2018 International Electrical Engineering Congress (iEECON). IEEE, Krabi, Thailand (2018). https://doi.org/10.1109/IEECON.2018.8712259
13. Sagara, M., Sediqi, M.M., Senjyu, T., Danish, M.S.S., Funabashi, T.: Voltage stability improvement by optimal active power and reactive power output control of storage battery system. In: Proceedings of 2016 IEEE Region 10 Conference (TENCON), pp. 2671–2674. IEEE, Singapore (2016). https://doi.org/10.1109/TENCON.2016.7848523

14. Wang, W., He, W., Cheng, J., Huang, X., Liu, H.: Active and reactive power coordinated control strategy of battery energy storage system in active distribution network. In: Proceedings of 2017 32nd Youth Academic Annual Conference of Chinese Association of Automation (YAC), pp. 462–465. IEEE, Hefei, China (2017). https://doi.org/10.1109/YAC.2017.7967453
15. Behera, M.P., Ray, P.K., Beng, G.H.: Single-phase grid-tied photovoltaic inverter to control active and reactive power with battery energy storage device. In: Proceedings of 2016 IEEE Region 10 Conference (TENCON), pp. 1900–1904. IEEE, Singapore (2017). https://doi.org/10.1109/TENCON.2016.7848352
16. Jana, A., Gayen, P.K., Dhara, P.K., Garai, R.: Simultaneous active and reactive power control of single-phase grid connected battery storage system. In: Proceedings of 2017 Devices for Integrated Circuit (DevIC), pp. 289–293. IEEE, Kalyani, India (2017). https://doi.org/10.1109/DEVIC.2017.8073954

A Novel Method for Cyber-Maintenance Prioritisation of Ageing Distributed Power Sources

Dalibor Gernhardt[1(✉)] and Ružica Kljajić[2]

[1] Croatian Defence Academy, Ilica 256b, 10000 Zagreb, Croatia
dalibor.gernhardt@morh.hr

[2] Faculty of Electrical Engineering, Computer Science and Information Technology Osijek, Kneza Trpimira 2B, 31000 Osijek, Croatia
ruzica.kljajic@ferit.hr

Abstract. Today's technology allows remote control of industrial processes, which can include distributed energy sources such as solar or wind power. As these remote control systems age, security updates for them will eventually cease. This means that any vulnerabilities found after end of service life cannot be patched, often requiring physical replacement of the equipment. The second problem is the fact that these systems are usually integrated with other devices and cannot be easily replaced. Looking at the lifespan of industrial systems, it is clear that IT devices used for remote control usually have a shorter lifespan than the rest of the system. Inspiration to investigate how poor maintenance of IT devices could be abused in the future was taken from the cyber-attacks on the Ukrainian power grid, where attackers took control of power switches relays equipment and gave the command to open circuit breakers, disconnecting loads from the power grid, resulting in regional blackouts. Using a similar approach, it was hypothesised that distributed power sources may become vulnerable to cyber-attacks as they age. Since the installed power of distributed power sources per unit is relatively small, it is plausible for the attacker to identify multiple closely located distributed systems and conduct a parallel cyber-attack by disconnecting them from the power grid simultaneously during peak demand periods, with the goal of disrupting the power grid. In this work, distributed power sources that are connected to the Croatian distribution network are identified, mapped and the spatial correlation of the sources was analysed using GIS software. The goal is to develop methodology which enables easy identification of power sources that are spatially close to each other and have similar age, which makes them a likely target for potential attacks. This information can be used to prevent such attacks in the future, as it would help identify critical groups of such sources and focus maintenance efforts.

Keywords: Cybersecurity · Legacy system maintenance · Remote access vulnerabilities

1 Introduction

How cyberspace and remote access to power facilities can be exploited is best illustrated by the case of two cyber-attacks in Ukraine that occurred in 2015 and subsequently in

H. Glavaš et al. (Eds.): OTO 2021, LNNS 369, pp. 205–216, 2022.
https://doi.org/10.1007/978-3-030-92851-3_15

2016, where attackers took control of power substations and disconnected loads from the power grid. Since these attacks, security has increased, and today, at least in the EU, critical infrastructure companies such as power plants or Transmission systems operators (TSO), are required by law to implement multiple cyber security measures. Unfortunately, this cannot be said for smaller companies that do not have the resources to implement such security measures.

It is a common paradigm that the risk of a coordinated cyber-attack on such companies is low. This may be the case for small businesses focused on manufacturing, but the ever-growing renewable energy sources (RES) could be exploited in a different way - by a simultaneous attack on multiple distributed systems in a given geographic area. Most of those facilities are automated and have an option for remote access and management via the Internet. This option appears to be the most cost-effective, as monitoring and control can be done remotely without the need for physical access; the downside of this approach is the introduction of additional risks.

The reason for this is that small distributed systems are not really a problem as their power output is not significant enough to make real impact on power grid stability. This may be true for photovoltaic systems, but the output of wind farms can be close to the balancing power in the system and a sudden outage of one wind farm may not be a problem. However, a parallel outage of multiple wind farms due to a coordinated cyber-attack may be a problem. Even a sudden and coordinated outage of even smaller units can cause problems in distribution networks.

As these renewable energy plants age, so does the IT equipment responsible for remote management. This equipment which is not without its flaws and as it ages, it becomes more likely that vulnerabilities will be found. It is important to emphasize that even specialized equipment has publicly known critical vulnerabilities (e.g. industrial gateway with critical vulnerability [1]) and must be patched regularly. If not properly maintained, they pose a security risk. The responsibility for patching lies with the people who maintain these systems, and it is reasonable to assume that some of updates will not be installed for fear of disrupting ongoing operations - “if it ain’t broke, don’t fix it”.

As mentioned earlier, most of those systems are not large enough to cause problems on their own, but a coordinated outage of multiple of such systems can be perceived as possible threat. In this work, a method for identifying such systems is presented, along with potentially vulnerable locations - clusters of locations with potentially outdated equipment. The aim is to make contribution to prioritization of maintenance and raise awareness by showing that this can be done with using only publicly available data, data that is also accessible to potential attackers.

Literature regarding power grid cybersecurity can be found in both, literature dealing with Industrial Control Systems (ICS) which are essential part of power systems and specialized cybersecurity literature of power grids. Most notable works dealing with concepts and potential vulnerabilities of ICS cybersecurity are [2, 3]. Secondly, literature is often further divided in cybersecurity of power grids as such and additional concerns dealing with smart grids security. Cybersecurity risks and challenges in power grids are discussed in [4, 5]. Most of recent works specializes in security concerns of smart metering and smart grids [6–10]. As visible, works are focused on either ICS or smart grids with, at least to author knowledge, no works dealing with vulnerabilities originated from usage of obsolete equipment and potential scenarios where attackers are trying to cause power grid stability attacks. Authors were able to identify only few works are dealing with cybersecurity topics in Croatia [11–13], none of these works focuses on cybersecurity of Croatian power grid.

This work is structured as follows: In Sect. 2 Croatian regulation on Renewable energy sources is described followed by examples of what attackers have done in past in Sect. 3. Section 4 describes methodology. Results are presented in Sect. 5 followed by discussion in Sect. 6 and conclusion.

2 Renewable Energy Sources in Croatia

Every year, RES increase their share in the structure of electricity production thanks to a global policy that encourages their use. The Republic of Croatia, at that time in the negotiations for accession to the European Union, and today a full member, is a signatory to the European Directive 2009/28/EC, thanks to which there have been positive developments in the field of RES.

According to the Ministry of Economy, in 2007 a total of 51,96 MW electric power produced from RES was installed in Croatia [14]. The year 2007 was taken as the starting year in data comparison, given that a number of laws defining the electricity market were adopted that year, i.e. sublaws related to RES. These sublaws are as follows: Ordinance on the usage of renewable energy sources and cogeneration (Official Gazette 67/07), Regulation on a mini-mum share of electricity produced from renewable energy sources and cogeneration in the electricity supply (Official Gazette 33/07), Regulation on the fee for the promotion of the electricity production from renewable energy sources and cogeneration (Official Gazette 33/07), Ordinance on the obtaining the eligible electricity producer status (Official Gazette 67/07) and Tariff system for the production of electricity from renewable energy sources and cogeneration (Official Gazette 33/07 [14]. These sublaws have led to the opening and liberalization of the electricity market and the emergence of new market participants. Electricity market, which was fully open on 1 July 2008 is now organized as bilateral market [15]. The incentive system has been replaced by a new law in the system of feed-in premiums [16], but at the moment a large part of these first privileged producers are still in the system of incentives. For this reason, we focused on the old law.

Table 1 shows statistical data on installed capacities for individual technology for years 2007 [14] and 2019 [17]. As it can be seen according to a report given by Croatian Energy Regulatory Agency, in Croatia there is 877,4 MW of installed power in 1347 power plants in incentive system. It can also be noted that there is greater variety of technologies compared to 2007.

This increase in numbers is significant as producers have to go through a long and complicated procedure to become eligible producers. Producers who were among the first to acquire this status, and which are covered by sublaws adopted in 2007, have concluded purchase agreements, during which the purchase of total energy produced at the price defined by the tariff system is guaranteed. For producers, it meant that anyone can become eligible producer after complying with conditions and regulations.

Table 1. RES installed power in 2007 and 2019 [14, 17]

	Installed power [MW]	
	2007	2019
Photovoltaic power plants	0,04996	53,4
Hydro power plants	32,76	5,9
Wind power plants	17,15	575,8
Biomass power plants	2	73,7
Geothermal power plants	0	10
Biogas power plants	0	42,7
Landfill gas power plant and gas from the wastewater treatment plant	0	2,5
Cogeneration	0	113,3
Other	0	53,4
Total	51,96	877,4

In 2017, the Law on renewable energy sources and high-efficiency cogeneration has been passed which led to forming Register of Renewable Energy Sources and Cogeneration and eligible producers (cro. Registar obnovljivih izvora energije i kogeneracije te povlastenih proizvodaca, OIEKPP), further known as Register OIEKPP [16, 18]. Register gives an overview on projects for production of energy from RES, together with the executed projects and projects in the planning phase and procedures for obtaining final permits. For producers this law meant departure from the previous system and made them more sensitive to market prices.

In bulk system, power flows are directed from the power plant, through the transmission grid to the distribution grid and finally to the consumer. Most of small RES are connected to distribution grid which results in bidirectional power flows. Other effect of RES penetration is increase in voltage. In long, loaded lines, the RES integration leads to an improvement in the voltage profile and increasing voltage at the end of the line. In every moment, produced active power needs to be equal to the required active power. In order to keep this balance, sufficient active power needs to be provided in the system. RES are stochastic so in each moment sufficient back-up needs to be ready in order to avoid consumer disconnection in the system. If RES reduces its output or disconnects, another power plant needs to deliver additional power to keep system stable. Also, RES integrated in distribution grid locally increases voltage. In rural locations this can be great measure to improve voltage profile for the consumers at the end of the line. However, unexpected failure due intentional or unintentional disconnection can cause voltage decrease. To avoid unwanted side effects, protection systems are installed and the whole system is monitored by using SCADA.

3 Cyber-Attacks on Power Grid

Knowing that eligible producers obtain mid-term contract, which guarantees purchase of electricity at a preferential price for the entire duration of the contract, the question of the obligations of producers with regard to the maintenance and improvement of their infrastructure is raised. Since duration of contracts for eligible producers is 12 to 14 years, this means that contracts for the first projects are finished. Unfortunately, there is no legal obligation to force them to engage in interventions that can improve their communication systems. Most of power plants built in mid and late 2000' have communication systems which are now outdated. In the new law, this topic remains uncovered, so any upgrade and maintenance of the system remains a matter of goodwill of the producer. The only reason that could force the producer to replace the equipment is the possible reduction in the efficiency of the system, which again remains the subject of economic cost-effectiveness analysis.

To emphasize possible consequences of cyber-attacks, two most notable cyber security involving energy sector that occurred in 2015 and subsequently in 2016 in Ukraine will be explained. These two attacks are very often equated in the literature, however the first attack took advantage of existing systems, while the second attack pushed the boundaries of what was considered possible and is often identified with the second Stuxnet attack [18, 20]. Although these examples are most notable, there are many similar examples and calls from experts that additional security measures are needed which are not discussed in this work [21, 22]. Generally, attackers potentially can cause damage in different ways, for example by disconnect loads from network, disabling safety equipment, changing various industrial parameters, disengage power generation units by sending various signals to switches and similar. In this example only few possible vectors of attacks are covered, additional cyber security risks in power grids are discussed in work [6].

3.1 First Ukraine Attack

Although attackers gave gained access to distribution companies enterprise and SCADA networks some 9 months prior in March, final stage of attack happened on December 23. 2015. Attackers, who have learned for months how daily operation of controllers in power distribution companies are conducted took over control over Human machine interfaces (HMI) and used existing software and procedures to disengage power lines overcurrent relays and disconnected loads from the power network. At the moment of final attack, dispatchers in three different control centres were locked out of their computers and could only observe how mouse moves on itself and sets switches to open position. At the end of attacks, controllers did not know extend of the damage because attackers have erased boot sectors form dispatchers' computers, disabled HMIs, UPS systems, and serial to ethernet converters. Attack itself lasted for 60 min, ant it took 6 h to set affected stations to manual operation mode [23, 24]. This event was not significant enough to cause major problems within stability of power grid network, but regional customers were left without access to electricity for hours.

3.2 Second Ukraine Attack

Second attack has occurred one year later. Although attack itself was unsuccessful, it showed what could have been done. In this attack, attackers have gained remote access by exploiting existing VPN connection to one power station and implanted malware that could directly interact with SCADA systems, to be more precise, malware supported following Industrial control systems (ICS) protocols: IEC 60870-5-101, IEC 60870-5-104, IEC 61850, OLE for Process Control Data Access (OPC DA). In attack malware had sent direct instruction to Siemens Siprotec relay and disengaged one power line. To make restoration of normal operation more difficult malware erased configuration of power station's SCADA network. Subsequent analysis of attack showed that malware has tried, but failed to put protective relay in state of update, a state in which no protection is active but display on relay itself shows normal operation. Without SCADA network, technicians could not know this and if attackers had managed in their goals upon manual switching of power, line would be left unprotected. It is assumed that attackers knew that Ukrainian operators will soon switch to manual operation and wanted to cause additional damage. According to available information malware had failed to do this due to bad implementation of IEC protocol. Additionally, malware had unused ability to put relay in flickering mode turning it on and off. Although this attack resulted in minor power loss it is difficult to assess what would happen if protection switches were disabled or put in flickering mode [18].

4 Methodology

The goal was to develop method for visualizing and identification of ageing cluster of renewable energy facilities that could be exploited by using only publicly available information. In this work, a method for identification of such, potentially vulnerable locations - clusters of potentially outdated equipment is introduced. Publicly available data on renewable sources connected to power grid in Croatia are used, put on map in order to try to identify clusters of aged, and possibly vulnerable equipment. The methodology can be divided as follows:

1. Identify publicly available information on renewable energy sources,
2. Match location name with coordinates,
3. Visualize results by showing facilities within one city as cluster of points,
4. Identify ageing facilities by visual methods.

Queries could have been used to group facilities with certain attributes, but they cannot give an answer if one city is close to another one.

In this paper, data obtained from the Register is used [18]. It can be noticed that there will be over 5 000 MW installed power in 4083 power plants. In addition, register contains data on: type of facility, power output, rough location – city name, and approximate age of facility can be extracted from license number where last two numbers contain a year.

Authors are aware that Register contains only approximate location of facility, and a year of a permit is not clear indicator of facility age, but this is only public information that could extract. Additionally, a search for each facility owner name and gain additional

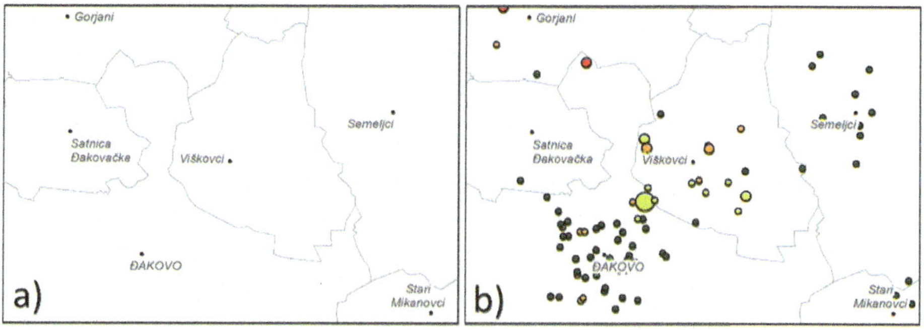

Fig. 1. a) Raw imported data, all facilities from one place are situated in single point, b) Facilities spaced apart to area 5 × 5 km so they can be visualized. Symbol size varies on power output, and colour on age with green indicates younger facilities.

Table 2. Number and installed power according to Croatian Energy Regulatory Agency and OIEKPP Register [14, 18]

	Eligible producers		Register OIEKPP	
	Number	Installed power [MW]	Number	Installed power [MW]
Photovoltaic power plants	1229	53,40	3782	307,53
Hydro power plants	14	5,90	49	1676,35
Wind Power plants	26	575,80	47	2021,45
Biomass power plants	39	73,70	119	218,32
Geothermal power plants	1	10,00	2	20,00
Biogas power plants	41	42,70	64	69,50
Landfill gas power plant and gas from the wastewater treatment plant	1	2,50	7	9,60
Cogeneration	6	113,30	12	1068,99
Other		53,40	1	1,00
Total	**1357**	**577,40**	**4083**	**5392,74**

information could be made, but due to number of queries (4 083) research was limited to city name.

To mitigate this, name of place in which facility is located have been joined with coordinates of that place by using Esri ArcMap software package “Join” function. This resulted in grouping of multiple facilities to single spot making visualization difficult as shown on Fig. 1a). To improve visibility, an option “spraying coordinates” was used. For each facility random number to coordinates was added in such manner to visualize them in area of 5 × 5 km instead of single point as shown in Fig. 1b). Parameter of 5 km was chosen after couple of iterations, and it seemed as suitable for visualization of Croatian data. This process resulted in fact that facilities are shown as randomly generated cluster of points instead of single point.

True location of facilities is unknown, only available data is place name so adding random number to coordinates significantly improves visualisation of data. Next, a symbols can be applied where color of symbol depends on facility age and size on power output. Division of levels and colors can depend on case by case scenario.

5 Results

Using GIS software data derived in previous step can be visualized in such manner that each facility symbol size varies on its power output, and color represents age of facility with green standing for young and unassumingly safe systems, and yellow to red indicated older systems which are more likely to contain old, unpatched equipment. As visible in Fig. 2, this method enables quick visualisation of both power output and age of facility. Second observation that can be made from both Fig. 2 and Table 2 is that power output of wind farms is orders of magnitude larger compared to other sources. Number of wind farms is relatively low and they can easily be analysed manually without necessity of this proposed method, to this reason in following sections focus will be on other renewable sources, sources whose grupation can make impact on power grid.

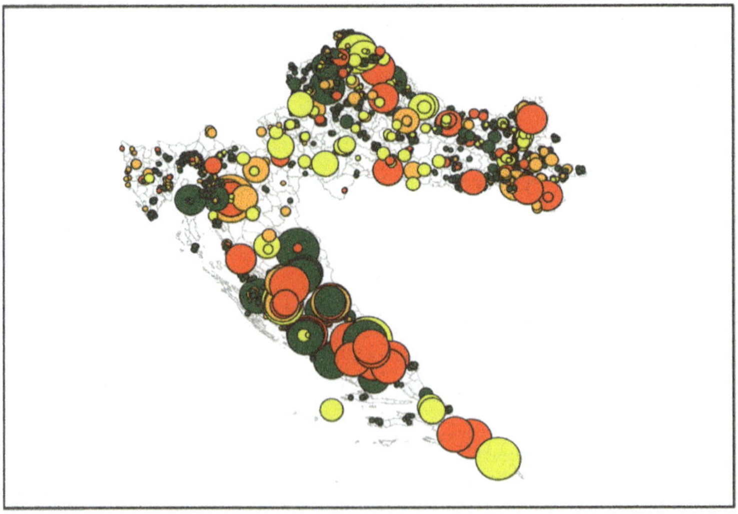

Fig. 2. Visualization of Croatian data on of all renewable sources depending on power output and age.

Examination of visualised data showed areas where ageing facilities are located in same places. For example, city of Slatina has 22 MW of installed power with average age of 11,6 years. Similar observations apply for places of Virovitica and Grubišno Polje. This data can be useful, but it can be derived without visualisation of data.

Upon close examination of results four clusters that meet following conditions are identified and can be easily identified using proposed methodology:

1. Contain facilities which are not situated in same place as this information can be extracted by query,
2. That are spatially close to each other,
3. Contain facilities with power output greater than 500 kW,
4. Contain facilities older than 3 years.

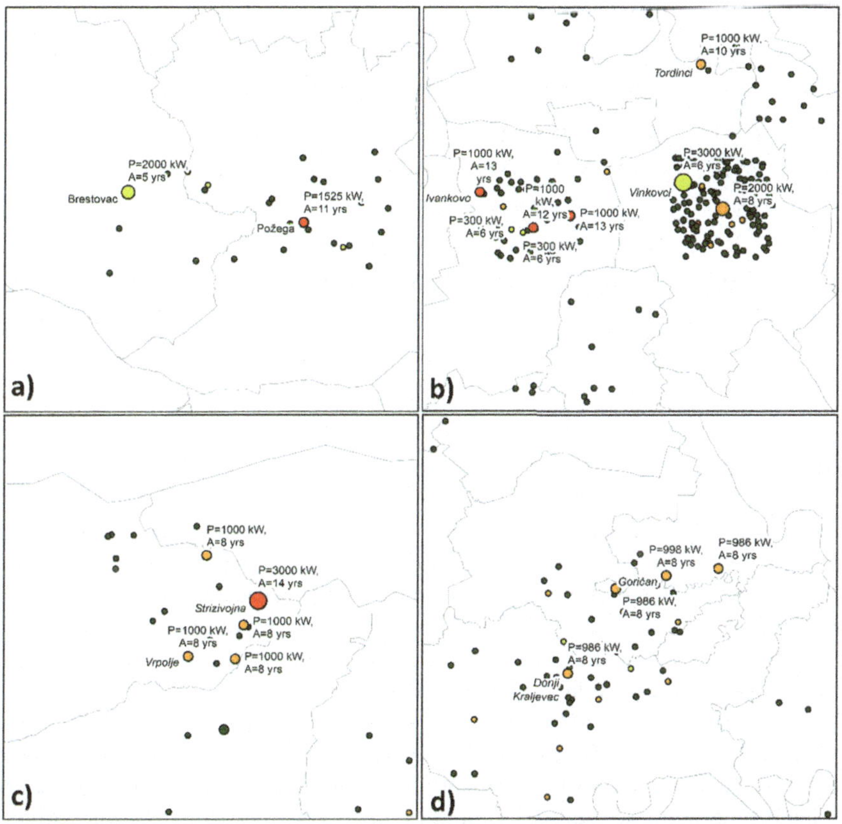

Fig. 3. Visualization of identified clusters of ageing distributed power sources with output less than 5 MW.

Identified areas containing clusters of potentially vulnerable facilities are presented visually in Fig. 3. These facilities are situated in different places and often in different counties. Without this type of spatial visualization their identification can be difficult. This data can be enriched with data presented in Table 3 which gives an detail overview of these location and provides an information on population of those areas. These clusters are 1. Brestovac & Požega, 2. Ivankovo & Vinkovci & Tordinci, 3. Strizivojna & Vrpolje, and 4. Donji Kraljevec & Goričan. Once again it should be emphasised that these facilities are not necessarily vulnerable to attack, but due to their age and grouping they could be identified as potential target.

Table 3. Identified clusters of ageing distributed power sources with output less than 5 MW. Number in brackets indicates population according to Census of 2011. Legend: BM - BioMass, BG - BioGass, PV - Photovoltaic, CG – CoGeneration

Strizivojna (2525), Vrpolje (3521)	Vinkovci (35312), Ivankovo (8006), Tordinci (2032)	Goričan (2823), Donji Kraljevec (4659)	Brestovac (3726), Požega (28248)
BM, 3 MW, 14 yrs	BG, 1 MW, 13 yrs	PV, 0,986 MW, 8 yrs	BM 1.5 MW, 11 yrs
BM, 1 MW, 8 yrs	CG, 1 MW, 13 yrs	PV, 0,986 MW, 8 yrs	BG, 2 MW, 5 yr
BM, 1 MW, 8 yrs	BG, 1 MW, 12 yrs	PV, 0,986 MW, 8 yrs	
BM, 1 MW, 8 yrs	BG, 1 MW, 10 yrs	PV, 0,986 MW, 8 yrs	
BM, 1 MW, 8 yrs	BG, 2 MW, 8 yrs		
	BM, 3 MW, 6 yrs		
SUM: 7 MW AVG age: 9.2 yrs	**SUM: 9 MW AVG age: 10,3 yrs**	**SUM: 3,956 MW AVG age: 8 yrs**	**SUM: 3,25 MW AVG age: 9 yrs**

6 Discussion

When dealing with thousands of distributed power sources, where each of them has it's owner, detection of cyber related risks is difficult. Outage of distributed power generation unit situated in large town or industrial area with great electric energy consumption does not have impact as one outage of same unit in rural area with weak power grid. This was main motive, how to identify relatively large clusters of distributed power sources, sources that pose greater threat compared to other ones.

Presented methodology showed current state of affairs based on publicly available data. As exact locations of facilities are not easily accessible this method shows that even imprecise data can be useful for prioritization of maintenance plans. This work does not go in detail on how these facilities should be maintained nor suggests that vulnerable remote connections are present on those location, only that these locations are likely to have an older and vulnerable equipment and if they would be attacked in conjunction they outage could impact grid in some manner.

As expected, results showed in Table 3 show that photovoltic power plants present smallest risk at this moment. At least in this test case made for Croatia, oldest facilities are Biogass, Biomass and Cogeneration plants. Reason for such results can be found in greater cost-effectiveness of such plants compared to other plant types. Secondly, this analysis is based with referent year of 2021, using this method it can be easily visualised how would map look in years to come. Facilities which are considered young and safe today will turn from green to yellow and will require additional cyber security checks and maintenance.

7 Conclusion

Distributed power sources are a reality today and their share will increase in the future. These systems are often automated and prone to cybersecurity risks. With each passing day, new, previously unknown vulnerabilities are detected meaning that old systems need to be patched or updated. If left unpatched, they pose a security risk, and one day someone could exploit these vulnerabilities and cause damage. Due to the number of distributed power sources, this task is becoming increasingly difficult, requiring prioritisation of efforts. As the power output of distributed power sources varies, so does the significance of that source. It is not the same if some hacker disables small unit in a large city or a biomass facility in a rural area. The same power outage can have different impacts in different areas.

In this work, the following contribution is made: a methodology that helps in prioritising maintenance activities and enables rapid identification of facilities whose outage can cause most significant impact. Using this methodology, by applying only publicly available data, it is possible to give an quick assessment of: age, power output, and clusters of such ageing facilities - locations where multiple ageing power generation units are located close to each other and can cause greater damage if shut down at the same time than if analysed separately. To put this method to the test, a Croatian power grid was analysed and four critical locations were identified. As time goes by and facilities age, the number of this locations is likely to increase.

References

1. Tenable.com: Teltonika Gateway TRB245 Multiple Vulnerabilities, Research Advisories (2020). https://www.tenable.com/security/research/tra-2020-48. Accessed 17 Aug 2021
2. Drias, Z., Serhrouchni, A., Vogel, O.: Analysis of cyber security for industrial control systems. In: Proceedings of the 2015 International Conference: Cyber Security of Smart Cities, Industrial Control System and Communications. SSIC 2015 (2015). https://doi.org/10.1109/SSIC.2015.7245330
3. McLaughlin, S., et al.: The cybersecurity landscape in industrial control systems. Proc. IEEE **104**(5), 1039–1057 (2016). https://doi.org/10.1109/JPROC.2015.2512235
4. Sun, C.C., Hahn, A., Liu, C.C.: Cyber security of a power grid: state-of-the-art. Int. J. Electr. Power Energy Syst. **99,** pp. 45–56 (2018). https://doi.org/10.1016/j.ijepes.2017.12.020
5. Kaster, P., Sen, P.K.: Power grid cyber security: challenges and impacts. In: 2014 North American Power Symposium, NAPS 2014, pp. 1–6 (2014). https://doi.org/10.1109/NAPS.2014.6965424
6. Martins, R., Knob, L.A.D., da Silva, E.G., Wickboldt, J.A., Schaeffer-Filho, A., Granville, L.Z.: Specialized CSIRT for incident response management in smart grids. J. Netw. Syst. Manage. **27**(1), 269–285 (2018). https://doi.org/10.1007/s10922-018-9458-z

7. Skopik, F., Friedberg, I., Fiedler, R.: Dealing with advanced persistent threats in smart grid ICT networks. In: 2014 IEEE PES The Innovative Smart Grid Technologies (ISGT) Conference, ISGT 2014 (2014). https://doi.org/10.1109/ISGT.2014.6816388
8. Skopik, F., Ma, Z., Bleier, T., Grüneis, H.: A survey on threats and vulnerabilities in smart metering infrastructures. Int. J. Smart Grid Clean Energy **1**, 22–28 (2012). https://doi.org/10.12720/sgce.1.1.22-28
9. Clements, S., Kirkham, H.: Cyber-security considerations for the smart grid. In:IEEE PES General Meeting, PES 2010 (2010). https://doi.org/10.1109/PES.2010.5589829
10. Acharya, S., Dvorkin, Y., Pandzic, H., Karri, R.: Cybersecurity of smart electric vehicle charging: a power grid perspective. IEEE Access **8**, 214434–214453 (2020). https://doi.org/10.1109/ACCESS.2020.3041074
11. Škiljić, A.: Cybersecurity and remote working: Croatia's (non-)response to increased cyber threats. Int. Cybersecur. Law Rev. **1**(1–2), 51–61 (2020). https://doi.org/10.1365/s43439-020-00014-3
12. Galinec, D., Možnik, D., Guberina, B.: Cybersecurity and cyber defence: national level strategic approach. Automatika **58**(3), 273–286 (2017). https://doi.org/10.1080/00051144.2017.1407022
13. Gernhardt, D.: Vulnerability assessment of the Croatian cyberspace to information warfare campaign via means of malicious websites comments. Strategos **4**(2), 39–65 (2020)
14. Energy in Croatia, Annual Energy Report 2007. Report, Republic of Croatia, Ministry of Economy, Labour and Entrepreneurship (2008). https://mingor.gov.hr/UserDocsImages/UPRAVAZAENERGETIKU/EnergijauHrvatskoj/EnergijauHrvatskoj2007.pdf. Accessed 10 Aug 2021
15. Croatian Transmission System Operator, Electricity market, https://www.hops.hr/en/electricity-market. Accessed 10 Aug 2021
16. Law on renewable energy sources and high efficiency cogeneration, Official Gazette of the Republic of Croatia, no. 100/2015., 123/16, 131/17, 111/18 effective from 20.12.2018
17. Croatian Energy Regulatory Agency. Annual Report on the Activities of the Croatian Energy Regulatoray Agency for 2019 (2020). https://www.hera.hr/en/html/annualreports.html. Accessed 10 Aug 2021
18. Ministry of Economics and Sustainable Development, Register of Renewable Energy Sources and Cogeneration and eligible producers. https://oie-aplikacije.mzoe.hr/pregledi/PopupIzvjestaj.aspx?ReportId=5b47346e-67aa-4df2-9603-fa83c47061e3#Pd91d581baf3245bdb28c6f1efc139f932173iT1R0x0. Accessed 10 Aug 2021
19. Slowik, J.: Crashoverride: Reassessing the 2016 Ukraine Electric Power Event as a Protection-Focused Attack, Dragos Inc. (2019)
20. Falliere, N., Murchu, L.O., Chien, E.: W32.Stuxnet Dossier. Cupertino (2011)
21. Seals, T.: Threatpost: solar, wind power utility disrupted in rare cyberattack (2019). https://threatpost.com/solar-wind-power-utility-cyberattack/149816/. Accessed 10 Aug 2021
22. Freudenberg, W.: DNV - Digital solutions: why windfarms need to step up cyber security (2020). https://www.dnv.com/article/why-windfarms-need-to-step-up-cyber-security-128082. Accessed 14 Aug 2021
23. Whitehead, D.E., Owens, K., Gammel, D., Smith, J.: Ukraine cyber-induced power outage: analysis and practical mitigation strategies. In: 2017 70th Annual Conference for Protective Relay Engineers (CPRE), pp. 1–8 (2017). https://doi.org/10.1109/CPRE.2017.8090056
24. Lee, R.M., Assante, M., Conway, T.: Analysis of the Cyber Attack on the Ukrainian Power Grid. Washington DC (2016)

Product Life Cycle Analysis on the Example of a Home Appliance

Domagoj Skorup[1], Goran Rozing[1](✉), Goran Heffer[2], and Ivan Vidaković[2]

[1] Faculty of Electrical Engineering, Computer Science and Information Technology, Kneza Trpimira 2B, 31000 Osijek, Croatia
goran.rozing@ferit.hr

[2] Faculty of Agrobiotechnical Sciences, Vladimira Preloga 1, 31000 Osijek, Croatia

Abstract. Today's industrial production, with the constant growth of the world's population, has a great impact on the environment loading and human health in the world. With the development of the industry, the need for large amounts of energy and materials have increased sharply, which has led to an increase in greenhouse gas emissions, which are the main cause of global warming on Earth. To reduce the impact of harmful gas emissions and global warming in the product life cycle, a new approach called LCA (*Life-cycle assessment*) has emerged. In the experimental part of the paper, the analysis of the impact of the product life cycle on the environment and human health was performed on the example of two household appliances using the software tool SimaPro. The conducted LCA analysis of two household appliances showed a significant impact of proper material selection at an early stage of construction on reducing the impact of production and products on the environment and human health.

Keywords: Industrial production · Life cycle analysis · Reduction of environmental impact

1 Introduction

The growth of the industry is significantly influenced by the development of new technologies and the modernization of existing industrial plants. This increase has resulted in faster production processes, reduced worker error rates, better material utilization, and improved the quality of finished product. On the other hand, the above mentioned development of industry has led to the need for larger amounts of energy and materials which has caused a sharp increase in emissions of harmful gases that are the main cause of global warming on Earth. To reduce the impact of greenhouse gas emissions on global warming caused by various production processes, a new approach called LCA (Life-cycle assessment) has been developed [1]. To reduce the harmful impact of man and industry on the environment, it is necessary to implement measures to reduce the consumption of energy and materials, but also to reduce the release of harmful substances into the environment. One of the most commonly used methods in the process of reducing the harmful impact of production on the environment is the LCA method. It is used to collect information,

H. Glavaš et al. (Eds.): OTO 2021, LNNS 369, pp. 217–227, 2022.
https://doi.org/10.1007/978-3-030-92851-3_16

document and monitor the life cycle of a product and its impact on the environment at all stages of life. The numbers obtained from product life cycle assessment data are called eco-indicators. Eco-indicator 99 is a method that greatly facilitates the work of designers, constructors, engineers and many others in creating new products. With the help of specific numbers obtained by Eco-indicators, it is much easier for designers to compare two or more products and decide for those production processes and materials that are less burdensome for the environment [2]. This paper briefly describes the LCA method and the Eco indicator-99 by which the ecological analysis is performed. On the example of two electrical household appliances (kettle), the most significant indicators of the impact of these products on the environment were compared. The software tool SimaPro 9.0 was used to determine the indicators.

2 LCA Method or Complete Balancing

Complete balancing or eco-balancing (LCA) is a method that enables monitoring and documentation of the impact of all phases of the product life cycle on the environment. The life cycle of each new product begins with the exploitation of the raw material from nature from which the product will be made and its design and construction "birth". Then the product comes to the stage "maturity" where it is distributed to end users who then use it for a certain time period. After a certain period of use of the product due to wear and tear or defect, it is disposed of or placed in the recycling process "death". Product life, origin, use and disposal are also called "cradle-to-grave" [3]. To effectively perform the LCA analysis, input and output data are collected for each of the above mentioned life cycle phases as shown in Fig. 1.

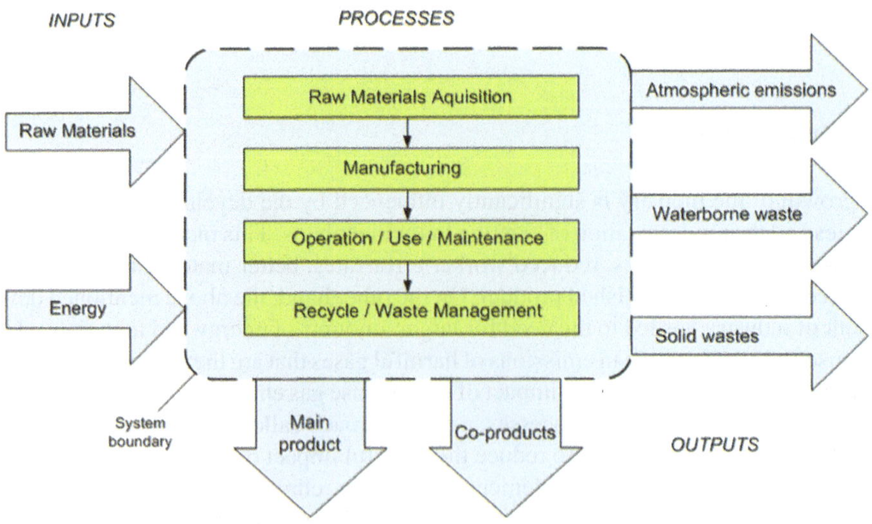

Fig. 1. Product life cycle input and output flows [4].

Products mostly have a negative impact on the environment due to the depletion of natural resources and energy during their creation and during use. However, there are also products that, in addition to the negative impact, also have a positive impact. Such an impact occurs on a product if it produces recycled material or energy during its life cycle.

2.1 Eco-indicators

Eco-indicators are indicators that take into account one or more areas of environmental impact, and the higher the amount of indicators, the greater the environmental impact. The values of eco-indicators have no dimension, so they are expressed as points or milli points. The basic meaning of eco-indicators is the comparison of two products or parts of products so that the unit of measurement is not crucial.

The eco-indicator can be understood as an upgrade of the LCA method. The foundation on which the eco-indicator 99 rests is the LCA method. The impact of products on the environment is observed through three areas where they can cause damage [5], which can be seen in Fig. 2. These three areas are: human health (this refers to the number and duration of diseases and the years lost due to premature death), ecosystem quality (implying impact on species diversity, effects included are hazardous substances emissions, acidity, eutrophication and soil exploitation), resources (especially the excess energy consumption that will be needed in the future to extract poorer mineral and fossil resources).

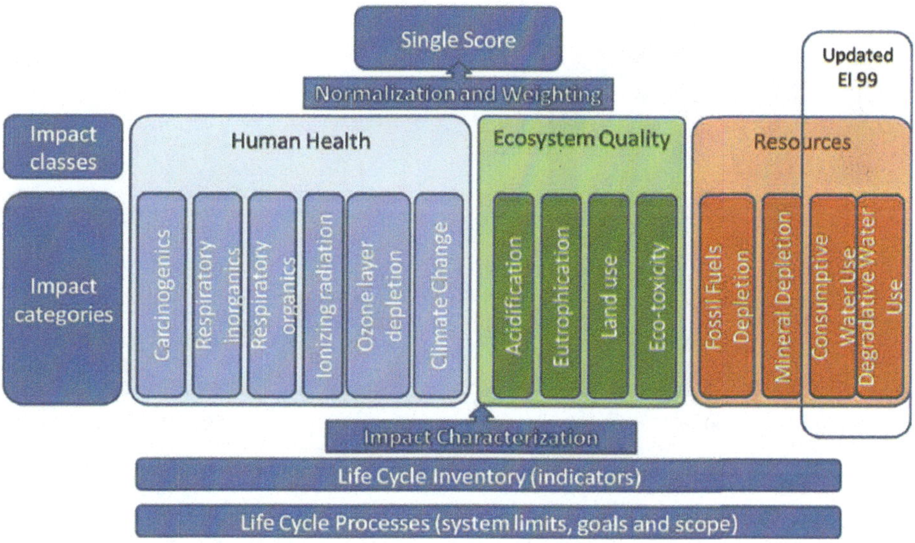

Fig. 2. Eco-Indicator 99, environmental impact categories [6].

To speed up the analysis phase, eco-indicators have been calculated in advance for individual product life cycle phases and are available in the databases of LCA analysis software tools.

Such pre-calculated indicators and eco-indicator points are available for: materials (including all stages of raw material production from ore extraction to the last stage of production and transport within the material production process), production processes (indicators refer to emissions of the process itself, but also all emissions to which occurs due to the generation of required energy), transport processes (refers to vehicle emissions, but also those arising from obtaining the required fuel or energy), energy production processes (indicators relate to obtaining and producing fuel, energy transmission and obtaining electricity), disposal (different products are disposed of in different ways, so it is necessary to think carefully when choosing the indicator which method of disposal is the most appropriate).

Although the purpose of standard eco-indicators is to facilitate the calculation of the impact of products on the environment, the process itself is quite complex and requires a good knowledge of materials and production processes. Many errors are possible because the procedure is based on a large number of assumptions.

3 LCA Analysis on the Example of Two Different Electric Kettles

In the first step before the LCA analysis in the SimaPro software tool, the procedure of disassembly of all assemblies and parts was performed, as well as the list of materials used in the production of each electric kettle. The analyzed electric kettles before disassembly are shown in Fig. 3.

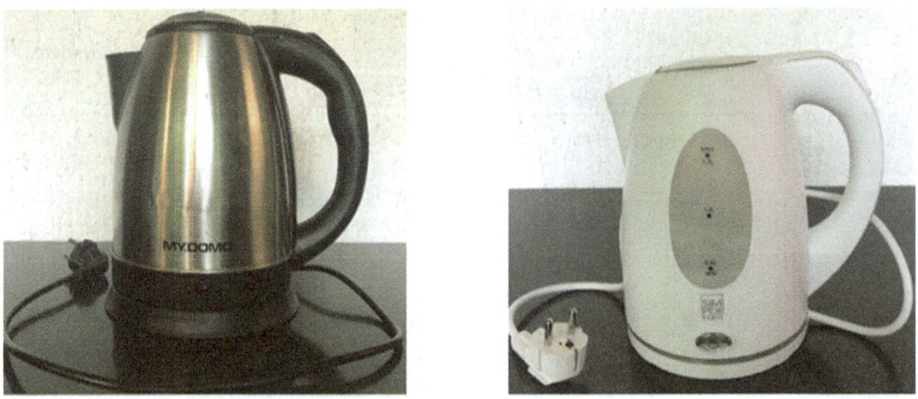

a) b)

Fig. 3. Electric kettle: a) MyDomo 631, b) Simpex Basic.

Both kettles: MyDomo, model 631 and Simpex, model Basic, have a nominal power of 2000 W, nominal voltage 220–240 V, frequency 50/60 Hz.

The analyzed MyDomo kettle is mostly made of stainless steel and polymer, with a low content of copper and brass, and the Simpex kettle is almost entirely made of polymer, with a lower proportion of copper, aluminum, and brass.

The measurement of the mass of the parts was also carried out in the disassembly analysis, and the portion of individual materials from which the kettles were made is shown in Table 1.

Table 1. Material content in the kettle MyDomo 631.

Material	Mass (grams)
Polypropylene	298.20
Steel	266.68
Copper	45.08
Brass	3.95
Aluminum	91.70
Polyvinylchloride	52.65

In Table 1, for the MyDomo kettle, a significant representation of two different types of material can be seen: polypropylene and steel from which the case parts of the analyzed kettle are made. The share of copper and brass is present in smaller quantities, from which the conductors in the power cable are made, as well as the conductors inside the kettle and the parts of the contacts. Polyvinylchloride (PVC) was used as the insulation of the power cable with the plug, while aluminum was used to make the heater case.

For the Simpex kettle in Table 2, it can be seen that it is mostly made of polypropylene (PP), and aluminum of which the heater case is made.

Table 2. Material content in the kettle Simpex Basic.

Material	Mass (grams)
Polypropylene	420.20
Steel	1.30
Copper	58.76
Brass	4.80
Aluminum	183.20
Polyvinylchloride	64.44

The share of copper and brass is present in smaller quantities, from which the conductors in the power cable are made, as well as the conductors inside the kettle and the parts of the contacts. Polyvinylchloride (PVC) was used as the insulation of the power cable with the plug.

3.1 Analysis of the MyDomo 631 Kettle Using the Simapro Software Tool

In the working interface of the SimaPro 9.0 software tool, it is necessary to select the databases used in the product life cycle analysis. After the selected databases, the input data are entered into the software tool. In the "Inventory" section under "Processes" it is possible to select the materials, production processes, type of transport, energy or waste disposal scenario required to perform the analysis. In the case of the analysis of the MyDomo kettle, all the materials of which it is composed are entered into the software tool, which includes the type of material, the total weight of each material and the process of their production.

After entering the input data, in the following steps, a "Process Tree" is formed, which is shown in Fig. 4, apropos a block diagram representing the network of processes and subprocesses that will be observed in the analysis.

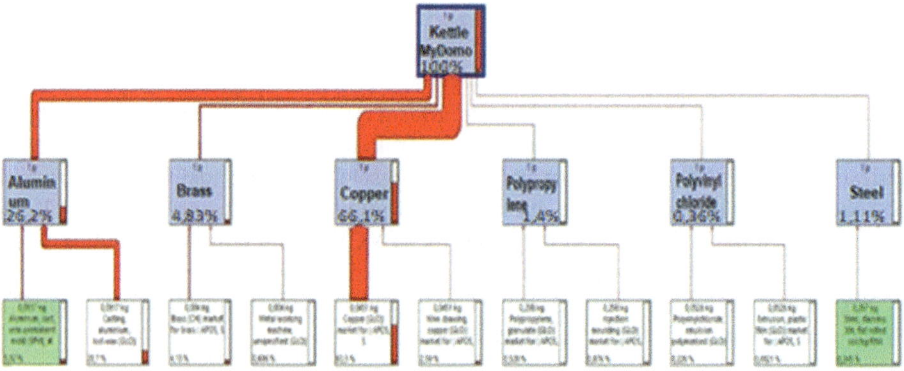

Fig. 4. MyDomo 631 kettle LCA analysis process tree.

In Fig. 4, it can be seen that the top node shows the kettle as a complete assembly, and from that node branch others representing a proportion of the individual material. In this case, they are aluminum, copper, steel, brass and various thermoplastics. Furthermore, processes for obtaining, shaping and processing these materials, such as casting, pressing, stamping, drawing, and other technological processes, branch out from these nodes. The red lines that connect the nodes create the branches of the process, and their thickness shows the share of influence in the overall process of product formation, the "thicker" the branch, the more significant the impact of environmental load.

Figure 5 shows a graphical representation of the impact of selected materials on human health, ecosystem and resources.

In the graph (Fig. 5), it can be seen that copper production and the technological process of copper processing, despite its small share in the kettle of only 5.94% of the total mass, has the greatest impact on human health amounting to 66% of the total share. the impact on the ecosystem is 60% of the total share.

Another material with a significant impact is aluminum, which affects human health by 26% of the total impact, while the ecosystem is 31%, and in the use of resources has a share of 63% of the total share.

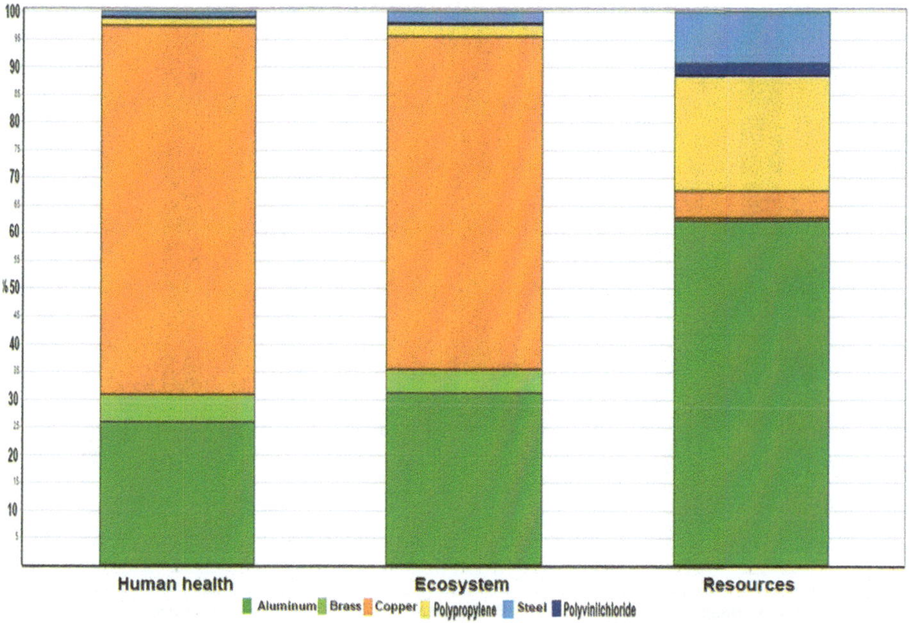

Fig. 5. Graphic representation of the influence of materials and processing.

Other materials (brass, steel, polypropylene, polyvinylchloride) have a much smaller impact on human health and the ecosystem compared to aluminum and copper, their total share is less than 5%, while the share in resource utilization is smaller, about 35%.

Figure 6 graphically shows the impact of each phase of the kettle life cycle on human health, ecosystem and resources.

The phase of using the stove has the greatest impact on human health due to the use of a large amount of electricity during 5 years of using the stove. Furthermore, it can be seen that transport has the greatest impact on human health due to emissions of harmful gases from the combustion of fossil fuels, and in slightly smaller quantities to the ecosystem and the depletion of natural resources.

It can also be seen on the graph that the reuse of materials has almost the same value in all three categories as the origin of the kettle, only with a negative result which shows us a positive impact on all categories. Equal values were obtained because this kettle is made of almost all recyclable materials.

3.2 Analysis of Simpex Kettle Using Simapro Software Tool

The life cycle analysis of the Simpex kettle was also performed using the SimaPro 9.0 software tool. The same databases were selected and the data entry was made the same as with the previous kettle.

After entering the input data, a "Process Tree" is formed, which is shown in Fig. 7, apropos a block diagram representing the network of processes and subprocesses that will be observed in the analysis.

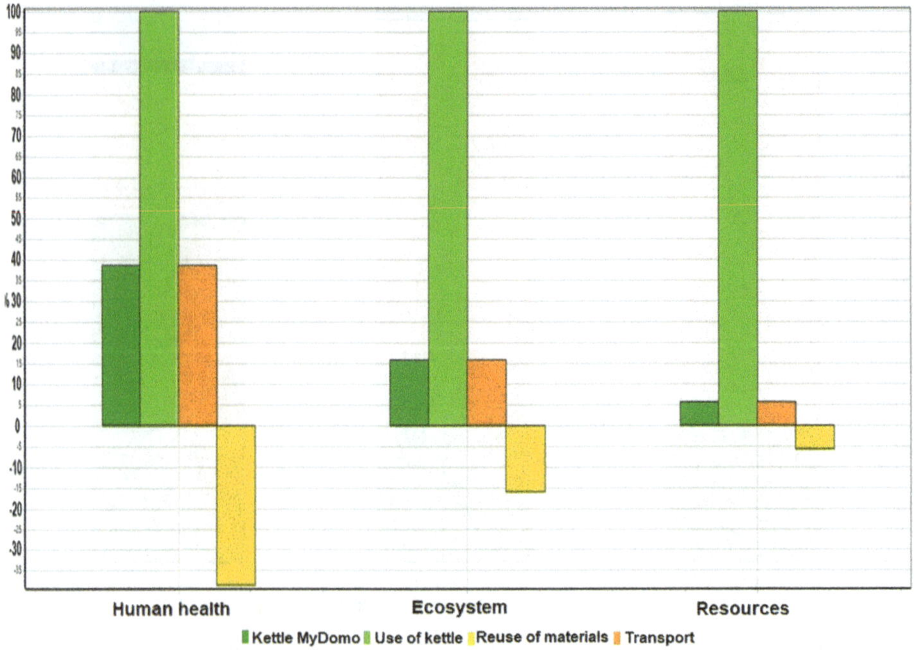

Fig. 6. Graphical representation of the impact of life cycle phases.

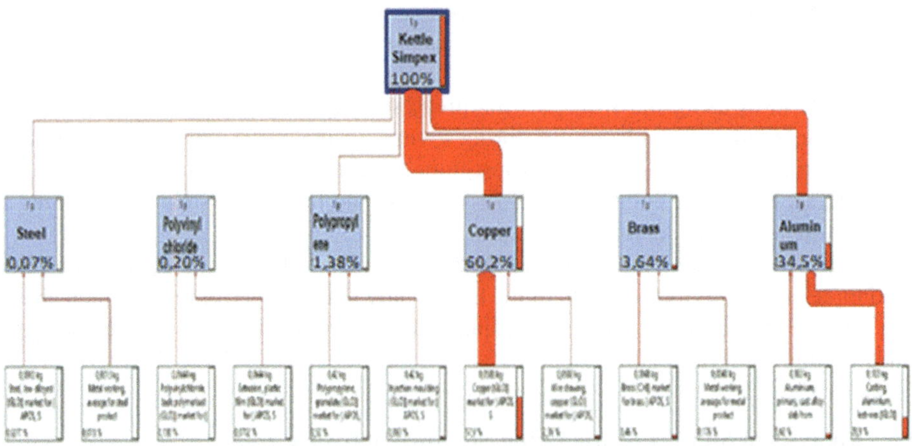

Fig. 7. Simpex Basic kettle LCA analysis process tree.

From the block diagram in Fig. 7, it can be seen that the uppermost node shows the Simpex kettle, as a complete assembly, and from this node branch others representing the share of individual material as well as the load of the environment. In the case of the Simpex kettle, copper comes first, followed by aluminum and, to a lesser extent, brass, followed by steel and various thermoplastics. Furthermore, processes for obtaining, shaping and processing these materials, such as casting, pressing, stamping, drawing, and many other technological processes, branch out from these nodes. The red lines that connect the nodes create the branches of the process, and their "thickness" shows the share of influence in the overall process of product formation, the thicker the branch, the more significant the impact of environmental load.

In Fig. 8, it can be seen a graphical representation of the impact of selected materials and their processes on the three categories.

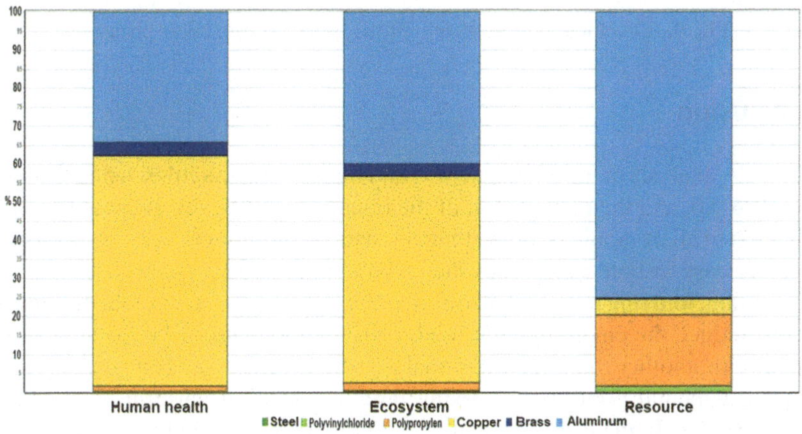

Fig. 8. Graphic representation of the influence of materials and processing.

In the "Human Health" category, the largest impact is contributed by copper with 60% of the impact, followed by aluminum with 34% of the total impact, brass with 4% of the impact and other materials with a total impact of 2%. In the category of impact on the "Ecosystem", we have approximately the same results as the impact on "Human Health". In the "Resources" category, aluminum has the largest impact, with a total impact of 66%, and copper with an impact of 19%.

The impact of individual phases of the life cycle of the Basic cooker on human health, ecosystem and resources is graphically shown in Fig. 9.

Figure 9 in graph shows that the stage of formation of the kettle and transport have the same effect on human health with a share of 47%, while the use of the kettle has a 100% share. The reuse phase of the material is 47% which is the same as the production, which would mean that the kettle is made of 100% recycled materials. The ecosystem category is most loaded by the kettle use phase, as well as the resource utilization phase.

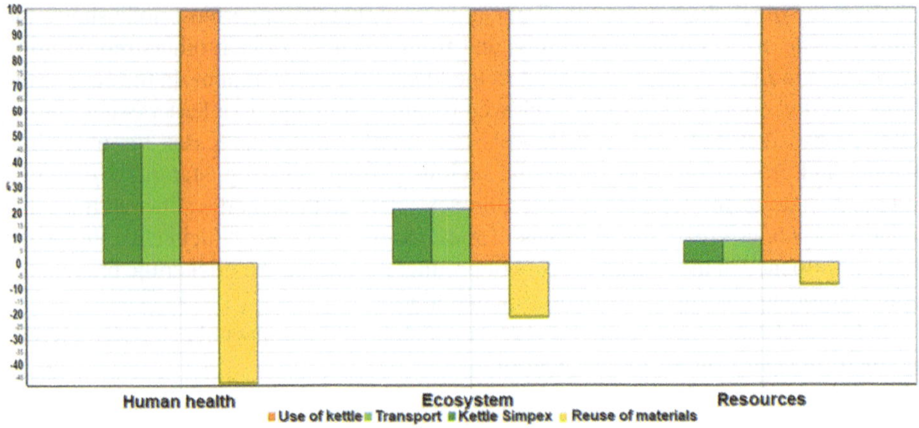

Fig. 9. Graphical representation of the impact of life cycle phases.

4 Conclusion

After the LCA analysis of the MyDomo 631 and Simpex Basic kettles, using the software tool SimaPro 9.0, and the comparison of the obtained results, several conclusions were reached. First of all, it should be noted that the share of aluminum used in the construction of the heater case in both kettles has the greatest impact on certain categories such as global warming, ionizing radiation, ozone depletion, eutrophication of the sea and ocean. In order to reduce the environmental load, consideration should be given to replacing aluminum with another material when making the heater case. Transport also greatly affects environmental pollution, releasing large amounts of harmful NOx (oxides of nitrogen) particles. In the near future, it is certainly to be expected that electric vehicles will be used, where the emissions of harmful gases are equal to zero. In addition to aluminum, the copper found in the power cable and conductors also creates a significant impact in certain categories of the analysis performed, however as far as cables and conductors are concerned there is no space for any improvement.

The main difference in construction from the aspect of material selection of the analyzed kettles is in the case in which the heating water is located. The MyDomo kettle case is made of stainless steel, and the Simpex kettles case is made of thermoplastic (polypropylene). This is why the MyDomo kettles puts a much greater load on the environment in terms of depleting natural resources from the earth due to the exploitation of iron ore.

The development and application of new and more advanced software tools greatly facilitates the work of scientists and engineers around the world, all with the aim of improving production and products that would ultimately have the least possible impact on the environment and human health.

1. References

1. Curran, M.A.: Life Cycle Assessment. CRC Press, Florida (2020)

2. Design life-cycle. http://www.designlife-cycle.com/. Accessed 20 May 2021
3. Curran, M.A.: Life Cycle Assessment Handbook. Wiley, Hoboken (2012)
4. Penn State. https://www.e-education.psu.edu/eme807/node/690lncs. Accessed 21 Nov 2016
5. PreSustainability. https://presustainability.com/leg-acy/download/EI99_Manual.Pdf. Accessed 17 June 2021
6. Comandrau, I.A., Barjoveanu, G., Peiu, N., Enel, S.A., Teodosiu, C.: Life cycle assessment of wine on water use impact assessment. Environ. Eng. Manage. J. **11**(3), 533–543 (2012)

The Maintenance of Agricultural Drainage Systems in the Conditions of Climate Change - The Jelas Polje Case Study

Siniša Maričić(✉) and Tatjana Mijušković-Svetinović

Faculty of Civil Engineering and Architecture Osijek, University Josip Juraj Strossmayer of Osijek, Vladimira Preloga 3, 31 000 Osijek, Croatia
{smaricic,tatjanam}@gfos.hr

Abstract. The paper presents the surface drainage of a part of Jelas polje, a field that is situated in the Croatian part of the Sava River valley. It is a complex hydrotechnical system with two safeguarding segments: the subsystem of reception and evacuation of runoff from hills, and the subsystem of flood risk management of the Sava River. The agricultural production in this part of the Jelas polje is mainly oriented towards field crops, which are most endangered by excess water during spring. According to farmers, in this century they are faced with problems they have not experienced before, situations where soil hold the water for an extended period of time and there is no adequate drainage. Direct consequence of such a situation is the decline of agricultural yields. All these problems farmers attribute to "bad and insufficient" maintenance of the channel network, which is under the jurisdiction of Hrvatske vode, legal entity for water management. The deeper hydrological and hydraulic analysis indicates extreme weather events and climate change, as well as the need for better adaptation to such a change. This paper analyzes the rainfall in the first vegetation months, April and May, during the long period from 1963 to 2020. In addition, these rainfalls are analyzed in correlation with the extreme water levels of the Sava River, when sluice gates are closed and drainage was provided by pumping stations. The performed analysis shows that in the last twenty years the amount of spring precipitations has significantly increased, as well as the danger of high Sava River water levels. Unlike in previous forty years, in this period there have been several coincidences of the threat from external and internal waters, which have adversely affected agricultural production for a long time. It all points out to a need for all stakeholders to join in, and with coordinated efforts find a better way to adapt maintenance of agricultural drainage system to the new climate conditions.

Keywords: Drainage · Rainfalls · Water levels · Climate change · Jelas polje

1 Introduction

The lowland area of Croatia has natural conditions for agricultural production and thus a long tradition of its development. For centuries, forested areas have been cleared, wetlands have been drained and plots have been adapted for agricultural work. Such processes also led to the establishment of water management services. By pooling resources

H. Glavaš et al. (Eds.): OTO 2021, LNNS 369, pp. 228–239, 2022.
https://doi.org/10.1007/978-3-030-92851-3_17

and organizing comprehensive works, better conditions for agriculture were created. Thus, since the middle of the last century, extensive work has been done in the Croatian Posavina on the development and management of the agricultural land drainage [1]. These are still the bases of local agriculture today. This paper describes the experience related to a part of the agricultural areas of Jelas polje, near to the settlement of Radovanje, not far from the town of Slavonski Brod. Local farmers suffered major damage in the spring of 2014 and saw the causes of that in poor maintenance of drainage channel networks. It all resulted in an expensive law-suit, thus attracting attention and encouraging writing articles about this subject. In the following chapters, first the existing field drainage system is presented, then the questionable elements of its maintenance are pointed out, followed by a hydrological and hydraulic analyses (of the condition when the problem occurred, but also of similar situations during the past sixty years).

2 Description of the Area and the Field Drainage System

Jelas polje is a left-bank part of the alluvial lowland of the Sava River in the Republic of Croatia, somewhere between its 360th and 400th kilometer. On the north side, Jelas polje is at the foot of Dilj gora (hillish area), and intersected by the international railway and motorway (A3), in parallel with the Sava River. The significant part of the area is used for fish farming. Figures 1 and 2 indicate the location of the respective agricultural land on the edge of the lowland, between traffic corridors (road embankments).

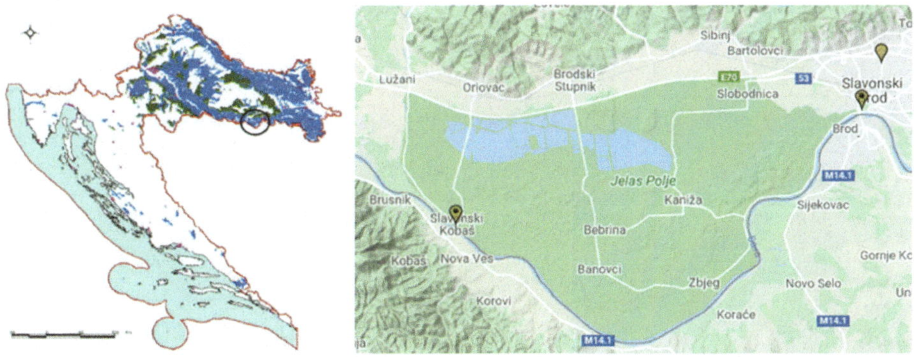

Fig. 1. Areas in the Republic of Croatia with a need for drainage of excess water [2] and the location of the Krčevina agricultural land, part of Jelas polje.

Construction and agricultural production are supported by two safeguarding subsystems. One of the subsystems consists of lateral channels and is a recipient of external surface waters originating from the southern slopes of Dilj gora. The eastern and western lateral channels of the Jelas polje, however, are directing torrential waters into the Sava riverbed, thus protecting the lowland area. The larger, western lateral channel is over 20 km long. Its course is from the mouth of the Sava River in the eastern part of Slavonski Brod to the north, and then from the mouth of the largest tributary of Sava, Glogovica River, turns to the west. The eastern lateral channel of Jelas polje is less than 5 km long, but collects all the streams between the settlement of Radovanja and the Orljava River. There are also some torrent streams (from the slopes of the Maglovac hill) beyond the scope of drainage channels, which drain a part of the settlement Radovanje and flow downstream through culverts under the railway. The second subsystem is the Sava River flood risk management system, encompassing embankments (around 80 km long) and hydrotechnical nodes (i.e. an assembly of a channel sluice gate and a pumping station). Flood risk management of the Jelas polje uses both a gravity and a pressure system, defending approximately 20000 ha large area. The system with four hydraulic nodes (with sluice gates and pumps) was built (between 1938 and 1990) for draining the defended area, at times when the Sava River water levels were high. Altogether, 14 pumps were installed, with total capacity of 32.4 m^3/s [3, 4].

The field drainage system consists of a network of detailed (land lot channels for immediate collection of waters from agricultural surfaces) and basic or main channels. For surface drainage of the area in question (agricultural area Krčevina) there are four long detailed channels (Fig. 2 and 3). On the right side, they transport water to the collecting receiver (the Topolovac River torrential stream), to which much shorter, detailed channels of the adjacent drainage area are connected, as well. The natural stream is regulated (channeled) and capacitated to receive and conduct catchment waters to a larger receiver, the Mrsunja River.

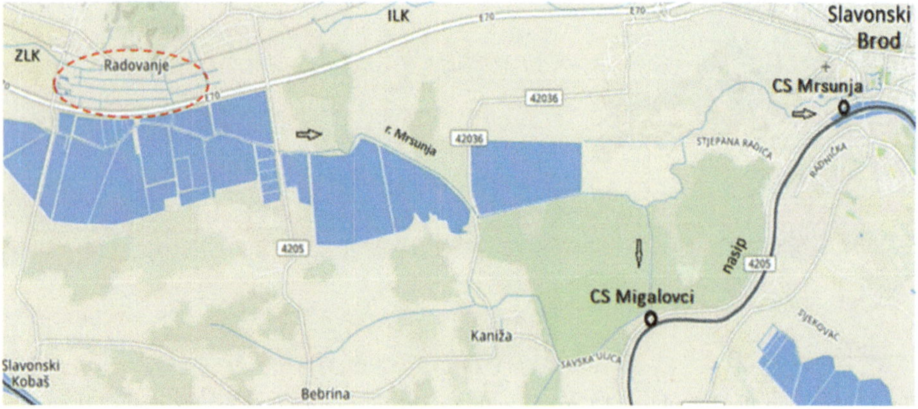

Fig. 2. The location of the Krčevina agricultural land (next to the settlement of Radovanje) in Jelas polje, with corresponding elements of the drainage system – lateral channels (ILK, ZLK), recipients and pumping stations (CS).

What is important for further drainage of the area in question is the Mrsunja riverbed as it allows the water to reach the nodes of Migalovci and Mrsunja, and from there into the Sava River [5]. The sluice gates of these hydrotechnical nodes must be closed when the water level on the Slavonski Brod (elevation 85.60 m above the sea level) water meter measures 380 cm. As the water level rises, the required number of individual pumps are turned on. There are four pumps (2 + 4 + 4 + 2 m^3/s) at CS Migalovci pumping station and as well as at CS Mrsunja pumping station ($2 + 2 + 2 + 2\ m^3/s$). The Water Act and its bylaws (i.e. rules and regulations related to the Sava River flood risk management system) regulate the commissioning and operation of the nodes [6, 7].

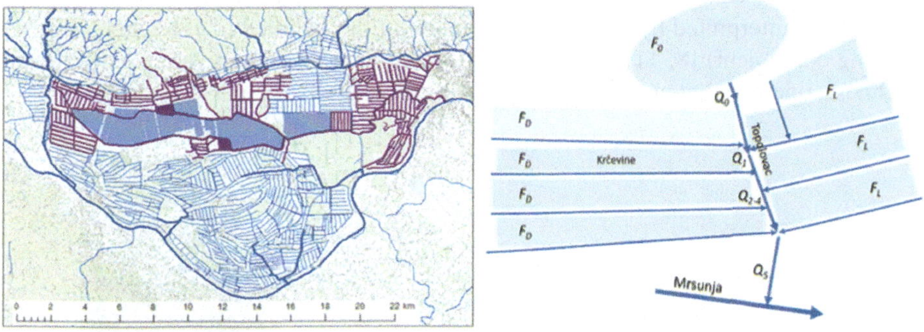

Fig. 3. Hydrographic network of Jelas polje with the part that flows into the Mrsunja River (right) and the calculation scheme of a part of the surface drainage system of the Krčevina land (left).

3 Maintenance of the Field Drainage System

The entire field drainage system needs to be maintained regularly, and this requires a legal, financial and organizational basis, as well as quality manpower and machinery. Stable socio-economic conditions support this, while every disturbance leaves consequences [8]. The economic downturn in the 1980s lead to a neglect of the maintenance of the field drainage systems, and due to the war circumstances during the 1990s, there was a complete halt. As a result, the systems lost their functionality and needed restoration [4, 9]. Then, the processes of privatization and restructuring were taking place, both in agriculture and water management. At the later stage, the Croatian laws were harmonized with the European Union legislation. However, despite all said misfortunate situations, the reconstruction and maintenance of hydrotechnical systems have begun again. For the area in question, and the time of occurrence of the agricultural plot damage (spring 2014) it was stated that a post-war reconstruction of the detailed channel had been carried out; that for a long period there was no silt removal from the riverbed of the first receiver (the Topolovac torrent stream); and that all the channels are regularly (once a year) mowed [5]. Some questions were raised when several road culverts of a small diameter (60 cm) were constructed on detailed channels, to enable simpler mowing. There were also objections that the plant cuttings were not removed from the riverbed, and as it was a swampy vegetation (Fig. 4), it posed a risk of channel congestion at locations with small culverts, in particular.

Referenced literature [8, 10] and the appropriate rulebook [11] indicate the need for mowing at least once a year, in late spring, under vegetative and hydrologically favorable conditions. As swamp vegetation was found in the respective area, it can be concluded that mowing should be even more frequent (2–3 times a year). There is a danger of partial or even complete congestion of culverts with small diameter, which indicates a need for a more frequent control and a more comprehensive cleaning job. The channel with overgrown grass creates an obstacle for water to flow through (for the same discharge value, the water level is higher and the velocity is lower), hence the longer duration of the excess water drainage. Grass overgrowth thus reduces the efficiency of the channel in view of the protection from excess water. In engineering terms, this is interpreted by change of roughness of the riverbed (e.g. by increasing the Manning coefficient) [8, 11]. By checking the runoff for different variants of rain and roughness, one may get an insight into the influence of a particular parameter [5].

Fig. 4. The state of overgrowth with swamp vegetation of one of the detailed channels of the Krčevina field drainage (upstream view), and the Topolovac receiver (view towards the highway downstream) in the spring of 2020.

4 Hydrological and Hydraulic Analysis of Runoff Circumstances

Relevant data needed to be established for these analyses. Rain and water levels data were recorded at the nearby measuring stations. The meteorological station of Slavonski Brod was considered as a representative one, for the climatological features of the entire area of interest. It is located about 20 km east of the observed field drainage system.

The properties of the climate were determined by statistically processing the rainfall data for the period from 1963 to 2018. The area in question is specific for a low annual amount of precipitation (annual sum of precipitation over a multiannual period is 769.8 mm). In addition, it is worth noting that the oscillations of precipitation are considerable in certain years. During 2005, there was significantly more rainfall, 1006.8 mm, than during 2011 when it was merely 433.1 mm. The months with most rain were July and August, while large deviations from average rainfall were measured for all other months. In 2014, the above average rainfall was recorded during five months, among which April, May and August were the rainiest ones (Fig. 5).

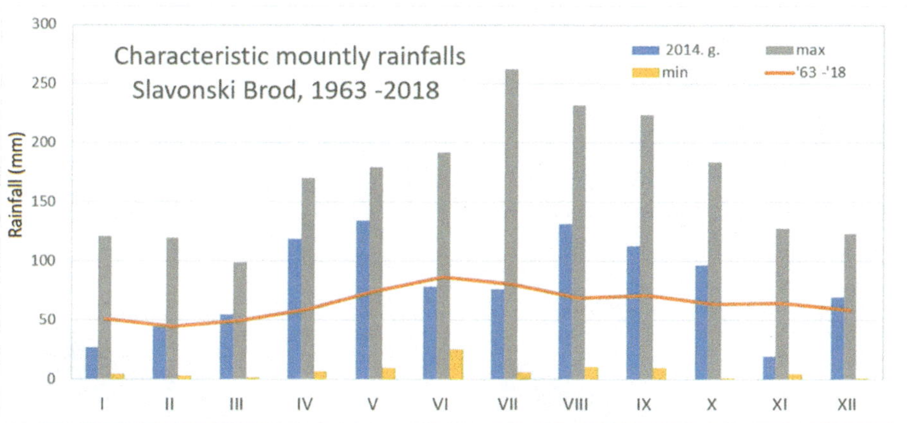

Fig. 5. Characteristic monthly rainfalls for a period of 56 years [12], for the meteorological station Slavonski Brod.

Figure 6 shows the results of the statistical analysis of daily (24-h) precipitation for the period from 1991 to 2010, in the area around Slavonski Brod. The highest rainfall occurred in the period from May to October, and the 24-h rainfall extreme was recorded at the level of more than 70 mm.

Humidity data for April and May 2014, and a multiannual average (1961–1990) for the surrounding area of Slavonski Brod indicate an extremely rainy April and a very rainy May. It was concluded that those were the months in which there was 200% and 183% more precipitation than average. The return period of such monthly precipitation was 100 years in April, and 10 years for May [13].

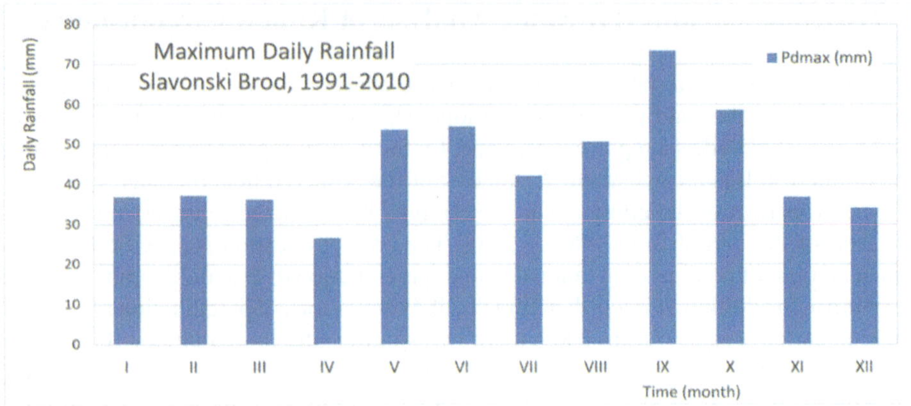

Fig. 6. Maximal daily rainfalls for a 20-year period [3], for the meteorological station Slavonski Brod.

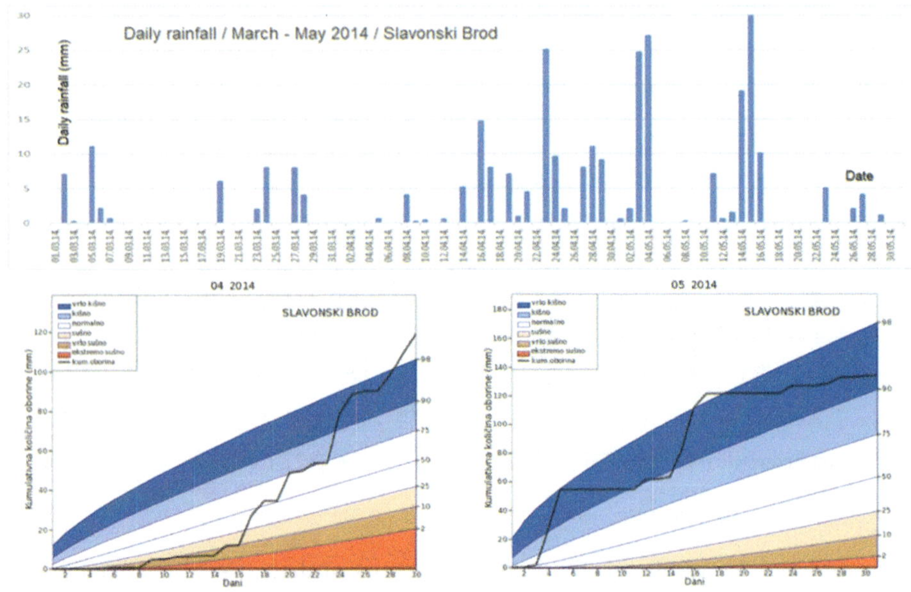

Fig. 7. The distribution of daily rainfall and an analytical presentation of the cumulative amounts of rain [14], for spring 2014 in the Slavonski Brod area.

Figure 7 shows that rain has soaked the soil for 10 days during March, and it intensified during the following two months. In the period from April 16 to May 16, there was 221.6 mm more rain per square meter, which falls into the category of the months with most rain ever (30 days), during the observed period from 1963–2018 (VII-262.2 mm/m^2; VIII-232 mm/m^2; IX-224 mm/m^2).

From the analytical statement (DHMZ [14]) of the cumulative amounts of precipitation during April and May 2014, with marked probability zones according to the multiannual average (1961–2000), it is noticeable that on three occasions during multiple days there was a state of great return period (25–100 years), i.e. greater than the 96th percentile.

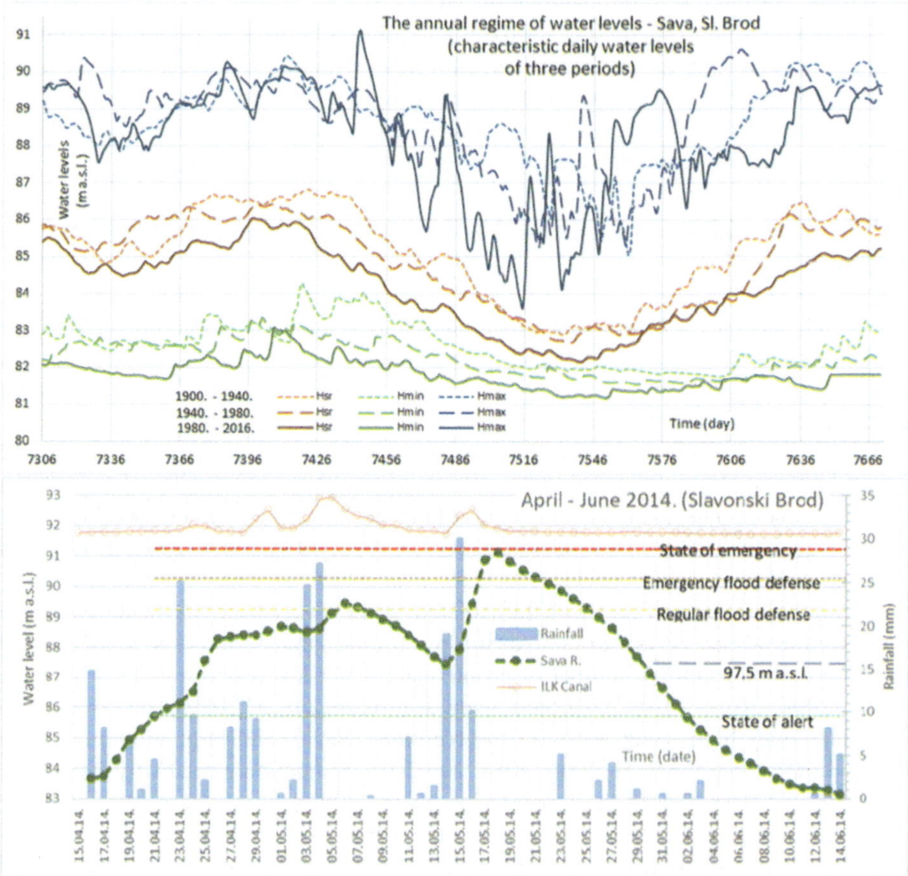

Fig. 8. Annual course of the Sava River water level [15], and the water wave in the spring 2014 at the gauging station Slavonski Brod, with marked reference (flood risk) water levels.

Along Jelas polje there are two gauging stations on the Sava River (Fig. 1), and they are located in Slavonski Brod (360.0 r.km) and Slavonski Kobaš (390.5 r.km). From the analysis of the annual characteristic water levels at the Slavonski Brod gauging station (Fig. 8, above [15]), it is evident that the highest water levels occur most often around April, and they can go up to over 90 m above sea level (or, even somewhat over 91 m above sea level). Water levels at this station are appropriate for managing a complex hydrotechnical system. The water level at which sluice gates close is at 85.69 m above sea level, while at the water level of 89.3 m above sea level a state of regular flood risk management is declared. In addition, by analyzing the cartographic data, a terrain altitude of 87.5 m a. s. l. was determined as a feature of low terrain in the agricultural area of Krčevina near Radovanje. The ratio of these elevations and the fluctuation of the Sava water level near Slavonski Brod in the spring of 2014 is presented in Fig. 8. It is obvious that at that time the problems of high water levels occurred simultaneously with the problems of heavy rainfall and that the possibility of draining the excess water with the existing system was significantly reduced.

The suitable runoff modelling for daily rainfall of 40 mm and 70 mm in conditions without the backwater of the Sava River, as well as with the backwater present, shows the inevitability of the occurrence of damage to crops in the weather conditions of spring 2014 in the area of the Krčevina. The question arose as to how often such conditions occurred, so further analyses were carried out.

5 The Coincidence of Spring Hydrological Extremes in Slavonski Brod

Prior to the analysis, it is necessary to emphasize that the spring conditions (April and May) have a strong influence on the development of field crops. Excess soil moisture can completely ruin yields in a span of 14 days, while in other stages of plant development it takes a few more days than that. That is why this period was further observed.

Firstly, a review of bimonthly spring precipitation (April + May) was made with the available data (monthly precipitation for the period 1963–2020). The data is shown in Table 1 and brings us to the conclusion that the "extreme" values (over 200 mm) appeared only eight times in almost 60 years, in 1975, 1980, 1989, 2004, 2010, 2014, 2017 and 2019.

Secondly, a review (Fig. 9) of the isolated water levels near Slavonski Brod was made (for the 90-day period around April/May) for the same period (1963–2020). With this process then a similar number of problematic water levels was noted: 1970, 1981, (1986), 2004, 2013, 2014, 2016, and again approximately in 2019. The coincidence of simultaneous high water levels of the recipient (Sava), and large amounts of rain occurred only three times, in 2004, 2014, and 2019, indicating a relatively rare occurrence. Nevertheless, during the last 40 years of the 20th century, there were 3 springs with abundant rain, and 2 or 3 occurrences the water levels were threatening, without occurring simultaneously. On the other hand, during the 20 years of this century there were already five conditions with heavy rainfall and high water levels, with unfortunate overlap during three of those situations. This, most likely, indicates climate change.

Table 1. Bimonthly (April, May) rainfall (mm) and its extremes (>200 mm) for Slavonski Brod during the period 1963–2020.

Bimonthly rainfall, Weather station Slavonski Brod					
Year	Mo. IV + V	Year	Mo. IV + V	Year	Mo. IV + V
1963	72.6	1982	99.2	2001	103.1
1964	171.2	1983	95.1	2002	134.0

(*continued*)

Table 1. *(continued)*

Bimonthly rainfall, Weather station Slavonski Brod					
Year	Mo. IV + V	Year	Mo. IV + V	Year	Mo. IV + V
1965	176.9	1984	173.6	2003	85.6
1966	114.2	1985	144.5	**2004**	**244.7**
1967	166.7	1986	108.5	2005	129.9
1968	96.8	1987	145.1	2006	176.1
1969	150.1	1988	75.7	2007	76.3
1970	114.0	**1989**	**251.5**	2008	140.1
1971	80.5	1990	61.5	2009	56.6
1972	121.6	1991	181.6	**2010**	**214.3**
1973	106.0	1992	71.1	2011	61.5
1974	93.9	1993	74.3	2012	173.1
1975	**215.2**	1994	89.9	2013	125.8
1976	131.3	1995	111.8	**2014**	**253.1**
1977	95.6	1996	180.1	2015	171.2
1978	191.5	1997	120.8	2016	107.4
1979	62.1	1998	106.6	**2017**	**246.0**
1980	**240.9**	1999	146.1	2018	122.5
1981	103.5	2000	79.8	**2019**	**235.8**

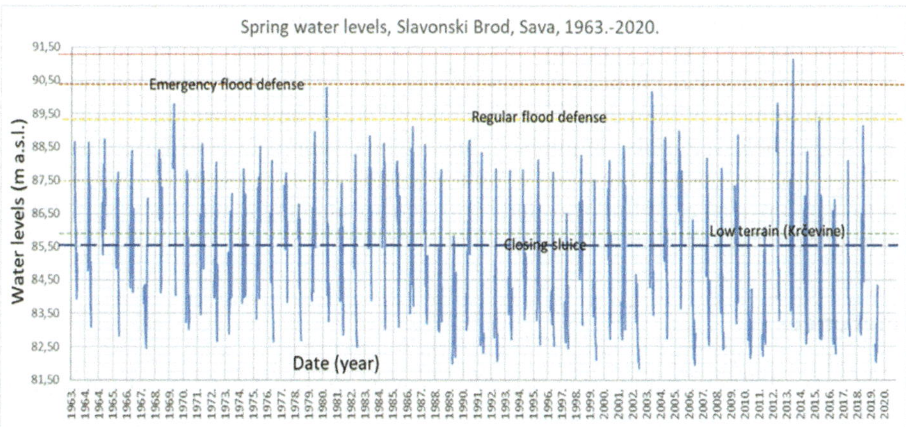

Fig. 9. The spring (15 April–15 June) water levels of Sava River (gauging station Slavonski Brod) in the period from 1963 to 2020 under boundary condition - closing sluice gates, low terrain of Krčevina and a flood risk management phase.

6 Conclusion

In the spring of 2014, Jelas polje was affected by large amounts of rain and high water levels of the Sava River. From the conducted analyses, it is evident that what we have been dealing with is above-average and long-lasting phenomena that have led to severe damage to agricultural crops. Hence, this was a natural disaster, which was then officially declared. The majority of the population is not familiar with the characteristics of the hydrotechnical system which improves the water-air regime of the soil and protects the local economy. Faced with a natural disaster, people also form wrong opinions about its causes. Many human dilemmas require expert clarification which encouraged the creation of this research paper.

Analyses have shown that two unfavorable effects on the water-air regime of the soil were a rare occurrence. Under such conditions, the efficiency of the existing drainage system is significantly lowered. Sluice gates need to be closed, and drainage has to be done with pumps. Pumping stations have a limited capacity, so the runoff is difficult and takes place with deceleration (higher water levels in the channel network). Then even the well-maintained (mowed) channels do not help the cause. However, a properly (according to the valid Rulebook) arranged network of channels preforms drainage much faster and shortens the time of wetting the soil in the rhizosphere (plant root zone). It is important that all elements of the hydrotehnical system are functional and in good condition (sluice gates, pumps, channels, culverts, …), and this is achieved with specified, regular and quality maintenance. In the shown example, remarks were made on the quality of maintenance, which should be considered, as well as it should be worked on respective improvements. It is also possible to control the system in a more complex way according to the weather and hydrological forecasts, which certainly needs to be comprehensively examined.

By performing additional analyses, a conclusion was drawn regarding the possible climate change. In comparison to the first two thirds of almost a sixty-year period to date, for its last third, it was observed that there was a significant increase in the occurrence of extreme rainfall events, and high Sava water levels, as well as the simultaneous occurrences of these two phenomena. Here is where we find the answer for the statements and the amazement of the population about how "We've never seen this before in our lives". Nevertheless, this also requires finding adequate approaches to agriculture under these new circumstances. In that sense, we recognize a need for all social actors to find a way to team up in an excelling way and contribute to adapting to these new agricultural circumstances through joint and coordinated efforts.

References

1. Piha, D.: WMBO for little basin "Brodska Posavina". J. HRVATSKA VODOPRIVREDA **213**(13), 24–33 (2015). (in Croatian)
2. Husnjak, S.: Croatian agricultural lands and the need for land reclamation measures. In: Reclamation Measures for the Improving Rural Areas, The Croatian Academy of Sciences and Arts (HAZU), pp. 21–37, Zagreb (2007). (in Croatian)
3. Tadić, Z.: Brod - Posavina County Irrigation Plan. Hidroing d.o.o. Osijek, Brod - Posavina County and Croatian Water, Osijek (2007). (in Croatian)

4. Budišić, D.: Condition of reclamation systems in the Sava River basin district. Croatian Water **45**(11), 465–477 (2003). (in Croatian)
5. Maričić, S.: Protection against excess water (part of Jelas Polje). In: Raos, P., et al. (eds.) 19th Natural Gas, Heat and Water Conference, 12th International Natural Gas, Heat and Water Conference, pp. 98–107. University of Slavonski Brod, Mechanical Engineering Faculty, Osijek (2021). (in Croatian)
6. Water Law (Official Gazette 53/2009, 63/2011, 130/2011, 56/2013 and 14/2014)
7. The Flood Defense Implementation Plan for defended areas, Sector D, the middle and lower part of Sava River. Croatian Water, Zagreb (2014). https://www.voda.hr/sites/default/files/clanak/bp_10_-_provedbeni_plan_obrane_od_poplava.pdf. Accessed 21 Aug 2021
8. Kos, Z.: Needs, principles and methods of maintenance of drainage hydromelioration systems. Manual for hydrotechnical reclamation, I. volume Drainage; Book 6. Maintenance, Croatian Society for Drainage and Irrigation, Zagreb, pp. 1–25 (1991). (in Croatian)
9. Šoštarić, J., Romić, D., Marušić, J., Josipović, M., Petošić, D.: The condition of melioration systems for drainage and irrigation in the Republic of Croatia. In: Pospišil, M., Vnučec, I. (ur.) 51st Croatian and 11th International Symposium on Agriculture, pp. 10–17. University of Zagreb, Faculty of Agriculture, Zagreb (2016). (in Croatian)
10. Lambert, K., Smedema, L.K., Vlotman, W.F., Rycroft, D.: Modern Land Drainage - Planning, Design and Management of Agricultural Drainage Systems, 2nd edn., pp. 420–425. CRC Press, London (2004)
11. Regulation on technical, economic and other conditions for the development of melioration drainage systems and bases for technical and economic maintenance of the systems (Official Gazette 4/98)
12. Croatian Meteorological and Hydrological Service, Whether Information for Slavonski Brod (1963–2000). https://meteo.hr/klima.php?section=klima_podaci&param=k1&Grad=slavonski_brod. Accessed 29 July 2021
13. Croatian Meteorological and Hydrological Service, Whether information for Slavonski Brod for April and May in 2014. https://meteo.hr/klima.php?section=klima_pracenje&param=klel&Grad=slavonski_brod&Mjesec=04&Godina=2014. Accessed 29 July 2021
14. Croatian Meteorological and Hydrological Service, Whether information for Slavonski Brod for April and May in 2014. https://meteo.hr/klima.php?section=klima_pracenje&param=spi&el=koa&Grad=SlavonskiBrod&Mjesec=05&Godina=2014. Accessed 29 July 2021
15. Maričić, S., Mijušković-Svetinović, T.: Water levels of the major rivers in Slavonia. In: Pelivanovski, P., Jovanovski, M., Petkovski, Lj., Donevska, K. (eds.) 16th International Symposium on Water Management and Hydraulic Engineering WMHE 2019, pp. 181–193. Ss Cyril and Methodius University, Civil Engineering Faculty - Skopje, Skopje (2019)

Graphic Analysis of the Environmental Influence on the Traditional Construction of a House in Órgiva, Granada, Spain

Antonio Martín-Cara[1] and José-Lázaro Amaro-Mellado[2,3](✉)

[1] Universidad de Coruña, Escuela Universitaria de Arquitectura Técnica, 15071 A Coruña, Spain
[2] Departamento de Ingeniería Gráfica, Universidad de Sevilla, 41092 Seville, Spain
jamaro@us.es
[3] Instituto Geográfico Nacional-Servicio Regional en Andalucía, 41013 Seville, Spain

Abstract. Software simulations allow the interior composition of buildings to be graphically analysed. Thus, it is possible to make an approximation related to very diverse parameters such as the comfort or earthquake resistance of the buildings. From the perspective of traditional architecture, this work aims to rigorously examine the characteristics of the construction systems used in Órgiva, a town of Las Alpujarras, in Granada, one of Spain's most seismically hazardous regions. Hence, using graphical interpretation, objective values are obtained to better understand the intrinsic performance of the materials. For this purpose, the simulation of a masonry wall is created by interpreting data related to heat and vapour flow, temperature, humidity and partial pressure. By means of internal analysis, and including the climatic conditions of the territory, we can observe how the construction systems affect the comfort of the users, causing condensation dampness in joints between walls and pathologies in the systems. In addition, the results are valuable for the maintenance of the building as they foresee future problems and guide the technician in the resolution of architectural encounters, benefiting the strong points and optimising the performance of the construction, even in terms of earthquake resistance.

Keywords: Graphic analysis · Software simulation · Traditional influence · Environmental influence · Earthquake resistance

1 Introduction

The popular architecture in La Alpujarra of Granada, southern Spain, has strong characteristics that adapt to the environment through physical and geometric composition. Thus, we can highlight the adaptation to the significant gaps and steep gradients [1] and the use of natural water channels from the rivers and tributaries that flow down from the Sierra Nevada [2].

The dwellings and buildings result from applying indigenous knowledge and the materials in the creation of habitable spaces such as rocks, clay, slate and various trees [3]. It should be noted that the use of the trees is not random; the inhabitants know which

H. Glavaš et al. (Eds.): OTO 2021, LNNS 369, pp. 240–252, 2022.
https://doi.org/10.1007/978-3-030-92851-3_18

is the most appropriate according to the intended use. As a result, a cultural and living landscape is created, reflecting a one-time use that still survives [4].

Besides, the Iberian Peninsula has a low-to-moderate seismic hazard. However, events of intensity VIII (heavily damaging) EMS-98 or higher have occurred in the province of Granada, specifically in the years 1431, 1531, 1806, 1886 or 1956 [5].

The main goal of this work is to analyse the risk factors that interfere with the structural stability of masonry walls, pointing out the possible causes of decoupling between ground motions [6]; in our case, this is a firm, rocky terrain immersed in the mountain.

Furthermore, there are non-destructive testing methods for building analysis such as rebound hammer, ultrasonic, pull-out, thermography, penetration resistance, or radiography [7] to help achieve the results. Nevertheless, we have used software simulation, creating analytical calculation models based on international standards [8] due to its low-nearly-zero cost. In earthen composite building systems, we can establish the balance between thermal mass and low conductivity, a scarce property in current materials [9]. First, however, we must evaluate which factors interfere with the systems' stability, such as material detachment, dampness, or erosion, to obtain the most vulnerable points and the effects of the environment on them, which is particularly valuable in the event of an earthquake.

2 Study Area

La Alpujarra, also called Las Alpujarras, is a rural region located on the southern face of the Sierra Nevada (southern Spain), within the framework of the Penibetic mountain range, in the south of the Betics. It straddles two provinces, Almería and Granada, and its most populated municipality is Órgiva (Fig. 1). The area has undergone a landscape restructuring concerning the mountainous areas. Hence, its soils were previously used for agriculture and, due to migration towards the capitals, have been reforested with holm oaks and oaks [10]. These wooded areas are maintained thanks to the network of irrigation ditches and water channels dating from the Andalusian period (between 711 and 1492 AD), which supply the fields [2]. The inhabitants of these mountainous areas are closely linked to the mining systems, where the extraction of copper, zinc, gold, silver, mercury and iron stands out [11].

Regarding climatology, it has a rainfall of 641 mm per year and an average temperature of 13.7 °C. Thus, it is classified as Csa (hot-summer Mediterranean climate) according to the Köppen-Geiger climate classification (https://es.climate-data.org/europe/espana/andalucia/orgiva-98709/); furthermore, under the DB-HE Energy Saving included in the Technical Building Code, Órgiva is in climate zone C4. The latter classification is established according to the global annual average daily solar radiation on a horizontal surface [12]. Hence, the classification determines the thermal transmittance values of the different elements that make up the building envelope, such as walls, windows, or roofs. In this case, there is an area where the sun shines most of the year but, being very close to the Sierra Nevada, temperatures can drop below 0 °C. Together with Ugíjar (to the east), these are the areas with the lowest rainfall aggressiveness. Nevertheless, the geomorphology is affected by other climatic, lithological and edaphological factors [13]

Fig. 1. Location map of Órgiva (highlighted in yellow), in southern Spain. Source: authors from https://www.ign.es/wms-inspire/mapa-raster. EPSG: 25830.

that humans have created for cultivation, presenting plains in the steep mountains that have eroded the natural terrain.

With regard to seismic hazard, Órgiva has an estimated peak ground acceleration (PGA) for a return period of 475 years of 0.19 g, which means that the maximum expected intensity EMS-98 is VII (almost VIII, since the threshold to consider it so is 0.196 according to the transformation suggested in IGN-UPM-Working Group [14]).

3 Materials and Methods

This chapter describes the steps taken to obtain the data derived from the different simulations and the application of criteria to the key points of the section we will study. Once the house "in situ" data has been gathered, it has been studied constructively and geometrically. Thereby, based on the construction techniques used when the house was built, we obtain a typical masonry wall section that we will explore later. Next, the standard values associated with the climate of the area are applied within the HTflux software (http://www.htflux.com/), with the indoor and outdoor climate properties being as follows (Table 1):

In addition, we apply the physical and chemical properties of the materials used at the time of construction described above to the system under study. The software provides the data and can be modified by adapting the composition and being as rigorous as possible. In our case, we will take the values from the software as they are verified by the UNE-EN ISO 10077–1:2020 standard in terms of thermal and energy performance (Table 2):

Table 1. Indoor and outdoor air climate data entered into the programme.

	Indoor climate	Outdoor climate
Temperature	20 °C	−10 °C
Relative humidity	50%	80%
Energy transfer resistance	0.13 $(m^2K)/W$	0.04 $(m^2K)/W$

Table 2. Energy performance data of the materials entered in the programme.

	Conductivity	Specific heat	Gross density
Sandstone	2.30 W/(mK)	1000 J/(kgK)	2600 kg/m^3
Sand/Gravel	2.00 W/(mK)	1000 J/(kgK)	2000 kg/m^3
Soft wood	0.13 W/(mK)	1600 J/(kgK)	500 kg/m^3
Limestone hard	1.70 W/(mK)	1000 J/(kgK)	2200 kg/m^3
Glass	1.00 W/(mK)	750 J/(kgK)	2200 kg/m^3

Once the above data is included, the program can calculate the factors and predict the behaviour of the wall in different situations. Firstly, the temperature path through the wall is analysed, obtaining the capacity to maintain the energy and taking advantage of it to reduce the demand as well as the power and cost of the heating and cooling equipment [15]. This factor helps to exactly locate the most singular points that present energy losses corresponding to thermal bridges, as this crucial element also affects aspects linked to the well-being of people, such as insulation and hygrothermal comfort of the spaces we inhabit [16].

Further on, heat flux allows finding the direction and energy loss of the wall itself as well as each material that composes it. Therefore, if the wall is made with stones that conduct heat better, there will be more serious energy losses if we do not have a retaining element. The most common problems with poor heat flux, apart from thermal bridges, would be the appearance of surface condensation, deterioration of the structure [17] and the detachment of the support material, in our case, compressed earth.

Next, concentrated relative humidity in buildings determines the possibility of fungal growth and diseases related to indoor thermal comfort affecting users' health. Besides, it hinders the conservation and maintenance of the building, creating the need to enhance interventions in the affected areas [18]. Therefore, corresponding to this case, the need for good natural ventilation to help alleviate the effects of humidity is established.

The composition of the building system also determines where the moisture escapes. This effect is called vapour flow or evapotranspiration. In addition, there must be an energy input provided by solar radiation or air circulation inside and outside the dwelling [19]. Fundamentally, the possibility of dampness is determined by the dwelling orientation and the air velocity in the area. However, poor execution of the system or choice of materials can cause the flow to extend from the exterior to the interior, resulting in loss of comfort and structural weakness.

Finally, about the partial pressure supported by the wall, we can define it as the pressure of the accumulated gases, which is determined by the Supporting Document to the Basic Document DB-HE 2 Energy Saving [20]. High partial pressure values lead to the accumulation of gases that do not find escape routes at certain points in the construction system. As with excess humidity, this factor damages the joints, especially those located on roofs, causing the weakening of the joints between the different elements and materials.

Once the simulations have been carried out, the programme determines at each point the incidence of the factors described above, creating isolines that mark the circulation and energy concentration based on the characteristics of the climate and the material used. With these outcomes, we can locate the most conflictive points with regard to comfort, evaluate the suitability of the materials and systems, and anticipate structural problems that may arise and affect the stability of the building in case of an earthquake.

4 Results and Analysis

This section presents and examines the results obtained after simulations for a series of parameters: temperature, heat flux, relative humidity, vapour flux and partial pressure. Furthermore, some of these outcomes are related to the possible impact on earthquake resistance.

4.1 Temperature

The first factor analysed is temperature. As shown in Fig. 2, the isolines in temperature values converge at the junction with the window. Thus, the execution of this point is crucial to the spatial segregation of the isolines and increasing indoor comfort. In addition, if there are remarkable temperature differences in a small space, condensation may appear on the entire face, affecting the upper and lower areas of the face where it meets the horizontal structure, weakening the points mentioned [17]. In this case, we see variations between 5 °C and 20 °C at the window frame and its junction with the wall. Nevertheless, the isolines dissipate at the base of the wall, not interfering with the wall's stability.

Looking at the upper part, at the junction between the facing and the roof, we see that the temperature is concentrated at this junction. The latter indicates the appearance of a thermal bridge and consequently the loss of energy. Although it does not affect the structural system, it is important to mention that the high-temperature variation in a system composed of earth could cause material detachment if the in-furniture is located in an area with a high rainfall index.

4.2 Heat Flux

Regarding heat flux, Fig. 3 illustrates how it is transmitted from the inside to the outside. Hence, the system's energy loss is described, especially at the junction points between

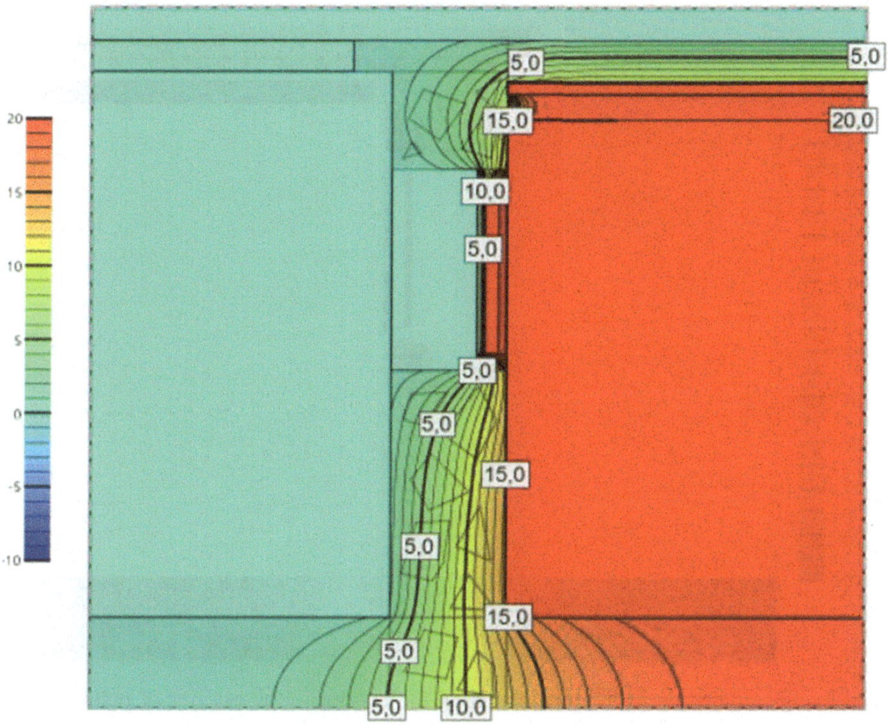

Fig. 2. Energy simulation of the temperature values transmitted at different points of the masonry wall. Source: authors from HTflux software.

the window and the wall. Certainly, the loss is due to the poor execution of the carpentry, joining it without any insulating element that separates it from the stone.

Focusing again on the most sensitive points, such as the wall-window or roof-wall joints, we can observe that the joints between the vertical and horizontal structure do not show any symptoms that would alert us to a possible problem of structural weakness. Up to now, energy loss due to the system's geometry is the main cause of the lack of thermal comfort. Also, we can see that the double glazing absorbs the thermal shock by creating an airtight air chamber and fulfils its function correctly.

4.3 Relative Humidity

The following parameter to be examined is relative humidity. In this respect, Fig. 4 shows a point where exists a high concentration of humidity in the junction area between the roof and the wall. This symptom warns of the appearance of fungus and material detachment. As it is a sandy material, this degradation may occur more quickly and require urgent intervention. Furthermore, we can notice how the problem extends throughout the roof, which is made up of wood, branches and sand.

At the base of the wall (Fig. 5), there is proportionality in the concentration of humidity; even the wall itself prevents the passage of humidity towards the interior,

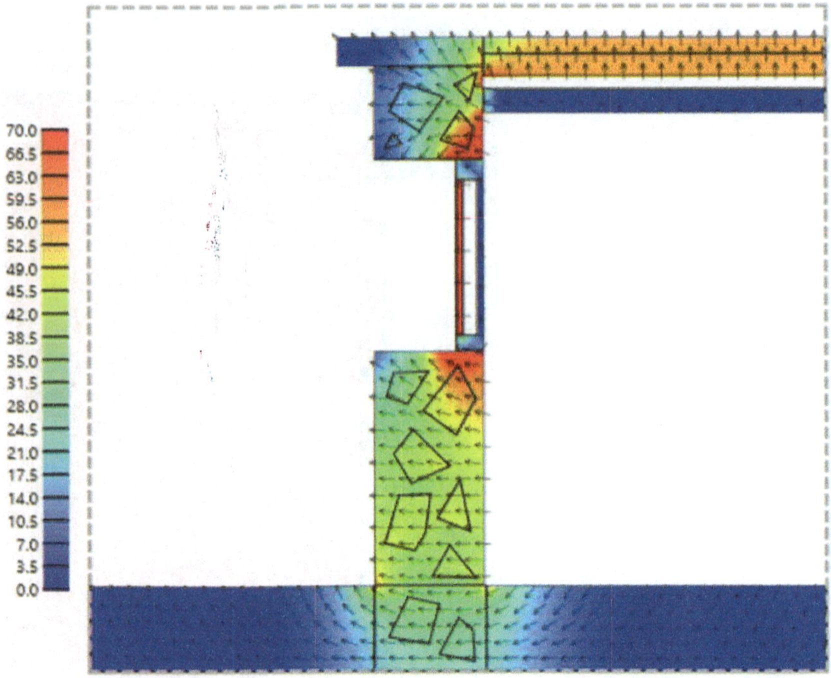

Fig. 3. Energy simulation of the heat flux from the interior to the exterior of building detail. Source: authors from HTflux software.

preventing condensation on the base of the dwelling and maintaining interior comfort. However, as mentioned above, the problem of the upper joint would cause damage to the health of the users.

4.4 Vapour Flow

Another studied parameter is vapour flow. Focusing on the vapour flow diagram linked to the relative humidity, the choice of material is of great importance in order to obtain the correct transpiration of the system. As can be seen in Fig. 6, the presence of stones prevents the moisture from dissipating correctly. However, the sandy material acts protruding along the wall, diverting excess vapour from the inside to the outside.

The deviation of excess humidity to the outside causes an increase in indoor air quality (IAQ), which favours user ergonomics, promoting passive air circulation and preventing contamination by biological agents indoors. Currently, there is great awareness of the importance of IAQ reflected in the NTP 243 standard of the Spanish National Institute of Occupational Safety and Health at Work (INSST).

In our case, as it is a stone floor without organic materials susceptible to the incubation of microorganisms, the vapour coming from it does not affect the base of the system, making the installation of a vapour barrier unnecessary. Nevertheless, if we examine the roof, it would be necessary to intervene and adapt the system to prevent the accumulation

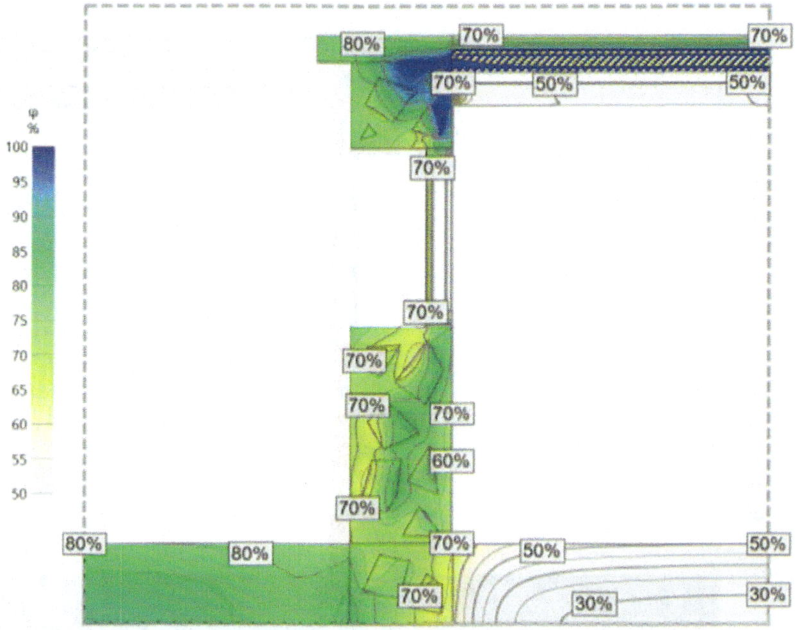

Fig. 4. Energy simulation corresponding to the relative humidity values in the selected building system. Source: authors from HTflux software.

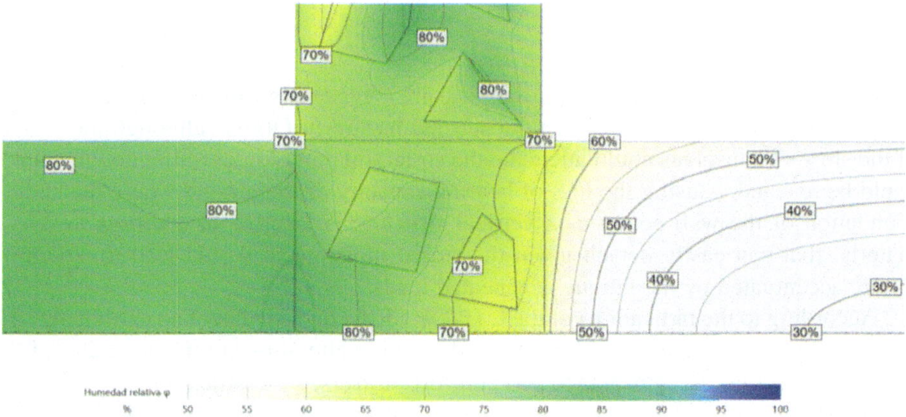

Fig. 5. Detail of the relative humidity values in the foundation base of the masonry wall. Source: authors from HTflux software.

of humidity. This accumulation is due to the lack of slope of the roof as well as the shortage of waterproofing material.

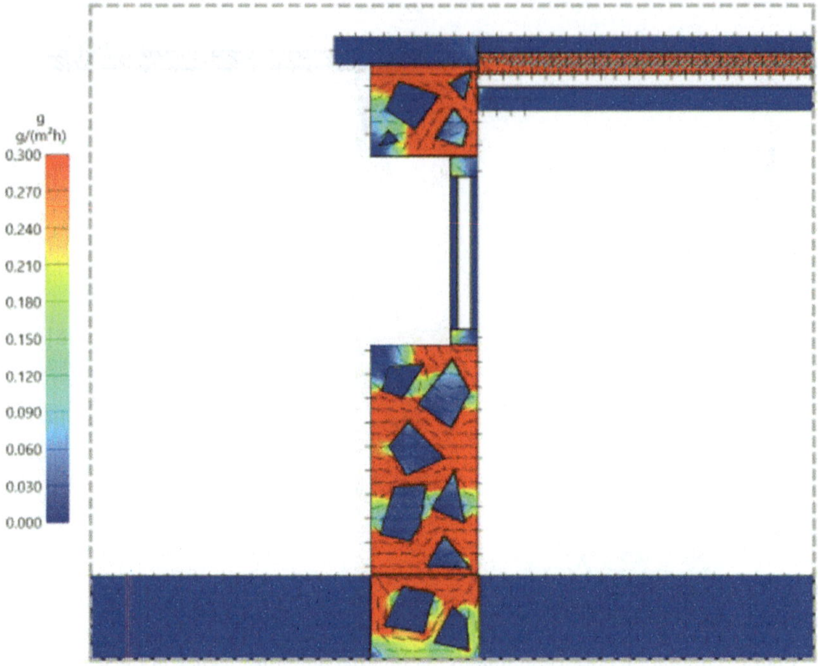

Fig. 6. Energy simulation of the vapour flow values transmitted at different points of the masonry wall. Source: authors from HTflux software.

4.5 Partial Pressure

Finally, the graphical analysis of partial pressure (Fig. 7) confirms the points where the structure is most vulnerable. Given the seismic hazard of Órgiva, although the sliding of the slopes themselves could attenuate the structural pathologies, ultimate limit states could be reached, causing the loss of bearing capacity of the ground where the sliding foundation of the wall acting as a vertical structure is located. In addition, as it is a material that can easily detach given the stored moisture, it would lead to structural failure accentuated by the fatigue of repeated loads.

According to the indications of the CTE DB SE-C, in the event of an earthquake, the serviceability limit states would be affected, causing alterations in the geometry of the building, affecting the appearance and comfort of users, generating problems with the installation systems and reducing functional efficiency. Therefore, it is crucial to evaluate the vulnerability of different buildings based on their geometry and composition.

4.6 Earthquake-Resistant Possible Implications

As a result of the partial pressure that runs through the wall, and the accumulation of relative humidity, the vapour flow is affected, thus impeding the correct transpiration of the wall. Besides, the composition of masonry walls and the effects of moisture in them affect the degradation of their resistance and their capacity to dissipate seismic energy

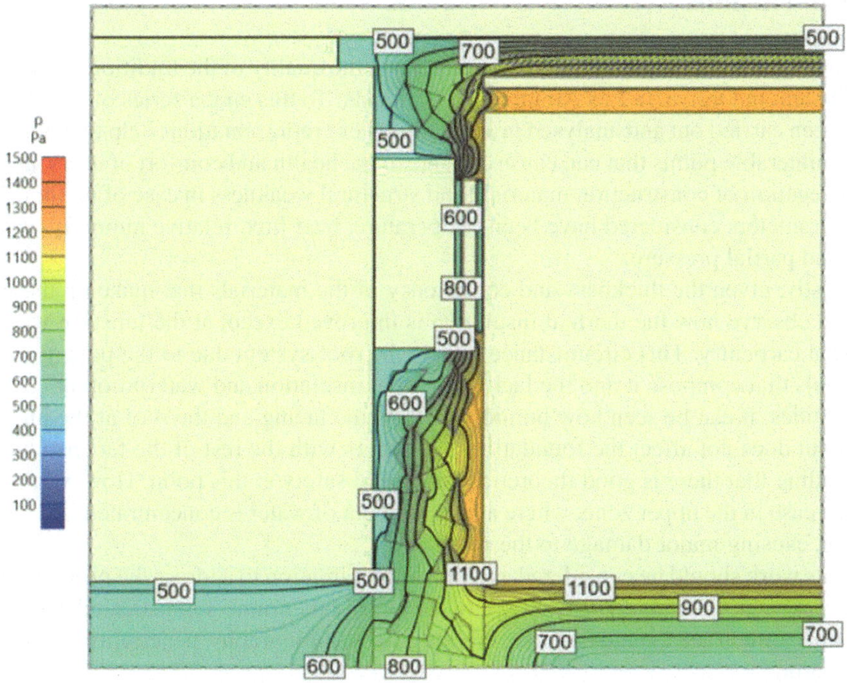

Fig. 7. Energy simulation of the partial pressure values transmitted at different points of the masonry wall. Source: authors from HTflux software.

[21]. Therefore, it is important to conduct an internal analysis of the wall. According to the results obtained and the area where it is located, it could be considered a vulnerable structure requiring intervention such as internal reinforcement and system strengthening [22].

As we can deduce from the results of the simulations carried out, the main points susceptible to loss of resistance capacity are the base of the wall and the crown of the wall. Research carried out with unreinforced masonry shows that the density of the walls has a direct influence on the seismic behaviour and the response of the construction system to lateral loads, which it absorbs with difficulty [23]. Together with the deficit caused by the detachment of material at sensitive points in the wall, these forces could lead to the building collapse. Furthermore, the seismic acceleration perpendicular to the plane of the wall, based on tests run in prone to earthquakes areas, indicates that the slenderness of the wall and the interruption of the wall (for the aperture of openings and windows) reduces the resistant capacity to absorb seismic accelerations to zero [24].

Given the signs of vulnerability, these buildings can be regarded as priority structures that require assessment and invasive testing to reliably determine the actual resistance to horizontal loads and identify the necessary measures to be implemented to reduce vulnerability [25].

5 Conclusions

This work has studied the factors that intervene in the quality of the traditional construction system in Órgiva, in Las Alpujarras of Granada. To this end, a series of simulations have been carried out and analysed graphically. These representations help to delimit the most vulnerable points that can cause damage to the health and comfort of users, as well as segregation of construction materials and structural weakness in case of earthquakes. The parameters considered have been temperature, heat flux, relative humidity, vapour flow and partial pressure.

Firstly, given the thickness and consistency of the materials that make up the wall, we can observe how the thermal insulation is improved except at the junction with the external carpentry. This circumstance affects the roof system due to the poverty of the materials that compose it and the lack of thermal insulation and waterproofing.

Besides, it can be seen how humidity affects the facing and the roof at the junction point but does not affect the foundations, as occurs with the rest of the factors studied, concluding that there is good theoretical structural safety at this point. However, this is not the case in the upper zone, where a large amount of water is concentrated and cannot escape, causing major damage to the roof.

This work should be considered as a first approximation to address the problem since a more exhaustive analysis would require the use of other more sophisticated and costly techniques, such as rebound hammer, thermography, radiography, penetration resistance, or so forth.

Acknowledgements. The authors would like to thank Universidad de Coruña for providing us with the HTflux software licences and the data related to the traditional construction analysed in the work.

References

1. Gómez Olazábal, L., Egido Orue, C.: La arquitectura popular de la Alpujarra. Narria Estud. artes y costumbres Pop. **3**, 4–5 (1976)
2. López Osorio, J.M., Lara García, J., Peñalver Martín, S.: Agua y arquitectura en la Alpujarra de Granada (España): Una aproximación a los conjuntos hidráulicos de las fuentes y los lavaderos de la comarca. J. Tradit. Build. Archit. Urban. (1), 395–410 (2020). https://doi.org/10.51303/jtbau.vi1.366
3. Casado Alcalde, A.: Introduction to the urbanism of Las Alpujarras. Gaz. Antroplogía. 6, Article 6 (1988)
4. Soroche Cuerva, M.Á.: Cultural landscape, town planning and traditional architecture. Cuad. Arte la Univ. Granada. **35**, 209–221 (2004)
5. Insitituto Geográfico Nacional: Catálogo de terremotos http://www.ign.es/web/ign/portal/sis-catalogo-terremotos. Accessed 15 Aug 2021
6. Tena-Colunga, A.: Simplified design method for base-isolated bearing-wall masonry structures. Rev. Int. Ing. Estructuras. **22**, 1–33 (2017)
7. Glavaš, H., Hadzima-Nyarko, M., Haničar Buljan, I., Barić, T.: Locating hidden elements in walls of cultural heritage buildings by using infrared thermography. Buildings **9**, 32 (2019). https://doi.org/10.3390/buildings9020032

8. Penner, D.E.: Chapter 1: cognition, computers, and synthetic science: building knowledge and meaning through modeling. Rev. Res. Educ. **25**, 1–35 (2000). https://doi.org/10.3102/0091732X025001001
9. Wieser, M., Onnis, S., Meli, G.: Desempeño térmico de cerramientos de tierra alivianada: posibilidades de aplicación en el territorio peruano. Rev. Arquit. **22**, 164–174 (2020). https://doi.org/10.14718/RevArq.2020.2633
10. Camacho Olmedo, M.T.: Ritmos temporales y dinámicas del paisaje de la alta Alpujarra Granadina mediante Sistemas de Información Geográfica. Geographicalia. 5–22 (2016). https://doi.org/10.26754/ojs_geoph/geoph.2005471327
11. Sánchez Hita, A.: El patrimonio minero en la Alpujarra granadina. Rev. PH. 68, 51 (2008). https://doi.org/10.33349/2008.68.2642
12. Ministerio de Fomento (Gobierno de España): DA DB-HE / 1. Zonificación climática en función de la radiación solar global media diaria anual. In: Código Técnico de la Edificación. Documento de Apoyo al Documento Básico DB-HE Ahorro de energía. Boletín Oficial del Estado (2011)
13. Goicoechea Acosta, M.: Aproximación al estudio de la influencia del clima en la erosión. El caso de la Alpujarra alta granadina. Cuad. Geográficos la Univ. Granada. 11, 153–182 (1981)
14. IGN-UPM-WorkingGroup: Actualización de mapas de peligrosidad sísmica 2012. Instituto Geográfico Nacional, Madrid, Spain (2013)
15. Alías, H.M., Jacobo, G.J.: Simulación del consumo de energía eléctrica para mantener el confort higrotérmico interior en viviendas con envolventes de madera y de mampostería . Av. en energías Renov. y medio Ambient. 8, 37–42 (2004)
16. Díaz Regodón, M.I., Tenorio Ríos, J.A.: Pérdidas de calor y formación de condensaciones en los puentes térmicos de los edificios. In: Científicas, C.S. de I. (ed.) I Jornadas de Investigación en Construcción (Instituto de Ciencias de la Construcción "Eduardo Torroja"), pp. 1131–1153, Madrid, Spain (2005)
17. Muñoz, C., Bobadilla, A.: Simulation and Evaluation of Thermal Bridges: Typical constructive solutions approved by the Thermal Norm for vertical elements in structure of wood and metalists in Zone 4. Simulations with Therm and Usai and evaluation with Method of Thermal Camera. Rev. la construcción. **11**, 92–111 (2012)
18. Salazar-Ceciliano, E.: Análisis de flujo de ventilación mediante software de CFD como mecanismo de conservación del patrimonio, aplicado a la Antigua Capitanía de Puerto Limón. Rev. Tecnol. en Marcha. **33**, 61–70 (2020). https://doi.org/10.18845/tm.v33i8.5509
19. Santos-Rufo, C., Lorite-Torres, I., Richard, A.: La evapotranspiración: concepto y métodos para su determinación. In: Mejora de la gestión de los recursos hídricos por medio de la integración de técnicas de teledección y modelos de simulación, pp. 19–25. Fundación Unicaja, Málaga (2008)
20. Ministerio de Fomento (Gobierno de España): DA DB-HE / 2. Comprobación de limitación de condensaciones superficiales e intersticiales en los cerramientos. In: Código Técnico de la Edificación. Documento de Apoyo al Documento Básico DB-HE Ahorro de energía. Boletín Oficial del Estado (2013)
21. Marinilli, A., Castilla, E.: Evaluación sismorresistente de muros de mampostería confinada con dos o más machones. Bol. Tec. Bull. **45**, 58–73 (2007)
22. Quintana Ramírez, H.S.: Reducción de la vulnerabilidad sísmica en viviendas por desarrollo progresivo con mampostería no armada. Designia. **3**, 99–119 (2015). https://doi.org/10.24267/22564004.150
23. Cornejo Vázquez, T., Henández Barrios, H.: Lateral resistance of confined masonry structures. Ingeniería. **17**, 179–190 (2013)

24. Jaramillo, J.D.: Mecanismo de transmisión de cargas perpendiculares al plano del muro en muros de mampostería no reforzada. Rev. Ing. Sísmica. **67**, 53–78 (2002). https://doi.org/10.18867/ris.67.205
25. Zora Mejía, F.N., Acevedo Jaramillo, A.B.: Índice de vulnerabilidad sísmica de escuelas del Área Metropolitana de Medellín, Colombia. Rev. EIA. **16**(32), 195–207 (2019). https://doi.org/10.24050/reia.v16i32.1035

Post-occupancy Evaluation of Higher Educational Buildings: Purpose, Obstacles and Benefits

Željka Jurković(✉) and Danijela Lovoković

Faculty of Civil Engineering and Architecture Osijek, Josip Juraj Strossmayer University of Osijek, Ulica Vladimira Preloga 3, 31000 Osijek, Croatia
zjurkovic@gfos.hr

Abstract. The post-occupancy evaluation (POE) is an overview of the architectural and technical performance of the building during its use. The purpose of POE is to improve future construction, based on previous experiences in the use of the building by immediate users. We primarily explored the architectural aspect of buildings (site, function, interior design) and the methods in which it should be valorized by users, primarily by a questionnaire. The article gives an overview of the benefits that investors and architects can have from POE and highlights obstacles to more POE procedure. We are focused on the long-term benefits that POE can provide if adopted as a standard method of valuing higher education buildings by users. POE focus in higher educational buildings is whether the physically constructed structure of the building with its architectural elements results in quality use by end-users. We found that the implementation of POEs procedures is the most necessary in educational buildings.

Keywords: Post-occupancy evaluation · Higher educational buildings · Architecture · Benefits · Questionnaire

1 Introduction

History of modern-day POE methods dates back to the late 1960s. The first publication with the term "POE" dates back to the 1975 but the terminology became more accepted after the textbook was written in 1988 [1] and after Technical report for US Federal Facilities Council was written in 2001 [2]. POE starts from the assumption that the performance of the building should be evaluated by the end-users at least one year after moving in. POE is defined procedure or process of systematically evaluating the characteristic of the buildings after they have been built and occupied after some time [3].

POE implies two basics aspects of buildings: technical performance and functional performance [4]. Technical performance includes: site performances, lighting, heating, cooling, ventilation, acoustics, indoor air quality, energy and water consumption, CO2 output, use of green energy sources, fire safety. Functional performance includes:

H. Glavaš et al. (Eds.): OTO 2021, LNNS 369, pp. 253–261, 2022.
https://doi.org/10.1007/978-3-030-92851-3_19

space arrangement, aesthetic value, amenity, life-cycle cost, maintenance [4, 5]. Consequently, there are two types of methods applied in POEs: objective methods (physical measurements) and subjective methods (questionnaires, interviews), measurable and non-measurable methods, i.e. qualitative and quantitative data.

The article primarily explores the functional parts of POE, those related to the architecture of the buildings. POE process can evaluate any purpose of building: commercial buildings, offices, residential buildings, educational buildings, medical buildings, transport buildings, etc. We chose higher education buildings because they are mostly built with public finances, they have significant building and maintenance costs, they have an area of several thousand m^2 and premises for various purposes (lecture halls, cabinets, laboratories, offices) and they must be nearly zero energy buildings (nZEB). Higher education buildings have a large number of different users (students, teachers, non-teaching employees) who spend majority of their time indoors. Spatially, because of their purpose and size, higher education buildings are important for the city in terms of urban position and they are often located in prominent city locations or on university campuses.

The purpose of POE is to improve future construction, based on previous experiences, actual performance of the building, feedback of the building, rather than project and calculations predicted performance. The goal is to learn from the experience of the building's users, determine has the building's performance met the expectations, how satisfied building users are with the environment that has been created [2]. The aim of POE is to correct possible mistakes and reduce maintenance costs, all in order to improve the next project. In this article, the focus is on the long-term benefits that POE can provide if adopted as a standard method of valuing higher education buildings by users.

2 POE Methods for Evaluating Architectural Elements

The general methodology of the POE process includes several phases that can be summarized in three: the preparatory phase (selection of POE team, what is to be inspected and evaluated, review of building documentation), implementation phase (total building observation tour, conducting interviews or questionnaires, data collection) and the POE reporting phase [5]. POE is an interdisciplinary process that involves the participation of architects and other engineers, together those who were involved in the designing, supervision, or construction of the building and those who were not involved. The participation of independent experts ensures the objectivity of the overall approach, objective analysis of information and obtained results. Royal Institute of British Architects (RIBA) defined POE as a systematic study of buildings in use to provide architects with information about the performance of their designs and building owners and users with guidelines to achieve the best out of what they already have [6]. RIBA also gave to the public Post Occupancy Evaluation Guidance to ensure that the profession and the public have access to POE-related methods and procedures [7]. Different characteristics of the building are examined by different methods. The functional characteristics of the building are examined mainly by subjective methods (surveys, questionnaires, interviews, building tour observation). POEs of functional performances of higher educational buildings often focuses on occupants' experience and use of the facilities and almost every POE procedure would use a questionnaire or an interview as the most common research method [8].

Basic architectural elements for POE of functional building performance are presented in Table 1 [5, 7, 9–12]. They include evaluation of site, architecture, functionality, environmental protection, security, interior and comfort of use. Table 1 displays the elements of the higher education building that need to be evaluated and the proposed established evaluation methods or POE feedback techniques.

Table 1. Basic architectural elements for POE of functional building performance.

Basic architectural elements for POE	Performance indicators	POE methods
Site, location context, zooning, surroundings	Urban presence and visibility, street frontage, access (pedestrian, bicycles, vehicles), walking distance to public transport, parking	Building tour, interviews, questionnaires
Overall architecture	Building form, building size, building volumes, number of stores, Flexibility and external envelope, fenestration, building phasing, potential for future expansion	Building tour, interviews, questionnaires
Functionality	Spatial configuration, spatial layout, grouping of spaces, spatial relationships, dimensions, proportions, heights of lecture halls, heights of laboratories, circulation, vertical and horizontal communications, accessibility	Building tour, interviews, questionnaires
Environmental protection	Flooding, earthquakes, wind and sun, erosion	Building tour, questionnaires
Interior	Furnishings, walls, doors, ceilings, materials, colors, finishes	Building tour, questionnaires
Safe and secure	Alarms, cameras, detectors, safety in use	Building tour, questionnaires
General occupant satisfaction	Space quality, Privacy and comfort, occupant's health and wellbeing response	Questionnaires

In addition to the listed standard evaluation elements (Table 1), higher education buildings, due to their specifics, require the evaluation of additional building characteristics as presented in Table 2. The tabular overview is systematized on the basis of research from the available literature [4–7] and research by the authors. Specific evaluation elements relate to the purpose of the building (higher education building), to different users (students, teachers, non-teaching employees) and the specific requirements that the building should have as a public building in which a large number of people gather, reside and work.

Table 2. Specific architectural elements for POE of functional building performance.

Specific architectural elements for POE	Performance indicators	POE methods
Aesthetics, image	Unique form, design innovations, inspiring architecture	Questionnaires
Creativity, innovations	Efficiency in teaching and student activities, choice of learning and teaching environment, workplace productivity, interactivity, collaborative work environments	Questionnaires, interviews
Occupancy	Capacity of lecture halls and laboratories	Schedules, interviews
Interior	Equipment	Building tour, questionnaires
Signability	Signposting, wayfinding, visible and clear visual communications	Building tour, questionnaires

It is evident that POEs are flexible procedures, which can be upgraded, depending on the specific requirements or new elements in the behavior of the buildings to be evaluated. Emergency issues that should be incorporated into POEs, for example, are the threat of terrorism or evaluating the buildings to prevent the spread of coronavirus or future pandemics, as presented in Table 3 [10].

Table 3. Emergency architectural elements for POE of functional building performance.

Emergency architectural elements for POE	Performance indicators	POE methods
Anti-terrorism protection	Barriers, alarms, detectors security equipment	Building tour, questionnaires
COVID-19 building response	Maintaining social distance, capacity of outdoor and indoor spaces, capacity of covered spaces, capacity of lecture halls and laboratories, natural ventilation, separate communications, adjustments for distance learning	Building tour, questionnaires

3 Benefits and Beneficiaries of POE for Higher Educational Buildings

POE is a diagnostic tool which allows investors and facility managers to identify and evaluate critical aspects of building performance systematically. POE projects are conducted for numerous purposes: to identify problem areas in existing buildings, to test new building prototypes and to develop design guidance and criteria for future buildings [8, 13]. POE provides the ability to compare projected building performance and actual building behavior in use. The most important benefit of POE is feedback, "lesson learned", valuable information how the building actually behaves in use, whether and how satisfied users are with the building.

There are three main beneficiaries of POE: investors (facility or building managers), architect (and other engineers) and users, as presented in Table 4.

We can also conclude that the general public or society as a whole is the most important beneficiary of POE because POE procedures lead to saving public money, ensure rise of architectural and building quality and provide knowledge for regulatory processes. Public buildings are particularly important in this context and thus buildings for higher education. Universities invest in more new buildings over the years and continuously renovate existing buildings, higher education sector is involved in numerous and often significant construction projects, sector must promote value for the money [11]. Higher education buildings are the ones that often guide construction standards and the sustainability of construction and influence national policy in regard to buildings and the environment [11, 12].

POE is a management aid, feed-back method for measuring building performance (technical performance) and functional performance (users interaction with built environment), POE leeds towards more sustainable production and consumption of built environment [20].

Table 4. Benefits and beneficiaries of POE

Beneficiaries of POE	Benefits from POE	References
Investors building managements	Improve the commissioning of subsequent buildings, improve management procedures, provides data for future buildings, provide procedures for quality monitoring, mclient gets more from their investment, cost and time savings, it supports fine-tuning of the building, it supports renovation of existing building and prioritize renovations, allow to build on success, not no repeat failures, POE Database	[2, 7, 15, 18–20]
Architects other engineers	Applying design skills more effectively, improve the design of subsequent buildings, learning from occupant response, access their work on a deeper level, allow to build on success, not to repeat failures, provide knowledge for design guides and regulatory processes, better understanding of the psychosocial aspects of buildings, POE Database	[2, 14–16, 18, 20]
Users	Improved fit between occupants and their buildings, evaluation of human (user) response, connect building with occupants' habits, occupant survey database	[2, 17, 20]

4 Obstacles or Why There is not Much POE in Practice?

POE has not become a norm in building industry, number of literary sources documented why POE is not carried out with regularity [8, 11, 14, 16, 20]. Although, POE has developed in last decade and continuing to rise. Many POEs protocols are in use in UK, USA, Canada, Australia and other countries [5, 7, 8, 11]. In the UKthere are two formal schemes underway which have been applied to the Higher Education sector. One (external) is PROBE (Post-occupancy Review of Buildings and their Engineering, from 1995) and the other (internal) to the higher education sector itself, is HEDQF (led by the Higher Education Design Quality Forum, from 1994) [11]. The PROBE project has demonstrated that a process of feedback from the original project team and the building users on key indicators can radically help improve building design [11]. More information about PROBE, the process, studies and conclusions, including downloadable reports may be found on the PROBE website [11, 21]. Multi-disciplinary HEDQF was set up to promote design quality and value for money ih higher education, bringing together professionals involved in design, construction, occupation and management of buildings. More information about HEDQF may be found on the HEDQF website [11, 22].

The question arises: why do we not see much of the POE in practice? The answer is somewhat paradoxical: those who should be beneficiaries of POE (see Table 4) are at the same time obstacles to the wider applications of POEs procedures.

POE takes time and resources and, above all, feedback culture. Investors, architect and other engineers have the realistic possibility to be harmed by POE [14]. Investors do not want to spend additional money for evaluating something that they had already procured and paid to professionals (architects, engineers). Who is certified to practice POE? And what if POE brings out some serious flaw in design? Who will pay for corrections? Who will pay for users' satisfactions? Architects and whole project team often think that they will be blamed for any problem, that their reputation will be ruined. POE is not part of regular or usual architect's service to their clients nor is it part of public procurement. Will only users have the power to determine the success of building? The notion of professional liability is the most significant contribution to the lack of POE, especially in a litigious society without feedback culture [11, 14].

POE is user-centered, but there is the absence of POE in the curriculum, we do not teach our students about user-centered design and POE methods [14, 19].

5 Discussion and Findings

Although many real and justified questions are asked about POEs, the longtime benefits of conducting POEs procedures (see Table 1,2,3) outweigh the concerns and fears of investors, architects, other engineers and construction industry.

Education is of the utmost importance to society as a whole, and so are education buildings. We found that higher education sector has the potential to be a leader in conducting POE procedures. Which sector is more scientifically prepared, more committed to constant learning and improving? Which sector, if not higher education, is more used to practice self-assessment methods and is committed to feedback culture?

POE is a tool for evaluating actual performance not just predicted performance. Considering the multiple advantages of POE, we found that the implementation of POE procedures is the most appropriate and necessary in educational buildings. Higher education buildings have a large number of different users, buildings are mostly built with public finances and have significant building and maintenance costs because they often have an area of several thousand m^2. Universities invest in more buildings over the years, continuously building new or renovating existing buildings. The implementation of POEs for educational building can be organized in cooperation with the academic community (faculties), within the framework of scientific research, so not to burden the architects and other design companies financially and in time.

6 Conclusion

POE ask questions and provide answers on how buildings actually work in architectural, technical, social and management terms for the users [23]. POE is a diagnostic tool which allows architects, investors and facility managers to identify and evaluate critical aspects of building performance systematically. POE provide lessons and feedback for the architect, investors and construction industry. There are numerous benefits of POE, including improved building design, improved procurement and cost savings. Various certificates, including awards in architectural competitions come before, in the design phase of construction, POE as a method of quality evaluation comes after moving into the building. We can conclude that POE bridge that gap between construction phase and the use of the building. POE is user-friendly method of evaluation. One of the common and most effective method of evaluating architectural elements is the questionnaire.

In the article, we were focused on the assessment of functional performance, i.e. the architectural elements of buildings, especially higher education buildings. Considering advantages of POE, we found that the implementation of POE procedures is the most appropriate and necessary in educational buildings and that higher education sector has the potential to be a leader in conducting POE. The implementation of POEs for educational building can be organized in cooperation with the faculties, within the framework of scientific research.

Further research of POE procedures should focus on linking POE and BIM technology in construction, with the aim of increasing quality and reducing construction and maintenance costs.

References

1. Preiser, W.F.E., White, E.T., Rabinowitz, H.Z.: Post-Occupancy Evaluation. Van Nostrand Reinhold, New York, NY, USA (1988)
2. Federal Facilities Council Technical Report No. 145: Learning from our Buildings, A State-of-the-Practice Summary of Post-Occupancy Evaluation. National Academy Press, Washington, DC, USA (2001)
3. Preiser, W.F.E.: Toward universal design evaluation. In: 17th Conference International Association for People-Enviromental Studies; Culture, Quality of Life and Globalization: Problems and Challenges for the New Millennium, p. 42. Corunna, Spain (2002)
4. Hassanain, M.A., Mathar, H., Aker, A.: Post-occupancy evaluation of a university student cafeteria. Architec. Eng. Des. Manag. **12**(1), 67–77 (2015). https://doi.org/10.1080/17452007.2015.1092941
5. Higher Education Funding Council for England (HEFCE): Guide to Post Occupancy Evaluation (2006). http://www.smg.ac.uk/documents/POEBrochureFinal06.pdf. Accessed 29 Aug 2021
6. Riba, R.S.G.: A research report for the architectural profession. In: Duffy, F.W. (Ed.) Architectural Knowledge: The Idea of a Profession, E. & F. N. Spon, London, UK (1991)
7. RIBA: Post Occupancy Evaluation Guidance https://www.architecture.com/knowledge-and-resources/resources-landing-page/post-occupancy-evaluation. Accessed 26 Aug 2021
8. Li, P., Froese, T.M., Brager, G.: Post-occupancy evaluation: state-of-the-art analysis and state-of-the-art review. Build. Environ. **133**, 187–202 (2018). https://doi.org/10.1016/j.buildenv.2018.02.024
9. Ahmed, H., et al.: Post occupancy evaluation of school refurbishing projects: multiple case study in the UK. Buildings **11**, 169 (2021). https://doi.org/10.3390/buildings11040169
10. Kennett, E.: Post Occupancy Evaluations. Whole Building Design Guide (2021), https://www.wbdg.org/resources/post-occupancy-evaluations. Accessed 29 Aug 2021

11. Enright, S.: Post-occupancy evaluation of UK library building projects: some examples of current activity. Liber Q. **12**(1), 26–45 (2002). https://doi.org/10.18352/lq.7665
12. People Places: Post-Occupancy Evaluation. https://www.sl.nsw.gov.au/sites/default/files/poe_survey_template.pdf. Accessed 29 Aug 2021
13. Preiser, W.F.E.: Post-occupancy evaluation: how to make buildings work better. Facilities **13**(11), 19–28 (1995). https://doi.org/10.1108/02632779510097787
14. Hadjri, K., Crozier, C.: Post-occupancy evaluation: Purpose, benefits and barriers. Facilities **27**(1–2), 21–33 (2009). https://doi.org/10.1108/02632770910923063
15. Whyte, J., Gann, D.M.: Closing the loop between design and use: post-occupancy evaluation. Build. Res. Inf. **29**(6), 460–462 (2001). https://doi.org/10.1080/09613210110072683
16. Hosey, L.: Going Beyond the Punchlist: Why Architects Should Embrace the Post-Occupancy Evaluations? Metropolis Magazine (2019). https://metropolismag.com/projects/architecture-post-occupancy-evaluations/. Accessed 30 Aug 2021
17. Tookaloo, A., Smith R.: International conference on sustainable design, engineering and construction, Post Occupancy Evaluation in Higher Education. In: Procedia Engineering, vol. 118, pp. 515–521 (2015)
18. HCM Architects: The Role of the Post Occupancy Evaluation in Architecture. https://hmcarchitects.com/news/the-role-of-the-post-occupancy-evaluation-in-architecture-2020-01-22/. Accessed 26 Aug 2021
19. Cranz, G., Morhayim, L., Lindsay, Sagan, J.: Post Occupancy Evaluation in architectural education and practice. Technol. Architect. + Des. **5**(1), 25–30 (2021). https://doi.org/10.1080/24751448.2021.1863663
20. Cooper, I.: Post-occupancy evaluation - where are you? Build. Res. Inf., Spec. Issue Post-occupancy Eval. **29**(2), 158–163 (2001). https://doi.org/10.1080/09613210010016820
21. https://www.usablebuildings.co.uk/UsableBuildings/ProbeListAll.html. Accessed 28 Aug 2021
22. https://www.hedqf.org/. Accessed 28 Aug 2021
23. Lorch, R.: Special issue on post-occupancy evaluation. Build. Res. Inf., **29**(2), 79–174 (2001)

The Effect of Insufficient Cover Thickness on Structural Performance of Reinforced Concrete Buildings

Ercan Işık[1(✉)], Ehsan Harirchian[2], Hüseyin Bilgin[3], Berfin Kaya[4], and İbrahim Baran Karaşin[4]

[1] Bitlis Eren University, 13100 Bitlis, Turkey
eisik@beu.edu.tr
[2] Bauhaus-Universität Weimar, 99423 Weimar, Germany
[3] Epoka University, 1032 Tirana, Albania
[4] Dicle University, 21000 Diyarbakır, Turkey

Abstract. Reinforcements used in RC structures lose their load-bearing capacity by suffering from various deteriorations due to external effects. Insufficient cover thickness is one of these reasons. The topic of this study consists of the concept of inadequate concrete cover. Concrete cover values with different thicknesses were chosen, and it was aimed to show how these values affect the performance of a building. Base shear forces, stiffness values and target displacement were obtained for each different concrete cover. A reduction in concrete cover resulted in an increase in the base shear force and stiffness values. Insufficient concrete cover layer will reduce the strength of the reinforcement over time. This will negatively affect the earthquake performance of the building. It should be kept in mind that the main function of a concrete cover is to prevent external effects which adversely affect the strength of the reinforcement. In this study, information was also provided about corrosion caused by the lack of concrete cover, and solutions to the corrosion were presented.

Keywords: Corrosion · Cover thickness · Net concrete cover · Performance

1 Introduction

Reinforced-concrete structures have emerged with the principle of withstanding of concrete and steel jointly against external loads. The general working principle of reinforced-concrete is such that the compressive stresses affecting a building are countered by the concrete, and the tensile stresses are countered by the steel. The position and arrangement of reinforcement used in the RC structural elements directly affect the calculations. The concrete cover used in the structural elements directly affects the effective depth value obtained according to the position of the reinforcement. The concrete layer, which ensures that the reinforcements used in the reinforced concrete bearing elements is completely contained in the concrete, is called the concrete cover. The aim of the concrete cover is to prevent the weakening of the reinforcement by preventing the corrosion of

H. Glavaš et al. (Eds.): OTO 2021, LNNS 369, pp. 262–277, 2022.
https://doi.org/10.1007/978-3-030-92851-3_20

the reinforcement used in the bearing element due to external effects. In addition to this, especially against effects like fire, it ensures that the reinforcement has a longer-lasting load-bearing capacity. If the concrete cover is not sufficient, the strength of the reinforcement may decrease, causing the load-bearing capacity to completely disappear over time. Corrosion may be defined as the loss of qualities of metals by degradation as a result of an electrochemical reaction caused by the environment [1–4]. Corrosion of reinforced concrete steel is one of the important problems in the reinforced concrete construction sector. Corrosion of reinforcement used in reinforced concrete causes loss of the cross-sectional area of the reinforcement, formation of expansion pressure caused by cracking of the concrete overcoming the concrete cracking strength, decrease of service life, and cracking and peeling of the concrete covering. As a result, the adherence between concrete and reinforcement is reduced, which degrades the strength of the building. In extreme cases, even the structural integrity may break down [5–10]. In general, different types of corrosion rates directly affect the degree of deterioration [11]. There are various studies on corrosion of reinforcements [12–15]. Therefore, it is important to follow and monitor corrosion in structures where corrosion is likely to occur. This strategy is even more important in buildings with significance. Therefore, building a database by conducting corrosion studies on structures makes traceability more effective. Different methods may be used for this purpose [16–19]. One of the factors that cause corrosion in reinforcements is severe climatic conditions. For this reason, it is recommended that certain precautions are taken in the specifications and regulations related to buildings to protect the concrete's strength by accounting for the calculation of the concrete mixture, material selection, and the concrete cover layer to prevent corrosion [20]. It is important to know the factors that will affect the corrosion rate to eliminate the damage caused by corrosion. In general, the concrete cover prevents the reinforcement used in reinforced concrete from being influenced by temperature differences and prevents the reinforcement from rusting by preventing the contact of reinforcement with water and air. Maintaining the material properties of the reinforcement used in load-bearing elements increases the strength and resistance of the structure against external loads. If the concrete cover is inadequate, reinforcements rust easily and lose their bearing property over time. As a result of this, the life of the building decreases and we end up with buildings which do not have enough strength under additional loads such as earthquakes. Reinforcements can corrode from place to place and cause loss of diameter and quality. There are many reasons why corrosion occurs. Corrosion in reinforced concrete structures means a condition in which the reinforcement loses its strength over time because of rusting. The most important of these is the lack of net concrete cover in concrete elements. The increase in moisture content alongside chlorine ions in the air especially in places close to seashores increases the significant of concrete cover in terms of corrosion. One of the reasons why buildings collapse in earthquakes is the use of concrete covers that are inappropriate and inadequate. The adhesion between the reinforcement and concrete will not be sufficient if the concrete cover is inadequate. Thus, the concrete and the reinforcement cannot withstand the load by themselves. Moreover, the reinforcement exposed to corrosion will lose all of its strength over time. Another function of the concrete cover is to protect the reinforcement from fire. Concrete is more resistant to fire in comparison to steel. The fire resistance of reinforced concrete building elements is directly proportional to the thickness of the concrete cover layer and the quality of the concrete. It is an important point that concrete covers should be emphasized in this regard.

The inadequacy of the concrete cover in RC structures is a problem of workmanship occurring during the construction phase. The inadequacy of the concrete cover becomes an issue due to the fact that people who manufacture the iron and form of reinforcements, which will be used in load-bearing elements, do not show the necessary care. They try to overcome this problem that arises by using a fabrication material like plaster. It should not be forgotten that the concrete cover is a part of the reinforced concrete element. In this study, different concrete cover values were chosen in a sample RC building. The results were compared, and recommendations were made. The purpose of this study was to show the importance of the concrete cover. In this situation, it was the calculation of how a building would react to an earthquake. In the first part of the study, information is given about the terms concrete cover, and net concrete cover as well as the values found in the regulations. Information is given on corrosion, a serious problem resulting from lack of concrete cover. In the study, information is presented about the pushover curves used in structural analyses. Then, analyses were carried out for the selected three-story reinforced concrete structure, using the analyses included in the software.

2 Methodology

Concrete cover (c) is expressed as the distance between the center of gravity of longitudinal reinforcements used in a structural element and the outermost concrete fiber. Net concrete cover (cc) is used to provide the required adherence to reinforcements and protect reinforcements from external effects. Net concrete cover is the distance between the outer surface of the outermost reinforcement and the outermost concrete fiber [21]. The representation of concrete cover and net concrete cover thicknesses in any load-bearing element is given in Fig. 1.

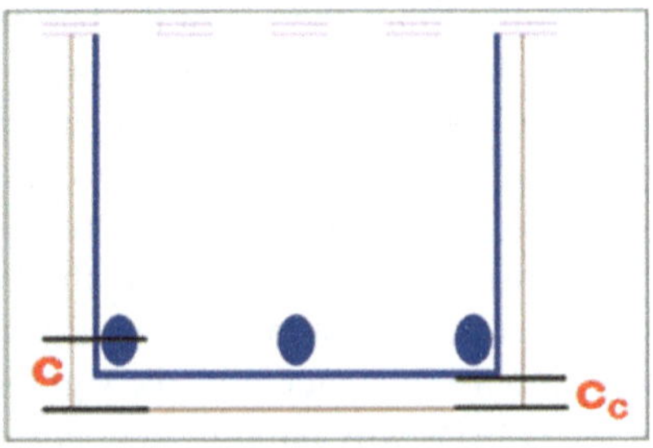

Fig. 1. Concrete cover (c) and net cover concrete (c_c)

As shown in Fig. 1, the relation between the thicknesses of concrete cover and net concrete cover is as follows.

$$c = \phi_1/2 + \phi_w + c_c \quad (1)$$

In this formula, ϕ_l is the longitudinal reinforcement diameter and ϕ_w is the transverse reinforcement diameter. The minimum net concrete cover thicknesses given in TS500 are presented in Table 1. When the effect of fire, corrosion and other harmful external factors may be excessive, the concrete cover thickness should be increased as much as necessary. This value is determined by the person who will design the building. There are no upper limitations in TS500.

Table 1. The required net concrete cover measured from the outside face of the outermost reinforcement [21].

In elements that are directly engaging with the ground	$c_c \geq 50$ mm
In columns and beams exposed to weather conditions	$c_c \geq 25$ mm
In columns and beams inside a structure, which cannot be exposed to external effects	$c_c \geq 20$ mm
In shear walls and floors	$c_c \geq 15$ mm
In shells and folded plates	$c_c \geq 15$ mm

One of the terms used while designing reinforced concrete (RC) structures is the term effective depth (d). Effective depth is defined as the distance from the center of gravity of the longitudinal reinforcement used in a building to the outermost concrete fiber in the compression region. Either of the thicknesses of concrete cover or net concrete cover may be used in the calculation of effective depth. Therefore, the insufficiency of concrete cover will directly affect the reinforced-concrete calculations of reinforced-concrete bearing elements. The effect of the amount of concrete cover to be applied on the calculation of effective depth should not be ignored.

The information regarding concrete covers in the TS500 Standard is rather incompatible with the equivalent European standards. EUROCODE 2 specifies the minimum concrete cover thicknesses according to environmental effect classes. Determining the thickness of concrete cover according to environmental effects is a necessary process for the service life of a building. It is more appropriate for designers to achieve this value as a result of calculation while choosing a concrete cover thickness. Concrete characteristic strength (f_{ck}), type of reinforcement, maximum diameter of reinforcement, environment exposure class, design working life member with slab geometry, concrete cast against uneven surface, special quality control of the concrete production, nominal aggregate size greater than 32mm and nationally defined parameters were taken into account as the input data for calculating the nominal concrete cover (c_{nom}) [22]. These data were used for calculation of structural class, concrete cover for durability ($c_{min,\ dur}$), and concrete cover bond ($c_{min,b}$). With these data, the minimum concrete cover (cmin) was calculated. The nominal concrete cover was obtained by adding the allowance for the deviation value (Δc_{dev}) to the obtained minimum concrete cover (c_{min}). The minimum cover may be reduced or increased for special conditions [22]. The concrete cover layer is obtained as a result of calculations in some parameters as well as limit values in EC2. In TS500, only the type of structural element is taken into consideration that given Table 1 for net concrete cover. There is no calculation method for concrete cover, therefore, the necessary modifications should be made on TS500 for the calculation-based thickness of concrete cover.

In this study, pushover analysis was used for determining the earthquake performance of building models. Pushover analysis is the common name for a type of procedure that uses simplified nonlinear analysis. Pushover analysis investigates the behavior of structures under earthquake loads in a non-linear manner. Sufficient information may be obtained about the seismic demands of structural systems and components resulting from design ground motion with pushover analysis. This method is a static procedure and practical way of showing a building's behavior in the inelastic region. The base shear force and peak displacement obtained from this analysis provide the capacity curve of the building. The pushover curve is a diagram obtained by geometrically combining the intersection points on an interaction diagram of the roof displacement values corresponding to the base shear forces under the applied load by increasing the structure from zero to unstable (Fig. 2). The pushover curve becomes meaningful by transforming it into modal capacity diagrams and calculating the maximum inelastic displacement capacity of the structure. Conventional pushover analysis is employed in the estimation of the horizontal capacity of structures implying a dynamic response that is not significantly affected by the levels of deformation incurred (i.e. the shape of the horizontal load pattern, which aims at simulating dynamic response, can be assumed as constant). It is automatically incremented by the program until it reaches the user-defined limit or a numerical error [23–34].

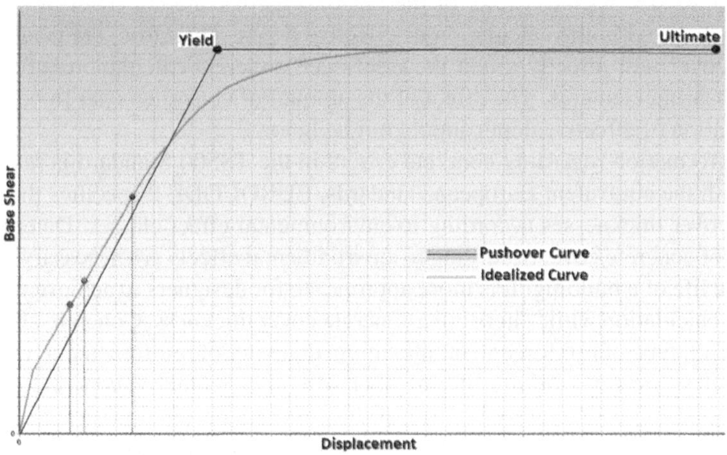

Fig. 2. Typical pushover and idealized capacity curves

In the study, a three-story reinforced concrete structure was selected and modelled three-dimensionally. The blueprint of the examined structure is shown in Fig. 3. A symmetrical building was chosen for ease of calculations.

Each story had an equal height and taken as 3 m. The material grades used for all structural elements was selected as C20 grade for concrete and S420 for steel grade in the software. All columns were selected as 0.40 * 0.50 m and beams were selected as

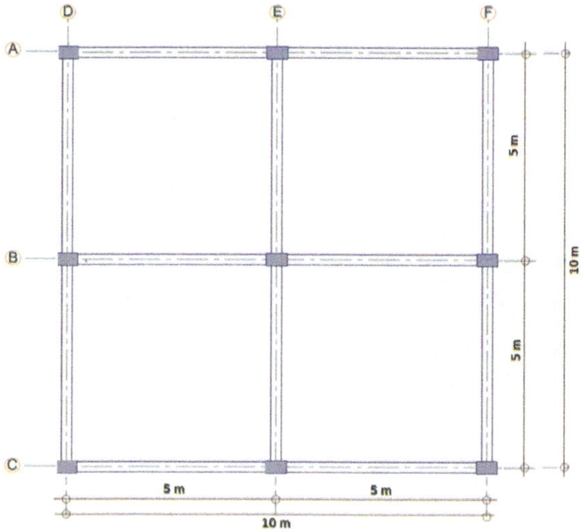

Fig. 3. The blueprint of the selected RC building.

0.25 * 0.60 m. The transverse reinforcements used in both elements were selected as ϕ10/10. The reinforcements used in the beams and columns were selected as ϕ14. The columns and beams used in the structure are shown in Fig. 4. The damping ratio was taken as 5% in all structural models. The ZA soil type which given in Turkish Building Earthquake Code (TBEC-2018) was chosen as the local ground class. The importance of structure was taken into consideration as Class II. Only the concrete cover thickness was selected as variable. All other values were the same in all models. The slabs were selected as rigid diaphragms.

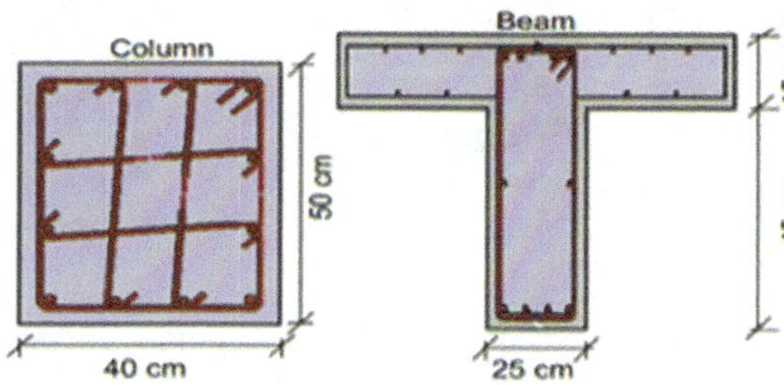

Fig. 4. The column's and beam's cross sections.

Different concrete cover thicknesses were chosen while making calculations about concrete cover. The calculations were made in the X direction by taking 0.001 m, 0.01 m, 0.02 m, 0.025 m, 0.03 m, 0.035 m and 0.040 m as the concrete cover values. It is not possible to select zero in the software. This value was chosen as the minimum possible value is 0.001 m. The Seismostruct software [35] was used to make the calculations. Permanent and incremental loads were applied to the structure in the software. Incremental load values of 11.11 kN and permanent load values of 5.00 kN were taken into consideration. The target displacement was selected as 0.18 m. All these values were taken as the same in all models that were used in this study. The three-dimensional model obtained in the software for the structure and the loads that were applied are given in Fig. 5.

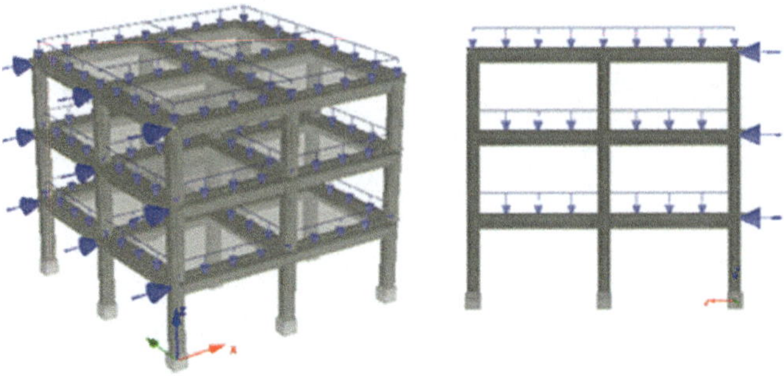

Fig. 5. The 3D and 2D models of the building

Three different situations were taken into consideration in this study. Firstly, the change of the concrete cover layer for a single column on the ground floor was examined. Secondly, only one beam on the ground floor was considered. Finally, different concrete cover thicknesses were considered in all columns and beams. The considered column and beam representations are shown in Fig. 6.

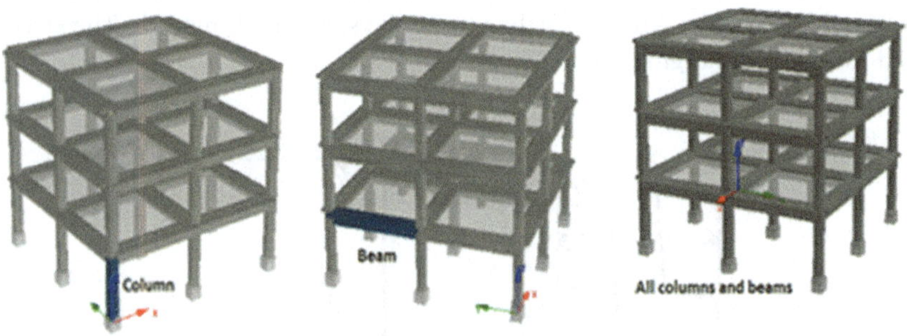

Fig. 6. The considered column and beam conditions in this study

The nonlinear concrete model [36] and steel model [37] were used for concrete and steel material. The stress-strain relationship of the material models considered in this study was given in Fig. 7.

Base shear forces for each construction model were calculated separately for both directions. Values were obtained for three different points on the idealized curve as displacement values. The first value refers to displacement at the moment of yield, the second value refers to the intermediate (d_{int}) displacement and the third value refers to the target displacement. The stiffness values of RC structural elements differ from the

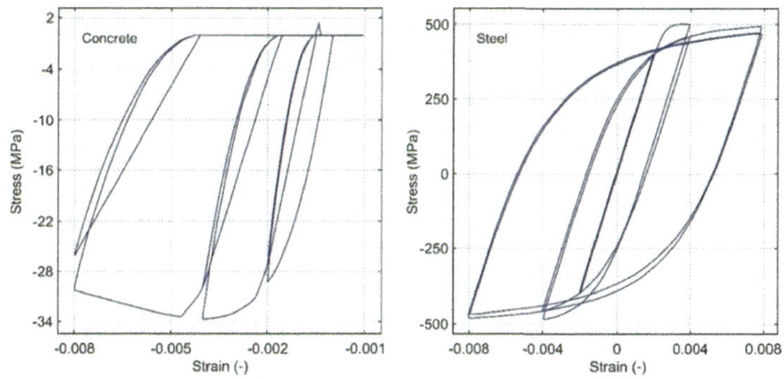

Fig. 7. The stress-strain relationship of material models for concrete and steel [35]

predicted stiffness values under the effect of an earthquake. Therefore, effective cross-sectional stiffness values are used in the design and analysis of these structural elements. The stiffness of cracked sections is taken into account to determine RC structural systems' performance under earthquake loads. The effective stiffness of cracked sections was obtained by using the prescribed stiffness reduction coefficients of the elastic stiffness value [38–40]. Elastic stiffness (K_elas) and effective stiffness (K_eff) values were also calculated separately for all models. It is crucial to determine the target displacements for damage estimation when certain performance limits of structural elements are reached in performance-based earthquake engineering. In the structural analysis, the limit states given in Eurocode-8 (Part 3) [41, 42] were taken into consideration for damage estimation used worldwide. All these displacements are shown in Fig. 8. Three different cases are stated for the damage cases in the software. These are considered as near collapse (NC), significant damage (SD) and damage limitation (DL). These values were calculated for all the structural models.

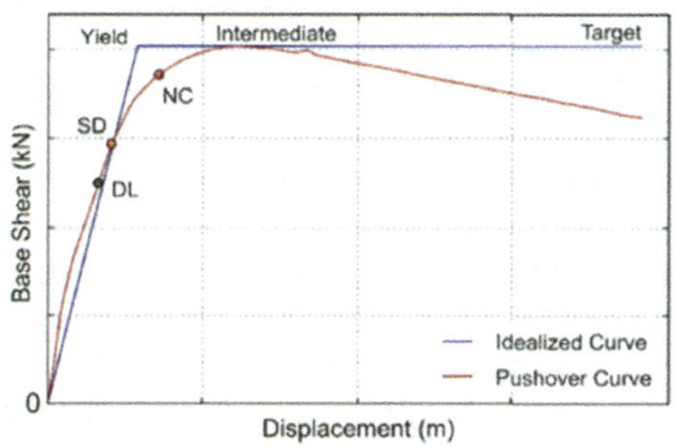

Fig. 8. Typical pushover and idealized curves and target displacements [43]

3 Results

For the selected sample reinforced concrete building, firstly, the change of the concrete cover for a single column located on the ground floor was considered. The pushover curve obtained due to the change of concrete cover on a single column is shown in Fig. 9 for the X direction.

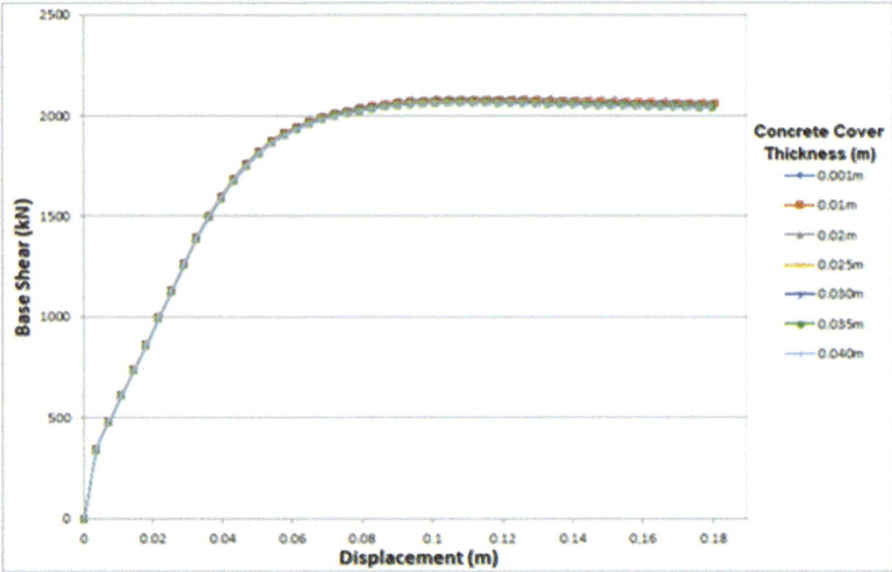

Fig. 9. Pushover curves obtained in the X direction due to concrete cover change for a single column on the ground floor

The comparison of all values obtained in the X direction as a result of the structural calculations for a single column is given in Table 2.

For the selected sample reinforced concrete building, secondly, the change of the concrete cover for a single beam located on the ground floor was considered. The pushover curve obtained due to the change of concrete cover on a single beam is shown in Fig. 10 for the X direction. The comparison of all values obtained in the X direction as a result of structural analyses for a single beam is given in Table 3.

As the concrete cover increased based on the analyses for changing the concrete cover on a single column or a beam on the ground floor, there is almost no difference in the base shear force. The period values were the same in all of these building models. An important change was not obtained on the elastic stiffness (K_elas) and effective stiffness (K_eff) values of the single column or beam models. The expected levels of performance are the same for all concrete covers. There was no change for the target displacement for performance criteria. The change in concrete cover was taken into account on all the columns and beams forming a concrete structure selected in the study. The pushover curves obtained due to the change of concrete cover on all the columns and beams are shown in Fig. 11 for the X direction. Table 4 shows the comparison of all values that were obtained in the X direction in case the same concrete cover value was selected for all columns and beams in the structure.

Table 2. Comparison of the results of analysis of concrete cover change for a single column

Concrete cover thickness (m)	Period (s)	Base Shear (kN)	Displacement (m)	K_elas (kN/m)	K_eff (kN/m)	DL (m)	SD (m)	NC (m)
0.001	0.2307	2086.05	0.0515	94835.83	40530.75	0.0131	0.0168	0.0292
			0.1086					
			0.1826					
0.010	0.2307	2080.20	0.0513	94832.83	40525.18	0.0132	0.0169	0.0293
			0.1084					
			0.1820					
0.020	0.2308	2074.04	0.0512	94755.05	40518.29	0.0132	0.0169	0.0293
			0.1081					
			0.1802					
0.025	0.2309	2071.00	0.0511	94694.87	40515.93	0.0132	0.0169	0.0293
			0.1080					
			0.1800					
0.030	0.2309	2068.54	0.0510	94555.34	40501.58	0.0132	0.0169	0.0293
			0.1080					
			0.1798					
0.035	0.2309	2065.53	0.0510	94421.69	40499.14	0.0132	0.0169	0.0293
			0.1078					
			0.1796					
0.040	0.2309	2063.45	0.0510	94418.32	40486.27	0.0132	0.0169	0.0293
			0.1077					
			0.1794					

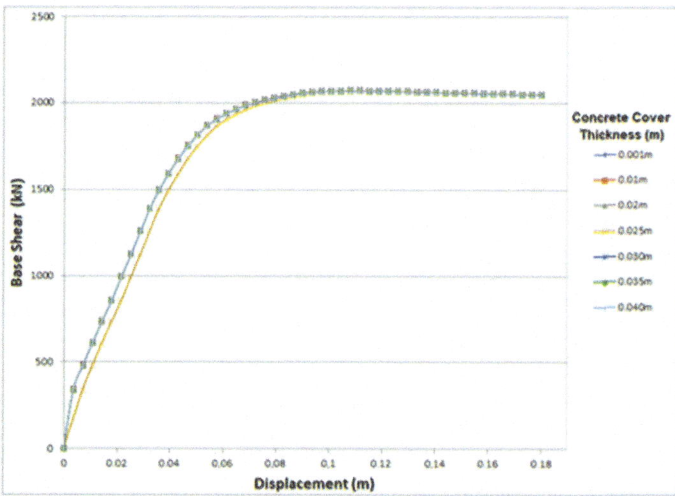

Fig. 10. Pushover curves obtained in the X direction due to concrete cover change for a single beam on the ground floor.

Table 3. Comparison of the results of analysis of concrete cover change for a single beam

Concrete cover thickness (m)	Period (s)	Base shear (kN)	Displacement (m)	K_elas (kN/m)	K_eff (kN/m)	DL (m)	SD (m)	NC (m)
0.001	0.2309	2071.00	0.0511	94907.59	40617.93	0.0132	0.0169	0.0293
			0.1080					
			0.1800					
0.010	0.2309	2071.13	0.0511	94827.68	40583.53	0.0132	0.0169	0.0293
			0.1080					
			0.1800					
0.020	0.2309	2071.13	0.0511	94471.08	40545.49	0.0132	0.0169	0.0293
			0.1080					
			0.1800					
0.025	0.2309	2071.00	0.0511	94694.87	40515.93	0.0132	0.0169	0.0293
			0.1080					
			0.1800					
0.030	0.2309	2071.00	0.0511	94658.52	40509.00	0.0132	0.0169	0.0293
			0.1080					
			0.1800					
0.035	0.2309	2071.00	0.0511	94633.71	40488.71	0.0132	0.0169	0.0293
			0.1080					
			0.1800					
0.040	0.2309	2071.00	0.0511	94591.15	40463.18	0.0132	0.0169	0.0293
			0.1077					
			0.1794					

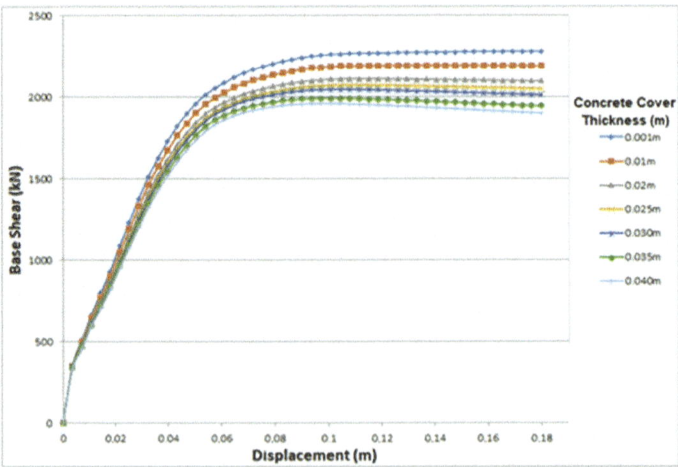

Fig. 11. Pushover curves obtained in the X direction for all columns and beams

Table 4. Comparison of the results of analysis of concrete cover change for all columns and beams

Concrete cover thickness (m)	Period (s)	Base shear (kN)	Displacement (m)	K_elas (kN/m)	K_eff (kN/m)	DL (m)	SD (m)	NC (m)
0.001	0.2267	2277.53	0.0539	97540.40	42244.05	0.0126	0.0162	0.0281
			0.1800					
			0.1800					
0.010	0.2285	2192.43	0.0523	96407.48	41938.20	0.0127	0.0163	0.0283
			0.1404					
			0.1800					
0.020	0.2305	2110.20	0.0516	95307.04	40907.65	0.0130	0.0167	0.0290
			0.1152					
			0.1800					
0.025	0.2309	2071.00	0.0511	94694.87	40515.93	0.0132	0.0169	0.0293
			0.1080					
			0.1800					
0.030	0.2317	2045.57	0.0507	94525.14	40368.05	0.0133	0.0170	0.0294
			0.1044					
			0.1800					
0.035	0.2324	1991.00	0.0499	93099.36	39928.57	0.0134	0.0172	0.0297
			0.1008					
			0.1800					
0.040	0.2334	1960.70	0.0499	91245.33	39267.17	0.0136	0.0174	0.0302
			0.9072					
			0.1800					

In case the value of concrete cover was the same for all the columns and beams of the building, the base shear force of the building was increased in the X direction due to the reduction of the concrete cover amount. When the cover was increased from 0.01 m to 0.40 m, the base shear force decreased by 11%. One of the reasons for this is that the effective height value decreases when the cover is increased. The increase in the effective depth was due to the small thickness of the concrete cover layer that was used. As the effective depth value increases, the amount of moment to be withstood by the structure will also increase. Considering the final damage conditions of the building models, it was determined that the amount of damage in the structure was increased as the concrete cover thickness decreases. This was due to the large base shear forces occurring at low concrete cover values. As the concrete cover thickness decreases, the stiffness value increases and depending on this increase, the period value has taken lower values. When the cover was increased from 0.01 m to 0.40 m, the period of building increased by 2%. This was due to the change in the effective depth value. Since the target displacement in the structure was chosen the same, this value was the same for all construction models. However, as the concrete cover layer increased, other displacement values in the structure decreased. A significant change was obtained on the elastic stiffness (K_elas) and effective stiffness

(K_eff) values of these models. When the concrete cover was increased from 0.01 m to 0.40 m, the stiffness of building decreased by 5%. This shows that the earthquake resistance of the structure decreases as the number of structural system element with insufficient concrete cover layer increases. With the increase of the concrete cover, the performance levels expected from the building also increased. As a result, the increase in the concrete cover layer decreases the effective depth and thus all structural calculations were affected. The selection of concrete cover values will directly affect the shear forces and damage conditions.

4 Conclusions

In this study, an attempt was made to show how changing the value of the concrete cover affects the performance of a building. For this purpose, a sample reinforced concrete structure was modelled with the help of a structural software. While the material class, sections, and loads remained the same, the thickness of concrete cover was chosen as variable. Pushover curves were obtained for each concrete cover value. The curves that were obtained were compared, the differences were observed, and the importance of the correct selection of the concrete cover thickness was emphasized. In this study, the importance of changing the concrete cover, which greatly affects the durability and performance of the structure, was clearly stated. Insufficient concrete cover reduces the life of a building and makes it vulnerable to earthquakes.

The minimum concrete cover thicknesses given in the TS500 standard are applied for all kinds of structures without any calculations. In Eurocode-2, the thickness of the concrete cover layer is decided according to the nature of the construction used and the environmental effects it will be exposed to. This study is important in that TS500, like Eurocode-2, should also set a calculation-based concrete cover thickness.

With TS500, which entered into force in 2000, the net concrete cover thickness has been expressed for the first time, and it has removed hesitations about the concrete cover. This added meaning to the basis of calculations of reinforced concrete structures. Any doubts that arise from the fact that the variations of transverse and longitudinal reinforcements have not been taken into consideration before have been removed. Studies should be carried out to determine the net concrete cover thickness, which is expressed independently of the transverse and longitudinal reinforcement, based on a calculation in the Turkish regulation.

In order for the concrete cover not to disappear, it is necessary to use fabricated elements instead of commonly used construction waste products. The concrete cover must be in sufficient quantity to fulfill the functions expected from it - to protect the reinforcement from fire, to prevent reinforcement corrosion and to ensure the adherence between concrete and concrete. Additionally, not using the concrete cover values specified on the project will change the obtained results to a large extent. This should not be ignored. The concrete cover directly affects the effective depth value used in the calculations of building elements. The elements used to provide the concrete cover are used to both protect the reinforcement against corrosion and increase the strength of the concrete against external effects. It is necessary for these elements to have the right material, shape and quantity depending on the place of use. If any of these are faulty and incomplete, problems arise during and after concrete casting. Therefore, the functions expected from the concrete cover layer will not be realized.

Adherence is one of the reasons why concrete and steel work together. An inadequate concrete cover layer affects adherence negatively. Therefore, the bond between the concrete and the steel will not be sufficient. This will cause loss of strain /stress transfer between the reinforcement and the concrete.

References

1. Smith, J.L., Virmani, Y.P.: Materials and methods for corrosion control of reinforced and prestressed concrete structures in new construction (No. FHWA-RD-00-081). Turner-Fairbank Highway Research Center (2020)
2. Broomfield, J.P.: Corrosion of Steel in Concrete: Understanding, Investigation and Repair. CRC Press, Boca Raton (2003)
3. Delikanlı, F.: The Corrosion Defects in Reinforced Concrete and Its Preventive Ways. İstanbul Technical University, Institute of Science and Technology, Thesis (M.Sc), İstanbul, Turkey (2001)
4. Cabrera, J.G.: Deterioration of concrete due to reinforcement steel corrosion. Cem. Concr. Compos. **18**(1), 1, 47–59 (1996)
5. Noh, H.M., Noor, N.M., Sarpin, N., Zainal, R., Kasim, N.: Structural effects of reinforced concrete beam due to corrosion. In: E3S Web of Conferences, vol. 34, p. 01024. EDP Sciences (2018)
6. Fernandez, I., Bairán, J.M., Marí, A.R.: Mechanical model to evaluate steel reinforcement corrosion effects on σ–ε and fatigue curves Experimental calibration and validation. Eng. Struc. **118**, 320–333 (2016)
7. Berrocal, C.G., Fernandez, I., Lundgren, K., Löfgren, I.: Corrosion-induced cracking and bond behaviour of corroded reinforcement bars in SFRC. Compos. B Eng. **113**, 123–137 (2017)
8. Rivera-Corral, J.O., et al.: Corrosion behaviour of steel reinforcement bars embedded in concrete exposed to chlorides: effect of surface finish. Constr. Build. Mater. **147**, 815–826 (2017)
9. Otunyo, A.W., Kennedy, C.: Effectiveness of resins/exudates of trees in corrosion prevention of reinforcement in reinforced concrete structures. Niger. J. Technol. **37**(1), 1, 78–86 (2018)
10. Zhu, W., François, R., Poon, C.S., Dai, J.G.: Influences of corrosion degree and corrosion morphology on the ductility of steel reinforcement. Constr. Build. Mater. **148**, 297–306 (2017)
11. Zhang, P., et al.: Steel reinforcement corrosion in Concrete under combined actions: the role of freeze-thaw cycles, chloride ingress, and surface impregnation. Constr. Build. Mater. **148**, 113–121 (2017)
12. Yu, B., Liu, J., Li, B.: Improved numerical model for steel reinforcement corrosion in Concrete considering influences of temperature and relative humidity. Constr. Build. Mater. **142**, 175–186 (2017)
13. Jin, X., Tong, J., Tian, Y., Jin, N.: Time-varying relative displacement field on the surface of concrete cover caused by reinforcement corrosion based on DIC measurement. Constr. Build. Mater. **159**, 695–703 (2018)
14. Ožbolt, J., Balabanić, G., Sola, E.: Determination of critical anodic and cathodic areas in corrosion processes of steel reinforcement in concrete. Mater. Corros. **68**(6), 622–631 (2017)
15. Hong, S., Wiggenhauser, H., Helmerich, R., Dong, B., Dong, P., Xing, F.: Long-term monitoring of reinforcement corrosion in concrete using ground penetrating radar. Corros. Sci. **114**, 123–132 (2017)

16. Dong, B., et al.: Monitoring reinforcement corrosion and corrosion-induced cracking by X-ray micro computed tomography method. Cem. Concr. Res. **100**, 311–321 (2017)
17. Li, W., Xu, C., Ho, S.C.M., Wang, B., Song, G.: Monitoring Concrete deterioration due to reinforcement corrosion by integrating acoustic emission and FBG strain measurements. Sensors **17**(3), 657 (2017)
18. Zhang, P., Liu, Z., Wang, Y., Yang, J., Han, S., Zhao, T.: 3D neutron tomography of steel reinforcement corrosion in cement-based composites. Constr. Build. Mater. **162**, 561–565 (2018)
19. El-Reedy, M.A.: Steel-Reinforced Concrete Structures: Assessment and Repair of Corrosion. CRC Press, Boca Raton (2017)
20. Aydın, Ö.: Investigation of measures for the prevention of corrosion of steel in concrete structures. Doctoral dissertation, Yıldız Technical University, Institute of Science, İstanbul Turkey (2012)
21. TS500: Requirements for design and construction of reinforced concrete structures, Ankara, Turkey (2000)
22. Walraven, J.C.: Eurocode 2: design of concrete structures EN1992-1-1. In: Symposium Eurocodes: Background and Applications, Brussels (2008)
23. Inel, M., Ozmen, H.B.: Effects of plastic hinge properties in nonlinear analysis of reinforced concrete buildings. Eng. Struct. **28**(11), 1494–1502 (2006)
24. Eslami, A., Ronagh, H.R.: Effect of elaborate plastic hinge definition on the pushover analysis of reinforced concrete buildings. Struct. Des. Tall Spec. Build. **23**(4), 254–271 (2014)
25. Biradar, M., Prasad, S.K.: Influence of soil stiffness on seismic vulnerability of irregular buildings. Indian J. Adv. Chem. Sci. **5**(1), 50–53 (2017)
26. Işık, E., Özdemir, M.: Performance based assessment of steel frame structures by different material models. Int. J. Steel Struct. **17**(3), 1021–1031 (2017). https://doi.org/10.1007/s13296-017-9013-x
27. Hadzima-Nyarko, M., Nyarko, E.K., Morić, D.: A neural network based modelling and sensitivity analysis of damage ratio coefficient. Expert Syst. Appl. **38**(10), 13405–13413 (2011)
28. Fajfar, P.: Capacity spectrum method based on ınelastic demand spectra. Earthquake Eng. Struct. Dynam. **28**(9), 979–993 (1999)
29. Chopra, A.K., Goel, R.K.: A modal pushover analysis procedure for estimating seismic demands for buildings. Earthquake Eng. Struct. Dynam. **31**(3), 561–582 (2002)
30. Freeman, S.A.: The capacity spectrum method as a tool for seismic design. In: Proceedings of the 11th European Conference on Earthquake Engineering, pp. 6–11 (1998)
31. Işık, E., Kutanis, M.: Performance based assessment for existing residential buildings in Lake Van Basin and seismicity of the region. Earthq. Struct. **9**(4), 893–910 (2015)
32. Hadzima-Nyarko, M., Morić, D., Španić, M.: Spectral functions of RC frames using a new formula for Damage Index. Tehnički Vjesnik **21**(1), 163–171 (2014)
33. Işık, E., Velioğlu, E.: A study on the consistency of different methods used in structural analysis. AKU J. Sci. Eng. **17**, 1055–1065 (2017)
34. Antoniou, S., Pinho, R.: Seismostruct–Seismic Analysis Program by Seismosoft. Technical Manual and User Manual (2003)
35. SeismoStruct v7: A Computer Program for Static and Dynamic Nonlinear Analysis of Framed Structures. Seismosoft (2018)
36. Mander, J.B., Priestley, M.J.N., Park, R.: Theoretical stress-strain model for confined concrete. J. Struct. Eng. **114**, 1804–1825 (1998)
37. Menegotto, M., Pinto, P.E.: Method of analysis for cyclically loaded RC plane frames including changes in geometry and non-elastic behavior of elements under combined normal force and bending. In: Symposium on the Resistance and Ultimate Deformability of Structures

Acted on by Well-defined Repeated Loads; International Association for Bridge and Structural Engineering: Zurich, Switzerland, pp. 15–22 (1973)
38. Caglar, N., Demir, A., Ozturk, H., Akkaya, A.: A simple formulation for effective flexural stiffness of circular reinforced concrete columns. Eng. Appl. Artif. Intell. **38**, 79–87 (2015)
39. Wilding, B.V., Beyer, K.: The effective stiffness of modern unreinforced masonry walls. Earthquake Eng. Struct. Dynam. **47**, 1683–1705 (2018)
40. Ugalde, D., Lopez-Garcia, D., Parra, P.F.: Fragility-based analysis of the influence of effective stiffness of reinforced concrete members in shear wall buildings. Bull. Earthq. Eng. **18**(5), 2061–2082 (2020). https://doi.org/10.1007/s10518-020-00786-3
41. EN 1998-3: Eurocode-8: Design of Structures for Earthquake Resistance-Part 3: Assessment and Retrofitting of Buildings; European Committee for Standardization: Bruxelles, Belgium (2005)
42. Pinto, P.E., Franchin, P.: Eurocode 8-Part 3: assessment and retrofitting of buildings. In: Proceedings of the Eurocode 8 Background and Applications, Dissemination of Information for Training, Lisbon, Portugal (2011)
43. Işık, E., Büyüksaraç, A., Ekinci, Y.L., Aydın, M.C., Harirchian, E.: The effect of site-specific design spectrum on earthquake-building parameters: a case study from the Marmara Region (NW Turkey). Appl. Sci. **10**(20), 7247 (2020)

Modelling the Durability of Cementitious Composites Elements by Means of Soft Computing Techniques

Slawomir Czarnecki(✉) and Lukasz Sadowski

Department of Materials Engineering and Construction Processes, Wroclaw University of Science and Technology, Wybrzeze Wyspianskiego 27, 50-370 Wroclaw, Poland
{slawomir.czarnecki,lukasz.sadowski}@pwr.edu.pl

Abstract. Proper modeling of the durability of cementitious composite elements is necessary to schedule maintenance and servicing of civil engineering structures. Knowing when the elements need their repair is a high-demand necessity in order to avoid sometimes very serious damage. It has been very often done using non-destructive testing (NDT) or other methods suitable for evaluation of the technical state of the buildings. However, the fact that there is an increasing number of research topics considering prediction of durability of cementitious composites structures using soft computing techniques indicates that this approach might also be effective solution for this purpose.

In this research, the authors would like to present a review of the latest research on the modelling of the durability of the cementitious composites with different admixtures using different machine learning algorithms. In this review, the most commonly used soft computing techniques for this purpose will be described and compared. Also, the main potential threats to loss of the durability of these elements such as wear of the material due to temperature (e.g. frost), mechanical (e.g. abrasion) and chemical (e.g. chloride concentration) will be recognised. Thanks to this review, an alternative approach to the evaluation of the technical state of buildings and its durability will be emphasized.

Keywords: Soft computing techniques · Durability of structures · Neural networks · Machine learning

1 Introduction

Concrete structures operate under adverse environmental and exploiting influences that can cause a decrease in their durability. It is stated that there are three important aspects that affect the durability of the structures. There are: proper design of cementitious composite mixture, erecting of the structure should be done fulfilling all of the requirements and during lifecycle of the structure it should be used according to its destiny and maintenance as well as periodic repair should not be neglected [1].

Unfortunately, it is a more difficult task than it looks, due to many variable factors related to the cementitious composite mixture and the influences during the life cycle [2].

H. Glavaš et al. (Eds.): OTO 2021, LNNS 369, pp. 278–287, 2022.
https://doi.org/10.1007/978-3-030-92851-3_21

In many cases, traditional techniques failed because they were costly, time-consuming, or sometimes impossible to use [3]. Thus, there is a necessity to propose other techniques which are competitive to e.g. laboratory or in-situ tests. With the acceleration of inventing novel machine learning algorithms, soft computing techniques have become a powerful device for solving many construction problems [4]. In this article the authors have focused on describing well-known usage of machine learning algorithms for prediction of the mechanical properties of cementitious composites and also for prediction of the potential threats connected with the environment in which the concrete structures might operate.

2 Soft Computing Techniques

For the purpose of modelling the durability of concrete structures, many different soft computing techniques are used. Analyzes using these techniques, for the purpose of finding a solution of the engineering problem, are performed during the training and testing processes. In this section, the short, schematic description of a few selected machine learning algorithms is presented.

2.1 Artificial Neural Networks

The most commonly used among others, the artificial neural networks imitating the biological neural connections in human brain, consist of fundamental units called neurons, which are grouped in layers. The first layer of the algorithm is an input layer composed of vectors that represent input parameters. The last layer called the output layer presents the solution of the scientific problem and consists of output vectors. The layers between these two are hidden layers consisting of hidden nodes. The layers between each other are connected and activated by activation functions. The training process is performed due to the learning algorithms and iterative adjustment of the weight between the nodes in neighboring layers. The most commonly used structure of the neural network is multi-layer perceptron which visual presentation is shown in Fig. 1.

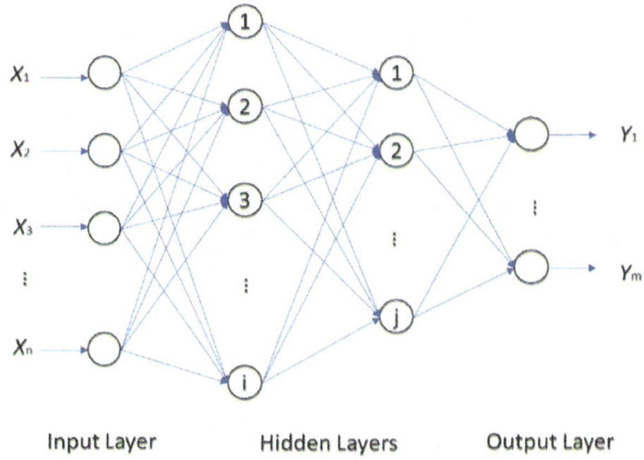

Fig. 1. The scheme of the multi-layer perceptron neural network.

2.2 Support Vector Machine

The Support Vector Machine is a soft computing technique used for analyzing classification and regression data, which is represented as a map of points in space and the solution is the hyperplane (lane in 2D, plane in 3D etc.) with the widest possible gap between the different classes. Each point in this space is described with a support vector (which is perpendicular to the hyperplane), as is shown in Fig. 2.

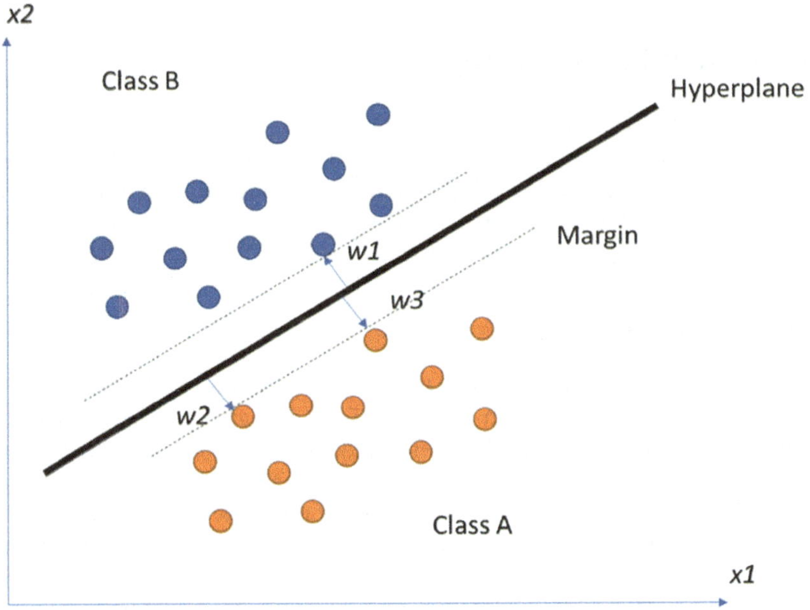

Fig. 2. The scheme of the support vector machine.

2.3 Ensemble Models

The ensemble models consist of decision trees which considered as supervised machine learning algorithm are able to solve both the regression and classification problems. The structure of such decision tree is consisting of nodes in which the binary decision is made, and this division is continued till the moment the algorithm will not be able to separate the data in the node. The most commonly used ensemble model algorithm is random forest, which consists of many decision trees and is schematically presented in Fig. 3.

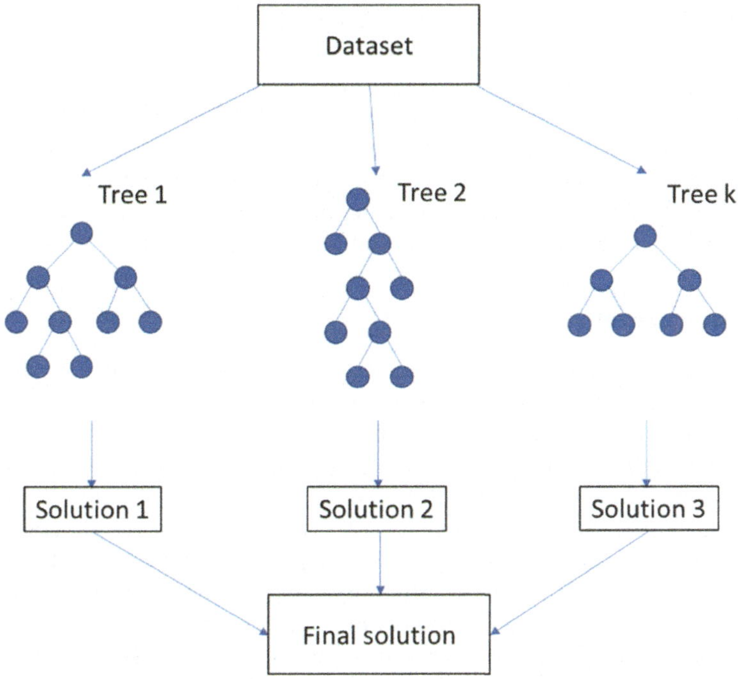

Fig. 3. The scheme of the random forest algorithm

3 Prediction of the Properties of Cementitious Composites

The aforementioned soft computing techniques are very often used to predict many important variables that are very connected or affect the durability of the entire structures. In this section, examples of such prediction of a few selected properties of elements made of cementitious composites are presented.

3.1 Compressive Strength

The most commonly predicted mechanical property of cementitious composites is compressive strength. This property is also considered to be the most important in the case of durability because the higher the compressive strength of the concrete, the better the durability. Therefore, lots of researches are performed in order to analyze this property by means of machine learning algorithms. They mostly investigate the compressive strength of concrete containing ordinary Portland cement using various algorithms such as neural networks [5], ensemble models [3], and support vector machine [7].

Recently, more often, self-compacting concretes have been subjected to studies [5]. In addition, the researchers tried to modify the mixture of concretes, these ordinary and self-compacting, by using eco-friendly admixtures. Such predictions were made successfully using neural networks for concretes containing ground granulated blast furnace slag [9], fly ash [10], silica fume [11], granulated rubber [12] and other materials that are wastes, but can be incorporated into cementitious composites as a admixture or substitute for cement, fine aggregate, or coarse aggregate.

Since the beginning of the XX century, the modeling of hardened concrete compressive strength plays a vital role in its technology development [13]. It is promising that after the laboratory inventions, the newest concrete mixtures, follow the numerical analyses, which accurately describes the properties of the designed mixtures. What is also important is the fact that in most of the aforementioned articles the parameters describing performance of these analyses confirm very high accuracy of these models.

3.2 Flexural Strength

Another commonly predicted property of cementitious composites is the flexural strength. It is mainly due to the fact that concrete is a brittle material and the potential threat of cracking could be dangerous to the safety of the structures [14]. Very often, the flexural strength is evaluated combined with the compressive strength due to the procedure of the strength tests. First, the beam samples are subjected to the 3-point bending test, and after the tests the remaining halves from the beams are subjected to the compression tests. It can be seen, as presented in [8] that using neural networks for the prediction of the flexural strength of self-compacting mortar can be as accurate as predicting the compressive strength. Recently, more research has been done to predict the flexural strength of reinforced concrete beams using the latest soft computing techniques such as gene expression programming [15]. However, even though the admixtures and reinforcements strongly affect the mechanical properties, the models presented in the literature are more often describing the prediction of the compressive strength instead of flexural strength as illustrated in Fig. 4.

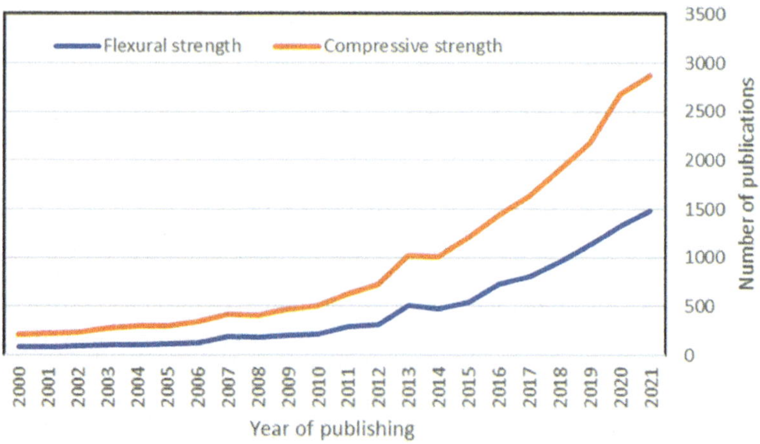

Fig. 4. The number of publications per year describing the prediction of the flexural strength of compressive strength in the Science Direct database.

According to Fig. 4 there are approximately twice more articles every year covering the topic of compressive strength prediction in comparison to the articles covering the topic of predicting flexural strength. It should be emphasised that when it comes to durability, fatigue loading during which tensile stresses occur is more dangerous due to potential threats of early corrosion [16].

3.3 Bonding Strength

The bonding strength of the cementitious composite was successfully predicted using a a neural network and information on the silica fume content in the mixture, compressive strength after 28 days and compressive strength after 90 days as input [17]. It has to be explained that bonding strength is not only the property of the material but also might be expressed as the adhesion between the two materials. Such prediction, of the two layers, cementitious composites have been made for newly constructed flooring elements where both layers were made of concrete [18] using neural networks. The interlayer bonding was also evaluated by means of neural networks for existing concrete elements with constant thickness of the layers[19], as well as for elements with variable thickness of the layers, the new one [20] and repaired [21].

In work [17], in which the most accurate model was obtained, the mechanical property evaluated was the bond strength of the subsurface layer, which is affected only by one material, not like in studies [18–21] where it is affected by the substrate and the overlay. It is a more difficult task to prepare such a model. It is also visible by the higher values of errors made by the machine learning algorithms, as shown in Fig. 5 for selected publish articles.

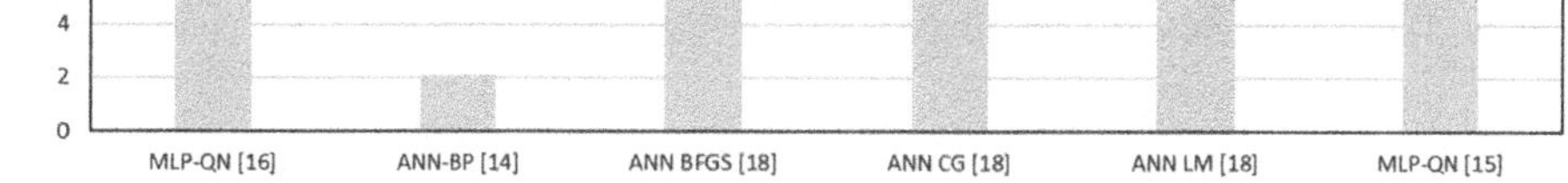

Fig. 5. Values of the mean percentage error to predict the bonding strength of cementitious composites elements.

The evaluation of the interlayer bonding is important from the durability point of view. The lower the adhesion, the lower the durability of the element. For this purpose there is a need to know the adhesion between the layers especially if the competitive method is destructive and cannot be performed in every place. What is also important is the fact that it is still possible to obtain more accurate results, compared to presented in Fig. 5, using other soft computing techniques.

3.4 Chemical Resistance

Cementitious composite elements are also exposed to different chemical influences, especially if they are located in coastal zones. These elements are highly affected by the penetration of chloride ions, and their failure in durability is mainly due to corrosion of steel reinforcements [22]. Thanks to the machine learning algorithms: neural networks and ensemble models there is a possibility to predict the chloride surface concentration [23] and based on experimental curves the depth of chloride ions penetration into the element can be estimated.

However, even for the most accurate model using gene expression programming [21], the values of the performance parameters, for example, the linear coefficient of correlation R^2 equal to 0.88, are not the highest among other properties investigated. Thus, there is still a need to improve the quality of the models presented in order to predict the chloride surface concentration. The performance of the selected models, in terms of linear coefficient of correlation, is presented in Fig. 6.

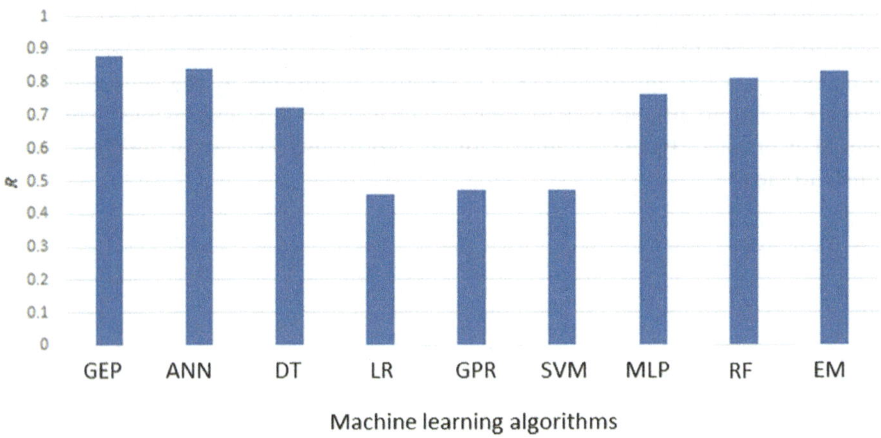

Fig. 6. The performance of selected models for predicting the chloride surface concentration.

The authors suppose that one of the main reasons why these models are not that efficient is the fact that chloride penetration into the concrete element is made by different mechanisms of transport of ions into the material in different zones.

3.5 Abrasion Resistance

In some cases the cementitious composites elements are exposed to various types of loading. One of these there is an abrasion which in some cases, e.g. concrete floors might be more important than compressive strength. This is due to the fact that the resistance to abrasion is primarily responsible for the durability of the concrete surface [25]. However, on the contrary, the compressive strength is strongly correlated with the resistance to abrasion and plays a vital role in the prediction of this property [26]. It is very promising that in [25] and [26] the values of the linear coefficient of correlation are very high, close to 1.0. It means that these models using neural networks accurately predict the abrasion resistance.

Although due to the fact that the abrasive wear of the surface of cementitious composites can be caused in many different ways, evaluating it only on the basis of the Boehme test and rotating cutter, could be slightly misleading, and it is the main limitation of these studies.

3.6 Hygrothermal Resistance

Due to different environmental influences in terms of temperature and humidity the concrete structures especially the concrete facades exposed outside are predestined to a physical and chemical damage. It is caused by the corrosion of reinforcing steel and freezing-thaw processes. In this case, it is important to evaluate, based on the data obtained from the hygrothermal monitoring system, the corrosion rate and other deterioration of the facade. This can be done by the recurrent neural network and the data collected consisting of hygrothermal values of temperature and relative humidity [27].

4 Conclusions

In this study, the review of the recent trends in the usage of soft computing techniques in order to predict the properties and influences affecting the durability of cementitious composite structures has been emphasized. In terms of the mechanical property, the most advanced models are presented for prediction of the compressive strength. This is mainly due to the fact that the quality of the concrete plays a vital role in durability and the higher the compressive strength, the higher the quality of the concrete. On the contrary, there are few very accurate models for such properties as bonding strength and chemical resistance of elements made of cementitious composites. The main reason for this is the fact that these cases are more difficult because more than one material is involved, and the phenomena, e.g. chloride ion transport, are difficult to describe. As the most common soft common technique used, the neural networks have been chosen. However, there are some other algorithms, e.g. gene expression programming, which in some cases perform better than neural networks.

Future research consisting of designing novel models in order to predict the durability of cementitious composites structures will be undertaken due to the constant necessity and their advantage in comparison with laboratory tests or in-situ tests.

Predicting many potential threats to these structures might be beneficial in avoiding unnecessary damage and costs related to their maintenance.

Thus soft computing techniques are competitive tools for evaluation of the durability of structures made of cementitious composites.

Funding. The authors received funding from the project supported by the National Centre for Research and Development, Poland [grant no. LIDER/35/0130/L-11/19/NCBR/2020 "The use of granite powder waste for the production of selected construction products."

References

1. Hooton, R.D.: Future directions for design, specification, testing, and construction of durable concrete structures. Cem. Concr. Res. **124**, 105827 (2019). https://doi.org/10.1016/j.cemconres.2019.105827
2. Qu, F., Li, W., Dong, W., Tam, V.W.Y, Yu, T.: Durability deterioration of concrete under marine environment from material to structure: a critical review. J. Build. Eng **35**, 102074 (2021). https://doi.org/10.1016/j.jobe.2020.102074
3. Farhat, H.: Chapter 7 - NDT processes: Applications and Limitations, Operation, Maintenance, and Repair of Land-Based Gas Turbines, Elsevier, Amsterdam pp. 159–174 (2021). https://doi.org/10.1016/B978-0-12-821834-1.00004-6.
4. Yu, Y., Nguyen, T.N., Li, J., Sanchez, L.F.M., Nguyen, A.: Predicting elastic modulus degradation of alkali silica reaction affected concrete using soft computing techniques: a comparative study. Constr. Build. Mater. **274**, 122024 (2021). https://doi.org/10.1016/j.conbuildmat.2020.122024
5. Nikoo, M., Moghadam, F.T., Sadowski, L.: Prediction of concrete compressive strength by evolutionary artificial neural networks. Adv. Mater. Sci. Eng., 849126 (2015). https://doi.org/10.1155/2015/849126
6. Asteris, P.G., Skentou, A.D., Bardhan, A., Samui, P., Pilakoutas, K.: Predicting concrete compressive strength using hybrid ensembling of surrogate machine learning models. Cem. Concr. Res. **145,** 106449 (2021). https://doi.org/10.1016/j.cemconres.2021.106449
7. Ling, H., Qian, C., Kang, W., Liang, C., Chen, H.: Combination of support vector machine and K-fold cross validation to predict compressive strength of concrete in marine environment. Constr. Build. Mater. **206**, 355–363 (2019). https://doi.org/10.1016/j.conbuildmat.2019.02.071
8. Nasr, D., Behforouz, B., Borujeni, P.R., Borujeni, S.A., Zehtab B.: Effect of nano-silica on mechanical properties and durability of self-compacting mortar containing natural zeolite: experimental investigations and artificial neural network modeling. Constr. Build. Mater. **229**, 116888 (2019). https://doi.org/10.1016/j.conbuildmat.2019.116888
9. Czarnecki, S., Shariq, M., Nikoo, M., Sadowski, L.: An intelligent model for the prediction of the compressive strength of cementitious composites with ground granulated blast furnace slag based on ultrasonic pulse velocity measurements. Measurement **172**, 108951 (2021). https://doi.org/10.1016/j.measurement.2020.108951
10. Song, H., Ahmad, A., Ostrowski, K.A., Dudek, M.: Analyzing the compressive strength of ceramic waste-based concrete using experiment and artificial neural network (ANN) approach. Materials **14**(16), 4518 (2021). https://doi.org/10.3390/ma14164518
11. Sun, L., Koopialipoor, M., Jahed Armaghani, D., Tarinejad, R., Tahir, M.M.: Applying a meta-heuristic algorithm to predict and optimize compressive strength of concrete samples. Eng. Comput. **37**(2), 1133–1145 (2019). https://doi.org/10.1007/s00366-019-00875-1
12. Hadzima-Nyarko, M., Nyarko, E.K., Ademovic, N., Milicevic, I., Sipos, T.K.: Modelling the influence of waste rubber on compressive strength of concrete by artificial neural networks. Materials **12**(4), 561 (2019). https://doi.org/10.3390/ma12040561
13. Karni J.: Prediction of compressive strength of concrete. Mat. Constr. **7**, 197–200 (1974). https://doi.org/10.1007/BF02473835
14. Sucharda O., Pajak M., Ponikiewski, T., Konecny P.: Identification of mechanical and fracture properties of self-compacting concrete beams with different types of steel fibres using inverse analysis. Construct. Build. Mater. **138**, 263–275 (2017). https://doi.org/10.1016/j.conbuildmat.2017.01.077
15. Murad, Y., et al.: Flexural strength prediction for concrete beams reinforced with FRP bars using gene expression programming. Structures **33**, 3163–3172 (2021). https://doi.org/10.1016/j.istruc.2021.06.045

16. Shen, J., et al.: Corrosion cracking process of reinforced concrete under the coupled effects of chloride and fatigue loading. KSCE J. Civ. Eng. **25**(9), 3376–3389 (2021). https://doi.org/10.1007/s12205-021-0768-4
17. Sancak, E.: Prediction of bond strength of lightweight concretes by using artificial neural networks. Scientific Research and Essay **4**(4), 256–266 (2009)
18. Sadowski, Ł, Hoła, J.: New nondestructive way of identifying the values of pull-off adhesion between concrete layers in floors. J. Civ. Eng. Manag. **20**(4), 561–569 (2014). https://doi.org/10.3846/13923730.2014.897642
19. Sadowski, Ł, Nikoo, M., Nikoo, M.: Hybrid metaheuristic-neural assessment of the adhesion in existing cement composites. Coatings **7**, 1–12 (2017). https://doi.org/10.3390/coatings7040049
20. Sadowski, Ł, Hoła, J., Czarnecki, S.: Wang, D: Pull-off adhesion prediction of variable thick overlay to the substrate. Autom. Constr. **85**, 10–23 (2018). https://doi.org/10.1016/j.autcon.2017.10.001
21. Czarnecki, S., Sadowski, Ł., Hoła, J.: Artificial neural networks for non-destructive identification of the interlayer bonding between repair overlay and concrete substrate. Adv. Eng. Softw. **141**, 102769 (2020). https://doi.org/10.1016/j.advengsoft.2020.102769
22. Ju, X., Wu, L., Lin, C., Yang, X., Yang, C.: Prediction of chloride concentration with elevation in concrete exposed to cyclic drying-wetting conditions in marine environments. Constr Build. Mater. **278**, 122370 (2021). https://doi.org/10.1016/j.conbuildmat.2021.122370
23. Cai, R., et al.: Prediction of surface chloride concentration of marine concrete using ensemble machine learning. Cem. Concr. Res. **136**, 106146 (2020). https://doi.org/10.1016/j.cemconres.2020.106164
24. Ahmad, A., Farooq, F., Ostrowski, K.A., Sliwa-Wieczorek, K., Czarnecki, S.: Application of novel machine learning techniques for predicting the surface chloride concentration in concrete containing waste material. Materials **14**(9), 2297 (2021). https://doi.org/10.3390/ma14092297
25. Malazdrewicz S., Sadowski L.: An intelligent model for the prediction of the depth of the wear of cementitious composite modified with high-calcium fly ash. Compos. Struct. **259**, 113234 (2021). https://doi.org/10.1016/j.compstruct.2020.113234
26. Gencel, O., Kocabas, F., Gok, M.S., Koksal, F.: Comparison of artificial neural networks and general linear model approaches for the analysis of abrasive wear of concrete. Constr. Build. Mater. **25**(8), 3486–3494 (2011). https://doi.org/10.1016/j.conbuildmat.2011.03.040
27. Taffese, W.Z., Sistonen, E.: Neural network based hygrothermal prediction for deterioration risk analysis of surface-protected concrete façade element. Constr. Build. Mater. **113**, 34–48 (2016). https://doi.org/10.1016/j.conbuildmat.2016.03.029

Maintenance of Building Elements to Fulfill the Basic Requirements Buildings

Daržislav Vidaković[1(✉)], Marijana Nyarko-Hadzima[1], Naida Ademović[2], and Silva Lozančić[1]

[1] Faculty of Civil Engineering and Architecture Osijek, Josip Juraj Strossmayer University of Osijek, Vladimira Preloga 3, 31 000 Osijek, Croatia
dvidak@gfos.hr

[2] Faculty of Civil Engineering, University of Sarajevo, Patriotske lige 30, 71 000 Sarajevo, Bosnia and Herzegovina

Abstract. Regulations define the basic requirements for buildings, which ensure the safety and comfort of users. Fulfillment of these requirements depends on the buildings' condition, i.e. its elements, which are affected by the use and various planned and unplanned effects. In addition to physical wear and tear, the service life often is shortened due to the lag of the building behind technological and economic progress. Therefore, proper and timely maintenance is required, and sometimes capital renewal and reconstruction. The article provides an overview of maintenance measures for the main building elements, from control to repairs and eventually to replacement. Availability of financial resources dictates how and when these measures will be conducted. In this respect, the costs of measures for maintenance and improvement of building properties in line with the new requirements are analyzed. Lifelong maintenance costs should be taken into account from the conception and design stage, because then the basic characteristics of the building are determined and, as a rule, increasing capital investment in construction can reduce the maintenance costs that will be required for numerous years. In some cases, major renovations on buildings that are already in use are also cost-effective. By time-linking the reconstruction and renovation with regular building maintenance multiplication of some works and the costs required would be avoided, leading to higher profitability of such projects.

Keywords: Prescribed requirements · Building elements · Degrading impacts · Service life · Maintenance · Improvement · Cost-effectiveness

1 Introduction – Prescribed Requirements for Construction and Maintenance

In Croatia, the Construction Law [1] stipulates that each building must be designed and constructed in such a way that it meets the basic requirements for the building and other prescribed requirements, i.e. conditions that affect the fulfillment of basic requirements, during its lifetime, depending on its purpose. The basic requirements are in line with the

H. Glavaš et al. (Eds.): OTO 2021, LNNS 369, pp. 288–309, 2022.
https://doi.org/10.1007/978-3-030-92851-3_22

principles of European harmonization of technical legislation and are elaborated by the Technical regulations.

Meeting the basic requirements of buildings is subject to change over the life of their use. In order for the condition of the building to remain satisfactory for use, even with constant physical degradation, it is necessary to take appropriate maintenance measures in a timely manner.

The need for maintenance during usage is influenced by the definition of the building during design and the quality of performance, and proper maintenance assures the designed duration or even extends it. Furthermore, maintenance prevents an increase in operating costs (for example, water supply costs since there will be no leakage and heating costs because the thermal insulation will be adequate), and in some buildings, it will increase the productivity of processes that are taking place in or on them. The main goal of building maintenance is to avoid the deterioration and misfunctioning of its component pieces, as well as their replacement to fit the intended purpose within the normal, planned duration. Figure 1 illustrates the benefits of building maintenance.

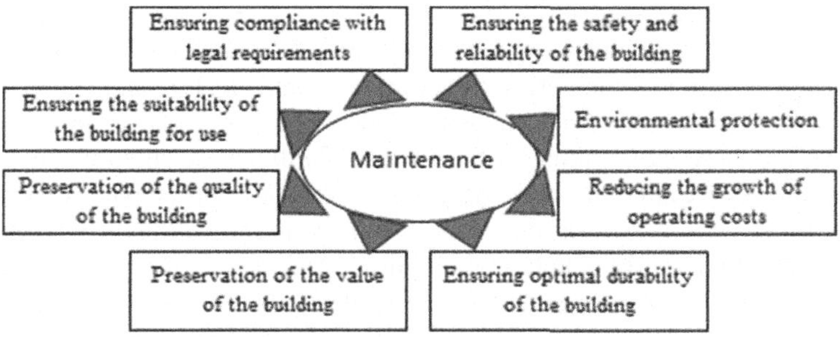

Fig. 1. Benefits of building maintenance

The contractor is obliged to compile a written statement on the performed works and conditions of maintenance of the building. The building owner is responsible for maintenance after construction and must preserve documentation on the maintenance and improvement of compliance with the basic requirements throughout its duration [1, 2]. The regulations established by the relevant standards, i.e. special rules prescribed by the Technical Regulation for various types of structures, are applied to the maintenance of buildings. The applied enforced standards at a specific time must be indicated in the design of the building structure, and there may be a technical specification that establishes stronger requirements than those from the standards [3].

New standards that develop during the use of a building usually impose stricter requirements, i.e., they demand that buildings have better properties. However, a building constructed in compliance with previously valid regulations is maintained in such a way that, during its lifetime, the technical properties are preserved and the requirements determined by its design and regulations are in compliance with which it was constructed are met [3].

Reconstruction of a building envisages the construction and other works on an existing building which, unlike maintenance, affect the fulfillment of basic requirements for that building or which change the building's compliance in respect to the location conditions under which it was built [1].

Published literature from 1994 to 2019 related to physical building conditions has increased approximately tenfold indicating the importance of the condition of buildings shows that a study of [4].

2 Impacts on the Condition and Durability of Buildings

2.1 Factors that Negatively Affect the Physical Condition of Buildings

Building's initial properties gradually deteriorate from the end of construction and the beginning of the use of buildings. For example, fatigue occurs with some materials, their strength decreases, deformations increase, metal corrosion, wood decay, surface delamination and flaking of concrete occur, air permeability and water permeability increase, etc. As a result, the structure of the building materials is destroyed, uneven deformations and settlement occur, as well as other negative consequences. Degrading changes are the result of the influence of many physical (e.g. freezing and thawing, wetting and drying), mechanical (static and dynamic loads), chemical (e.g. action of chlorides and sulfates on concrete, carbonization of concrete and mortar, etc.), and electrochemical factors (such as reinforcement corrosion). The intensity of these processes varies in a fairly wide range, and often there is a combined effect of several influences with a synergistic effect on the building. The result of the action depends on the state of the environment, the level of use (especially if some work processes take place in the building), the importance, and construction quality of the buildings. Figure 2 depicts the different influences from air and ground on a building, which degrade its elements. Other external influences, in addition to those indicated in Fig. 2, can be other, such as fire, earthquake, and other catastrophic events or demolition and construction of nearby buildings, especially during underground works.

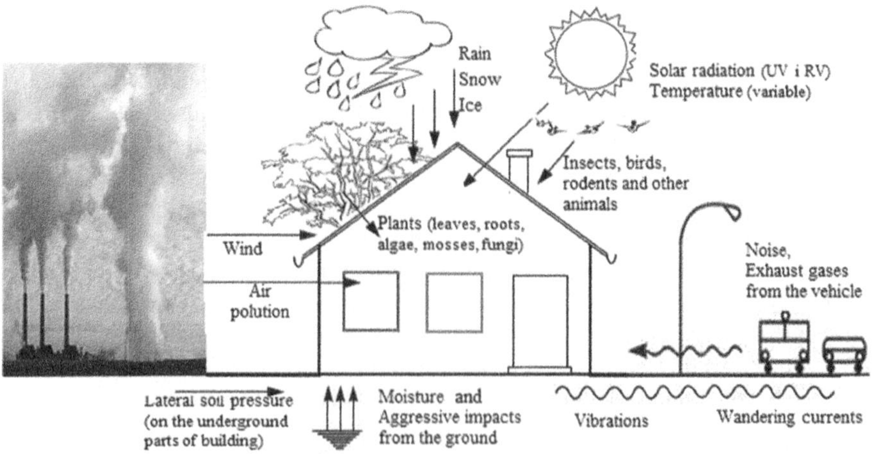

Fig. 2. Negative effects on the building which requires the maintenance of its elements.

One of the most dangerous and common negative effects on buildings is moisture. As the humidity increases, the thermal conductivity of the material increases, and the thermal protection properties deteriorate. In addition, with humidity change, the bulk density of the material increases, sand repeated wetting disrupts their structure and reduces durability [5]. Moisture in buildings (in the air, on the surface of walls, and condensed on windows) causes rot, fungi, and mold, which are harmful to health. Humid air is also more difficult to heat, resulting in higher heating costs. (If the humidity is twice as high as the ideal 45%, it takes almost 50% more energy to heat.) Moisture can come from a variety of sources; during construction, atmospheric precipitation (particularly if the building is poorly waterproofed), condensation of water vapor, groundwater, electroosmosis and capillary lifting, leakage from the water and sewage system, and floods. (Research of defects in wet areas of buildings has shown that the biggest problem is water leakage through joints [6].) Water leaks out of the water supply or sewerage system due to failures, feeding the foundation soil of buildings, causing erosion and, as a result, loss of stability [7]. Inadequate or unmaintained drainage of rainwater (discharge of rainwater from roof surfaces without gutters or vertical gutters discharging water adjacent to walls, damaged or clogged gutters, pipes, and sewers, etc.) is a common cause of softening and settlement of foundation soil, resulting in deformations and cracks in buildings. Poor, unmaintained surface drainage can also cause excessive softening of the foundation soil over time (neglect of drainage channels, overgrown terrain, unfavorable terrain slope). An increase in moisture can induce swelling and consequent damage to the building in soils that are particularly sensitive to shrinkage and swelling [8]. The influence of moisture increases with oscillations of temperature and humidity, as well as when the environment is polluted by aggressive impurities. Pollution from industrial plants enters groundwater, which spreads over large areas and often reaches the underground parts of buildings, destroying their foundations and the ground beneath them. Increases in the intensity of transportation flows, as well as the carrying capacity of machinery and rolling stock, results in increased vibration and shock strengths, which, when combined with changes in soil structure, can be extremely harmful. It is the increase in soil moisture content that leads to an increase in the rate of vibration propagation, leading to a decrease in the damping properties of the soil, and ultimately, to stronger dynamic effects on buildings.

Table 1 summarizes the main groups of factors of physical deterioration of buildings, some of which are random in nature and others which are caused by organizational and technological factors. In addition to the above, the weakening of some material properties and degradation of building elements occurs with aging (change in physical and/or chemical properties of materials regardless of the time of use [9]) and wear through normal use, in proportion to its intensity. Dilapidation or wear can lead to the failure of certain elements, which means the cessation of the ability to perform the required function.

These factors may have different degrees of influence in each specified situation (building). In 2017, Ledenyov and Yartsev [10] conducted a statistical analysis of the damage pattern and misfunctions of building structures, finding that 50% of all of them were caused due to construction errors, 20% due to the usage phase, 18–20% due to design errors and shortcomings, and 10–12% due to geological research.

Table 1. Classification of causes of building elements deterioration.

Causes of damage	Explanation - examples
Design and construction	Errors, i.e. omissions in the project documentation and during construction (especially with poor supervision) lead to insufficient load-bearing capacity and durability, and thus to damage to buildings. (For example, due to insufficient insulation of the attic floor, heat flows there from the heated floor below, causing snow and ice to melt on the roof and ice to form along the eaves.)
Violation of usage regimes and inadequate maintenance	Non-compliance with the rules of use and content of the building, which are provided by the project, lead to unplanned actions (loads, etc.) for which the building does not have sufficient capacity to withstand them (e.g. irregular, uneven load when filling the silo, excessive traffic load, unforeseen, unprofessional changes on the building or use of inappropriate heating fuel). Damage and an increase in damage are also results of improper and untimely repairs
Technological activities	Aggressive, polluted environment, mechanical actions, etc. from technological processes that take place in/on the building
External actions	Natural influences and human action (strong wind, sea salt in the air, groundwater, machine vibrations, vehicle shocks, and many other impacts)

2.2 Decay and Obsolescence of Buildings and Their Durability

Physical (material or technical) deterioration means partial or complete loss of the original qualities of the building, which can be sudden or gradual. The technical condition has steadily deteriorated since the completion of construction, although there may be some more immediate, more severe disruptions, such as floods, landslides, fractures, and so on. Susceptibility of building elements to physical wear depends on the quality of the materials from which they are made, usage, and technological processes in/on the building and external conditions. A building's physical deterioration cannot be reversed in full, but it can be controlled through maintenance to keep it above the project's compliance level.

Durability is the calendar time after construction during which the basic requirements and usability properties above the permitted level are met for buildings, with possible repair breaks. Figure 3 shows how without regular (planned) maintenance durability and the service life is reduced, and how service life can be prolonged further with the additional capital renovation and reconstruction interventions.

All elements of the building have a certain duration, and the duration of the building as a whole is primarily determined by the service life of the main load-bearing structures, i.e. elements that are not replaceable during major repairs. For example, in most countries, stone and concrete foundations are expected to last 100–150 years [7]; buildings with reinforced concrete (RC) frame structures, concrete and masonry with vertical ties, and massive floor structures are expected to last 120 years; and buildings with the same floor structure and walls of brick, concrete, concrete blocks, or stone without vertical ties are expected to last 100 years [14].

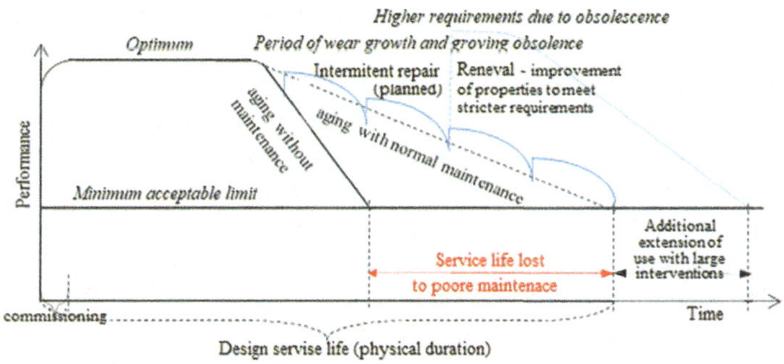

Fig. 3. Impact of maintenance on the expected service life of the building (combined according to [11–13]).

Predicting the duration of building elements is a complex procedure and there are deterministic and stochastic methods that can be applied [15]. It is most often calculated deterministically using factors defined by the standard ISO 15686:8 - Part 8: Reference service life and service-life estimation [16]. Factors categorized into groups from A to G provide the ability to take into account a lot of specific conditions [16] (quality of the material, level of workmanship, frost days index, temperature, relative humidity, heavy precipitation index, insects, surrounding activities, type of use, type of ownership, budget limitations, ease of maintenance, etc.). Loli et al. [17] believe that the planning of expected interventions on the renovation of buildings, especially buildings that are historical heritage, will be more accurate if long-term climate change is taken into account, i.e. their impact on building materials. The service life can change significantly depending on the method of use and the method of maintenance. Different scenarios of using one particular house show that the duration varies from −15% to 20% compared to the reference 63 years, depending on whether it is converted for commercial purposes (more people come and more damage), whether it is rented and not maintained according to actual needs or the owner uses it himself and additionally invests in the maintenance of the facade [17]. For this reason, Chow [18] provided data in 1990 for the durability of individual building elements and their renewal of cycles according to the purpose of buildings (condominiums, retail, public housing, hotels, office, and airports), and the differences between the duration of some types of elements in different buildings are multiple. Design Life is the period of time during which a building is predicted to remain in operation. Such durability should be achieved by the building, according to regulations and standards adopted by the project, with regular maintenance, but without unforeseen effects on the building (and unforeseen costs) and without large, expensive interventions (time due to durability/structural failure [19]).

All elements of the building must resist the negative influences and degrading processes described in Sect. 2.1. Their durability and reliability depend on the strength and duration (frequency) of these actions and the properties obtained by the design and execution. However, while some building elements are long-lasting (such as concrete and stone foundations, brick and RC walls, and RC floor slabs), some have much shorter durability than the building of which they are a part, so they need to be replaced or renovated one or more times during the service life of the building. For example, the expected service life for galvanized sheet metal cover is 25–40 years, for PVC floors 15–25 years, for softwood windows 20–30 years, for elevators 20–30 years, etc. [14, 20, 21]. While some buildings have a much longer lifetime (often with a change of original purpose) than the duration expected by the project, others, due to poor functionality and competitiveness compared to newer buildings of the same type, do not remain in use as long as they could due to structural elements and installed materials.

Physical lifetime is the period of the actual existence of a building in a functionally capable state before its demolition [22], but the actual service life does not always last as long as possible according to physical predispositions (especially in RC and steel buildings). There are cases in the world where buildings aged 15–20 are demolished to build new buildings on the same site [23]. This is because the duration of use depends on the obsolescence of the building and its technical equipment and on the conditions around the building (new or missing contents, which may be attractive or unsuitable) and the economic and political situation (living standards, taxes, tourism, market situation, large migrations, war, etc.). Obsolescence is the deviation of buildings from current standards and in general from current expectations for use (they change during long-term use). Obsolescence can be functional (lack of more favorable conditions for users, such as surface and spatial layout, sanitary facilities, elevators, insufficient thermal and sound insulation, etc.) or technological (when there is a mismatch between the service life of the building and progress in the development of technological equipment, which in its dimensions, weight, etc. does not fit into the existing building). With proper maintenance, buildings usually become obsolete faster than physical decay, and according to UNESCO, residential buildings have become obsolete every 8 years [7]. In Croatia, most buildings were built more than 40 years ago [24], and the housing stock is very old in most other European countries. These buildings continue to fulfill the function for which they were built, but at levels below modern standards. The global trend of awareness of the need to preserve the environment and energy and the recent earthquakes in Croatia and other countries have especially highlighted the requirements for improving energy efficiency and seismic safety. Even today some buildings are still in use in which some materials that are known to be dangerous today (such as asbestos and mercury-containing paints) have been installed, so it is imperative to replace them.

Figure 4 shows how the obsolescence of a building increases in a very steep stepwise manner as requirements, i.e. standards, change. In reality, there are micro-steps in the process of losing the physical and mechanical characteristics of the material (if no catastrophe occurs), but this can be approximated by a curve. Due to aging, buildings have a smooth change in the state of wear, while changes in working conditions and external influences contribute to a more sudden change in physical condition. Physical deterioration increases over time, but the parameters of physical wear are less pronounced (reduced) as a result of maintaining the building elements in normal technical condition. The curves of the growth of wear and tear of the building and its elements can be convex, concave, or combined depending on the conditions of maintenance (as shown in Fig. 4). The results of the research show that the wear of buildings, i.e. their individual elements occurs more intensively in the first 20–30 years of use and after 90–100 years [7], which is shown in Fig. 4.

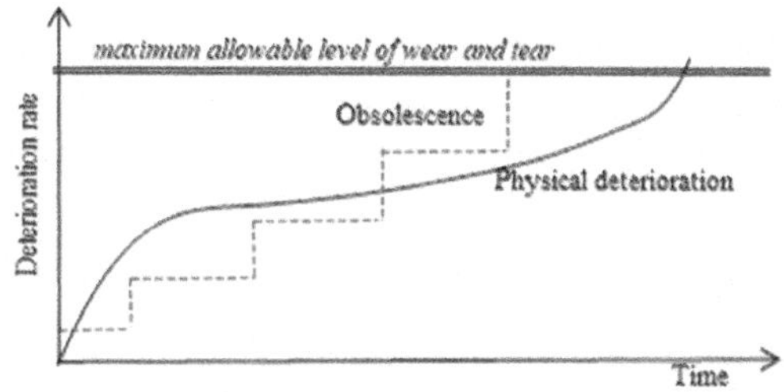

Fig. 4. Physical deterioration and obsolescence of buildings over time.

Obsolete buildings are less attractive and lose value, so it is questionable how profitable it is to invest in larger interventions just to prolong their physical life. It is possible to have a positive effect on the obsolescence of building the same as in the case of physical condition (i.e. physical lifetime) [22], i.e. it is possible to modernize them, by improving the characteristics compared to those envisaged before construction.

3 Building Maintenance Activities

3.1 Maintenance System

In addition to performing direct maintenance activities on the building, the maintenance system must include maintenance preparation (planning, contracting) and supervision of maintenance implementation, recording and reporting on maintenance and construction (databases), monitoring and analysis of costs of all maintenance activities, and development of measures for optimization of total operating/use and maintenance costs of the building.

HRN EN 13306 interprets maintenance in general as a combination of all technical, administrative, and managerial procedures during the life of an element with the aim of retaining or returning the element to a condition in which it can perform the required function [25]. Such an interpretation, in a broader sense than the Law on Construction, implies a complex system that performs a series of procedures and planned and unplanned activities, depending on the adopted maintenance strategy. These are inspections and monitoring to determine the condition of buildings, replacement of worn-out materials and elements, and repairs to restore the physical condition and preserve the physical life of buildings. For regular maintenance, cleaning, and maintenance of passable parts of the building are prescribed when the fulfillment of the basic requirements for it (e.g., for roads) or its durability depends on it [2, 26]. More recently, authors (e.g., Asaul [22], Olanrewaju et al. [27], Ihuah and Fortune [28] on building maintenance, and Radović et al. [29] on road maintenance) expand the maintenance activities on construction facilities improvements to meet current standards.

Building maintenance activities refer to the underground and aboveground elements of the building, to the elements of finishing works, installations, and equipment and devices [30]. In addition, this usually includes maintenance work on the building's surroundings, including external installations (roadside vegetation is mandatory on roads [26]).

According to the ISO 15686 standard [16], maintenance can be planned (according to the determined time period), preventive (to avoid malfunction, meaning to prevent the loss of the function envisaged by the project), corrective (reactive - removal of the malfunction after it has occurred), based on the condition of the elements (preventive, based on performance, parameter monitoring, and follow-up activities), based on prediction, delayed, urgent and unpredictable, and direct or indirect [9, 16].

Unplanned maintenance is a short-term corrective action necessary to return the building to service condition (associated with unforeseen maintenance costs). Planned maintenance can be prevention, corrective, and refurbishment [28]. For road facilities, preventive maintenance is usually applied if they are in good condition and have a long remaining service life, while reconstruction (work to improve or replace the entire part of the road) is undertaken when the structural efficiency of the existing facility is compromised [31]. For maintenance, it is necessary to make appropriate plans (different levels) for predictable activities, optimized according to determined priority criteria, available resources, and the lowest possible costs. Long-term, medium-term, and annual plans are made for road maintenance, and the operational maintenance program determines the time and schedule of individual works, their scope and technological procedure, the deadline for execution, as well as other conditions that are important for their execution [32]. Records must be kept on the implementation of plans and all performed activities, costs, and findings.

Maintenance planning should begin already at the stage of conception and design of the building when the characteristics important for later maintenance are determined (defining the location, position, dimensions, materials and equipment, availability of elements, etc.). With modern systems that help in the management process, such as Cloud computing, Internet of Things, Building automation system, and Building information modeling (BIM) further expands the application [33]. BIM's databases have been continuously expanded and improved since the graphic design (3D), and the last, the "seventh dimension" (7D) of BIM is the life cycle and maintenance (planning and management of maintenance operations throughout the building's lifecycle). This lifelong approach, with advanced computer support, effectively improves the quality and efficiency of building operation and maintenance, from construction to the removal of the building. There is a possibility of introducing BIM models also in the reconstruction projects. Then it is necessary to collect the data for existing buildings (mostly built a long time ago) so that they can be used in future situations to save money and time, and for such Perhavec and Vidaković [34] are showing an example of Knowledge management systems and Rules and Legislation Database and Building Database.

3.2 Inspection and Monitoring of the Condition of Buildings

In addition to the mentioned numerous factors that affect buildings and their elements, with a longer period of use, appropriate, professional control of their condition is required. An inspection is an assessment of the conformity of an element by observing, measuring, examining, or evaluating their relevant characteristics, while during monitoring the condition the parameters are constantly monitored over a period of time [9] (e.g., settlement, crack propagation, etc.). The actual loads determined during the use of the building can differ significantly from those envisaged by the project, both in size and in place of action. Therefore, in some cases, systematic monitoring of operating loads should be developed, being careful that they do not exceed the prescribed values [5].

Inspections of buildings are required in order to determine the need for certain maintenance measures (from emergency to delayed) to maintain safety, but also to prevent a significant deterioration (growth of spread) of damage due to delayed repairs. Inspections are focused on the physical condition of the building elements and its technical equipment - the degree of change in material properties, structural defects, i.e., wear, damage, defects, and malfunctions/faults. In addition, the causes of unplanned major deterioration should always be discovered.

During the inspections, certain parameters are detected in different ways, with standard methods (sometimes a combination of several of them) and suitable equipment, i.e., instruments, depending on the building element and its availability (inaccessibility of some parts of the building makes them much more difficult). Data important primarily for safety, but also for the efficiency and quality of use and durability of the building are checked, such as longitudinal dimensions and cross-section, stresses, deformations and displacements (deflections, slopes), cracks and other damage, material properties (strength, homogeneity, toughness, stiffness, humidity, etc.), impermeability (joints, wall panels, covers, etc.), cover thickness (as in concrete reinforcement and thermal or hydro insulation), vibrated property, humidity in the room, composition and airflow, resistance to heat transfer, functioning/malfunctions of technical equipment and compliance of devices (e.g. for heating) and equipment with the design of the building, etc. [2, 5, 7]. It is also necessary to examine the environmental conditions (physical, chemical, and biological).

Building inspection diagnostics are often based on a comparison of existing values with some reference or initial one, i.e., a deviation from the designed condition. Testing can be done by non-destructive and destructive methods. It can be visual (e.g. according to color, without instruments, or with a camera), touch, acoustic methods (it can be searching for cavities by knocking), scraping or notching (to check the strength, because the crushing of the material indicates poor hardness), measuring electrical resistance (for reinforcement corrosion), the radiographic testing of welded joints, pressure test (for chimneys and gas and plumbing installations), chemical analysis and other laboratory tests (e.g. swabs, sawdust or samples taken by sounding).

To help in obtaining a more complete picture of the building, *in situ* inspection is preceded by acquaintance with the technical documentation and other materials (documentation on previously performed inspections), if available. If such documentation does not exist or is incomplete, the initial (general) inspections will have to be more extensive, with complete measurements of the dimensions of the structures, determination of the construction scheme, and drafting. The entire work on the inspection is carried out at different levels of expertise, in stages, as shown in Fig. 5.

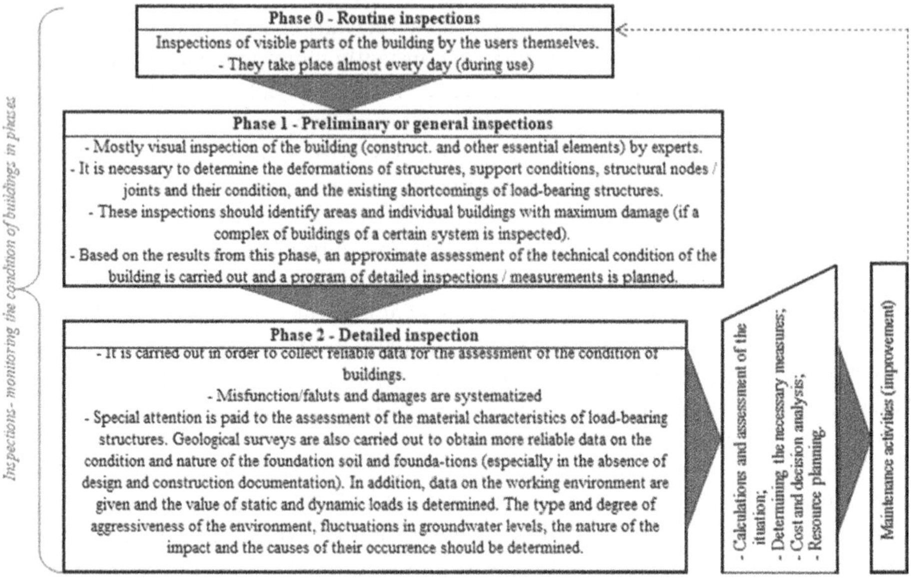

Fig. 5. Contents of all phases of testing the condition of buildings and further action.

A detailed inspection requires a lot of time and resources, so the need for it should be proven by a preliminary examination. If necessary, after a preliminary inspection, measures are taken to temporarily strengthen the structures. A detailed inspection can be selective (only individual elements) or complete (checking all essential elements). If during the examination of the sample a significant reduction in the load-bearing capacity of structures (inhomogeneous material properties, different load conditions, etc.) is found or significant deviations from the design are detected, a complete inspection is performed [7]. A complete measurement is also carried out in cases when there is no technical documentation. When performing all types of reviews, it is necessary to record the identified data because they are used for further calculations and databases for decision-making and planning.

According to the Croatian regulation, inspections can be [3]:

– regular - basic (every year or more often), main (at an interval of 5 years for bridges, towers, and other engineering structures and 10 years for buildings), and additional

(then the basic inspection identifies deficiencies that may affect the requirements of mechanical resistance and stability and fire resistance) or
- extraordinary (as soon as possible after natural or technical disasters or extraordinary loads).

In addition, an inspection of the buildings must be made in the following cases [5]:

- when the reconstruction of the building is required, ie a change that changes the spatial-planning solution, load, or working conditions or
- accidents on similar buildings.

Road inspections on the regular basis are carried out [26]:

- by the professional patrol service (at least once a month visual inspection of road facilities, primarily elements and parts related to stability, functionality, durability, and safety of traffic);
- seasonal (usually after the winter period and in the fall to check the condition of the roads and determine the extent of damage);
- annual (steel and wooden road structures are inspected at least once a year, other road structures are inspected at least once every two years);
- main (at least once in six years all parts of the facility are inspected and specialist measurements and tests are carried out).

Laws, regulations and norms, and a written statement of the contractor after the completion of the construction prescribe the content and inspection periods for some elements and equipment of the buildings. For installed equipment, devices, and installations, the service plan and deadlines are prescribed in the guarantees of their manufacturers [2]. For example, residential and public buildings require a series of inspections related to lightning protection and electrical installations, fire protection systems, chimneys and heating systems, air conditioners, elevators, etc. (usually for a period of 1 or 2 years).

The need for some inspections is dictated by the time of year, so the facades of buildings should be inspected in general by a professional in autumn (ie before winter) and in spring (together with that at least every 5 years in detail) [35], and chimneys always before the heating season.

For some buildings (e.g., bridges, waterworks, pipelines, steel structures, etc.) it is possible to apply the Risk-Based Inspection strategy. However, although these procedures are elaborated on a theoretical level, in practical use they become a significant problem in connection with complicated modeling of the structure (necessary before risk analysis). This is modeling in terms of dividing the building structure, according to different criteria (e.g., sensitivity, i.e., vulnerability, state of the structure, state of the environment, etc.), into different areas that should be treated as a whole according to the maintenance procedure [36].

After testing on the building, data processing and calculations of the condition and impact of damage on the reliability and load-bearing capacity of structures as a whole follow. Based on the test data, it is possible to define the degradation curve as it has been so far and assume its progress in the future [37] and determine the remaining service life of structures according to it. The determined level of physical degradation gives an objective idea of the technical condition of the structural elements and through them the building as a whole. The study of the overall building condition is an extremely important basis for building management and decision-making on maintenance [38] - on repairs and replacement of elements (urgency and scope of work and the necessary cost of restoration), reconstruction, modernization, or abandonment/demolition buildings.

The classification of condition assessments can be different, adapted to the specific type of building, but as a rule, it always goes from a good or fully usable condition, where all the requirements of the building are met (suitable for use with normal maintenance), to a condition that is not for use (if urgent actions are not taken there will be a collapse). Yacob et al. (2016) suggest the following types of maintenance activities depending on the five categories of building condition [38]:

- Very Good – Preventive maintenance;
- Good – Condition Based Maintenance;
- Fair – Repairs;
- Poor – Rehabilitation;
- Very poor – Replacement.

3.3 Repairs and Replacement of Elements and Improvement of the Condition of Buildings

If the technical properties of the building structure do not meet the requirements of regulations and rules according to which it is designed and constructed, it is necessary to replace individual elements, conduct periodical repairs, conduct repair or reconstruct by which the building structures to a level that meets the minimum requirements or remove the building [3]. Defining the type of maintenance work according to the Ordinance on maintenance of buildings is shown in Fig. 6.

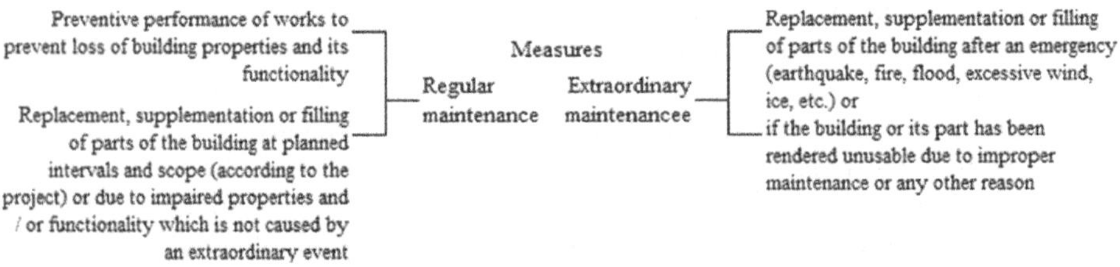

Fig. 6. Measures of regular and extraordinary maintenance.

Repairs are considered to be works on restoring worn-out, i.e., decaying, deteriorated, or in other ways damaged elements to a normal usable condition (the function provided by the project is returned to them). Repairs differ in the scope of work and the impact on the building. Minor repairs to buildings are, for example, replacement of individual roof tiles or battens, placing of fallen bricks and fixing with mortar, grouting, occasional repairs of plaster, etc. These are ongoing, routine repairs of a shorter duration (usually 1–2 days). They can be planned and in accordance with regular inspections or unplanned when the need for them is revealed by extraordinary inspections. Such repairs allow for further use and prevent deterioration and increasing of subsequent maintenance costs. Buildings require major repairs when satisfactory properties for use cannot be provided by small, ongoing repairs. The lost properties of the individual constructive building elements are selectively restored. Larger repairs extend the physical life but do not increase the economic value of the building because they do not improve it compared to the initial requirements. The application of new materials and technologies can significantly increase the reliability and durability of structures and buildings as a whole. In the complete renovation (overhaul) and reconstruction of buildings, some long-lasting structural elements, and the system of technical equipment are replaced as needed (it is modernized with modern equipment). This achieves the fulfillment of modern requirements and improves usability. Modernization extends the economic life of the building (duration until when the usage is profitable). For such large interventions, project documentation must be prepared in advance and special resources (contractors) are usually hired for them. Due to the high cost of these works, a quantitative assessment of the obsolescence of the building and verification of the cost-effectiveness of the project is required to justify them.

The urgency of eliminating the existing damage, i.e., detected defects in the building depends on their importance. The identified damages can be of different weight [7]:

- insignificant damage, which does not reduce the load-bearing capacity of structures and can be easily removed by routine (ongoing) repairs;
- significant damage, which reduces the load-bearing capacity and usability of buildings and their physical durability;
- critical (alarming) damage to structural elements due to which further use is unsafe and the condition is extraordinary.

In case of damage to the building, which endangers the life and health of people, environment, nature, and other buildings, as well as soil stability on the surrounding land, the owner of the building, is obliged to take urgent measures to eliminate the danger and mark the building as dangerous until such damage is eliminated [1]. In some cases when the usability reliability and durability are impaired, action is necessary, but if it is possible to use the building for some time and in that condition (no emergency intervention is required) repairs can be postponed until the planned date within the existing maintenance program. This is the so-called delayed maintenance. Repairs can also be temporary if they only allow the worn-out element to perform its function until the necessary repair is performed in full [9]. It is necessary to recognize whether the misfunction is primary or secondary, or caused by another misfunction or defect (which is quite common in buildings), and then be sure to eliminate the primary cause.

Long-lasting structural elements are not subject to replacement, but due to physical wear, they undergo regeneration procedures (e.g., bearing renewal), which restore their correctness and the ability to meet the necessary basic requirements. Elements, as well as parts of building equipment whose durability is less than the design life of the building when they were worn out, are replaced by similar ones [22]. Nowadays, improving the properties of residential buildings usually includes replacing windows and doors with better thermal insulation properties, increasing the thermal envelope of external walls, roofs, and ceilings to unheated space, replacing heating systems with more energy-efficient ones, replacing plumbing and other technical equipment and installation lines, etc. [7, 39].

Buildings protected as a cultural asset always require an individual approach and cost-effectiveness is not primary. Their protected monumental properties must not be disturbed [2] and therefore works on them may only be designed and carried out by a company with a permit to work on the heritage asset property [40]. In some cases, original materials that are more difficult to obtain on the market and more expensive must be used, and today it is very difficult to find craftsmen who know some of the necessary traditional works (e.g., restoration of decorations).

The main reason for the high degree of physical deterioration of the housing stock is the untimely execution of repairs and restorations, which is the result of the limited availability of funds [7]. The degradation of load-bearing structures in buildings generally goes from the stage of crack formation at the places of stress concentration, through their slow development to sudden destruction when critical stresses and deformations are reached. From that, it can be seen that longer interruptions in the necessary maintenance activities lead to emergencies and shorten the physical duration of buildings. Their restoration when the critical stage has already occurred requires very large works and the cessation of use until their completion. That means tens of times more costs than they would be for timely action. (They can even exceed the value of the structure, i.e., building a new one, so sometimes the renovation is abandoned [7].) Irregular maintenance of the water supply and sewerage network is also dangerous, and damage to it (due to water leakage) can lead to the deterioration of the structural elements of the building. One should always try to avoid situations when repairs should be performed immediately and inevitably because then it is the most expensive.

In conditions of limited financial resources and at risk of developing deterioration of the building, maintenance based on the risk of failure or multi-criteria decision-making is recommended, which takes into account some other criteria in addition to the risk. For this, it is convenient to use the Analytic Hierarchy Process method to determine the priority of maintenance activities (there are computer programs, such as various versions of *Expert Choice*, which greatly simplify the process). Such a need is particularly pronounced when managing the maintenance of large buildings or a multi-building complex, and the available maintenance budget should be allocated to a lot of different necessary replacements, repairs, and remediation of certain elements (not urgent interventions or major renovations). Cerić and Katavić [41] proposed the status of the building, physical condition, significance for use, impact on users, and impact on the building as criteria for such decisions when maintaining buildings. First, the criteria should be compared with each other to determine their weight, and then for each criterion, maintenance interventions are compared with each other as variants that should be ranked according to priority.

In some cases, for organizational and technological reasons and due to the same one-time costs of preparation and safety measures at work, it is more convenient to repair or renovate the entire building element or the entire structure (e.g. a reinforced concrete wall damaged by corrosion), although there is not the same degree of damage everywhere (e.g. if a scaffolding has already been installed or a crane has been hired, if the maintenance team is already on the site, it is better to do the whole job than to work part-time, etc.).

4 Analysis to Minimize the Total Cost

Maintenance costs are one of the Life-cycle costs (LCC) groups that are part of Whole-life costing [16]. Along with them, there are the costs of Construction, Operation, Occupancy, and End of life, and there is a connection between all these costs. Maintenance affects all other costs after construction (e.g., the costs of required energy and cleaning), and Construction, Operation, Occupancy affect the required maintenance (explained in the introduction). A review of research in different countries shows a large share of operating, maintenance, and refurbishment costs in LCC buildings [42]. This share is higher the longer the service life (according to Boussabaine et al. [43] more than 80%).

The structure of building maintenance costs according to [30] is shown in Table 2.

Table 2. Maintenance cost by type of activities.

Costs of expert inspections	Costs of preventive maintenance activities	Costs of replacing worn materials, elements, and techn. equipment	Reactive (corrective) maintenance costs
– the costs of periodic inspections prescribed by law for the purpose of taking measures necessary for the safety, health, and life of people. (These are various tests and inspections of individual parts of the building, devices, and installations.)	– the costs of those activities that are repeated at relatively the same time intervals in order to maintain a defined level of easement of the building	– costs of activities on changes of certain worn-out materials and elements of the building (according to the planned schedule or inspection determined by the condition)	– costs of repair and replacement of materials and elements and equipment in interventions after failures and damage. (The magnitude of these costs is unpredictable because it is impossible to predict all possible faults/malfunctions that may occur due to extraordinary impacts.)

Performing maintenance and other work on existing buildings can be very expensive. The reason for this is work in conditions of lack of space, which makes it difficult to engage optimal mechanization and a lot of manual work, sometimes very professional and in particularly dangerous conditions. Special work equipment may be necessary, and small amounts of individual works (relative to new construction) increase the unit price due to the share of fixed, one-off costs (mainly indirect labor costs that must be made regardless of the amount of work on the building). A special organizational problem is a work in the conditions of leaving the building in use, otherwise, there are costs of suspension of use (relocation of tenants, alternative traffic routes, etc.) and speeding up the completion of works as soon as possible, which also increases costs, (such as work in multiple shifts).

The required maintenance costs mainly follow the dynamics of the physical deterioration of the building. They become the largest in a period of intensive deterioration of a building, and then one should consider how economically justified the investment in further maintenance is, compared to the demolition and construction or purchase of a new building.

Cost analysis is aimed at minimizing them, through long-term (lifelong) planning of maintenance and other related activities of realization and use of the building. In order to analyze the costs, it is necessary to include all their types and carriers and determine the dependence with other activities on the construction and with the costs they require. The process of analysis and decision-making (costs are often the most important criterion) about the building is shown in Fig. 7.

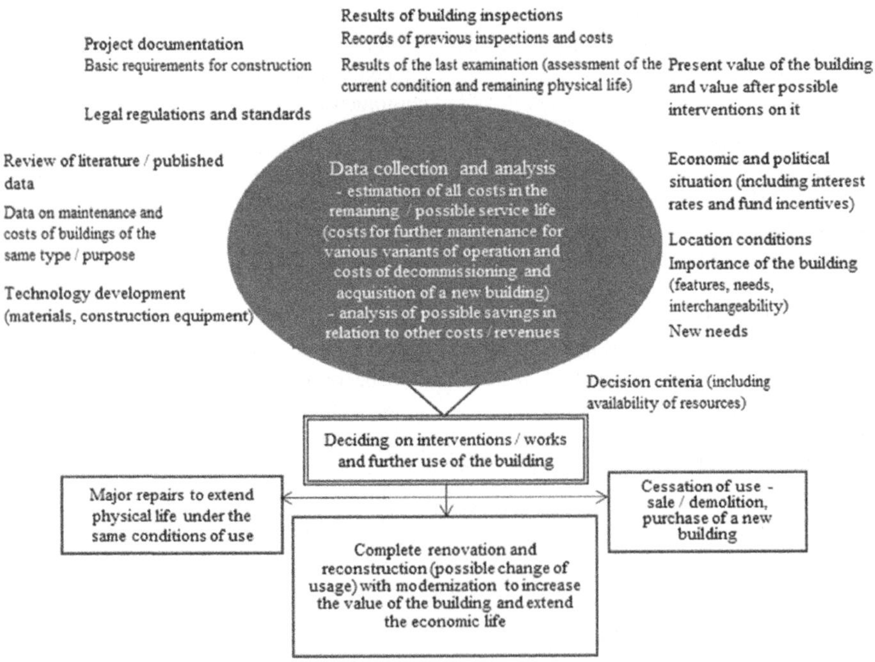

Fig. 7. Cost analysis and therefore decision making on building.

In the cost analysis, it should not be neglected that the maintenance of buildings has direct and indirect costs. Direct costs are the costs of materials, the labor of workers and machines, and all work equipment. Maintenance may involve an own workforce (from the one who manages the building) or it can be outsourcing to external companies/trades (especially for consultants, for highly specialized activities, or large amounts of work that is rarely done). Indirect costs are all costs of the maintenance team, logistics costs (e.g. IT systems, education, etc.), and penalties (inability to use the building or degradation of quality/performance) [30]. Reduced usability or unavailability of a building may result in many associated costs (depending on the type or purpose of the building and compensation for discontinuation of its use), such as the cost of providing alternative services, warranty costs, liability, lost revenue, and also with this can also affect the bad reputation of the building owner [9].

It has been shown that at the design stage operating, maintenance, and refurbishment costs can be predicted [43] and that these costs can be well influenced over the life cycle even before construction. Suitable location and design solutions, including materials that will require less maintenance, it can save significantly more over its useful life than it does to increase initial investment in construction. Today, there are many structural elements of reinforced concrete structures that are negatively affected by environmental factors, and the appropriate composition of concrete greater resistance can be achieved. For example, in concrete sidewalks, the addition of macrofibre reduces the potential cracking due to shrinkage and limits the movement of cracks, and the base gravel reduces the risk of lifting due to frost. According to the analysis of Czarnecki and Poon [19], the service life without macrofibre and substrate is 15 years. Appropriate composition of concrete and substrates increases the construction costs by 20–30%, but thus the service life is extended to 35–50 years, and the costs of necessary renovation and replacement are reduced. Preventive maintenance at 40% of assumed service life further reduces rehabilitation costs.

Improvements to meet the requirements of existing buildings should also be considered. As with replacing elements with better properties when the old ones expire lives, the cost-effectiveness is always higher in situations where some of the costs are reduced or avoided. This can be achieved if work on the complete reconstruction (overhaul) or improvement of the building's properties takes place in connection with another project (which requires the construction site formation, scaffolding, crane delivery, removal of some parts of the building, etc.). Thus, for example, the addition of floors or rearrangement of the attic into a living space can be combined with the complete renovation or replacement of some structural elements (e.g., the roof structure can be further strengthened and the ceiling with wooden beams replaced with reinforced concrete) and the installation of better equipment or installations. Likewise, the reconstruction of a building after an earthquake or fire can be well combined with energy efficiency improvements or interventions that increase the level of compliance with the basic requirements for resistance and stability and safety in the event of a fire. Table 3 highlights the types of costs of investors, i.e., building owners, and the costs of contractors that can be saved by such project conjugation.

With the increase in the volume of construction work, indirect costs (e.g., the establishment and management of the construction site, for cranes that serve everything, etc.) are distributed over a larger amount, so in terms of the unit of work, the costs are lower. Even greater savings are possible if maintenance works and other works are carried out simultaneously on several buildings that are close and well-coordinated according to the work schedule (if the types of works on the buildings are the same, then best tactfully).

Table 3. Costs that can be reduced by combining the necessary maintenance activities with the project of reconstruction and modernization of the building.

Investor/owner costs	Contractor costs
- for testing the condition of the building (harder to reach elements) - for the procedure of selection of contractors and contracting - for supervision of the execution of works - payment of the contractor (the price of works) - temporary unusability or reduced level of usability of the building (losses because it is not operational)	- Indirect construction site costs – preparatory works, dismantling and insurance, safety measures at the worksite, construction site management, internal supervision of works and communication with the investor, transports - Direct realization costs - hours of work of workers and machines, and there may be less material (In some cases they are reduced due to avoidance of double cleaning, demolition, and closing of the structure, due to the existing availability of elements with (insured for other works) and higher utilization of machines and workers for a larger amount of work on the construction site (especially universal workers)

This also increases the amount of work for the contractor, and the necessary resources can be easily exchanged (for which the requirements on individual projects are usually variable).

5 Conclusion

Many buildings show at first glance that their owners do not allocate enough funds for satisfactory maintenance. There can be many reasons for neglecting maintenance, but this has consequences whose correction in the end almost always exceeds the cost of omitted maintenance measures. In addition, failure to monitor the condition of buildings and failure to take appropriate maintenance measures can lead to catastrophic accidents, as exemplified in both the poorest and most developed countries. In addition to meeting the requirements of safety and reliability, the maintenance of buildings has other important goals, including financial and quality of use and regulation of service life.

Maintenance decisions are crucial for successful building management, and one of the fundamental decisions is choosing the most appropriate strategy. The maintenance strategy depends primarily on the importance and purpose of the building and the processes that take place through it and the risks that can lead to dismissals, and the available resources should be taken into account (including the availability of funds and people with relevant knowledge).

Maintenance activities should be planned (although some unforeseen activities will always be required), taking into account long-term, i.e., lifelong costs. Only by analyzing all maintenance costs and related building costs, it is possible to optimize the total costs of the owner and the management of the value of the property. That is why it is important to constantly collect data on maintenance and related costs (costs, causes of problems, measures taken, etc.), especially when one owner/manager has several buildings of the same type, and especially of a similar type (by construction, materials, etc.)

Savings are possible if improving the fulfillment of basic requirements and modernizing the properties of buildings (e.g., energy efficiency, better sound insulation, higher earthquake resistance, etc.) is associated with the implementation of major repairs (e.g., constructive renovation of buildings after earthquakes). Each construction project is unique and the possibility of avoiding the double performance of part of the construction and other works and procedures, and thus the size of the savings always depends on the specific case (must therefore be analyzed separately).

In order to ensure optimal maintenance, looking to minimize all costs throughout the life of the building and taking into account all the benefits that maintenance brings, even in the pre-construction stages, the issue of future maintenance of buildings should be considered. Lifelong maintenance care is provided by BIM, but its application requires a very serious interest in the "fate" of the building since the beginning of the development of thc construction project.

Today, buildings are being built for a future of maybe 100 or more years, so the planning of their maintenance should be flexible and with the adoption of the latest knowledge and technologies.

References

1. Construction Law, Official Gazette 153, Croatia (2013). (in HR)
2. Ordinance on maintenance of buildings, Official Gazette 122, Croatia (2014). (in HR)
3. Technical regulations for building structures, Official Gazette 17, Croatia (2017). (in HR)
4. Faqih, F., Zayed, T., Soliman, E.: Factors and defects analysis of physical and environmental condition of buildings. J. Build. Pathol. Rehabil. **5**(1), 1–15 (2020). https://doi.org/10.1007/s41024-020-00084-0
5. Kazarnovsky, V.S., Grigoriev, P.Ya., Neustroev, A.Ya.: Technical Exploitation of Buildings and Facilities of Railway Transport. Training and Methodological Center for Education in Railway Transport, Moscow (2006). (in RU)
6. Chew, M.Y.L.: Defect analysis in wet areas of buildings. Constr. Build. Mater. **19**, 165–173 (2005)
7. Afanasyev, A.A., Matveev, E.P.: Reconstruction of Residential Buildings. JSC Center for Design Products in Construction, Moscow (2008). (in RU)
8. Jurić, A., Vidaković, D., Minažek, K.: Measures of preparation and maintenance in the construction and rehabilitation of buildings. In: Proceedings of the 6th Conference Maintenance 2020, pp. 105–114. University of Zenica, Faculty of Mechanical Engineering, Zenica (2020)
9. Frković, D., et al.: Maintenance and property management. ESUO manual. Croatian Maintenance Society, Zagreb (2016). (in HR)
10. Ledenev, V., Yartsev, V.P.: Inspection and Monitoring of Building Constructions and Structures. Publishing House TGTU, Tambov (2017). (in RU)
11. Building Research Board: Commission on Engineering and Technical Systems: National Research CouncilFourth Dimension in Building: The Fourth Dimension in Building: Strategies for Minimizing Obsolescence Studies in Management of Building Technology. National Academies Press, Washington, D.C. (1993)
12. Thomsen, A., van der Flier, K.: Understanding obsolescence: a conceptual model for buildings. Build. Res. Inf. **39**(4), 352–362 (2011)
13. Alexander, M.G.: Service life design and modelling of concrete structures – background, developments, and implementation. Revista ALCONPAT **8**(3), 224–245 (2018)

14. Željko, Ž.: Real Estate Sales; Real Estate Valuation. Handbook, Croatian Chamber of Commerce, Zagreb (2004). (in HR)
15. Ortega Madrigal, L., Serrano Lanzarote, B., Fran Bretones, J.M.F.: Proposed method of estimating the service life of building envelopes. J. Constr. **14**(1), 60–68 (2015)
16. ISO 15686-8:2008: International Organisation for Standardization, Geneva (2008)
17. Loli, A., Bertolin, C., Kotova, L.: Service life prediction of building components in the times of climate change. IOP Conf. Ser. Mater. Sci. Eng. **949**, 1–9 (2020)
18. Chow, K.F.: The Construction Agenda. Construction Industry Development Board, Singapore (1990)
19. Czarnecki, B., Poon, B.: Concrete sidewalk design analysis and optimization for improved life cycle and sustainability. In: Conference of the Transportation Association of Canada St. John's (2017)
20. Aničić, D.: Planning the service life of buildings - translation of ISO 1586 series standards. Građevinski godišnjak 03, HDGI, Zagreb (2004). (in HR)
21. Akmatov, A.K., Ryspaev, J.A., Ordobaev, B.S.: Technical Exploitation of Buildings and Construction. Faculty of Architecture, Design and Construction, Bishkek (2017). (in RU)
22. Asaul, A.N., Abaev, H.S., Molčanov, Y.A.: Management, Exploitation and Development of Property Complexes. Humanism, St. Petersburg (2007). (in RU)
23. Čaušević, A., Rustempašić, N.: Reconstructions of Masonry Buildings. Faculty of Architecture, Sarajevo (2013). (in BA)
24. Long-Term Strategy to Encourage Investment in the Renovation of the National Building Fund of the Republic of Croatia, Ministry of Physical Planning, Construction and State Assets, Zagreb (2020). (in HR)
25. HRN EN 13306, HZN e-Glasilo 12 (2017). (in HR)
26. Ordinance on road maintenance, Official Gazette 90, Croatia (2014). (in HR)
27. Olanrewaju, A.A., Khamidi, M.F., Idrus, A.: Appraisal of the building maintenance management practices of Malaysian Universities. J. Build. Apprais. **6**, 261–275 (2011)
28. Ihuah, P.W., Fortune, C.J.: Toward a framework for the sustainable management of social (public) housing estates in Nigeria. J. US-China Public Admin. **10**(9), 901–913 (2013)
29. Radović, N., Šešlija, M., Peško, I.: Expert project analyses in the process of road maintenance management. Građevinar **65**(7), 641–652 (2013)
30. Marenjak, S., El-Haram, M.A., Horner, R.M.W.: Analysis of overall project costs for the building industry. Građevinar **54**(7), 393–401 (2002)
31. EU criteria for green public procurement for the design and construction of office buildings and their management. EU Commission staff working document, Bruxelles (2016)
32. Road Law, Official Gazette 84, Croatia (2011). (in HR)
33. Fang, T., Zhao, Y., Gong, J., Wang, F., Yang, J.: Investigation on maintenance technology of large-scale public. Sustainability **13**, 1–18 (2021)
34. Perhavec, D., Vidaković, D.: Is it time for new approach on historical building preservation or conservation project? In: Proceedings of the REHABEND 2016 Euro-American Congress on Construction Pathology, Rehabilitation Technology and Heritage Management, pp. 1411–1418, University of Cantabria, Burgos (2016)
35. Building facade maintenance – Legal liability and damage limitation. A guide for building owners and occupiers. www.bre.co.uk/pdf./facademaintenance.pdf. Accessed 10 Oct 2016
36. Li, Y., Vrouwenvelder, T., Wijnants, G.H.: Deterioration and maintenance of concrete bridges based on spatial variability of structural properties. In: Proceedings Risk-Based Maintenance of Civil Structures, pp. 86–93, Serie Workshop Proceedings no. 8, Faculty of Civil Engineering and Geosciences, Delft University of Technology, Delft (2004)
37. Gaspar, P.L., de Brito, J.: Assessment of the overall degradation level of an element, based on field data. In: Proceedings of the 10th International Conférence on Durability of Building Materials and Components on Proceedings, TT8-68, Lyon (2005)

38. Yacob, S., Shah Ali, A., Cheong Peng, A.: Building condition assessment: lesson learnt from pilot projects. In: Proceedings of the 4th International Building Control Conference. MATEC Web of Conferences. Vol. 66, 00072, pp. 1–7 (2016)
39. Hećimović, D., Vidaković, D.: Analyses cost effectiveness energy reconstruction family houses in Osijek-Baranya county. In: Proceedings of the 27th International and Scientific Professional Conference 'Organization and Maintenance Technology' – OTO 2018, pp. 97–104, Faculty of Electrical Engineering, Computing and Information Technology, Osijek (2018)
40. Law on Protection and Preservation of Cultural Heritage, Official Gazette 69 (1999). (in HR)
41. Cerić, A., Katavić, M.: Building maintenance management. Građevinar **53**(2), 83–89 (2001)
42. Petrović, B., Zhang, X., Eriksson, O., Wallhagen, M.: Life cycle cost analysis of a single-family house in Sweden. Buildings **11**(2015), 1–20 (2021)
43. Boussabaine, A., Kirkham, R.: Whole Life-Cycle Costing Risk and Risk Responses. Blackwell Publishing Ltd., Oxford (2004)

Maintenance Condition and Seismic Vulnerability of Buildings in Rural Areas – A Case Study of Two Rural Settlements in Osijek-Baranja County

Marijana Hadzima-Nyarko[1](✉), Vedran Stojnović[2], Gordana Pavić[1], and Borko Bulajić[3]

[1] Faculty of Civil Engineering and Architecture Osijek, University J.J. Strossmayer Osijek, Osijek, Croatia
mhadzima@gfos.hr

[2] Geoprem d.o.o., Trg Lava Mirskog 1, 31000 Osijek, Croatia

[3] Faculty of Technical Sciences, University of Novi Sad, Trg Dositeja Obradovića 6, 21000 Novi Sad, Serbia

Abstract. The December 29, 2020, Petrinja earthquake highlighted the significance of maintenance of buildings in seismically active regions. Many family houses in rural parts of central Croatia were damaged by the Petrinja earthquake. During the earthquake, the insufficient response of the buildings in the rural regions resulted in the majority of human injuries and fatalities. These deaths and catastrophic property losses in rural areas have underlined the importance of earthquake prevention and mitigation in these communities. A study of the seismic performance of structures in rural areas, as well as an evaluation of seismic risk and susceptibility, were conducted with the aim to improve the overall capability to prevent seismic disasters in rural areas. The structural characteristics, construction method, and application of seismic codes and building rules of two rural villages in the Osijek-Baranja County were researched, and they can be considered representative of the continental area of Croatia and surrounding countries. GIS technology is used to process and present information on the location, distribution, and other characteristics of structures.

Keywords: Seismic vulnerability · Building exposure model · Rural areas · Geo-graphic information system

1 Seismo-Geological Features of the Research Area

Croatia is situated in the northeastern section of the area of permanent collisions between the Adriatic and Eurasia continental blocks, in the compression zone [1–5]. The devastating 2020 Petrinja earthquake struck near the border of Croatia's two main geological provinces: the Pannonian basin and the Dinarides mountain range. The Dinarides range is the product of an Alpine collision between the Adriatic and European tectonic plates,

H. Glavaš et al. (Eds.): OTO 2021, LNNS 369, pp. 310–323, 2022.
https://doi.org/10.1007/978-3-030-92851-3_23

whereas the Pannonian basin is a large depression in the interior of the arcuate Carpathian Mountain chain [6].

Slavonija and Baranja County, in the southern central section of the Pannonian Basin, is where the analyzed villages in this research are located. This is an intraplate seismicity region with a low frequency of major earthquakes, which are more common in Croatia's central and coastal regions. Recent microzonation studies for the city of Osijek showed that deep geological sediments in this area have a significant impact on the severity of surface ground motion during earthquakes [7, 8]. In Slavonija and Baranja County, seismic activity has been detected in the past near the towns of Đakovo, Slavonski Brod, and Beli Manastir, as well as in the mountains of Dilj Gora, Požeška Gora, Krndija, Papuk, and Bansko Brdo. The area around the mountain Dilj Gora experienced the most earthquakes.

As published by the United States Geological Survey [9], Fig. 1 depicts the epicenter locations of regional earthquakes with $Mw \geq 2.5$ that occurred between 1900 and 2017. The greatest historical earthquake near Osijek-Baranja County occurred on April 13, 1964, with an epicenter ten kilometers northwest of Slavonski Brod at Dilj Gora (see the right plot in Fig. 1), a hypocentral depth of around 15 km, and a magnitude of 6.0 [9]. The cities of Slavonski Brod, Đakovo, Požega, Našice, and Vukovar and the surrounding villages suffered the most damage. The first reports of damage in the area of the city of Đakovo and surrounding villages mention 442 demolished and almost 3300 damaged buildings. In the western part of the Đakovo region, in the villages of Musić, Lapovci, Hrkanovci, Ovčara, Paučje, Slobodna Vlast, Levanjska Varoš, and in the villages of Vučevci in the east and Ivanovci in the west of the Đakovo region, the greatest damage was caused by earthquakes.

The largest historical earthquake near Osijek occurred on November 24, 1922, with the epicenter 20 km north of Osijek (see Fig. 1). It had a magnitude of $Mw = 4.6$, and a hypocentral depth of 18 km. The intensity in the epicentral region was VII–VIII (°MCS) and around VII (°MCS) in Osijek. The destructive earthquake struck central Croatia on December 29, 2020, near the town of Petrinja and around 40 km south of Croatian Capital Zagreb, and with a magnitude of 6.4 [10].

Figure 1 depicts the epicenters of all $Mw \geq 2.5$ earthquakes in the region from 1900 to 2017 [9], as well as a recent devastating Mw6.4 Petrinja earthquake, 1964 Slavonski Brod earthquake, and 1922 Osijek earthquake.

Although seismologists predict a low probability of stronger earthquakes for the analyzed area, increased seismic activity in northwestern Croatia in recent times, especially the 2020 Petrinja earthquake, have indicated numerous problems with the resilience of the existing buildings stock. Due to the similarity of application of building regulations, construction, and maintenance of buildings in the analyzed area, we can expect a detrimental effect on the load-bearing capacity and stability of the construction fund even in the case of weaker seismic events.

2 Building Stock Vulnerability in Rural Parts of Sisak-Moslavina County

On December 29, 2020, a shallow severe earthquake struck Croatia's Central region, causing the ground to shift 45 cm in the line-of-sight direction [10]. According to the

USGS, the earthquake had a magnitude of 6.4 on the moment magnitude scale, and the epicenter was near the town of Petrinja (40 km south of the Croatian capital, Zagreb). Due to the west-side slippage of the fault, lying below the NW-SE mountain ridge that runs west of Petrinja, the fault ruptured for around eight kilometers during the Petrinja earthquake [10, 11].

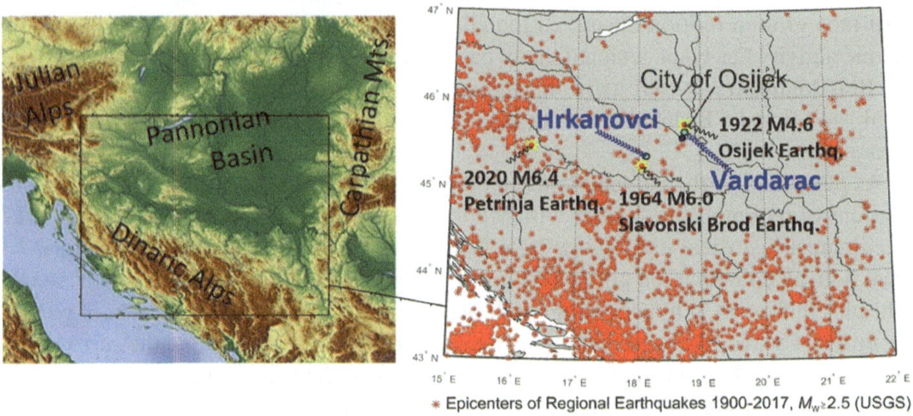

Fig. 1. Epicenters of regional earthquakes in the period between 1900 and 2017 [9]. Locations of the two rural settlements (Vardarac and Hrkanovci (Đakovački)) for which an exposure model was developed are also shown.

The contemporary building stock in rural areas of Sisak-Moslavina County was constructed using masonry, adobe, and timber structural frames. Masonry structures usually have a timber floor, whereas the most recent structures have a reinforced concrete slab. Adobes are sun-dried composites of water, mud, and straw/chaff that are put in rectangular molds. Composite components have previously been used in rural areas due to their low cost and easy availability, as well as their simple manufacturing process. They are strong enough to carry gravitational loads and provide adequate acoustic and thermal insulation, but they are not waterproof and have poor mechanical properties. In the rural parts of Sisak-Moslavina County, the third most common type of structure is a timber frame structure filled with reeds or planks and a mixture of mud and chaff. Modern timber structures are filled with adobe or masonry.

After examining the damage caused by the earthquake, it was discovered that many of the structures, both new and older, had maintained better resistance and stability than what the standard analysis of their load-bearing capacity would have suggested. However, ignoring the maintenance of non-structural elements (roofs, chimneys, gable walls, and so forth) caused considerable damages during the earthquake. Non-structural elements of individual buildings that were improperly built and maintained have also caused damages to the adjoining structures that were well constructed and with the required earthquake resistance. For example, the roof of house No. 11 on Gajeva Street in Petrinja collapsed due to insufficient maintenance, causing damage to the roof and house No. 13, which would otherwise have been unaffected by the earthquake (Fig. 2).

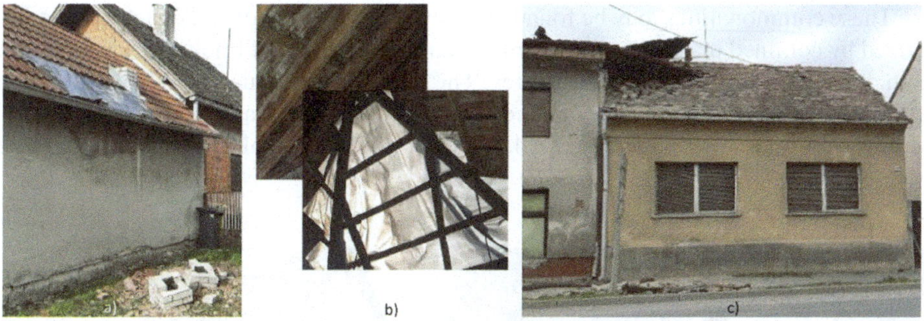

Fig. 2. Inadequate maintenance of chimney (a) or roof (b) can lead to the damaged roof of the adjacent house (c) – Houses No. 11 (left) and No. 13. (right) on Gajeva Street in Petrinja.

When comparing the damages to outbuildings that are not maintained and often built without considering the regulations (see Fig. 3), with the damages to buildings of approximately equal age and structural system, the impact of complying with building regulations during construction and subsequent maintenance during use on the seismic resistance of buildings in rural areas is particularly visible. Outbuildings frequently experienced major structural damage and collapsed, whereas residential buildings sustained minor damage, which primarily consisted of reductions in the load-bearing capacity of non-structural materials.

Fig. 3. Damages to a house in Vinogradi IV settlement, Petrinja – an example of construction without complying with any building codes and with poor maintenance.

3 Characteristics of Buildings Stock in Rural Areas of Osijek-Baranja County

When comparing the characteristics of structures in rural parts of Osijek-Baranja County to those in Sisak-Moslavina County, significant similarities in construction methods and materials were discovered.

These commonalities may be found not only in modern buildings, which are characterized by confined masonry and unreinforced masonry made of brick or brick blocks in both counties but also in older structures. The load-bearing structure of village houses from the beginning of the twentieth century from Osijek-Baranja County and Sisak-Moslavina County is depicted in Fig. 4, which consists of a timber frame and an adobe or brick filling.

Fig. 4. Timber frame with infill of brick or adobe in a) Sisak-Moslavina County, b) Osijek-Baranja County.

Two settlements in the rural areas of Osijek-Baranja County, which were significantly damaged by earthquakes in the past, were chosen for analysis Hrkanovci Đakovački, damaged by the 1964 Slavonski Brod earthquake, and Vardarac, damaged by the 1922 Osijek earthquake.

Even today, in the Slavonia and Baranja region, the most prevalent type of village is a typical line-road village with a central square or a square as an extension of streets. The spatial concept of rural housing estates has been affected by centuries of development, functional adaptations, and local climatic conditions. To retain spatial concepts, elongated plots perpendicular to street corridors and the orientation of the gables of the buildings to the street side are still employed today.

In the villages of Osijek-Baranja County, traditional building techniques and materials were used to construct residential and commercial structures, and they are still used today. Some villages still have old houses made of compacted earth, timber frame structures filled with reeds or planks coupled with a mixture of mud and chaff, or wooden frame structures filled with adobe or brick. Since the mid-nineteenth century, houses have been made with bricks. Although bricks are still widely used as a building material, brick blocks are gaining popularity as a result of their faster and less expensive construction, particularly in commercial buildings.

3.1 Data Collection and Processing Methodology

Because the data in this study is based on spatial data, a GIS-based approach for data collection and processing in the office and the field was adopted. Longley et al. [12] define a geographic information system as a computer system that allows for the collection, modeling, storage, access, exchange, manipulation, and display of georeferenced data. It is made up of six basic components: software, hardware, users (people), data, and procedures. A geoinformation system's data represent features on Earth in a simplified, digital format for a specific location. The spatial distribution of objects is described by modeling georeferenced spatial data, which also includes attribute data that describes their qualities [13]. The geometric component of a GIS can be organized as a raster or vector data set.

A vector approach was employed in this research for data processing, i.e. the data is arranged as geometric objects (buildings) for which the following set of attribute data is defined:

street and house number,
- purpose of a building,
- year of construction,
- position in the block,
- regularity of the floor plan,
- regularity in elevation,
- number of floors,
- wall material,
- ceiling construction design,
- vertical reinforcement,
- construction system,
- the current state of preservation,
- roof construction design,
- a comment.

In addition to attribute data that describe the qualities of an object, features are geometrically described as georeferenced polygon features which define their dimensions, spatial distribution, and mutual relationship.

GIS data processing has a significant benefit over traditional data processing methods in that it allows for the visual component of study objects, positional distribution, and faster navigation in space. Geometric elements in the GIS are defined in the ground plan using the most recent orthorectified aerial imagery that are publicly available (DOF5 2019/20) [14]. Attribute data on the year of construction of the buildings are taken from the 1991 census, and other attributes are manually collected and processed. A system of automated controls with visible warnings has been built up within the established system to avoid human errors during the process of entering and changing data. This assures data integrity for each object and prevents inhomogeneities and redundancy of data. This study project needed data gathering in the field for specific locations, in addition to data analysis and collection in the office. The system was built using QGIS [15], which is free and open-source software for working with GIS (Fig. 5).

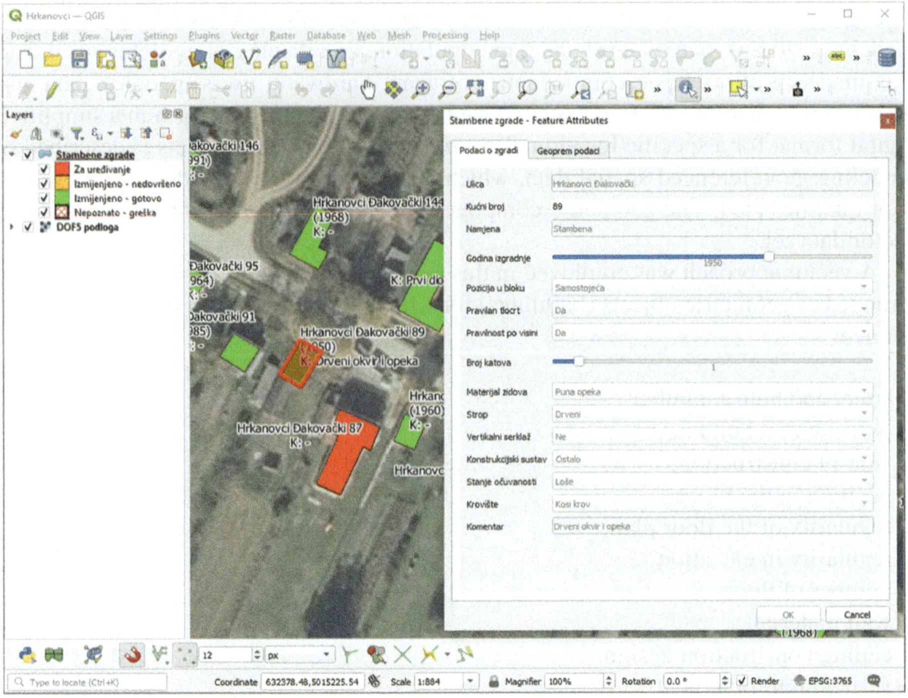

Fig. 5. Data processing in QGIS – an example for Hrkanovci Đakovački village.

We used a mobile version of QGIS software - QField [16] (Fig. 6) for mobile data collection since devices with built-in GNSS receivers - mobile phones - are widely available and their positioning precision satisfies the criteria of this research. The exchange format – GeoPackage – was chosen to meet the job requirements: it had to be easily portable, support raster and vector formats, and work in conditions where there was no Internet access (offline) in the field. This format has been developed recently specifically to meet all these requirements [17].

3.2 State of Preservation of Buildings in Selected Villages

The basic data that the database of buildings should contain are data on the geographical position of buildings, spatial orientation, purpose, area, geometric and structural characteristics, and the state of maintenance [18–20]. Because Google Street View does not show individual streets, detailed forms with each address have been created for a more efficient town inspection.

On a worldwide scale, the preservation, usability, and stability of houses are questionable. Many older houses, as well as many newer ones, are in poor condition.

The replacement of roof tiles, carpentry, and gutters, as well as the restoration or strengthening of chimneys, broken parts of facades and walls, and repairing visible cracks that are caused most likely by subsidence and/or material degradation, are all necessary maintenance tasks on most older houses.

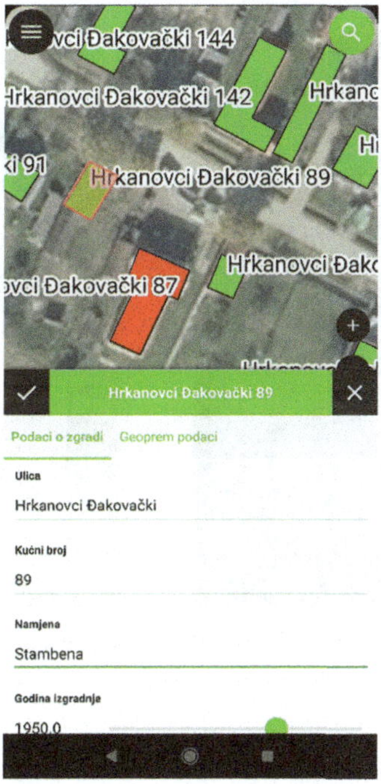

Fig. 6. Field data processing using the QField application – an example for Hrkanovci Đakovački village.

According to locals, after the 2020 Petrinja earthquake, some existing cracks widened, and the walls of several abandoned houses that had wider cracks or where the mortar had degraded collapsed, often resulting in the entire building collapsing.

In natural disasters, insufficient and unprofessional maintenance is frequently the primary cause of destruction and loss of cultural and architectural heritage. Figure 7 depicts a house in Hrkanovci Đakovački that collapsed following the earthquake in Petrinja. The left picture shows the house ten years ago and the right picture shows the same house today.

During the field study, it was also discovered that many older and newer houses lack a facade, which leads to faster wall disintegration and moisture absorption. The parapet masonry is particularly wet and damaged, most likely due to a lack of foundation or insulation between the foundation and the masonry, as well as inadequate or unresolved drainage of rain and wastewater (sewage) that frequently ends up around the house.

4 Results

4.1 Statistical Analysis of Main Characteristics of Buildings in Selected Villages

After collecting buildings data of the selected villages and developing exposure models (building databases), the following building characteristics were selected for comparison: construction typologies, condition of maintenance, rigidity, and regularity of the floor plan.

Fig. 7. State of the house in Hrkanovci Đakovački: a) Google Street view in November 2011, b) photo of the same house taken in August 2021.

The findings of the statistical study for selected building characteristics in the villages of Vardarac and Hrkanovac Đakovacki are presented in Fig. 8.

Unreinforced masonry with a rigid floor is the most common structural typology in Vardarac, and unreinforced masonry with a flexible floor is the most common structural typology in Hrkanovci Đakovački. In Hrkanovci Đakovački, wooden buildings account for 31% of the total number of buildings, while there is only one wooden building in Vardarac.

The authors' opinion, which is based on knowledge and observations gained during fieldwork on data collection for buildings in the subject villages, the distribution of structural typologies statistics has been influenced by the adopted architectural tradition and the financial possibility of each village.

Better maintenance statistics for Vardarac can be explained by the village's improved economic status (which we associate with the recent development of tourism). In contrast to Hrkanovac, where the number of well and poorly-maintained dwellings is approximate, 2/3 of the houses in the Vardarac village are well-maintained.

The statistics on the rigidity of the floor plan show that only 1/3 of the buildings in Hrkanovci have a rigid floor plan, while it is the opposite for Vardarac, where 2/3 of the buildings have a rigid floor plan.

The statistics on the regularity of the floor plan of the houses in both villages are the same and show that regular floor plans predominate in both villages.

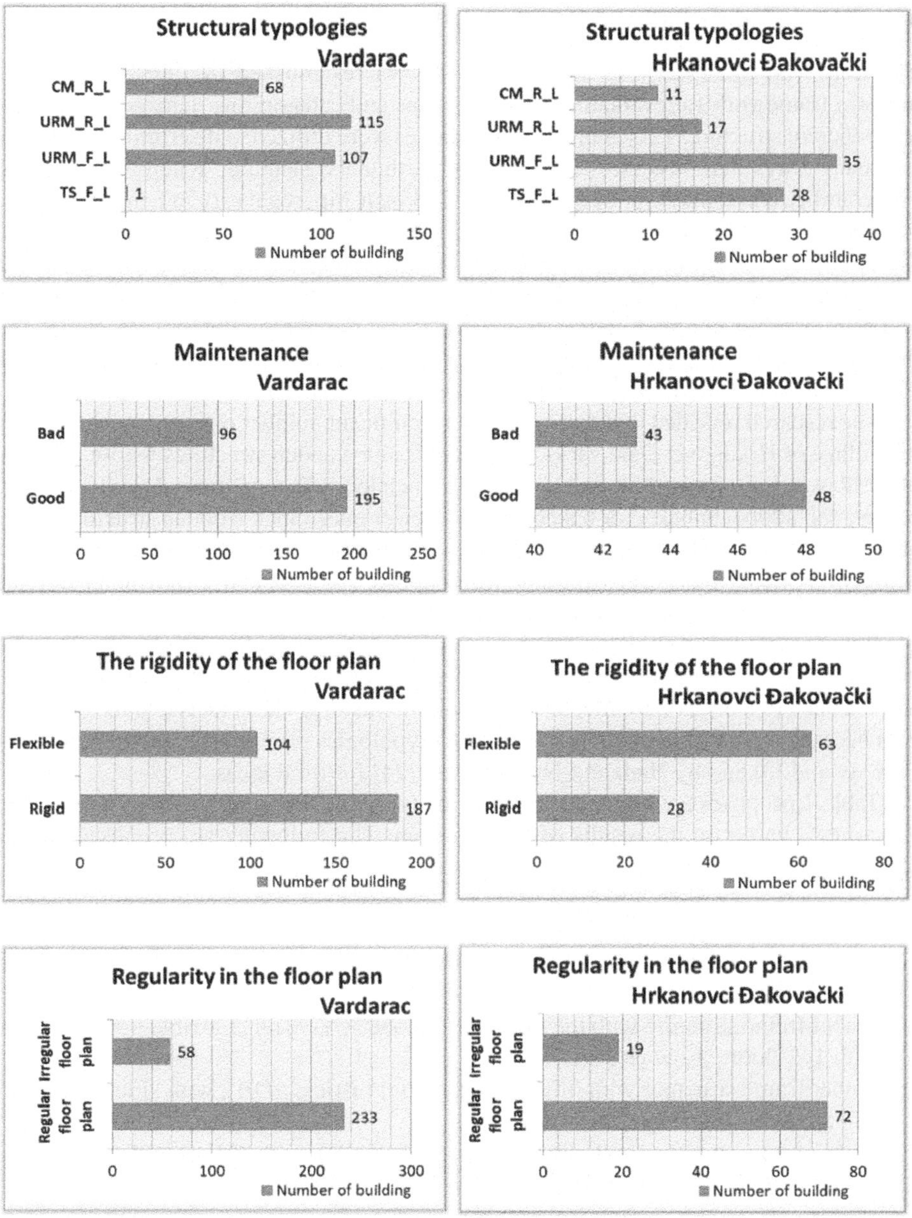

Fig. 8. An examination of the key characteristics of buildings in selected villages.

4.2 Seismic Vulnerability of Buildings in Rural Areas

The European Macroseismic Scale (EMS-98) was established in 1998 to provide a uniform framework for evaluating vulnerabilities and categorizing damage to masonry and concrete structures [21]. Vulnerability classes were utilized to discriminate between the many ways that structures can react to earthquake shaking. When assessing the vulnerability class, the level of quality and workmanship, regularity in height and plan, state of preservation, and other vulnerability impacting variables must all be considered [22]. EMS-98 distinguishes between rubble stone, adobe, simple stone, massive stone, two forms of unreinforced buildings: with manufactured stone units and RC floors, and reinforced or confined masonry constructions. For the rural exposure model and the adopted design typologies, vulnerability classes were adopted in accordance with EMS-98, which were developed in [23] and shown in Table 1.

The research revealed that there are still a significant number of wooden buildings in the village of Hrkanovci Đakovački. These are frame wooden structures whose filling in older houses is composed of reed or planks associated with a mixture of mud and chaff binder, and adobe, unbaked or baked brick in newer ones. Such wooden structures were also found after the 2020 Petrinja Earthquake in the earthquake-affected villages of Sisak-Moslavina County, and are also present in the exposure model of other Slavonian-Baranja villages, which is currently being produced.

Private owners and public institutions are increasingly deciding to rebuild and maintain these old wooden buildings recently, owing to the growth of tourism in rural areas and the need to conserve architectural and cultural history. These facts and the collected data necessitate the adjustment of the building typologies proposed for urban regions by Pavić et al. [23] and by Hadzima-Nyarko et al. [24] for rural areas.

Table 1 proposes a classification of the analyzes timber buildings into "Timber old structures", based on the results of this research. The following are proposed for the class "Timber old structures" (TOS):

1. timber frame structure with the filling of "Reeds or Planks" (RP) associated with a mixture of mud and chaff binder, with "Flexible floor" (F), and with 1–2 floor: "Low-rise" (L)
2. timber frame structure with the filling of "Adobe" (A), with "Flexible floor" (F), and with 1–2 floor: "Low-rise" (L)
3. timber frame structure with filling made of "Old Brick" (OB), with "Flexible floor" (F), and with 1–2 floor: "Low-rise" (L)
4. timber frame structure with the filling of "Brick" (B), with "Flexible floor" (F), and with 1–2 floor: "Low-rise" (L).

Table 1. Proposed vulnerability classes for timber and masonry buildings – extended table from [23].

Masonry building typology		Proposed vulnerability class (EMS-98)	Masonry building typology		Proposed vulnerability class (EMS-98)
Unreinforced masonry	URM_OB_F_L URM_OB_F_M URM_B_F_L URM_B_F_M URM_B_R_L URM_B_R_M	B A B A C B	Confined masonry	CM_R_L CM_R_M CM_R_H	D C B-A
Timber building typology		Proposed vulnerability class (EMS-98)			
Old timber structures	TOS_RP_F_L TOS_A_F_L TOS_OB_F_L TOS_B_F_L	A A A A			

5 Conclusion

The devastating earthquake in Petrinja in 2020 revealed the shortcomings of the existing buildings stock in central Croatia, especially rural buildings. Similar typologies of buildings exist in other parts of Croatia and adjacent countries, so this requires a detailed reassessment of seismic risk regardless of seismic hazard.

In this research, an exposure model was developed for two rural settlements in Osijek-Baranja County, which due to structural characteristics, construction, and implementation of seismic codes and construction standards, can be considered representative of the continental area of Croatia and neighboring countries.

During the development of the exposure model, the characteristics of old timber buildings were observed, which were not considered by the existing typology of timber buildings. Therefore, the paper proposes the classification of the analyzed timber structures as "Old timber structures" with 4 different typologies.

GIS technology was used to collect and present data on the location, distribution, and other characteristics of the buildings of selected rural settlements. By using the GIS approach, the process of data collection was very practical, especially in the field, which made the whole process much faster and decreased the number of errors to a minimum. After the initial data collection, GIS technology provided visual inspection and control of the entire dataset, resulting in easier data processing, and obtaining statistical values.

The findings of this study and the proposed typology can be considered as criteria for currently higher accuracy of damage estimates in the analyzed parts of Croatia, and damage estimates obtained based on actual input data after the 2020 Petrinja earthquake struck Sisak-Moslavina County may be due to similar typologies, materials, applied construction techniques and possible earthquake intensity.

Acknowledgment. The research described in this paper has been supported by the Croatian Academy of Sciences and Arts under a scientific research project entitled "Preliminary research in the field of seismic risk in rural areas of eastern Croatia".

References

1. Anderson, H., Jackson, J.: Active tectonics of the Adriatic Region. Geophys. J. Roy. Astron. Soc. **91**, 937–983 (1987)
2. Grenerczy, G., Sella, G., Stein, S., Kenyeres, A.: Tectonic implications of the GPS velocity field in the northern Adriatic region. Geophys. Res. Lett. **32**(16) (2005)
3. Benetatos, C., Kiratzi, A.: Finite-fault slip models for the 15 April 1979 (Mw7.1) Montenegro earthquake and its strongest aftershock of 24 May 1979 (Mw6.2). Tectonophysics **421**, 129–143 (2006).
4. Bennett, R.A., et al.: Eocene to present subduction of southern Adria mantle lithosphere beneath the Dinarides. Geology **36**(1), 3–6 (2008)
5. Métois, M., et al.: Insights on continental collisional processes from GPS data: dynamics of the peri-Adriatic belts. J. Geophys. Res. Solid Earth **120**, 8701–8719 (2015)
6. Tondi, E., Blumetti, A.M., Čičak, M. et al.: 'Conjugate' coseismic surface faulting related with the 29 December 2020, Mw 6.4, Petrinja earthquake (Sisak-Moslavina, Croatia). Sci. Rep. **11**, 9150 (2021)
7. Bulajić, B.Đ, Hadzima-Nyarko, M., Pavić, G.: Vertical to horizontal UHS ratios for low to medium seismicity regions with deep soil atop deep geological sediments—an example of the City of Osijek. Croatia Appl. Sci. **11**(15), 6782 (2021)
8. Bulajić, B.Đ, Hadzima-Nyarko, M., Pavić, G.: Horizontal UHS amplitudes for regions with deep soil atop deep geological sediments—an example of Osijek. Croatia. Applied Sciences **11**, 6296 (2021)
9. USGS: Earthquake catalog for all earthquakes with Mw≥2.5 in the period 1900–2017 for the geographic region between 40.5°N and 47.5°N, and 12.5°E and 24.5°E (2017). https://earthquake.usgs.gov/earthquakes/search/, Accessed 23 Feb 2017
10. Ganas, A., Elias, P., Valkaniotis, S., Tsironi, V., Karasante I., Briole, P.: Petrinja earthquake moved crust 10 feet. Temblor (2021). https://doi.org/10.32858/temblor.156. Accessed 14 July 2021
11. Miranda, E., et al.: PRJ-2959/StEER-EERI: Petrinja, Croatia December 29, 2020, Mw 6.4 Earthquake. In: Kijewski-Correa, T. (ed.) Field Research, Joint Reconnaissance Report, DesignSafe, p. 206. Oakland, CA, USA (2021). https://www.designsafe-ci.org/data/browser/public/designsafe.storage.published/PRJ-2959/#details-4541797621822058005-242ac118-0001-012, Accessed 15 July 2021
12. Longley, P.A., Goodchild, M.F., Maguire, D.J., Rhind, D.W: Geographic Information Systems and Science, John Wiley & Sons, Ltd., Chinchester (2015)
13. Neteler, M., Mitasova, H.: Open source GIS – A GRASS GIS Approach. Springer, New York (2008)
14. Croatian State Geodetic Administration, Publicly available datasets and services. https://geoportal.dgu.hr/#/menu/podaci-i-servisi. Accessed 22 July 2021
15. QGIS software. https://qgis.org. Accessed 29 Aug 2021
16. QField software. https://qfield.org. Accessed 29 Aug 2021

17. Stojnović, V., Miler, M.: Determining possibilities of GeoPackage file format for exchanging cadaster and land register data in Republic of Croatia. In: 13. Symposium of Chartered Geodetic Engineers - Transformation of Geodetic Profession Through the Goals of Sustainable Development (Virtual Conference), Croatian Chamber of Chartered Geodetic Engineers, Zagreb, Croatia, pp. 9–17 (2020)
18. Pavić, G., Hadzima-Nyarko, M., Bulajić, B.: A Contribution to a UHS-based seismic risk assessment in Croatia—A case study for the City of Osijek. Sustainability **12**(5), 1796 (2020)
19. Pavić, G., Bulajić, B., Hadzima-Nyarko, M.: The vulnerability of buildings from the Osijek database. Front. Built Environ. **5**, 66 (2019)
20. Hadzima-Nyarko, M., Lešić, M., Morić, D. Seismic vulnerability assessment for residential buildings in Osijek, Croatia. In: Proceedings of the 16th World Conference of Earthquake Engineering, pp. 1–12, Chilean Association of Seismology and Earthquake Engineering Board Directors ACHISINA, Santiago (2017)
21. Grünthal, G., Musson, R.M.W., Schwarz, J., Stucchi, M. (eds.): European Macroseismic Scale 1998, vol. 15, Cahiers du Centre Européen de Géodynamique et de Séismologie, Luxembourg (1998)
22. Hadzima-Nyarko, M., Pavić, G., Lešić, M.: Seismic vulnerability of old confined masonry buildings in Osijek, Croatia. Earthq. Struct. **11**(4), 629–648 (2016)
23. Pavić, G., Hadzima-Nyarko, M., Bulajić, B., Jurković, Ž: Development of seismic vulnerability and exposure models—a case study of Croatia. Sustainability **12**(3), 973 (2020)
24. Hadzima-Nyarko, M., Pavić, G., Bulajić, B., Plaščak, I.: Seismic vulnerability assessment of small rural areas in Eastern Croatia. In: Meguro, K. (ed.) Proceedings of the 17th World Conference on Earthquake Engineering. Japan Association on Earthquake Engineering JAEE, Sendai, pp. 1–12 (2021)

Material Consumption Estimation in the Construction of Concrete Road Bridges Using Machine Learning

Miljan Kovačević[1] (✉) and Borko Bulajić[2]

[1] Faculty of Technical Sciences, University of Pristina, Knjaza Milosa 7, 38220 Kosovska Mitrovica, Serbia
miljan.kovacevic@pr.ac.rs

[2] Faculty of Technical Sciences, University of Novi Sad, Trg Dositeja Obradovića 6, 21000 Novi Sad, Serbia
borkobulajic@uns.ac.rs

Abstract. The state-of-the-art techniques of machine learning for assessing material consumption in the construction of prestressed concrete (PC) road bridges are described and analyzed in this paper. For model training and evaluation, a database of material consumptions and design features for 75 PC bridges is compiled. The achieved accuracy of the model for estimating prestressed steel consumption per m^2 was 6.55%, calculated as the mean absolute percentage error. Based on the proposed model, a MATLAB-based software with a graphical user interface was developed to allow for the input of basic variables and the calculation of estimated prestressing steel quantities. The use of such software to determine the ideal amount of steel will, in most cases, result in the construction of higher-quality structures that require less maintenance.

Keywords: Material consumption · Bridges · Prestressed concrete bridges · Machine learning

1 Introduction

In general, using modern algorithms to estimate the ideal amount of steel in a bridge or other RC structure will result in the construction of a higher-quality structures that will require less long-term maintenance. Early resource planning and optimization gives a better foundation for later maintenance that is easier and less expensive.

There are many examples of such algorithm used for bridges. Marcous et al. [1] used a neural network model to model optimal concrete volume and the weight of prestressing steel for 22 prestressed reinforced concrete bridges over the Nile River in Egypt. The result was an error size of 7.5% in terms of predicting the amount of concrete and 11.5% in terms of predicting the amount of prestressing steel. Fragkakis et al. [2] used regression analyses in modeling the consumption of steel and concrete for bridge foundations in order to consider the costs in the conceptual phase of the project. Mučenski et al. [3] used a neural network model to estimate the required amounts of reinforcement and

H. Glavaš et al. (Eds.): OTO 2021, LNNS 369, pp. 324–343, 2022.
https://doi.org/10.1007/978-3-030-92851-3_24

concrete in multi-storey buildings based on data from 115 major projects of residential buildings. They obtained the best results using the network trained by Broyden–Fletcher–Goldfarb–Shanno algorithm with an average error of 12.49%. Fragkakis et al. [4] defined regression equations for estimating the consumption of concrete and steel per meter of culvert, based on the database of 104 culverts on the Egnatia highway in Greece. They tested the models using tenfold cross-validation and evaluated against MAPE criteria. The result showed the accuracy of the estimate in terms of consumed concrete and steel, which is 13.78% and 19.79%, respectively. Marineli et al. [5] created a model to estimate the material consumption in bridge superstructure using a multilayer perceptron neural network model based on the data from 68 bridges in Greece. The models were evaluated via the correlation coefficient R. The construction costs and consumption of steel and concrete in underpasses were analyzed by Antoniou et al. [6]. Their database included data from 28 underpasses in Greece. The paper points out the average values per m^2 of material consumption and the actual costs of construction of underpass bridges. Dimitriou et al. [7] analyzed a model for estimating the consumption of steel and concrete in the construction of road bridges using linear regression models and neural network models, based on the data on 68 bridges in Greece. The value of MAPE and R^2 have been used to assess model accuracy. In the prediction of concrete consumption only for the bridge superstructure, the value of MAPE ranged from 11.48% to 16.12% depending on the bridge span, while the values for R^2 range from 0.979 to 0.995. In terms of consumption of concrete and steel per column, the obtained values of MAPE and R^2 were, respectively, 37% and 0.962 for concrete and 31% and 0.962 for steel.

In this paper, we will develop, examine, and verify a number of prediction models for early estimation of prestressed steel consumption per m^2, based on artificial intelligence approaches. For the assessment model, a database of design attributes for 75 PC bridges in Serbia is constructed. All models will be trained and tested using cross-validation under the same conditions.

2 Methods

2.1 Multilayered Perceptron Artificial Neural Network (MLP-ANN)

A multilayer perceptron is a forward signal-propagating neural network having minimum of three neurons layers: the input, the hidden, and the output. As shown in Fig. 1, every neuron of one layer is connected to every neuron of the next layer.

The number of neurons in the hidden layer and the activation type have a major impact on the network's features. When MLP-ANN is applied as an approximator, activation functions are usually chosen to be continuous and differentiable. A linear activation function is most typically used in the output layer. If the case of sufficient number of neurons in the hidden layer [8], an arbitrary multidimensional function can be approximated for a given data set using the MLP-ANN model with a single hidden layer whose neurons' activation function is bipolar sigmoid while for the output layer the activation function is linear.

The number of neurons in the hidden layer is established through trial and error, with certain guidelines taken into account. The following formulas [8–11] can be used to estimate the maximum of the number of hidden layer neurons N_H:

$$N_H \leq 2 \times N_i + 1 \tag{1}$$

$$N_H \leq \frac{N_s}{N_i + 1}. \tag{2}$$

where N_i is the number of inputs, and N_s is the number of instances used for training. Less than two acquired values for the number of neurons are accepted utilizing these formulas based on the obtained values for the number of neurons.

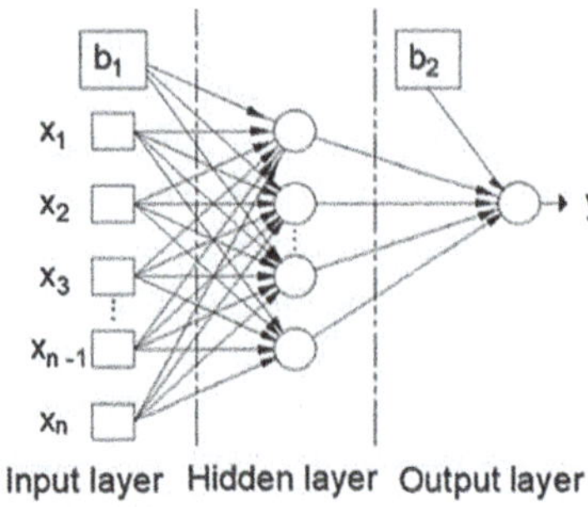

Fig. 1. Schematic presentation of a multilayer perceptron artificial neural network [8].

2.2 Regression Tree Ensembles

The binary tree depicted in Fig. 2 represents an example of a regression tree model, in which the entire set of data is started at the top of the tree, and observations are divided on each node based on whether or not they meet specified conditions in that node. Individual regions (R_1, R_2, R_3, R_4, R_5, R_6, R_7, R_8) correspond to the tree's end nodes or leaves.

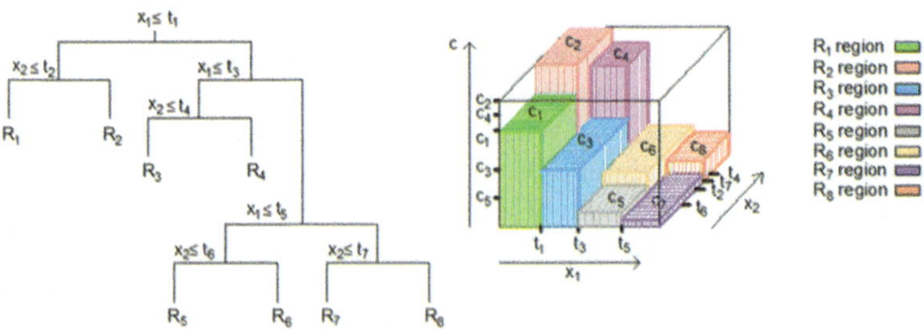

Fig. 2. Examples of space segmentation (left) and 3D regression surface in the regression tree (right) [11].

The regression model assigns a constant value c_m to each region based on the defined input variables:

$$\hat{f}(X) = \sum\nolimits_{m=1}^{8} c_m I\{(X_1, X_2) \in R_m\} \tag{3}$$

where the constant c_m represents the mean of the output variable for region R_m, and $I\{(X_1, X_2) \in R_m\}$ is an indicator function that has a value of one for exact statements and zero for all other statements.

Assuming that space segmentation is performed in M domains $R_1, R_2, \ldots, R_M$, the output model with a constant value of c_m for each region in this case has the following form

$$f(x) = \sum\nolimits_{m=1}^{M} c_m I(x \in R_m). \tag{4}$$

The optimal value $\hat{c}_m$ is the mean value y_i for the region R_m, as determined by the criterion of minimizing the sum of the squares $\sum (y_i - f(x_i))^2$:

$$\hat{c}_m = ave(y_i | x_i \in R_m). \tag{5}$$

where *ave* is the mean value.

A Greedy algorithm [12, 13] is used to find the input space's binary segmentation point. When using this algorithm, only the best results from each step of the process are considered, with no consideration for subsequent steps.

Let us consider the variable j on which the split will be conducted, as well as the value or split point s, and define a pair of half-planes starting with the entire set of data:

$$R_1(j, s) = \left\{X | X_j \leqq s\right\} \text{ i } R_2(j, s) = \left\{X | X_j > s\right\}. \tag{6}$$

The variable j, as well as the value of the split point s, must be determined to meet the following expression:

$$\min_{j,s}\left[\min_{c_1} \sum\nolimits_{x_i \in R_1(j,s)} (y_i - c_1)^2 + \min_{c_2} \sum\nolimits_{x_i \in R_2(j,s)} (y_i - c_2)^2\right]. \tag{7}$$

By using the mean y for the region R_1 and the mean y for the region R_2 as follows, the value in square brackets is minimized:

$$\hat{c}_1 = ave(y_i | x_i \in R_1(j, s)) \text{ and } \hat{c}_2 = ave(y_i | x_i \in R_2(j, s)). \tag{8}$$

The split point can be determined by examining all the model's input variables. The operation continues with further splitting of the space until a specified stop criterion is met after finding the optimal point (which corresponds to the smallest value of the formula (7)). A similar approach is used for multidimensional input variables, with the exception that it is a matter of space segmentation in so-called hyperboxes in that case. There is a chance that the produced regression tree has very good performance on the training set but poor generalization on the test data set if the technique for constructing a regression tree model is followed.

The bootstrap sampling method [14–16], i.e., repeated sampling for the same training dataset, can solve the aforementioned difficulty. Each of these sets can be used to train a regression tree on its own. The ensemble is made up of the trees that have been generated in this manner.

All input variables are considered potential splits in the Treebager method, but just a subset of the input variables is used for tree splitting in the Random Forest method. If the prediction function for the b-th bootstrap training set is $\hat{f}^{*b}(x)$ at the point x, then an ensemble model [17] is created by averaging all B models (Fig. 3), with the prediction function specified by the expression:

$$\hat{f}_{bag}(x) = \frac{1}{B}\sum\nolimits_{b=1}^{B}\hat{f}^{*b}(x). \tag{9}$$

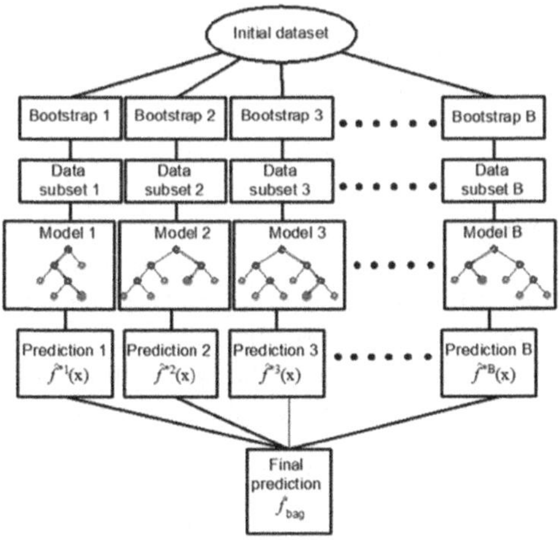

Fig. 3. Regression tree ensembles with Bootstrap aggregation-bagging [11].

The Boosting approach employs sequential training of models, with new regression trees that are added increasing the preceding tree collection's performance [17]. For quadratic error function (Gradient Boosting method), each sub-sequent step (Fig. 4) adds a new sub-model to the basic model that best guesses the residuals of the preceding model. By applying an iterative approach to the addition of a model, a definite model is defined. Such model is an ensemble of the models that were obtained in previous steps.

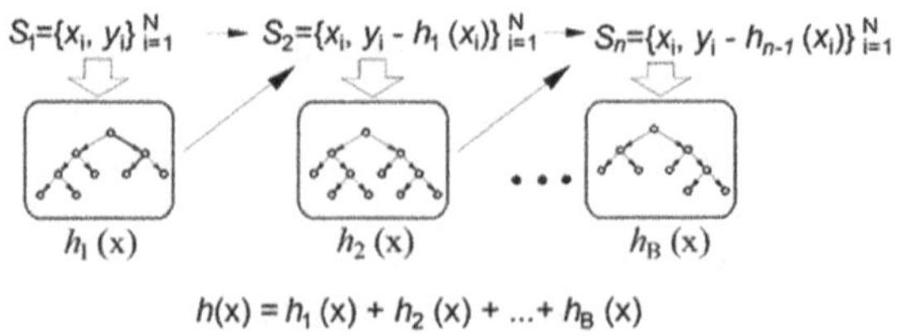

Fig. 4. Regression tree ensembles using gradient boosting [11].

2.3 Support Vector Regression (SVR)

Consider the following training datasets: $\{(x_1, y_1), (x_2, y_2), \ldots, (x_l, y_l)\} \in R^n \times R$, where $x_i \in \mathbb{R}^n$ is the n-dimenzional vector expressing the inputs and y_i are the responses. The approximation function can be written as:

$$f(x) = \sum\nolimits_{i=1}^{l} \left(\alpha_i^* - \alpha_i\right)\mathbf{K}(\mathbf{x_i}, \mathbf{x}) + b. \quad (10)$$

The empirical risk function needs to be minimized. The problem of minimizing of the empirical risk function can be solved by using the Vapnik's linear loss function with ε-insensitivity zone (Fig. 5), which is defined as follows [18, 19]:

$$|y - f(\mathbf{x}, \mathbf{w})|_\varepsilon = \begin{cases} 0 & \text{if } |y - f(\mathbf{x}, \mathbf{w})| \le \varepsilon \\ |y - f(\mathbf{x}, \mathbf{w})| - \varepsilon & \text{otherwise.} \end{cases} \quad (11)$$

As a result, the problem can be simplified to minimization of the equation below:

$$R = \frac{1}{2}\|\mathbf{w}\|^2 + C\sum\nolimits_{i=1}^{l} |y_i - f(\mathbf{x_i}, \mathbf{w})|_\varepsilon. \quad (12)$$

By introducing slack variables ξ and ξ^*, which are shown in Fig. 5, minimization of R is equivalent to the minimization the following:

$$R_{w,\xi,\xi^*} = \frac{1}{2}\left[\|\mathbf{w}\|^2 + C\left(\sum\nolimits_{i=1}^{l} \xi + \sum\nolimits_{i=1}^{l} \xi_*\right)\right], \quad (13)$$

To implement the model, the LIBSVM software [20, 21] was utilized within the MATLAB program.

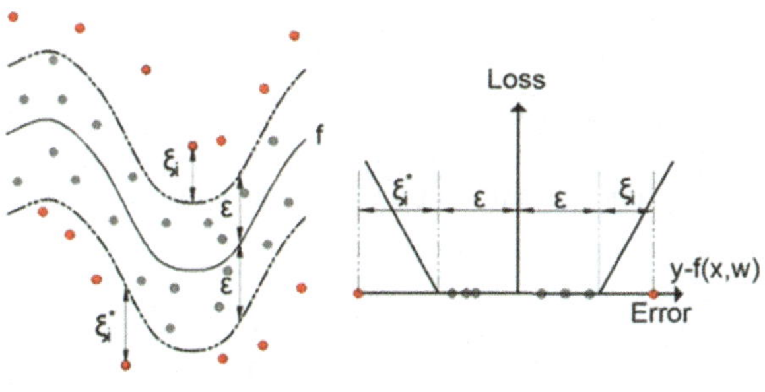

Fig. 5. Nonlinear SVR with the ε-insensitivity zone [11].

2.4 Gaussian Process Regression (GPR)

Limited collections of the random variables are characterized by a multivariate normal distribution. In other words, a finite linear combination of random variables is normally distributed in a Gaussian process. Gaussian process regression model is a probability distribution over alternative functions, which fit a set of observed points. Let us consider the following nonlinear regression problem:

$$y = f(x) + \varepsilon, \varepsilon \sim \mathrm{N}\left(0, \sigma^2\right). \tag{14}$$

where $f(\cdot) : R^n \rightarrow R$ is an unknown function that must be estimated, y is the target and x are the input variables while ε is an additive noise that is normally distributed.

The Gaussian process-type regression [22] implies that $f(\cdot)$ has a Gaussian distribution with a mean function $\mu(\cdot)$, and a covariance function $k(\cdot, \cdot)$. Here, the observations in an arbitrary data set $\mathbf{y} = \{y_1, \ldots, y_n\}$ may always be thought of as a sample from a multivariate Gaussian distribution:

$$(y_1, \ldots, y_n)^T \sim N(\boldsymbol{\mu}, \mathrm{K}), \tag{15}$$

where $\boldsymbol{\mu} = (\mu(x_1), \ldots, \mu(x_n))^T$ is the mean vector, K is the $n \times n$ covariance matrix with the (i, j) element defined as $K_{ij} = k\left(x_i, x_j\right) + \sigma^2 \delta_{ij}$, and the Kronecker delta function is represented by δ_{ij}. Let x^* be any of the test points and y^* be the response. The joint distribution $(y_1, \ldots, y_n, y^*)$ is a $(n + 1)$ variate normal distribution $(y_1, \ldots, y_n, y^*) \sim N\left(\boldsymbol{\mu}^*, \sum\right)$, where $\boldsymbol{\mu}^* = (\mu(x_1), \ldots, \mu(x_n), \mu(x^*))^T$ and covariance matrix:

$$\sum = \begin{bmatrix} K_{11} & K_{12} & \cdots & K_{1n} & K_{1*} \\ K_{21} & K_{22} & \cdots & K_{2n} & K_{2*} \\ \cdots & \cdots & \cdots & \cdots & \cdots \\ K_{n1} & K_{n2} & \cdots & K_{nn} & K_{n*} \\ K_{*1} & K_{*2} & \cdots & K_{*n} & K_{**} \end{bmatrix} = \begin{bmatrix} K & K^* \\ K^{*T} & K^{**} \end{bmatrix} \tag{16}$$

where $K^* = (K(x^*, x_1), \ldots, K(x^*, x_n))^T$ and $K^{**} = K(x^*, x^*)$.

Given $\mathbf{y} = (y_1, \ldots, y_n)^T$, the conditional distribution of y^* is $N\left(\hat{\mathbf{y}}^*, \hat{\boldsymbol{\sigma}}^{*2}\right)$ with

$$\hat{\mathbf{y}}^* = \mu(x^*) + K^{*T} K^{-1} (y - \mu) \tag{17}$$

$$\hat{\boldsymbol{\sigma}}^{*2} = K^{**} + \sigma^2 - K^{*T} K^{-1} K^*. \tag{18}$$

Using the automatic relevance determination (ARD), hyperparameters may be utilized to define the inputs that are more relevant than the others. Let us consider the squared exponential covariance function and variable length scale parameters used for every input (ARD SE):

$$k\left(x_p, x_q\right) = v^2 exp\left[-\frac{1}{2} \sum\nolimits_{i=1}^{n} \left(\frac{x_p^i - x_q^i}{r_i}\right)^2\right] \tag{19}$$

where r_i is the covariance function's length scale along input dimension i. The input is reduced if r_i is very large [10, 17].

The maximum likelihood technique can be used to estimate the noise variance σ^2 and the hyperparameters of covariance functions. The training data's log-likelihood is given by:

$$L\left(v, r_1, \ldots, r_n, \sigma^2\right) = -\frac{1}{2} \log \det K - \frac{1}{2} \mathrm{y}^T K^{-1} \mathrm{y} - \frac{n}{2} \log 2\pi. \tag{20}$$

3 Dataset

The proposed material consumption methods are based on the creation of a data collection that comprises project and contract documentation for prestressed concrete bridges built in Serbia's Corridor X (Fig. 6) [10]. On the eastern and southern branches of Corridor X, the bridge data set comprises complete data on 75 PC bridges (prefabricated or cast on site). The choice of input variables is critical in predicting material consumption in the construction of PC bridges because the bridge design is often influenced by a number of variables.

Fig. 6. Eastern and southern legs of Corridor X in Serbia [10].

To estimate the amount of prestressed steel, the following factors were used as independent variables in the model: x_1 – maximum individual bridge span, x_2 – average span, x_3 – total bridge span, x_4 – bridge width (Table 1).

Table 1. Mean, minimum and maximum values of variables in the model used to estimate the prestressed steel consumption per m^2 of the bridge superstructure

Variable	Average value	Minimum value	Maximum value
Max. individual bridge span [m]	31.58	18.00	49.00
Average bridge span [m]	30.74	17.60	44.91
Total bridge span [m]	161.33	21.20	628.74
Bridge width [m]	12.81	8.40	17.80

Based on project documents, the dependent variable is y - mass in kg of prestressed steel per m^2 [23] of the bridge superstructure. The properties of prestressing steel ropes are listed in the Table 2.

4 Evaluation and Performance Measures

To assess the model quality, the following errors were analyzed: RMSE (root mean square, MAE (mean absolute), R (Pearson's Linear Correlation Coefficient), and MAPE (mean absolute percentage). The RMSE, described by Eq. (21), is a measure of the model's overall accuracy, and is given in the same units as the modeled quantity:

$$RMSE = \sqrt{\frac{1}{N}\sum\nolimits_{k=1}^{N}(d_k - o_k)^2} \tag{21}$$

where d_k is the actual value (the target value), o_k is the modeled output (forecast), and N is the number of the used training samples.

Table 2. Geometrical and mechanical rope characteristics [23].

Abbreviated name	Class	Nominal values			Guaranteed values			Maximum relaxation at a force of 0.7 Fpk after 1000 h [2.5%]
		Diameter Ø [mm]	Tensile strength [N/mm^2] fpk	Elastic modulus [kN/mm^2] E	Section area [mm^2] Apk	Characteristic breaking force [kN] Fpk	Characteristic 0.1% proof-stress of prestressing steel [kN] Fp0.1k	
Y1770S7	A	15.2	1770	195	140	248	213	2.5
Y1860S7	B	15.2	1860		140	260	224	
Y1770S7	A	16.0	1770		150	265	228	
Y1860S7	B	16.0	1860		150	279	240	

The MAE is also a measure of the model's absolute accuracy. It is used to estimate the model's mean absolute error as follows:

$$MAE = \frac{1}{N}\sum\nolimits_{k=1}^{N}|d_k - o_k| \tag{22}$$

The R coefficient, defined by expression (23), is a relative criterion for the evaluation of the model's accuracy:

$$R = \sqrt{\left[\sum\nolimits_{k=1}^{N}\left|(d_k - \overline{d})(o_k - \overline{o})\right|\right]^2 \times \left[\sum\nolimits_{k=1}^{N}\left|\left(d_k - \overline{d}\right)^2 (o_k - \overline{o})^2\right|\right]^{-1}} \tag{23}$$

where $\overline{o}$ is the predicted mean that is obtained by the related model, while $\overline{d}$ is the mean target value.

The MAPE criterion, which is defined by the Eq. (24), is a measure of the relative accuracy of the model prediction.

$$MAPE = \frac{100}{N}\sum\nolimits_{k=1}^{N}\left|\frac{d_k - o_k}{d_k}\right| \tag{24}$$

The model cross-validation approach is used in the analysis. The approach was chosen because it lowers the bias associated with random data partition into training and training models, which should, in theory, have the same statistical features. The ten-fold cross-validation procedure, for example, must be performed by dividing the data set randomly into ten disjoint sections of the same size (10 folds) [24].

5 Results and Discussion

Because it is a regression problem, the output layer neurons number is one in the MLP model, which has an architecture with the same number of input layer neurons as predictors, i.e., with four input layer neurons. For the hidden layer, the number of neurons was determined experimentally by observing the upper value obtained using Eqs. (1) and (2). The Levenberg-Marquardt algorithm [25] was employed to train the MLP-ANN that had one hidden layer. The maximum epochs number of 1000, the minimum gradient magnitude of 10^{-5}, or the mean square error of 0 were used to determine when the training should be stopped. Prior to training, the normalization of the input data was done in the range $[-1, 1]$.

In this situation, network architectures with up to 8 neurons in the hidden layer must be examined. In terms of accuracy criteria, two similar models can be seen (Fig. 7). According to the RMSE and R criterion, the model with three neurons within the hidden layer is superior, while the model with four neurons within the hidden layer is better in terms of MAE and MAPE (Table 4).

The parameters of the model were adjusted using the cross-validation procedure for developing a model for estimating the amount of prestressed steel utilizing ensembles of regression trees, and in order to attain appropriate predictive performance. The following methods were used to generate the model:

- Bagging method (TreeBagger),
- Random Forests (RF) method,
- Boosted Trees method.

Alternative values of adjustable model parameters were analyzed when using the TreeBagger (TB) approach, as follows:

- The number of the generated trees, B. The maximum number of generated trees is limited to 500. The number of created trees is set to 100 by default in the MATLAB software that is used to apply the Bagging method. The method of bootstrap aggregation was employed to construct trees, in which samples of the same size as the original samples were formed, i.e., a total of 75 samples were formed throughout each iteration to generate a tree model.
- Within the tree, the min leaf size, i.e., the minimum number of samples that are assigned to the leaf. Values ranging from one to ten samples were evaluated, with a step of one per tree leaf. When using the Bagging method in MATLAB, the default setting takes 5 samples per tree leaf for regression problems. Within this study, a somewhat greater range for the number of data per tree leaf was explored, as well as the impact of the number of data on the model's generalization.

The RF method randomly selects a subset of input variables that are split in a tree. Different values of adaptive model parameters were analyzed when this method was implemented, as follows:

- The number of variables in the tree that are split. According to L. Breiman's paper Random Forests [26], the subset *m* of the variables subjected to the splitting should be approximately $\sqrt{(p)}$ or p/3 predictors, depending on whether the task is a classification or regression problem. The number of input variables or predictors in this study is 4, therefore when the second regression criterion is used, the subset of variables on which splits can be is 1 or 2. In this study, a slightly broader range of models is examined (Fig. 8), namely models with two, three, and eventually four variables as a subset of the variables subjected to the splitting. The TB method is obtained after analyzing the splitting with all four variables.
- Within the tree, the min leaf size, i.e., the minimum number of samples that are assigned to the leaf. We considered the values from 1 to 10 samples (Fig. 8) with the step of 1 per tree leaf. When using the RF method in MATLAB, the default configuration is to take 5 samples per tree for regression issues. In this study, it is analyzed a slightly greater range for the number of data per tree leaf.

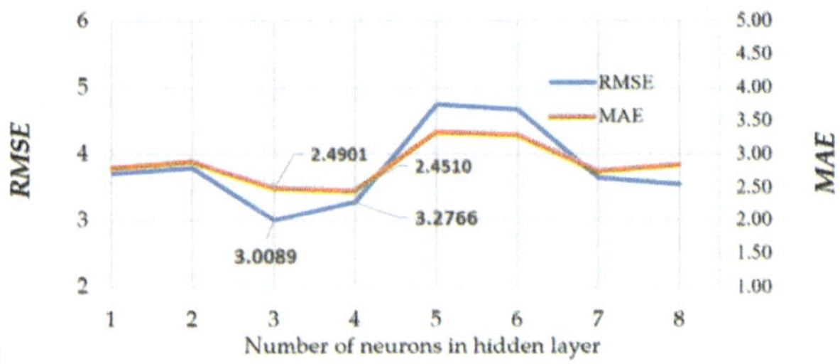

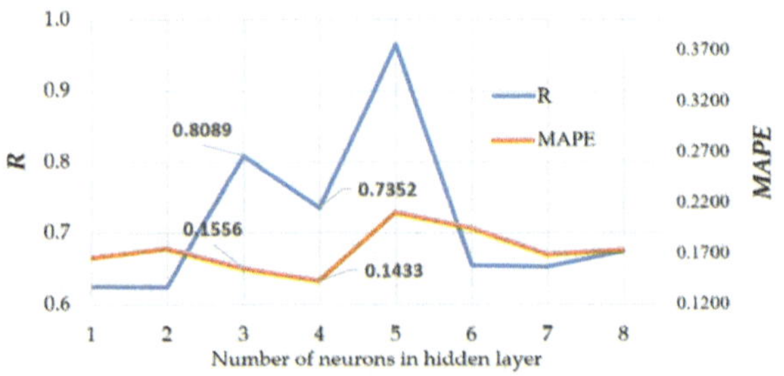

Fig. 7. Performance measures compared using MLP-ANNs with varying number of neurons within the hidden layer: a) the RMSE and MAE criteria, b) the R and MAPE criteria.

The parameters of the TB and RF procedures were defined using the grid-search method. The impact of the number of splitting variables and the minimal number of data for terminal leaf were investigated. The ideal model (Fig. 8) was found to be the RF model, in which two input variables are separated and the number of the data per leaf is kept to a minimum of one.

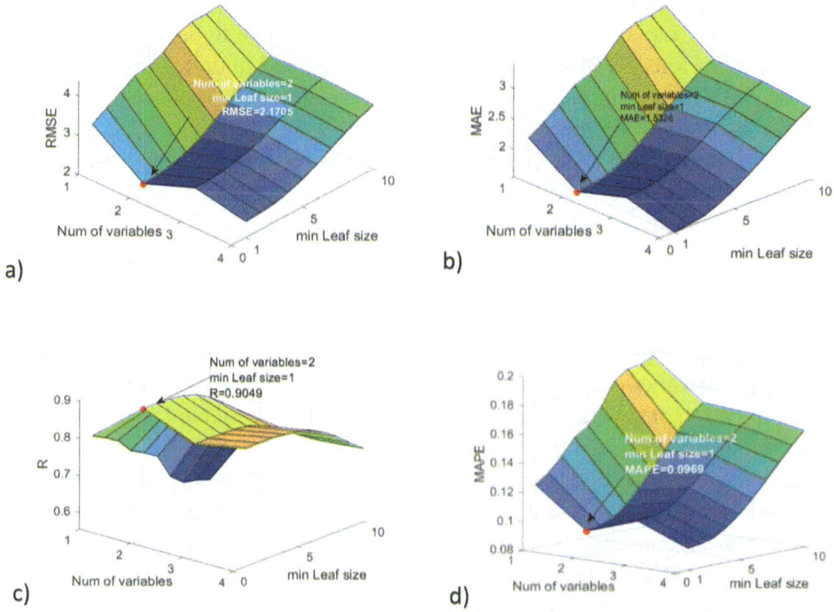

Fig. 8. Comparison of performance measures for tree-based ensemble methods a) RMSE, b) MAE, c) R, d) MAPE.

In both the Boosting and other methods, the cross-validation procedure was utilized to identify the parameters of the best model. The model parameters for the Boosting Trees approach are as follows:

- The number of generated trees, B. When using the Boosting approach, there is a risk of model overtraining if too many trees are formed. Since the number of the analyzed models in this research is large, the number of trees (base models) in the ensemble is limited to 100.
- Reduction parameter λ (learning rate). The λ parameter defines the model's training speed. Although the usual values are 0.01 and 0.001, we analyzed several values: 0.001; 0.01; 0.1; 0.5; 0.75, and 1.0.
- The number of splits in a tree, d. The values of the number of splits with an exponential increase were analyzed. Tree models were created with a number of splits limited to $2^0 = 1, 2^1, 2^2, 2^3, 2^4, 2^5, 2^6 = 64$.

Figure 9 shows the influence of the adopted parameters on the model's accuracy, i.e., the influence of the maximum number of splits and reduction parameters on the model's accuracy in terms of the MSE value obtained by cross-validation. The boundary cases are modeled with maximum number of splits equal to 64. The optimal model with the lowest MSE value for the evaluated ensemble of 100 base models is shown in Fig. 9 and is represented by a yellow line.

The following parameter values were determined for the optimal model: a total of 100 trees, learning rate of 0.1, and a number of splits limited to 8.

The use of various SVM kernel functions is examined in this study to discover the optimal one. SVR models using RBF, linear, and sigmoid kernels were investigated. Before training and testing the model, all input data were transformed into the range (0, 1) via normalization. For all kernels, the best model was found using the grid search technique ($C = 0.6208$, $\varepsilon = 0.0642$ for the linear kernel; $C = 1.2263$, $\varepsilon = 0.0302$, $\gamma = 7.9800$ for the RBF kernel; $C = 243.9853$, $\varepsilon = 0.0656$, $\gamma = 0.0028$ for the sigmoid kernel).

A comparison examination of several SVR models reveals that the models have different accuracy for different kernel functions. According to several criteria, models with the linear and sigmoid kernels exhibit equivalent accuracy. Considering all criteria functions, the SVR model with the RBF kernel has significantly highest accuracy.

The application of several covariance functions was investigated when analyzing the consumption of prestressed steel per m^2 of superstructure using GPR. The data used in modeling are standardized. The so-called Z-score standardization [17] procedure was performed on each column. Models with a constant base function were analyzed. The parameter values are defined by maximization of the log marginal probability.

Table 3 shows the model's parameters for different covariance functions. Table 4 shows the parameters of the model with ARD covariance functions. Table 5 shows the results of different machine learning models.

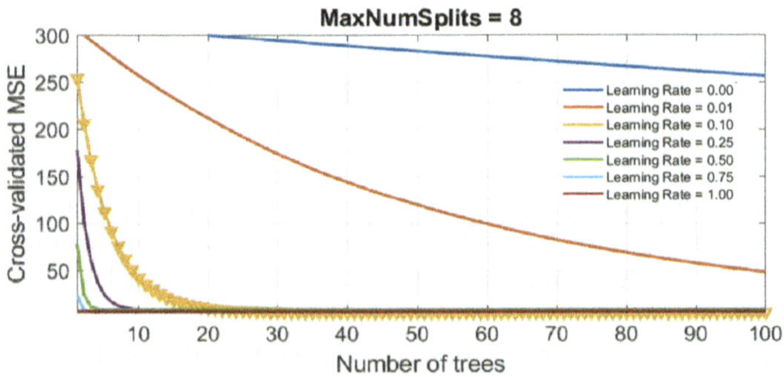

Fig. 9. Dependence of the MSE value on the reduction parameter λ and the number of trees (base models) used in the Boosted Trees method when the maximum number of splits is limited to 8.

Table 3. The GPR model covariance functions parameters

GP model covariance function	Covariance function parameters		
Exponential	$k((x_i, x_j \mid \Theta)) = \sigma_f^2 exp\left[-\frac{1}{2}\frac{r}{\sigma_l^2}\right]$		
	$\sigma_l = 5.4002$		$\sigma_f = 9.9766$
Squared exponential	$k((x_i, x_j \mid \Theta)) = \sigma_f^2 exp\left[-\frac{1}{2}\frac{(x_i - x_j)^T(x_i - x_j)}{\sigma_l^2}\right]$		
	$\sigma_l = 2.4565$		$\sigma_f = 10.7092$
Matern 3/2	$k((x_i, x_j \mid \Theta)) = \sigma_f^2\left(1 + \frac{\sqrt{3}r}{\sigma_l}\right)exp\left[-\frac{\sqrt{3}r}{\sigma_l}\right]$		
	$\sigma_l = 5.1291$		$\sigma_f = 13.3442$
Matern 5/2	$k((x_i, x_j \mid \Theta)) = \sigma_f^2\left(1 + \frac{\sqrt{5}r}{\sigma_l} + \frac{5r^2}{3\sigma_l^2}\right)exp\left[-\frac{\sqrt{5}r}{\sigma_l}\right]$		
	$\sigma_l = 3.6042$		$\sigma_f = 11.8095$
Rational quadratic	$k((x_i, x_j \mid \Theta)) = \sigma_f^2\left(1 + \frac{r^2}{2a\sigma_l^2}\right)^{-\alpha}$		
	$\sigma_l = 2.6978$	$a = 2.8872$	$\sigma_f = 11.3713$

where $r = \sqrt{(x_i - x_j)^T(x_i - x_j)}$.

In a model, the relevance of specific predictors or variables can be seen utilizing ARD covariance (Automatic Relevance Determination) functions. Higher ARD covariance function parameter values indicate less significance of the variable to which they refer. Models using an ARD covariance function have higher accuracy than models with only one length scale parameter, as shown by the values of the criteria (Table 5).

The study of the length scale parameters values may be used to see the significance of individual variables (Fig. 10) for the model with the highest accuracy. This means that our analysis will be performed on a model with an ARD exponential covariance function.

The variable x_1, which indicates the maximum individual span in the model with ARD exponential function, has the largest value (the value of the variable from the aspect of the analyzed model is inversely proportional to the length scale parameter) when the values of the length scale parameters are examined. This can be explained by the fact that bridges are usually made up of similar individual spans with little variation, so the variable x_2, which represents the mean value of the spans, implicitly contains some information about the maximum span with little variation, and this information is hence included in the model.

Table 4. The GPR ARD model covariance functions parameters.

Parameters of covariance function			
σ_1	σ_2	σ_3	σ_4
ARD exponential:			
$k((x_i, x_j\|\Theta)) = \sigma_f^2 \exp(-r); \sigma_F = 7.9176; r = \sqrt{\sum_{m=1}^{d} \frac{(x_{im}-x_{jm})^2}{\sigma_m^2}}$			
327822.2661	**0.4200**	**8.4734**	**170.7902**
ARD squared exponential:			
$k((x_i, x_j\|\Theta)) = \sigma_f^2 exp\left[-\frac{1}{2}\sum_{m=1}^{d} \frac{(x_{im}-x_{jm})^2}{\sigma_m^2}\right]; \sigma_f = 8.4546$			
3,3532	0,8216	2,8218	11,5585
ARD matern 3/2:			
$k((x_i, x_j\|\Theta)) = \sigma_f^2 \left(1 + \sqrt{3}r\right) exp\left[-\sqrt{3}r\right]; \sigma_f = 5.7735$			
0,5812	0,9767	71499.7664	0,1014
ARD matern 5/2:			
$k((x_i, x_j\|\Theta)) = \sigma_f^2 \left(1 + \sqrt{5}r + \frac{5r^2}{3}\right) exp\left[-\sqrt{5}r\right]; \sigma_f = 7.1089$			
1,8478	0,3117	4243.2352	4,8180
ARD rational quadratic:			
$k((x_i, x_j\|\Theta)) = \sigma_f^2 \left(1 + \frac{1}{2\alpha}\sum_{m=1}^{d} \frac{(x_{im}-x_{jm})^2}{\sigma_m^2}\right)^{-\alpha}; \alpha = 0.1673; \sigma_f = 8.8932$			
1,6509	0,3324	7441.4850	5.2814

where $r = \sqrt{\sum_{m=1}^{d} \frac{(x_{im}-x_{jm})^2}{\sigma_m^2}}$.

Table 5. Results of different machine learning models

Model	RMSE	MAE	R	MAPE/100
NN 4-3-1	3.0089	2.4901	0.8089	0.1556
NN 4-4-1	3.2766	2.4510	0.7352	0.1433
TreeBagger	2.4138	1.5349	0.8736	0.1433
Random forest	**2.1705**	1.5326	0.8849	0.0969
Boosted trees	2.3822	1.6650	0.8896	0.1025
SVR-Lin. kernel	3.2536	2.5971	0.7906	0.1498
SVR-RBF kernel	2.4980	1.8839	0.8762	0.1234

(*continued*)

Table 5. *(continued)*

Model	RMSE	MAE	R	MAPE/100
SVR-Sig. kernel	3.2484	2.5866	0.7890	0.1490
Exponential	2.6536	1.4834	0.8565	0.0912
ARD-exponential	2.3800	**1.3302**	**0.8870**	**0.0815**
Squared exponential	2.8456	2.0747	0.8396	0.1293
ARD-Sq. exponential	2.4239	1.7913	0.8822	0.1099
Matern 3/2	3.6520	2.2463	0.7113	0.1331
ARD-matern 3/2	3.0418	1.9661	0.8087	0.1254
Matern 5/2	3.1981	2.2096	0.7851	0.1327
ARD-matern 5/2	3.1297	1.9085	0.7937	0.1147
Rational quadratic	2.9855	2.0910	0.8180	0.1289
ARD rational quadratic	2.6568	1.7374	0.8565	0.1069

The length of the bridge, x_3, is variable with a larger value of the length scale parameter or less relevance in comparison to the variable x_2, which can be explained by the fact that the output variable reflects prestressed steel consumption expressed in m^2. The significance of the variable x_3 would be great if the output variable was, for example, the total amount of prestressed steel. For the variable x_4, i.e., the influence of bridge width, a similar explanation can be given. Steel consumption per m^2 of bridge superstructure is between 12.5 and 22 kg/m^2 in most investigated bridges, which is consistent with the approximate values stated in the literature [27]. Figure 11 depicts the modeled values in respect to the model's target values. Figure 12 shows a regression plot of the modeled and target values in the optimal model. The model that uses the

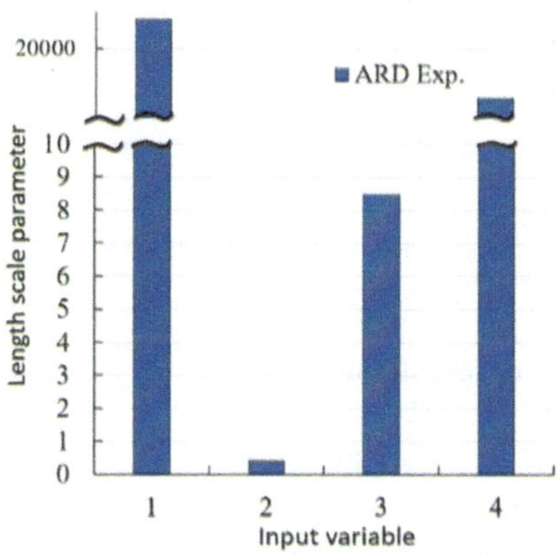

Fig. 10. Length scale parameter values for models with ARD exponential covariance function for prestressed steel consumption models.

most relevant variables is considered further in the analysis. In some circumstances, a model with the same or superior accuracy can be obtained by employing a narrowed set of variables.

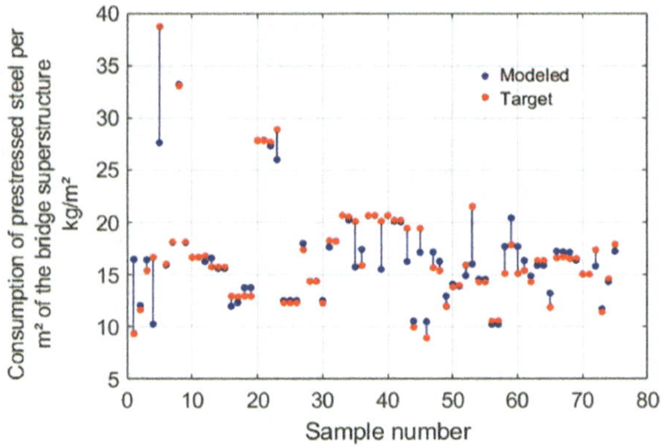

Fig. 11. The GPR ARD Exponential model – target and modeled values.

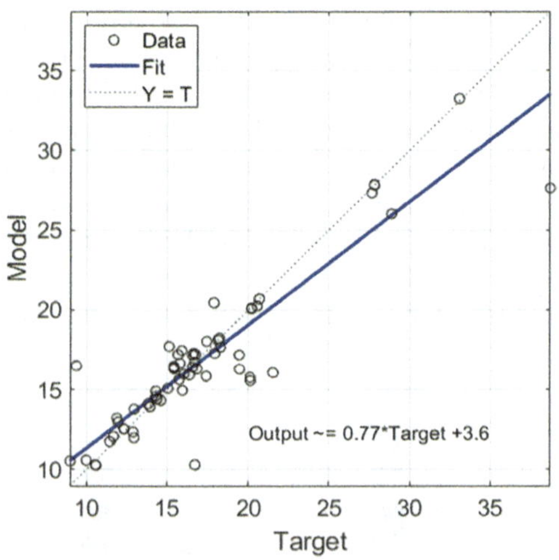

Fig. 12. The GPR ARD Exponential model – the target and modeled values' regression plot.

Three more models were investigated further:

a. Variables 2, 3, and 4 are included in the model.
b. Variables 2 and 3 are included in the model.
c. Only variable 2 is included in the model.

Table 6 compares alternative sets of input variables to the model of prestressed steel consumption per m^2 of bridge superstructure. In the table, a binary value (0 or 1) indicates that a variable is used in the model. By analyzing Table 5, it can be seen that by reducing the number of input variables in a particular scenario, the model can be simplified while also improving its accuracy. In the case of estimating prestressed steel consumption per m^2 in Model 2, improved values for the accuracy criteria RMSE, MAE, MAPE, and R were obtained by eliminating variables x_1 and x_4 from a total of four studied input variables. The values of RMSE and MAE indicate that the model is accurate, with the value of RMSE being significantly higher than the value of MAE, indicating slightly lower accuracy in predicting extreme values of steel consumption. The R and MAPE values also imply that the model is accurate enough.

Table 6. Comparative analysis of the model of prestressed steel consumption per m^2 of the bridge superstructure with alternative sets of input variables

Model	x_1	x_2	x_3	x_4	RMSE	MAE	R	MAPE
1	0	1	1	1	2.8232	1.3881	0.8368	8.2593
2	**0**	**1**	**1**	**0**	**2.1676**	**1.0993**	**0.9096**	**6.5537**
3	0	1	0	0	3.0318	1.7853	0.8088	10.0922
4	1	1	1	1	2.3800	1.3302	0.8870	8.1541

6 Conclusions

In order to decide whether or not to begin construction of the PC bridge, an early techno-economic analysis must be conducted during the preparatory phase of the investment project, which must also contain the planned resources. The early planning and optimization of the resources provides a better basis for later easier and less-costly maintenance.

A number of prediction models for early estimation of prestressed steel consumption per m^2 were developed, examined, and verified in this paper, based on artificial intelligence approaches. A database of design characteristics for 75 PC bridges was created for the assessment model. Cross-validation was used to train and test all models under same conditions.

Individual neural network models were used, and the results were unsatisfactory in terms of accuracy across all criteria. Hence, the employment of regression tree ensembles was proposed. The Bagging, Random Forests, and Boosting methods were investigated. For each of the approaches, hyperparameters are defined, as well as a procedure for optimizing them.

A comparison of multiple SVR models revealed that, depending on the kernel function, the models have varying accuracy for alternative criteria. The procedure for parameter optimization has been established. The grid search procedure produced an optimal answer by optimizing parameters through a rough and detailed search. In terms of all criterion functions, the model with RBF kernel function showed much superior accuracy. The results indicate that the RBF kernel would be our recommendation for solving this and similar regression problems.

The application of the Gaussian process to the problem of forecasting prestressing steel consumption is also discussed. The usage of various covariance functions or kernel functions is examined in this context. ARD covariance functions, which apply different length scale parameters to input variables, have proved to be superior to covariance functions with a single distance parameter. The results show that using the ARD exponential covariance function is the optimum solution for the problem.

An accuracy of RMSE = 2.1676 was achieved using the ARD exponential covariance function, together with MAE = 1.0993, R = 0.9096, and MAPE = 6.55%, which is the best result in terms of accuracy of all analyzed models. The application of the ARD covariance function allows the relevance of individual input variables to be considered. The model's achieved accuracy can be considered satisfactory.

Based on the suggested model, a MATLAB-based software was created with a graphical user interface for the input of basic variables, from which estimated prestressing steel quantities can be derived. The use of such software for the estimation the optimal amount of steel will in a general case lead to the construction of higher-quality structures that will require less maintenance on a long-term basis.

References

1. Marcous, G., Bakhoum, M.M., Taha, M.A., El-Said, M.: Preliminary quantity estimate of highway bridges using neural networks. In: Proceedings 6th International Conference on the Application of Artificial Intelligence to Civil and Structural engineering, Stirling, Scotland (2001)
2. Fragkakis, N., Lambropoulos, S., Tsiambaos, G.: Parametric model for conceptual cost estimation of concrete bridge foundations. J. Infrastruct. Syst. **17**(2), 66–74 (2011). https://doi.org/10.1061/(ASCE)IS.1943-555X.0000044
3. Mučenski, V., Peško, I., Trivunić, M., Dražić, J., Ćirović, G.: Optimizacija neuronske mreže za procenu potrebnih količina betona i armature u višespratnim objektima. Građevinski materijali i konstrukcije **55**(2), 27–46 (2012)
4. Fragkakis, N., Marinelli, M., Lambropoulos, S.: Preliminary cost estimate model for culverts. Procedia Eng. **123**, 153–161 (2015). https://doi.org/10.1016/j.proeng.2015.10.072
5. Marinelli, M., Dimitriou, L., Fragkakis, N., Lambropoulos, S.: Non-parametric bill of quantities estimation of concrete road bridges superstructure: an artificial neural networks approach. In: Proceedings 31st Annual ARCOM Conference, Lincoln, United Kingdom (2015)
6. Antoniou, F., Konstantinitis, D., Aretoulis, G.: Cost analysis and material consumption of highway bridge underpasses. In: Eighth International Conference on Construction in the 21st Century (CITC-8), Changing the Field: Recent Developments for the Future of Engineering and Construction, Thessaloniki, Greece (2015)
7. Dimitriou, L., Marinelli, M., Fragkakis, N.: Early bill-of-quantities estimation of concrete road bridges - an artificial intelligence-based application. Public Work Manage. Policy **23**(2), 127–149 (2018). https://doi.org/10.1177/1087724X17737321
8. Kovačević, M., Ivanišević, N., Dašić, T., Marković, L.: Application of artificial neural networks for hydrological modelling in Karst. Građevinar **70**, 1–10 (2018). https://doi.org/10.14256/JCE.1594.2016
9. Beale, M.H., Hagan, M.T., Demuth, H.B.: Neural Network Toolbox. The Mathworks, Inc (2010)
10. Kovačević, M., Ivanišević, N., Petronijević, P., Despotović, V.: Construction cost estimation of reinforced and prestressed concrete bridges using machine learning. Građevinar **73**, 1–13 (2021). https://doi.org/10.14256/JCE.2738.2019

11. Kovačević, M., Lozančić, S., Nyarko, E.K., Hadzima-Nyarko, M.: Modeling of compressive strength of self-compacting rubberized concrete using machine learning. Materials **14**, 4346 (2021). https://doi.org/10.3390/ma14154346
12. Black, P.: Dictionary of Algorithms and Data Structures, NISTIR (1998). http://www.nist.gov/dads. Accessed 11 Oct 2021
13. Cormen, T.H., Leiserson, C.E., Rivest, R.L., Stein, C.: Introduction to Algorithms. The MIT Press, London (2009)
14. Hastie, T., Tibsirani, R., Friedman, J.: The Elements of Statistical Learning. Springer, Heidelberg (2009)
15. Breiman, L., Friedman, H., Olsen, R., Stone, C.J.: Classification and Regression Trees. Chapman and Hall/CRC, Wadsworth (1984)
16. Breiman, L.: Bagging predictors. Mach. Learn. **24**, 123–140 (1996). https://doi.org/10.1007/BF00058655
17. Kovačević, M.: Model for forecasting and assessment of construction cost of reinforced-concrete bridges, Doctoral dissertation, University of Belgrade, Faculty of Civil Engineering, Belgrade, Serbia (2018). https://doi.org/10.13140/RG.2.2.24025.65129
18. Vapnik, V.: The Nature of Statistical Learning Theory. Springer, New York (1995)
19. Kecman, V.: Learning and Soft Computing: Support Vector Machines, Neural Networks, and Fuzzy Logic Models. The MIT Press, Cambridge (2001)
20. Chang, C.C., Lin, C.J.: LIBSVM: a library for support vector machines. ACM Trans. Intell. Syst. Technol. **2**(3) (2011). https://doi.org/10.1145/1961189.1961199
21. LIBSVM-A Library for Support Vector Machines. https://www.csie.ntu.edu.tw/~cjlin/libsvm/. Accessed 21 Feb 2021
22. Rasmussen, C.E., Williams, C.K.: Gaussian Processes for Machine Learning. The MIT Press, Cambridge (2006)
23. IMS: Sertifikat sistema kvaliteta broj 170101, SPB SUPER Sistem za prenaprezanje. IMS, Centar za prednaprezanje, Beograd (1999)
24. Chou, J.S., Lin, C.W., Pham, A.D., Shao, J.Y.: Optimized artificial intelligence models for predicting project award price. Autom. Constr. **54**, 106–115 (2015). https://doi.org/10.1016/j.autcon.2015.02.006
25. Hagan, M.T., Demuth, H.B., Beale, M.H., Jesus, O.D.: Neural Network Design. Oklahoma State University, Stillwater (2014)
26. Breiman, L.: Random forests. Mach. Learn. **45**, 5–32 (2001). https://doi.org/10.1023/A:1010933404324
27. Radić, J., Šavor, Z., Puž, G.: Tipizacija mostova za autoceste. Građevinar **52**(6), 321–330 (2000)

Importance of Bedrock Depth Knowledge in Basins: Çanakkale (Dardanalles) Case History

Aydın Büyüksaraç[1(✉)], Öznur Karaca[1], Onur Eyisüren[2], Özcan Bektaş[3], and Ercan Işık[4]

[1] Çanakkale Onsekiz Mart University, 17020 Çanakkale, Turkey
absarac@comu.edu.tr
[2] Çanakkale Muncipality, 17000 Çanakkale, Turkey
[3] Sivas Cumhuriyet University, 58000 Sivas, Turkey
[4] Bitlis Eren University, 13100 Bitlis, Turkey

Abstract. It is not correct to produce the necessary information for structuring, especially in environments such as Çanakkale, which exhibit a basin structure, without determining the bedrock or the strict ground conditions in bedrock. This approach is the basis of earthquake resistant building design. In this study, which was carried out to determine the bedrock/seismic foundation depth for the central settlement of Çanakkale and to define the basin structure to a certain extent, microgravity measurements were taken on a large scale, and the study area was modelled in three dimensions based on the obtained gravity data. By taking long-term microtremor measurements, one-dimensional depth-shear wave (Vs) velocity models were obtained using the Rayleigh ellipticity method. A depth map of the engineering bedrock was created, in which the velocity Vs reached to 2500 m/s.

Keywords: Soil · Bedrock · Microgravity · Rayleigh ellipticity · Vs velocity · Seismic foundation

1 Introduction

Accurate definition of soil-structure interaction has become very important for in terms of both earthquake damages and environmental safety. How seismic waves propagate, change and transfer to structures in the rock and soil environment remains an important research topic [1, 2]. Earthquake codes, on the other hand, are updated periodically by evaluating the new information and findings obtained through this research and the lessons of damage caused by earthquakes [3]. In the Building Earthquake Code [4–7] published in Turkey, the acceleration spectra required for the earthquake design of the structures, the desired earthquake level (ground motion) and the average shear wave velocity at the top 30 m, (Vs)30, SPT impact number, it defines it with six local soil classes ranging from (N60)30 or undrained shear strength (cu)30 to ZA-ZF [6, 7]. Accordingly, in the presence of certain thicknesses of fill, loess, peat, high plasticity clay units, very thick, soft/medium solid clay deposits or liquefiable soils in the environment, site-specific research should be carried out by defining the local soil class as ZF, soil

H. Glavaš et al. (Eds.): OTO 2021, LNNS 369, pp. 344–352, 2022.
https://doi.org/10.1007/978-3-030-92851-3_25

behavior (response) analysis of the site-specific acceleration spectrum and requires the use of a design supervisor in this process. Since TBEC-2018 came into force on January 1, 2019, many soils behavior analyses have been carried out in many regions, especially in Çanakkale, where thick alluvial and liquefiable ZF soils are found. Shallow soil investigation is made in the investigations. The location of the engineering bedrock, the size and shape of the basin are not known. Therefore, the basin and basin-corner effect are not defined, the absence of microzonation studies, the Vs profile of the units up to the bedrock cannot be fully defined, and the lack of experiments revealing the nonlinear dynamic behavior of rocks emerged as important shortcomings.

As it can be understood from here, the necessary information cannot be reached in the soil behaviors analyses with the soil survey studies carried out on the parcel basis. Moreover, it is not possible to determine the topography and characteristics of the sedimentary basin, if any, while it is not attempted to find the bedrock depth with site-specific surveys. However, on the parcel basis, the depth information called $Z_{2.5}$ along with Vs30 and at which Vs = 2.5 km/s is reached should be measured [8, 9]. The deepness of the bedrock and the presence of soft-medium solid clay deposits or liquefiable soils with low shear wave velocity on them and the inclusion of basin effects increase the PGV values and extend the period of the acceleration spectra. In this case, ductile structures such as high-period structures, multi-story buildings and bridges may be adversely affected. The location and topography of the bedrock in Çanakkale, which is in a basin with very high earthquake hazard, thick alluvium, and liquefiable soils, was obtained using geophysical methods.

2 Seismicity of the Region

The main seismotectonic structures of this region are the North Anatolian Fault Zone (NAFZ) in the north, the Yenice-Gönen Fault Zone (YGFZ) in the east, and the Ganos Fault (GF), which is the continuation of the NAFZ in the northwest. In fact, the Biga Peninsula forms the southwest of the NAFZ. The most important active fault that may affect the study area is the Ganos Fault along the Gaziköy-Gölcük-Kavakköy line and the SW-NE trending Yenice-Gönen faults, one of the most active fault systems in Southern Marmara. The distance of these faults to the study area is approximately 50 km (Fig. 1) [10]. On the other hand, Çanakkale is affected by all earthquakes in the Aegean Sea. When the distribution of earthquakes affecting Çanakkale is examined, four sub-seismic zones can be mentioned. The outer focus distribution of the earthquake with a magnitude of $M \geq 3$ that occurred between 1900 and 2021 is given in Fig. 2 [11]. As can be seen from this map, Çanakkale is in the affected zone of faults that produce intense and destructive earthquakes. For this reason, information such as soil thickness and soil behaviors are much more important. Acceleration distribution is also given on the map, and Çanakkale central settlement is seen as the region where 0.3–0.4 g maximum acceleration can occur.

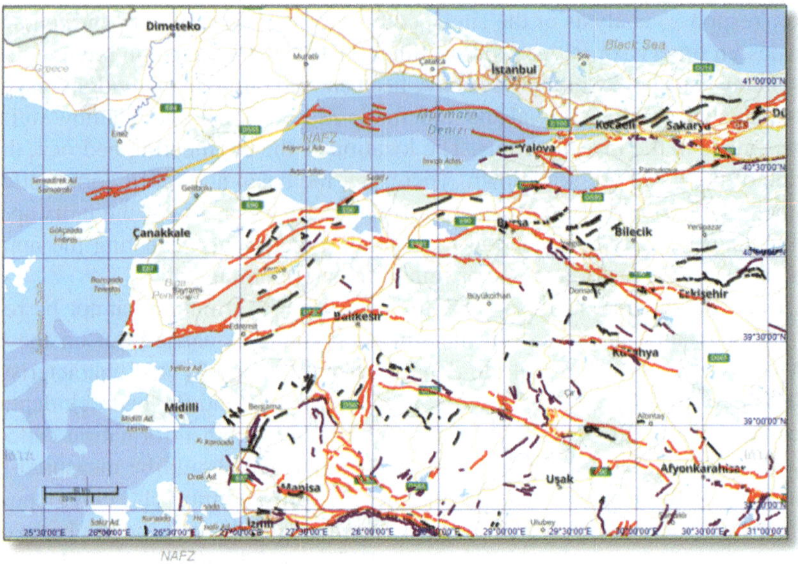

Fig. 1. Active Fault map of the study area and its surroundings [10] Here, the red lines are the Holocene Faults; black lines, Possible Quaternary Fault or lineaments; purple lines, Quaternary Fault; yellow lines indicate Surface Fracture.

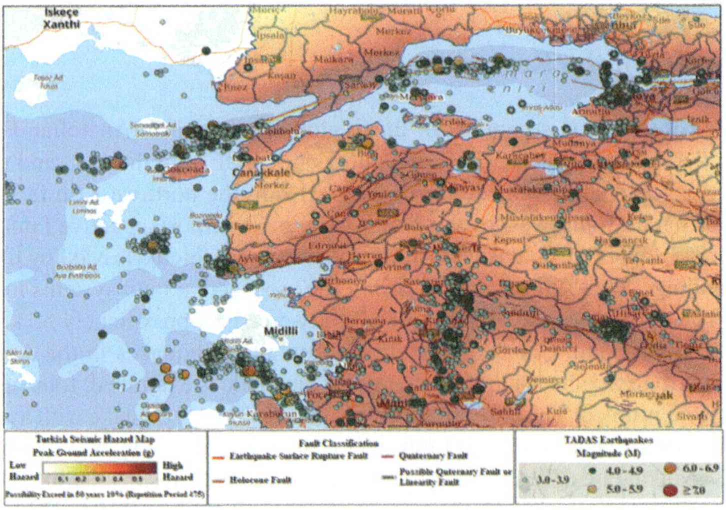

Fig. 2. Active faults, epicentres (M $\geq$ 3.0) and acceleration distribution of Biga Peninsula and its surroundings [11]

3 Soil Conditions of the Region

When the soil structure of Çanakkale is examined, it is widely exposed on both sides of the Dardanelles Strait and consists of four litho-stratigraphic units. The lowest levels of the Çanakkale soils are composed of the Gazhanedere formation, which was deposited in a terrestrial environment and represented by conglomerate, sandstone and mudstone. On top of this unit, the Kirazlı formation consists of fine-coarse grained sandstone and lesser amounts of fine conglomerate, siltstone and mudstone. At the upper level of the Kirazlı Formation, there is the Çamrakdere formation consisting of mudstone-claystone units. The Alçıtepe Formation, which starts with the clastics of the Çamrakdere Formation, transitionally, with moderately bedded, pebbly and sandy carbonates, generally consists of mudstone, marl, siltstone, sandstone, calcarenite and occasionally fine conglomerate and is located on top of the Çamrakdere Formation. At the top are alluviums, which are very commonly unconsolidated sediments. These sediments were defined as three separate units according to their formation types and the locations of the lithology types that compose them (Fig. 3). These; a) Sand, clayey sand, partly gravel channeled sand unit; b) Sand, silty sand (with occasional clay lenses) unit; c) Block, gravel and sand (current stream bed and floodplain sediments) unit.

4 Rayleigh Ellipticity Curve Inversion

Many surface wave analysis studies have been carried out to obtain the structure of the earth from surface wave distribution analysis. The surface wave distribution varies along the propagation path depending on the nature of the place where these waves pass. Therefore, surface wave distribution analysis is a good tool to study the most important features of ground structure. For example, in the Iberian Peninsula, various slip rate models of the crust and upper mantle have been described by Corchete et al. (1993, 1995) from the analysis of the Rayleigh wave distribution [12, 13]. This analysis consists of filtering and inverting the Rayleigh wave distribution to obtain the variation in shear wave velocity versus depth. Sexton et al. (1977) showed that the Rayleigh wave ellipticity is mainly dependent on the local shell structure and does not exhibit azimuth dependence in the range of 10–50 s [14]. Therefore, for this period range, the observed ellipticity is primarily controlled by the local crustal geology below the seismic station and does not depend on the propagation path of Rayleigh waves. This essential feature of ellipticity means that ellipticity analysis is a very useful tool for obtaining local crustal models that can be used to determine site and/or local effects in seismic risk and/or seismic design studies [15, 16]. It is known that the observed ellipticity of a Rayleigh wave can be calculated using the formula (Eq. 1) [17, 18].

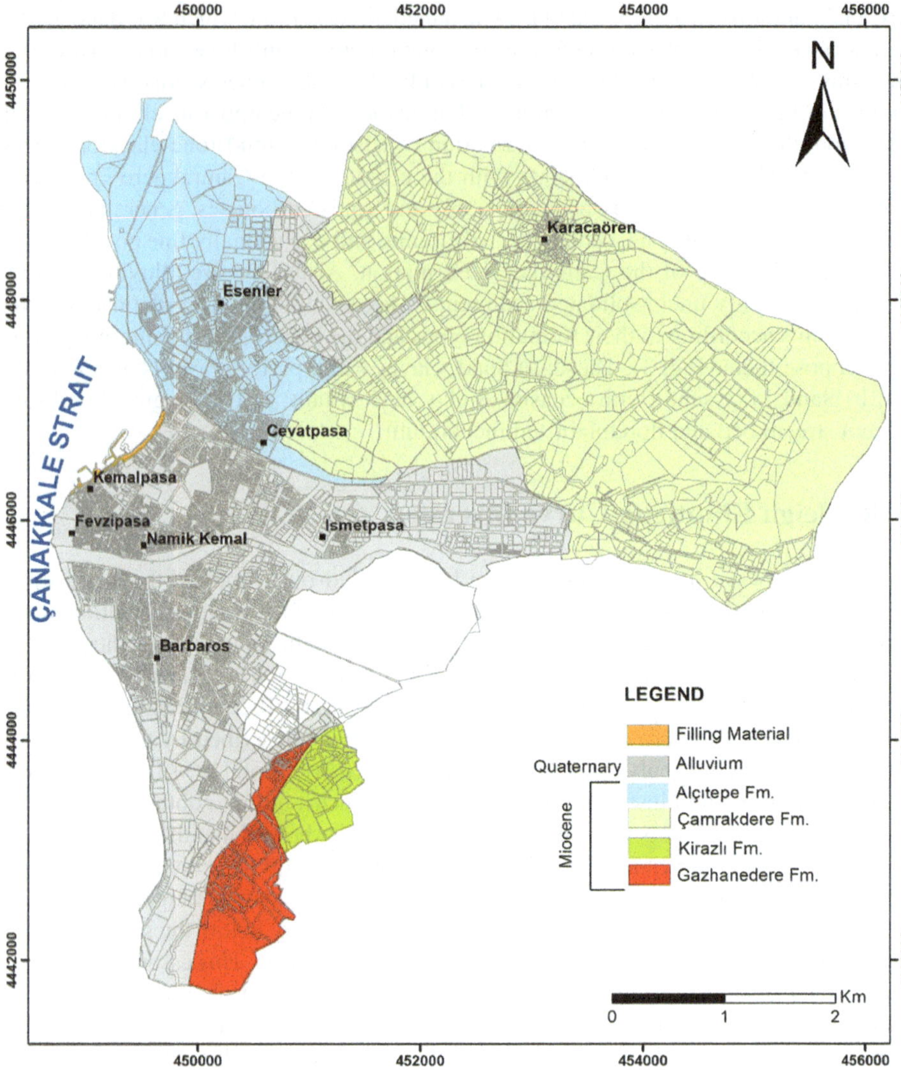

Fig. 3. Soil distribution map of the study area

$$\varepsilon(T) = A_L(T)/A_Z(T) \quad (1)$$

where $A_L(T)$ and $A_Z(T)$ are the instrument corrected spectral amplitudes of the longitudinal (also called radial) and vertical seismograms for the period T [14]. The inversion of the ellipticity curves of Rayleigh waves (Fig. 4) is done using Dinver software from the Geopsy package [19] based on the neighborhood algorithm (Fig. 5).

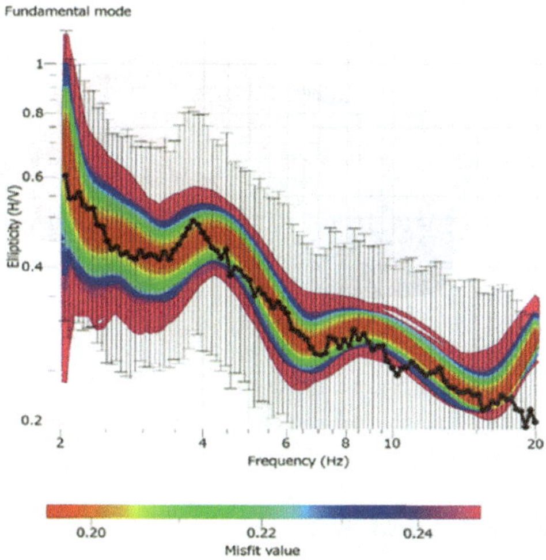

Fig. 4. Rayleigh ellipticity graph

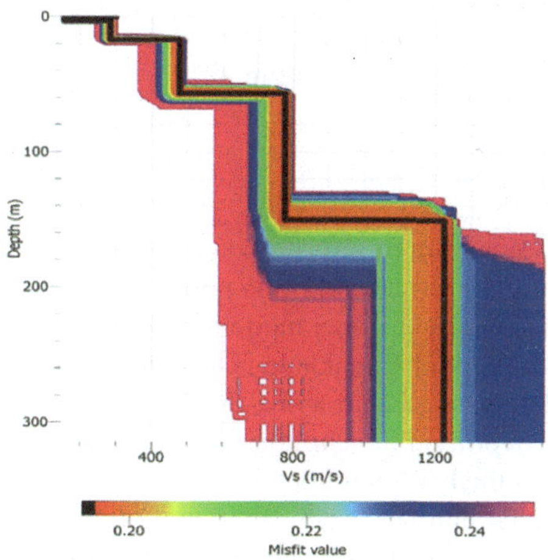

Fig. 5. Obtaining Vs velocity values

5 Microgravity

Microgravity data were collected with the grid network created in an area of approximately 10 × 10 km, which aims to reach the borders of the Çanakkale basin, which includes the study area and its surroundings. The distance between profiles and measurement points was equal and taken as 1 km. The 3D modeling application suggested by Cordell and Henderson (1968) [20] was applied to the Bouguer anomalies obtained from the microgravity measurements made in the study area.

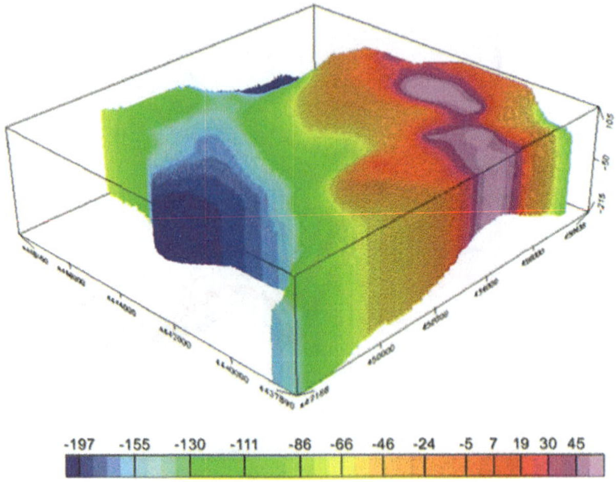

Fig. 6. 3D prismatic gravity depth map

As seen from the 3D depth model, the thickness of the young alluvial unit in the study area was determined to be approximately 300 m at most. The prismatic representation of the three-dimensional model is given in Fig. 6.

6 Results

It is known that there is not enough information about the depth of the bedrock under the influence of shallow groundwater. For this purpose, it is aimed to obtain a bedrock depth map by creating three-dimensional modeling of the plain regime settlement area, in which the central settlement of Çanakkale is located. Thus, the effects of the results, which will be achieved in an area with a one-dimensional wave shallow surface depth (15–30 m) and bedrock depth, on the earthquake soil behavior will be observed. Since the shear wave velocity of the basement rock is almost constant, the possible earthquake damage, in other words the unit shear deformation that will occur, will be determined by the acceleration in the basement rock and the dominant frequency and amplification value of the ground above it [21, 22]. In the proposed project, it is aimed to obtain the detailed earthquake site behavior of the study area, to determine the bedrock depth of the Çanakkale city center settlement, which is not well defined and mostly uncertain, and to be a case study of the most appropriate method selection for future studies by using more than one method together. Considering the geological units defined in the deep soundings up to 240 m across Çanakkale, a clear unit that is the bedrock could not be reached. However, this statement does not coincide with compression, which may be a seismic basis. In other words, a clear distinction could not be made within the definition of loose and tight material of the basin. Equation 2 was developed by Campbell and Bozorgnia (2013) [23] to determine the level at which the generally accepted ground reaches the shear wave of 2500 m/s.

$$\ln Z2.5 = 6.510 - 1.181 \times \ln Vs30 \tag{2}$$

The Vs2.5 velocity distribution based on the Vs velocity values obtained by the Rayleigh Ellipticity method from microtremor measurements made in Çanakkale given in Fig. 6. Accordingly, the slowness limit of the earthquake waves that will affect the structures starts at 320 m in the Çanakkale Basin and shows significant amplification effects as it approaches the surface. A bedrock depth of up to 2 km is expected in the south of the area (Fig. 7).

The estimated $Z_{2.5}$ map in Fig. 7 can be used as very useful information in seismic hazard studies for further measurements. Further geophysical investigations are required to determine $Z_{1.0}$ and $Z_{2.5}$ to provide additional constraints and refine these models in the future. At Çanakkale, efforts are underway to measure $Z_{2.5}$ depths and integrate the geophysical view of the basin. The necessary working algorithm for these studies, which is expected to be an example for other basin-based practices throughout Çanakkale, will also be completed.

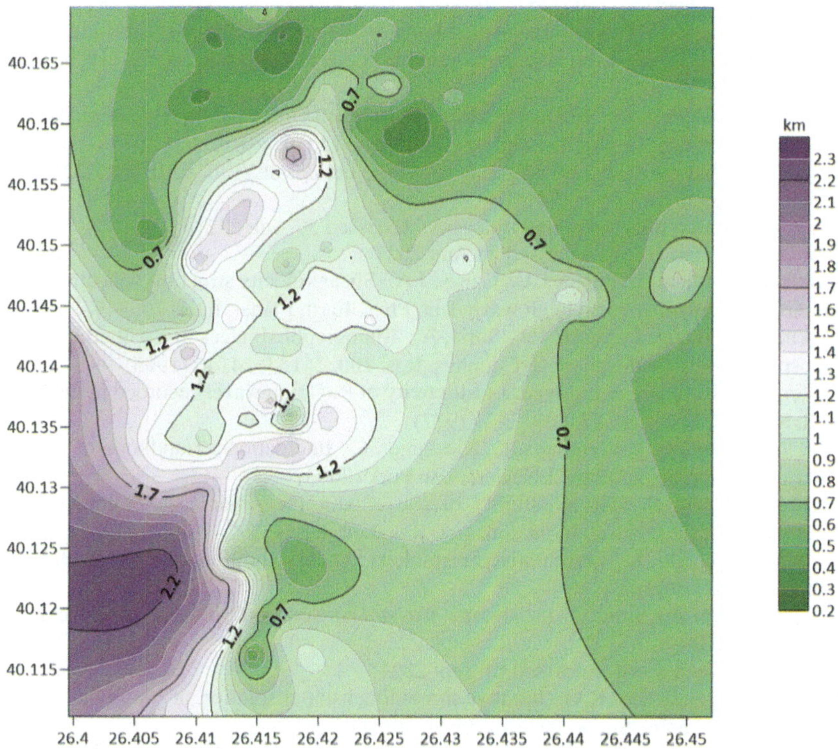

Fig. 7. Depth distribution up to 2.5 km/h (Z2.5)

Acknowledgments. This project was supported by ÇOMÜ BAP Unit as an independent research project numbered FBA-2020-3307.

References

1. Büyüksaraç, A., Bektaş, Ö., Yılmaz, H., Arısoy, M.Ö.: Preliminary seismic microzonation of Sivas city (Turkey) using microtremor and refraction microtremor (ReMi) measurements. J. Seismolog. **17**(2), 425–435 (2013)
2. Büyüksaraç, A., Över, S., Geneş, M.C., Bikçe, M., Kaçın, S., Bektaş, Ö.: Estimating shear wave velocity using acceleration data in Antakya (Turkey). Earth Sci. Res. J. **18**(2), 99–105 (2014)
3. TBEC: Turkish Building Earthquake Code; T.C. Resmi Gazete: Ankara, Turkey (2018)
4. Büyüksaraç, A., Işık, E., Harirchian, E.: A case study for determination of seismic risk priorities in Van (Eastern Turkey). Earthq. Struct. **20**(4), 445–455 (2021)
5. Işık, E., Ekinci, Y.L., Sayıl, N.L., Büyüksaraç, A., Aydın, M.C.: A time-dependent model for earthquake occurrence and effects of design spectra on structural performance: a case from the North Anatolian fault zone Turkey. Turk. J. Earth Sci. **30**, 215–234 (2021)
6. Işık, E., Büyüksaraç, A., Ekinci, Y.L., Aydın, M.C., Harirchian, E.: The effect of site-specific design spectrum on earthquake-building parameters: a case study from the Marmara Region (NW Turkey). Appl. Sci. **10**, 7247 (2020)
7. Işık, E., Aydın, M.C., Büyüksaraç, A.: 24 January 2020 Sivrice (Elazığ) earthquake damages and determination of earthquake parameters in the region. Earthq. Struct. **19**(2), 145–156 (2020)
8. Rukstales, K.S., Petersen, M.D.: Data Release for 2018 Update of the U.S. National Seismic Hazard Model: U.S. Geological Survey data release (2019)
9. Simpson, A.R., Louie, J.N.: Measurements and Predictions of Vs30, Z1.0, and Z2.5 in Nevada (2.0), Zenodo (2020)
10. http://yerbilimleri.mta.gov.tr/anasayfa.aspx. Accessed 30 Aug 2021
11. AFAD. www.afad.gov.tr. Accessed 28 Aug 2021
12. Corchete, V., Badal, J., Pujades, L., Canas, J.A.: Shear velocity structure beneath the Iberian Massif from broadband Rayleigh wave data. Phys. Earth Planet. Inter. **79**, 349–365 (1993)
13. Corchete, V., Badal, J., Serón, F.J., Soria, A.: Tomographic images of the Iberian subcrustal lithosphere and asthenosphere. J. Geophys. Res. **100**, 24133–24146 (1995)
14. Sexton, J.L., Rudman, A.J., Mead, J.: Ellipticity of Rayleigh waves recorded in the Midwest. Bull. Seismol. Soc. Am. **67**, 369–382 (1977)
15. Lee, W., Kanamori, H., Jennings, P., Kisslinger, C.: International Handbook of Earthquake and Engineering Seismology. Elsevier, New York (2003)
16. Rastogi, B.K., et al.: The possibility of site effects: the Anjar case, following the past earthquakes in the Gujarat, India. Seismol. Res. Lett. **82**(1), 692–701 (2011)
17. Richards, P.G., Aki, K.: Quantitative Seismology: Theory and Methods, vol. 859. Freeman, San Francisco (1980)
18. Ben-Menahem, A., Singh, S.J.: Seismic Waves and Sources. Springer, New York (1981). Broomfield, J.P.
19. https://www.geopsy.org. Accessed 01 Sept 2021
20. Cordell, L., Henderson, R.G.: Iterative three-dimensional solution of gravity anomaly data using a digital computer. Geophysics **33**(4), 596–601 (1968)
21. Nakamura, Y.: Clear identification of fundamental idea of Nakamura's technique and its application. In: Proceedings of the XII World Conference Earthquake Engineering, Auckland, New Zealand (2000)
22. Nakamura, Y.: On the H/V spectrum. In: The 14th World Conference on Earthquake Engineering, Beijing, China (2008)
23. Campbell, K.W., Bozorgnia, Y.: NGA-West2 Campbell-Bozorgnia Ground Motion Model for the Horizontal Components of PGA, PGV, and 5%-Damped Elastic Pseudo-Acceleration Response Spectra for Periods Ranging from 0.01 to 10 s, Berkeley: Pacific Earthquake Engineering Research Center, University of California (2013)

Conceptual Design of Smart Furniture: A Case Study

Damir Blažević, Mario Bukovčan(✉), Tomislav Keser, and Krešimir Nenadić

Faculty of Electrical Engineering, Computer Science and Information Technology, Ul. Kneza Trpimira 2b, 31000 Osijek, Croatia
{damir.blazevic,mario.bukovcan}@ferit.hr

Abstract. Today design industry relies heavily on stylish and feature-rich designs applied to everyday applications. That is especially emphasized in the furniture industry where new trends are rising thus making it, so-called, smart furniture. This paper deals with paradigms and conceptual ideas in the enrichment of furniture with added value but with a stressed focus on furniture longevity and maintainability. The following case study will provide an insight into current smart furniture solutions and key characteristics that ensure the longevity of smart furniture.

Keywords: Furniture design · Industrial design · Maintenance · Smart furniture

1 Introduction

In order to maintain their competitiveness, furniture manufacturers are trying to adapt to the growing desires and needs of users. With the development of IoT technology, artificial intelligence and the desire to stand out from other manufacturers, manufacturers are becoming more innovative, which leads to the emergence of new trends in the furniture industry. As a consequence, the classical concept of furniture has changed from just basic objects that humans use to furniture units that are able to sense, identify and interact with users to meet their needs and desires [1]. Smart furniture with numerous functionalities reduces human effort and increases the quality of life, which is why it is increasingly used in private and business premises. Given the difference in the price of smart furniture and ordinary classic furniture, it is expected that smart furniture will last as long as possible. In order to meet human needs for some time, this paper proposes key characteristics to be considered in the production of smart furniture.

2 Smart Furniture

Given that the idea of "Smart Furniture" is relatively new and still in development, there is no explicit definition of so-called "Smart furniture". The first patent in which the term "Smart furniture" has occurred dates back to 1998. when innovator proposed integration of computer and card reader into a desk. Since then, technology has advanced

H. Glavaš et al. (Eds.): OTO 2021, LNNS 369, pp. 353–363, 2022.
https://doi.org/10.1007/978-3-030-92851-3_26

rapidly, so the term "Smart furniture" today has a different meaning. Based on analyzed research projects, studies and patents applications, the authors of the article [2] proposed a description of smart furniture in the context of smart homes, smart devices, smart environments and users as shown in Fig. 1.

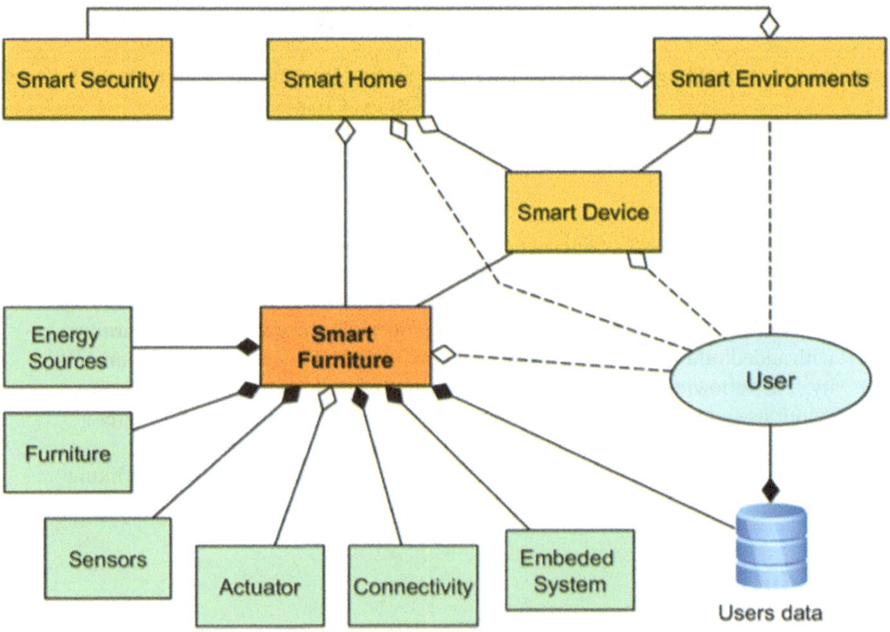

Fig. 1. Role and position of smart furniture within the Smart City [2].

A Smart Home is a house that uses numerous types of information technology to observe the environment, manage electric appliances and communicate with the outer world. Smart devices are described as critical resources that enable the delivery of numerous smart service systems to users [3]. Users should be able to interact with smart furniture, but also with other components of the smart home. By knowing the user's data and individual user preferences, smart furniture can be customized to the needs and wishes of end-users.

Different authors propose different definitions, according to one of those proposals, "Smart furniture" is defined as a "Physical structure which is composed of hardware such as computers, sensors, and appliances, and developing software technologies for realizing smart hot-spot services" [4].

According to the article [5], "Smart furniture is furniture which brings added value, functionality, comfort and elegance to fit every personalized requirement issued by the user".

Smart Furniture can also be defined as furniture which instantaneously changes the legacy non-smart space into a smart space where location-based context-aware services, service roaming, personalized services and connectivity to the Internet are provided.

Since the Smart Furniture is equipped with embedded networked computers, sensors and various I/O devices, it can provide various services in open public and/or private spaces [6].

Given the numerous definitions of "Smart Furniture", the authors of the article [2] provided an explicit definition of the term "Smart Furniture" based on patent databases and the frequency of keywords used in the literature. They proposed the following: "Smart Furniture is designed, networked furniture which is equipped with an intelligent system or controller operated with the user's data and energy sources. It can communicate and anticipate the user's needs using sensors and actuators inside the user's environment, resulting in a form of user-adapted furniture or an environment that satisfies the user-declared needs and non-declared needs for the aim of improving their quality of life in a smart world".

3 Current Trends and Characteristics of Longevity

The authors of the article [7] implemented an Arduino-based system in the chair to scan the user ID and check the status of the chair occupancy. As shown in Fig. 2, the data is then sent to a cloud server where it is stored. An authorized user can access this data at any time using a mobile device and thus, for example, check the presence of a student.

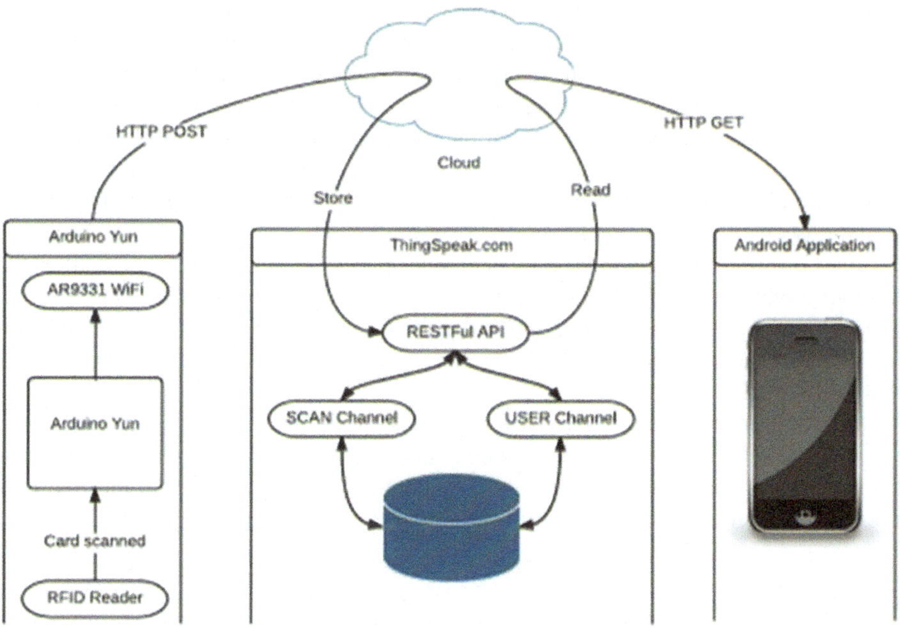

Fig. 2. System architecture of intelligent chair [7].

Office employees perform their work by sitting most of the time. Sitting longer than recommended can lead to serious health problems and musculoskeletal disorders. Smart chairs, equipped with sensors in the seat surface and backrest, monitor the sitting behaviour of the user [8]. If a certain time is exceeded, the user can be provided with a tactile feedback signal. Another solution for providing feedback to users is by sending notifications to a PC, smartphone or smartwatch. Sensors embedded in a smart chair can also be used for the detection of improper posture.

By implementing various sensors in the chair, some vital parameters that are important for human health can be detected. The authors [9] and [10] suggest the implementation of piezoelectric sensors in the chair to obtain heart rate or monitor children who have attention-deficit hyperactivity disorder (ADHD) what would help to determine the required therapy.

A large number of companies strive to increase the satisfaction of their employees and improve their work environment. Nowadays, almost every person owns a smartphone, so wireless chargers are being implemented into office desks to reduce the need for each employee to carry their own charger. The further enrichment of the desk could be the possibility of adjusting the slope or height to improve comfort and the health conditions for the eyes and spine [11].

The development of smart furniture can especially contribute to the health of the elderly. With built-in sensors, the smart bed could monitor an individual's health during sleep and automatically notify the central system in the event of a heart attack [11]. The authors of the article [12], propose the concept of a smart paramount bed system as shown in Fig. 3. It would continuously monitor physical conditions such as breathing and heart rate while the user simply lies on the bed with nothing attached to his body. It could also determine a user current state, such as general motion, being away from the bed, sleeping, and awakening. Additionally, the system could read vital signs such as blood pressure and body temperature by linking with measuring. The information measured and acquired in these ways could be shared with a bedside terminal, health care facility and electronic medical record system.

Smart furniture is also widely used in private spaces. The most common features that are smart tables equipped with are charging ports, Bluetooth speakers, LED lights, food and beverage refrigerators. With a touch-sensitive surface and the ability to connect to other devices such as TVs, smart tables make a perfect example of multifunctional smart furniture.

In order to meet the human needs for smart furniture in the long run, it is necessary to consider what are the most important characteristics that end-users expect from smart furniture. According to the research of the authors from the article [5], the five most important characteristics that furniture should have are as follows:

- Functionality,
- Design,
- Safety in Use,
- Customization and
- Structural Design.

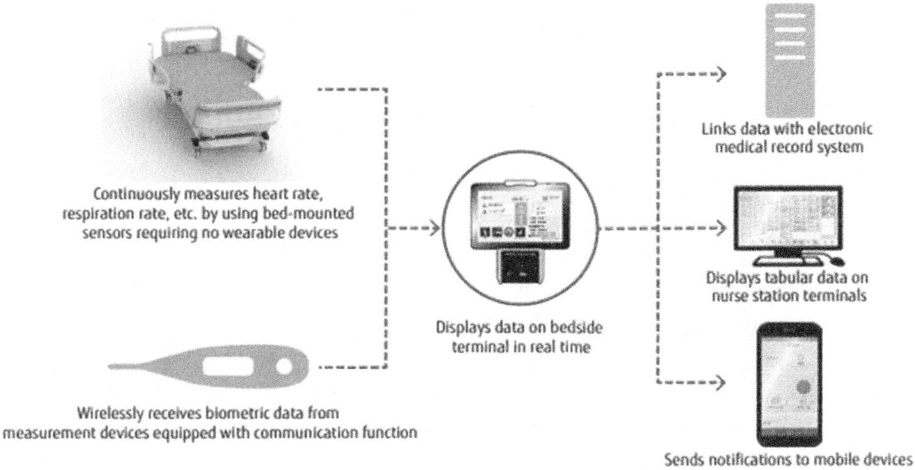

Fig. 3. The basic configuration of smart bed system [12].

The research was conducted according to the relative importance of each individual characteristic for a certain customer profile. The following Table 1 shows the results of research for young single adults.

Table 1. Importance of characteristics for young single adults.

Customer profile Young single adult	
Characteristics	Importance
Functionality	13.0%
Space saver	12.6%
Price	12.2%
Multiple functions	11.5%
Modularity	7.3%
Safety in use	6.3%
Design	5.7%
Durability	5.4%
Eco-friendly	5.0%
Structural design	4.9%
Ease of assembly	4.7%
Customization	4.6%
Easy to clean	4.3%
Type of material	2.6%

A graphical representation of results is shown in Fig. 4.

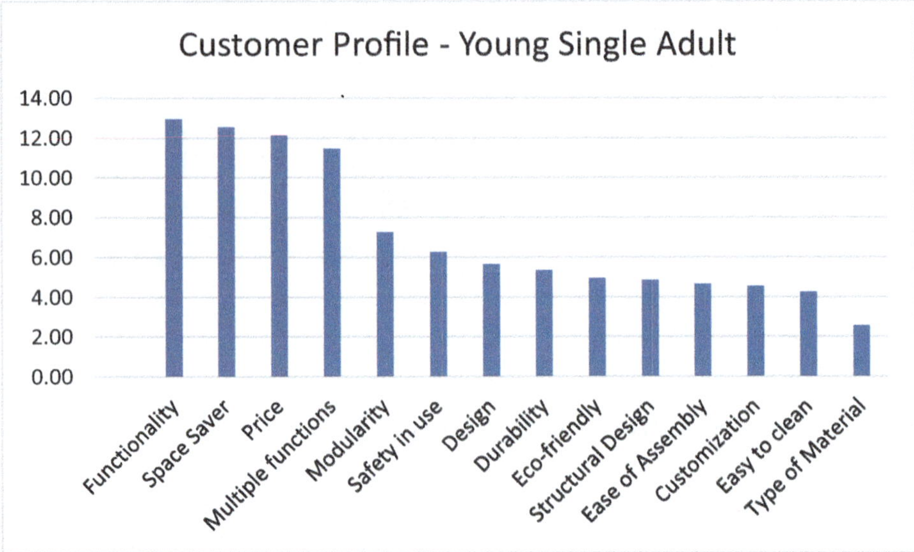

Fig. 4. Importance of characteristics for young single adults – graphical representation.

Due to low incomes, young people want to have as much functionality as possible provided by furniture at the lowest possible price. Low incomes lead to smaller housing spaces where young people live, so space-saving is expected to be another important characteristic of young adults. The following Table 2 shows the results of research for families with small children.

As can be seen in Fig. 5, for families with children the most important feature is Safety in Use. Unlike the results of a survey of young people, families are able to set aside more money for furniture that will meet their needs for a longer period of time. The last group of respondents consisted of wealthy users, the results are given in Table 3.

As can be seen in Fig. 6, for wealthy, upper-class customers the most important furniture features are design, customization and type of material. According to research results, various groups of people have different desires in terms of the possibilities that furniture can provide them, so before production, it would be good to research the market for which the furniture is planned to be produced.

Complex systems like smart homes should consist of different devices that have the ability to communicate with each other, [13] so the possibility of communication is one of the crucial device functionalities for furniture that need to be controlled.

Since the consumer should also be able to communicate with the device, one way of doing this is by voice control which enables a hands-free way to control the device.

When it comes to communication, it is always important to keep in mind possible device compatibility issues as well as security issues, to ensure privacy and safe usage of equipment. With machine learning, intelligence can be added to furniture to learn

Table 2. Importance of characteristics for families with small children.

Customer profile Families with small children	
Characteristics	Importance
Safety in use	14.6%
Functionality	11.7%
Durability	11.2%
Structural design	9.4%
Easy to clean	8.1%
Customization	6.4%
Modularity	5.3%
Design	5.2%
Price	5.2%
Type of material	5.2%
Multiple functions	5.2%
Space saver	4.9%
Eco-friendly	4.4%
Ease of assembly	2.9%

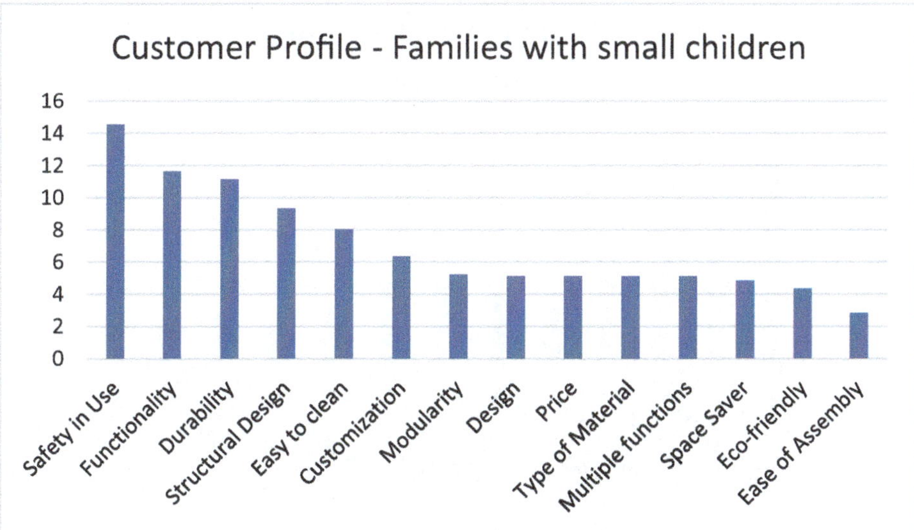

Fig. 5. Importance of characteristics for families with small children – graphical representation.

Table 3. Importance of characteristics for wealthy, upper-class customers.

Customer profile Wealthy, upper-class customers	
Characteristics	Importance
Design	16.9%
Customization	12.5%
Type of material	10.7%
Structural design	10.0%
Safety in use	8.0%
Functionality	7.7%
Durability	6.9%
Eco-friendly	6.5%
Easy to clean	4.3%
Modularity	3.8%
Multiple functions	3.7%
Ease of assembly	3.2%
Price	3.0%
Space saver	2.8%

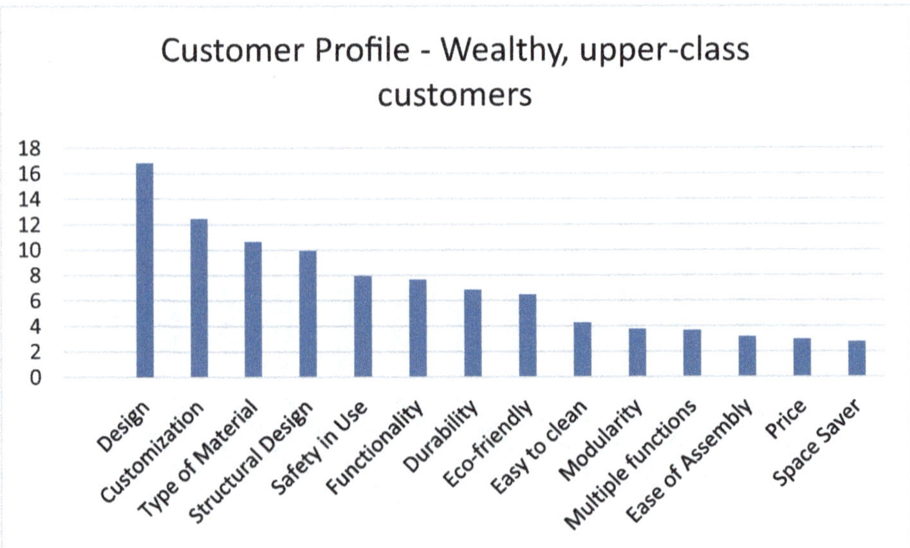

Fig. 6. Importance of characteristics for upper-class customers – graphical representation.

individual user preferences. By knowing individual user preferences, the environment can be personalized and act as an intelligent support system to improve user comfort, health and productivity [14]. Because of population growth and urbanization, the number of small houses and apartments is increasing [15]. To make the best use of space, space-saving design and the possibility of transforming furniture such as wall beds are becoming increasingly important features. Some other key characteristics regarding smart furniture longevity are shown in Fig. 7.

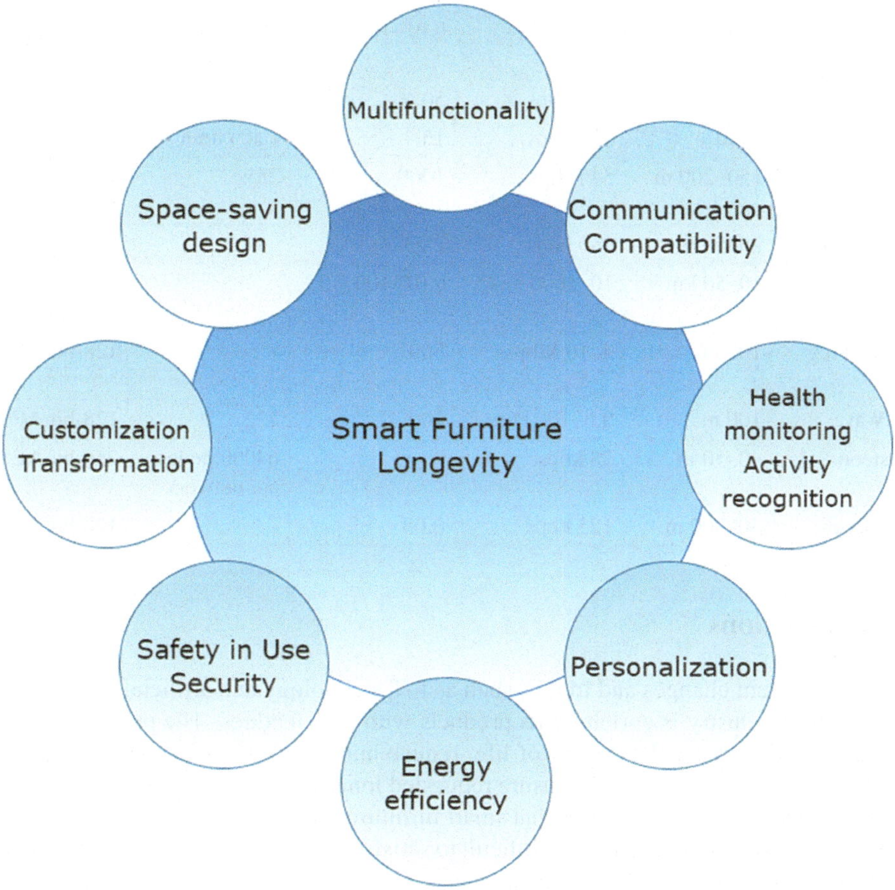

Fig. 7. Key characteristics for smart furniture longevity.

As mentioned earlier, smart devices should be able to communicate with each other. This interconnection of devices as well as the connection of the physical and virtual environment can be wired or wireless. Wired systems are more reliable concerning interference because of shielded cables, but wireless systems are generally more prevalent due to lower costs. Another advantage is the ability for devices that use this method of communication to be easily transferred from location to location. The most common wireless communication standards are shown in Table 4, [11].

Table 4. Some key smart home communication technologies.

Wireless technology	Range	Data rate	Power (mW)	Max nodes	Security
RFID	0–3 m	64 kbps	200	1 at a time	N/A
Bluetooth	1–100 m	1–3 Mbps	2.5–100	1 M + 7 S	56–128 bit key
BLE	1–100 m	1 Mbps	10	1 M + 7 S	128-bit AES
ANT	30 m	20–60 kbps	0.01–1	65533 in one channel	64-bit key
ZigBee	10–100 m	250 kbps	50	65533	128-bit key
NFC	5 cm	424 kbps	15	1 at a time	AES
WiFi	150–200 m	54 Mbps	1000	255	WEP, WPA, WPA2
6LoWPAN	25–50 m	250 kbps	2.23	-	128-bit key
Sigfox	10–50 km	10–1000 bps	0.01–100	-	No default encryption
HomePlug GP	100 m	4–10 Mbps	500	-	128-bit AES
Z-Wave	100 m	9.6–100 kbps	100	232	128-bit AES
Insteon	40–50 m	38 kbps	-	64000 nodes per network	256-bit AES
EnOcean	30–300 m	125 kbps	0.05	-	128-bit AES

4 Conclusions

Based on current changes and trends, such as IoT technology and artificial intelligence, the furniture industry is enriching its products with new features. The purpose of such products is to increase the quality of life, reduce human effort and ensure that human needs are met for some time. To ensure requested longevity, the study listed some basic functionalities and characteristics that smart furniture should possess. Due to individual differences among consumers, it is difficult to satisfy everyone's needs with one product. And while some characteristics such as energy efficiency and safety are desirable to all, other characteristics, such as design that vary from person to person, should be able to be customized to each individual. With all this in mind, the production of smart furniture should take into account the characteristics that will affect its longevity, maintainability, and thus consumer satisfaction.

Acknowledgements. This work results from implementing research activities on the project "Investing in research activities with the aim of developing a new product line" (KK.01.2.1.02.0029). The stated project is funded by European Regional Development Fund.

References

1. Mohamed, A.M.S., Ibrahim, N.F., Hassan, M.R.: The impact of smart furniture design on meeting the needs of a limited-space housing. J. Archit. Art Hum. Sci. **6**, 25 (2021). https://doi.org/10.21608/mjaf.2020.23392.1499
2. Krejcar, O., et al.: Smart furniture as a component of a smart city-definition based on key technologies specification. IEEE Access 1 (2019). https://doi.org/10.1109/ACCESS.2019.2927778
3. Lim, C., Maglio, P.P.: Data-driven understanding of smart service systems through text mining. Serv. Sci. **10**(2), 154–180 (2018). https://doi.org/10.1287/serv.2018.0208
4. Ito, M.: Smart furniture: improvising ubiquitous hot-spot environment. In: 23rd International Conference on Distributed Computing Systems Workshops, pp. 248–253. IEEE, Rhode Island, USA (2003). https://doi.org/10.1109/ICDCSW.2003.1203562
5. Vaida, C., Gherman, B., Dragomir, M., Iamandi, O., Banyai, D.: Smart furniture -Quo Vadis. In: 3rd International Conference on Quality and Innovation in Engineering and Management (2014)
6. Tokuda, H.: Smart furniture: a platform for creating context-aware ubiquitous applications everywhere. In: Yang, L.T., Guo, M., Gao, G.R., Jha, N.K. (eds.) EUC 2004. LNCS, vol. 3207, pp. 1112–1112. Springer, Heidelberg (2004). https://doi.org/10.1007/978-3-540-30121-9_108
7. He, J., Atabekov, A., Haddad, H.M.: Internet-of-things based smart resource management system: a case study intelligent chair system, In: Proceedings of the 2016 25th International Conference on Computer Communication and Networks (ICCCN), pp. 1–6, Waikoloa, HI, USA (2016). https://doi.org/10.1109/ICCCN.2016.7568533
8. Roossien, C.C., et al.: Can a smart chair improve the sitting behavior of office workers? Appl. Ergon. **65**, 355–361 (2017). https://doi.org/10.1016/j.apergo.2017.07.012
9. He, Z., et al.: A heart rate measurement system based on Ballistocardiogram for smart furniture. In: 2018 IEEE Asia Pacific Conference on Circuits and Systems (APCCAS), pp. 151–154 (2018). https://doi.org/10.1109/APCCAS.2018.8605567
10. Yang, R.C., et al.: Objective evaluation of therapeutic effects of ADHD medication by analyzing movements using a smart chair with piezoelectric material. Appl. Sci. **11**(12), 5478 (2021). https://doi.org/10.3390/app11125478
11. Shidiq, D.D., Chico Hermanu, B.A., Fauzan, A.M., Ubaidillah, A.: Development of adjustable desk with wireless charging features. AIP Conf. Proc. **2097**, 030021 (2019). https://doi.org/10.1063/1.5098196
12. Majumder, S., et al.: Smart homes for elderly healthcare-recent advances and research challenges. Sensors **17**(11), 2496 (2017). https://doi.org/10.3390/s17112496. PMID: 29088123; PMCID: PMC5712846
13. Nakajima, R., Sakaguchi, K.: Service vision design for Smart Bed System™ of paramount bed. Fujitsu Sci. Tech. J. **54**, 9–14 (2018)
14. Zdankin, P., Weis, T.: Longevity of smart homes. In: 2020 IEEE International Conference on Pervasive Computing and Communications Workshops, Austin, USA (2020). https://doi.org/10.1109/PerComWorkshops48775.2020.9156155
15. Aryal, A., Becerik-Gerber, B., Anselmo, F., Roll, S.C., Lucas, G.M.: Smart desks to promote comfort, health, and productivity in offices: a vision for future workplaces. Front. Built Environ. **5**, 76 (2019). https://doi.org/10.3389/fbuil.2019.00076
16. Husein, H.A.: Multifunctional furniture as a smart solution for small spaces for the case of Zaniary towers apartments in Erbil City. Int. Trans. J. Eng. Manag. Appl. Sci. Technol. **12**, 1–11, (2020). https://doi.org/10.14456/ITJEMAST.2021.8

Assessment of Wireless Signal Propagation in an Obstacle Rich Environment and Maintenance Communication Capability

Marko Božić, Tomislav Keser(✉), and Andi Bašić

Faculty of Electrical Engineering, Computer Science, and Information Technology, JJ. Strossmayer University of Osijek, Kneza Trpimira 2b, Osijek, Croatia
marko.bozic@ferit.com, tomislav.keser@ferit.hr

Abstract. There is a high and constant thrive to utilize a variety of wireless communication systems to achieve some sort of remote monitoring and/or controlling processes and systems overall. The wood processing industry is no exception but there is still an ultimately low presence of such systems, either in automated or semi-automated form, but their rise and presence has become inevitable and is constantly growing. Such an exemplary system is set up and adjusted in the Spačva wood processing industry where wireless devices are used for constant monitoring of environmental parameters and wood structural moisture in a wooden plank during the phase of naturally aspired drying process and forced one in a controlled environment.

However, because of fact that such devices must be tightly incorporated into a pack of planks a new problem arises regarding wireless signal propagation that becomes significantly affected. This paper deals with the problem in a way that we describe and propose a technique to strengthen measurement device and receiving unit wireless signal propagation to maintain quality communication link. Also, we define some practical guidelines for keeping good signal and their maintenance propositions.

Keywords: Wireless signal propagation · LoRaWAN · Radio wave propagation

1 Introduction

Propagation channel is one of the key areas of research and analysis for the design and implementation of secure and successful wireless communication systems between transceiver and receiver [1].

Wireless channel causes the transmitted signal to lose power as it propagates from transmitter to receiver. To address these issues there are several elements to consider during design, development and eventual deployment of communication stack in real-time environments.

Project was developed for outside contractor "*Drvna industrija Spačva d.d. Vinkovci*" with the special purpose of measuring humidity in wood during drying. The transmitter is designed to be placed in special housing which then transmits measured data to the receiver.

H. Glavaš et al. (Eds.): OTO 2021, LNNS 369, pp. 364–377, 2022.
https://doi.org/10.1007/978-3-030-92851-3_27

Obstacles and several key factors like reflections, diffraction, and scattering [2] create multiple propagation paths between the transmitter and the receiver, each with a different delay. This leads to a loss of signal power, crates noise in the communication channel, slowing or significantly disrupting the communication channel so the transmitted message is either received at a significant loss or not at all.

During project development project designer needs to be aware of these issues and model communication accordingly. To better illustrate when we became aware of these issues and how did we overcome them we put forward some of our findings.

In medium, to large industrial environments, consumer-grade communications protocols are limited by small propagations characteristics since high bandwidth is a priority as well as serving multiple clients with data packets with variable sizes. Data can be something as simple as text and complex as video streaming. Data arrival and time for a signal to reach the receiver is also not a priority. Most of these priorities are normal and consumers won't mind as long data arrives in a reasonable amount of time. In industrial settings, priorities are different and a lot stricter. Data has to have a fixed size, data time of arrival needs to be within specified margins. The communication channel needs to be active for long periods and needs to have extremely small downtime.

This paper has the following structure. In the section *State of the art survey and problem statement* characteristics of multipath propagation, channels are talked about and what causes them. In Sect. 3, we briefly describe important factors affecting wireless propagation channels to explain the need for several different channel models. Our proposed solutions and results are analysed and put forward in Sect. 4.

2 Signal Propagation Mechanisms

There are several key propagation mechanisms for a transmitted signal to reach the receiver for any given communication system. In this paper, we are using *LoRaWAN* [3] as low power wide area networking protocol in a communication system for structured measurement data delivery.

To model such a system we present few key mechanisms, each potentially associated with a different propagation path (see Fig. 1).

Propagation Along the Line-of-Sight

Propagation along the line-of-sight (LOS) path is when a signal is sent from transmitter to receiver by the receiver without suffering any reflections, diffractions, or scattering. A LOS has the smallest time delay and is usually the strongest signal to be found on any given communication system. To be able to call signal LOS signal obstructions to need to be far away or non-existent between transmitter and receiver. Fresnel zones are used to model such a system. In this project, we measured the signal without any obstructions between transmitter and receiver.

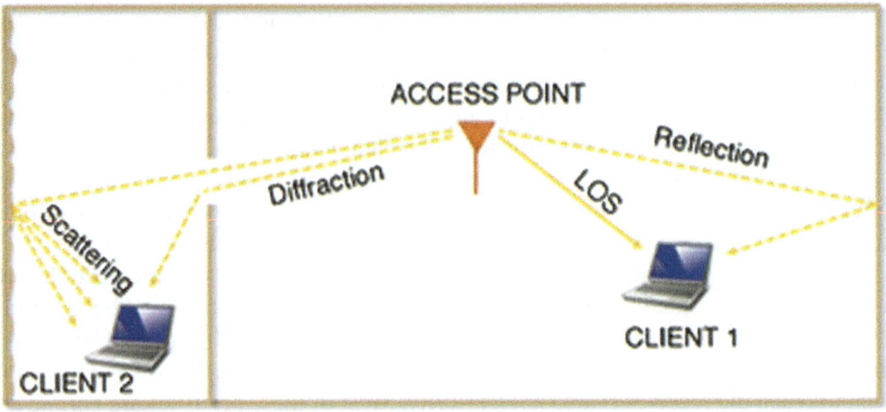

Fig. 1. LoRaWAN network in an indoor environment. The LOS path is unobstructed between the access point and clients. That client 1 receives weaker signals because of a reflection off a wall. Client 2 receives signal through diffraction in the doorway and scattering off a rough wall.

Non-Line-of-Sight Propagation

When a transmitted signal travels through communication paths other than the LOS path then we have non-line-of-sight (NLOS) propagation. Transmitted signal reaches the receiver via one or more indirect paths, each having different frequency attenuations and delays. It's used for signal coverage in dense and obstacle-rich environments like a city where a lot of buildings and other obstructions are present. Those obstructions are responsible for propagation mechanisms like reflection, scattering, and diffraction which are responsible for most of the communication to be made.

Reflection occurs when a wave impinges on a smooth object. Reflection is accompanied by refraction (transmission of the wave through the object). The strength of the reflected and refracted waves depend on the type of material. The angles and indices of reflection and refraction are given by Snell's law [4]. In our project, this type of mechanism was important for transmitter setup which is discussed in later chapter measurement and results.

Signal regularly hits rough and uneven objects this splits transmitted waves and crates smaller waves. This is called scattering and usually comes with energy loss for signal. Split wave is spread over a larger, wider area which in turn comes at a cost for other signals that need to pass as it crated noise in the communication channel. The signal is received from multiple communication paths and all with different time delays. This property is extremely important during project development for communication congestion during high network traffic.

Diffraction is the "bending" of waves around sharp corners as in over the tops of buildings, around street corners, and through doorways. Diffraction is one of the main ways that it is possible to provide cellular coverage in cities and busy industrial settings where every signal is important.

3 Assessment of Wireless Signal Propagation

LoRaWAN network is used in this project for its reliability in harsh industrial environments, high signal penetration properties at low frequencies such as 32 kHz up to 500 kHz, and wireless connectivity at a wide-open area (see Fig. 2).

Fig. 2. Area for storing and drying cut wood before going to production. It's used for testing signal penetration and coverage. Red dot symbol signifies placement of Gateway on lamppost with general direction of antenna described with arrow. (Color figure online)

LoRa based networks have low energy consumption, configurable frequency as well as other signal configuration properties. It allows us to set up data packets in sizes that are suitable for the desired task. Those tasks are usually simple measurement data as is in our project. It allows us to model projects based on extremely low power data exchange. The exchange needs to be quick with minimal hardware and software assistance to save on power consumption. Data needs to be transmitted quickly and reliably. Messages between the LoRa node device and Gateway need to be quick and acknowledged on the software layer. Up to 100 connected devices are connected at any given moment.

3.1 LoRaWAN Specification

LoRaWAN stack can be separated into two layers. The first layer starts with LoRa enabled device commonly called a node. This device can be a custom printed circuit board (PCB) or prefabricated device from other manufacturers. To reduce cost custom PCB was developed with 868 MHz sx1276 module onboard LoRa needs outside device, concentrator to exchange messages with the server. Concentrator called Gateway is a device with several LoRa modules on board with one or more channels of communication.

In the project we are using a single channel LoRa enabled device from Dragino.com [5]. The device uses Atmega328p [6] from Microchip on the LoRa side for communication with other devices and on another side, it uses an ARM chip with OpenWRT [7] Linux distribution preinstalled for communication with a server.

LoRa Parameters and Requirements

Several parameters allow us to balance obstacle penetration in densely populated environments with speed and coverage (see Fig. 3). LoRa Spread Factor or shortly SF [4] represents the number of symbols sent per bit of information. LoRa can trade-off data rate for sensitivity.

Fig. 3. Parameter tuning and calculation for LoRa communication protocol. Depending on communication requirements and constraints for a given data size or a payload length it is shown how long it takes to transmit messages from client to access point.

Lower SF means more encoded data per second while higher SF means fewer data encoded per second. With lower SF more data can be sent over time which reduces power consumption. This is useful in long-distance, low populated areas with a partial or clear line of sight. It can be useful for connecting long-range sensors if messages are not too big.

The benefits of high SF are that more extended airtime gives the receiver more opportunities to sample the signal power which results in better sensitivity. In addition, to the already mentioned, LoRa uses forward error correction coding to improve resilience against interference.

LoRa network was set up in steps during testing depending on project requirements, network constraints, and various environmental use cases.

Project requirements are:

- LoRa client needs to send 42 bytes of data
- Up to 5 retries for data exchange
- Data exchange time needs to take less than a second and not more than two seconds
- Needs to work reliably in high humidity and temperatures

Network connectivity cases were made:

- Wide area network connectivity in line of sight
- Network connectivity in a designated testing area
- Data transmission times for two sets of frequencies 125 kHz and 250 kHz and spreading factors from 9 to 12 [5].

Access node or Gateway in LoRaWAN specification needs to be set up in a relatively high place. It helps if it's set up in such a way that it's the highest point where obstructions are not present. Initially, during design and testing, we placed gateway high on lamppost together with fixed antenna directly to the transmitter.

Gateway was paired up with two different 868MHz omnidirectional antennae. A smaller Omni antenna was used for small to medium indoor places and under the right condition for larger open areas. This was done for project cost reduction and to assess signal coverage in the designated testing area for future reference. At a later date, we replaced a smaller antenna with a larger version (see Fig. 4) which is used for higher frequency use cases like GSM and 3G connectivity.

The client or node had a limited time to deliver the data packet to the gateway given the obstacles and that LoRa has limited bandwidth at higher spreading factor values [6].

Clients were placed in the middle of wood pallets (see Fig. 5). Depending on weather conditions, the number of active connections, placement of nodes inside the wood pallet time of delivery can vary as well as signal quality and strength.

Node has a small helical antenna inside. Measurement data was logged and placed inside the CSV file which was later analyzed (see Fig. 6).

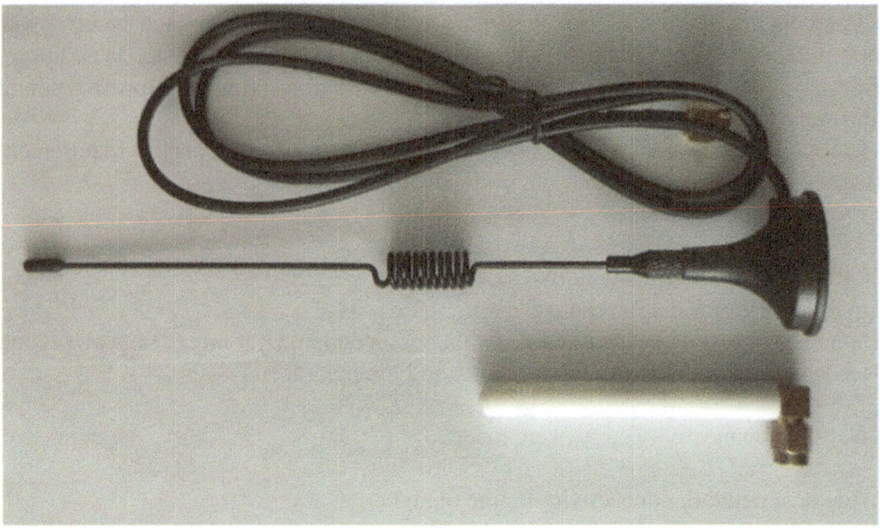

Fig. 4. Antennas used in project testing. Small 'white' antenna initially used for reference testing of LoRa signal coverage and large 3G/GMS omnidirectional antenna used later for wide area communication on open field and for better signal quality during message exchange with large amounts of obstructions.

Fig. 5. Placement of communication device inside of stacked wood pallets. Communication device sends up to 10 messages to access point. When the device is taken out number of successfully sent messages is checked. Process is repeated for all stacks of wood pallets until all messages are unsuccessful.

Fig. 6. On-filed analysis of collected data samples, number of successfully exchanged messages between client and access point. Laptop shows printouts of results while measurement device in red casing sends messages. (Color figure online)

4 Measurement and Results

Testing revealed several issues in the communication channel. During the normal day with temperatures exceeding 28 °C LoRa parameters need to reflect a higher data rate with corresponding spreading factor and frequency so that most measurement data in close vicinity of a gateway can pass as fast as possible.

This was mostly due to the thermal throttling of devices with prolonged exposure to high temperatures. Due to high temperatures signal strength (RSSI) decreases over time (see Fig. 7) which can be seen from dates on 7.8.2021 to 15.8.2021.

Congestion on communication channels can make issues with a higher number of active clients. To mitigate those issues, we implemented two-step solutions. If the client can't send the message during the first cycle it saves data for the next one where two

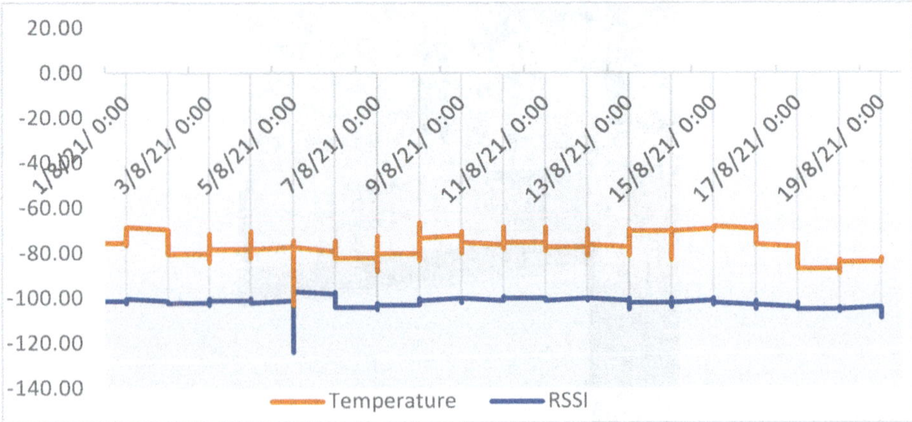

Fig. 7. Signal strength in relation to temperature depending on the date of measurement during testing. From 4.8.2021 to 10.8.2021 there is reduction in signal strength due to high temperature conditions of open field area for drying wood pallets.

messages with data will be sent. The second solution is to encode messages on the client side. Encoding data packets means getting to know what data types are going to be sent and what are their respective value ranges. The third solution is to set the number of possible retries during the transmission cycle.

Client-Side Analysis

On the receiver side, we implemented a small circular queue that accepts messages for a limited amount of time. This method of polling the clients before responding adds flexibility when it's time to respond to received messages and when it's time to pass messages further over the network if needed. We implemented three sets of processes that are active only a certain amount of time and can be exited sooner if the right conditions are met.

Distance from transmitter to client plays a big role in getting the messages on time. The further the client is the more time is needed to transmit the message and even more energy is lost to transmit it.

Antenna Placement and Effects on Signal Strength

Getting the right antenna and adding additional boost or gain can make a difference from 5 dBm up to 15 dBm in our testing. Changes in gain don't notably affect clients closest to the receiver. To cover longer distances higher frequency was chosen to reduce noises in the communication channel and additionally speed up data transmission rates.

With higher spreading factor values we additionally reduced noise in the communication channel for modules that are furthest away.

Weather conditions can affect data transmission in a significant way especially humidity or simple water vapor. Since our node was placed in the middle of wood pallets. Those wood pallets are affected by outside water conditions such as snow, hail, fog, or simply rain. Water can cause mobile signal problems by impeding or scattering radio waves and therefore causing bad signal reception. Water conducts electricity, which interferes with a signal wave, but it also reflects and/or refracts radio waves and absorbs energy from radio waves, and converts the energy to heat, causing the propagation delay effect.

Wood is especially affected by water. It can absorb water up to 50% from 10% in just a couple of minutes or hours. The absorbed water is then released during days with temperatures turning it to water vapor. The combination of water vapor and heat affects electronics considerably and can reduce signal significantly.

From 30.7.2021 to 31.7.2021 almost 10% of humidity is gained back (see Fig. 8). Since all that vapor can't escape from the same part of the wood pallet at the same time vapor settles further inside until the next cycle as shown from 31.7.2021 to 3.8.2021.

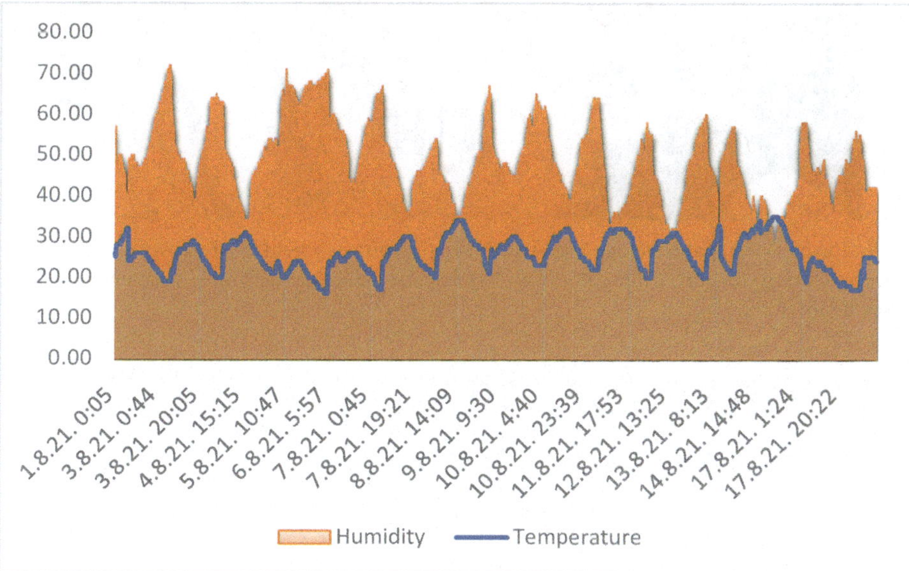

Fig. 8. Relationship of humidity and temperate. On the graph temperature spikes during the day are shown and how much it affects the release of humidity in wood during measurement cycle. With higher temperature more moisture is released which causes high humidity conditions with high signal and energy losses.

Depending on signal coverage the right antenna must be chosen. For testing purposes, we choose a designated testing site where signal coverage and penetration are going to be measured (see Fig. 9).

For distance testing large area of the open field was chosen (see Fig. 10). Where the green line represents where the signal with the receiver was achieved. As shown from points 2 to 3 of the wide testing area there is no communication signal. This is mostly due to large amounts of obstructions. From points 1 to 2 the closer we get to point 2 the smaller signal quality and strength were observed while from points 3 to 4 only a small area of signal coverage was achieved.

This represents the only valid point of wide-area coverage with obstructions. This points to the fact that choosing the right testing area for signal penetration wide signal

Fig. 9. Results of signal coverage and strength testing on open field. Red arrows point to places where measuring was done while color of arrows denotes frequency used for LoRa communication. (Color figure online)

Fig. 10. Birds eye view of field where wood pallets are placed for drying and conditioning. Open field testing was done along the drawn lines where the green color of the line denotes successful signal connection and red where signal connection losses occurred. (Color figure online)

testing is necessary before all other tests are made. Wide area testing was done with a small omnidirectional antenna. It was replaced after it failed to adequately cover all designated areas for signal coverage. A bigger antenna with a gain of 2.5 dBm was placed instead of a smaller one with few exceptions. It was placed 1 m from the lamppost on a metal stand (see Fig. 11). The additional extended cable needed to be used. This ads signal loses due to transmission impedance from longer cable. To medicate those issues signal power gain needed to be raised which usually comes at the expense of power consumption which is acceptable for a gateway that remains stationary inside the box [7].

Fig. 11. Placement of LoRa receiver (Gateway) on the lamppost. Antenna is at least one meter from the lamppost standing in upright position. Lamppost is located on field for drying wood pallets.

The antenna must be set up in a high place like a lamppost (see Fig. 12). During testing, it was observed that the antenna must be at least one meter from any object especially if that object is made from metal. We observed noticeable signal reflection which created no-signal gaps in the testing area.

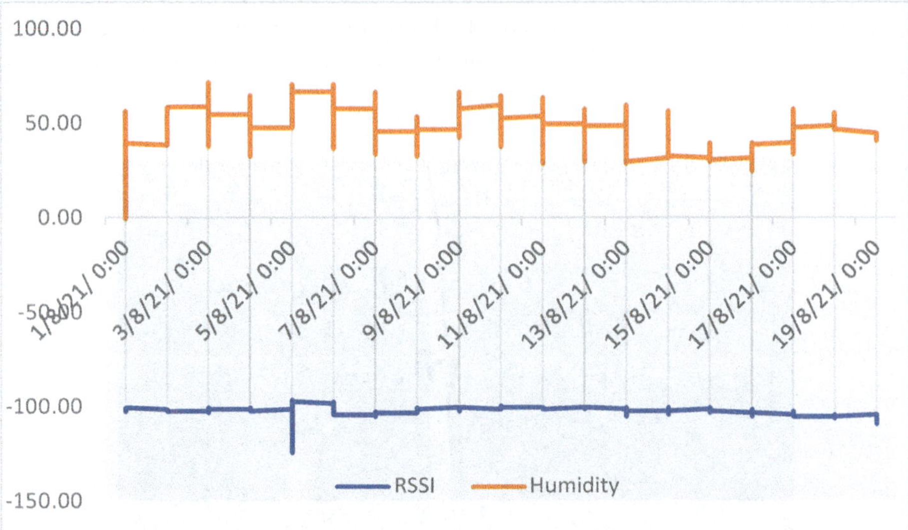

Fig. 12. Relationship of humidity and signal strength over time. During high humidity cycles signal strength drops as shown from 3.8.2021 to 8.8.2021.

5 Conclusion

In this paper, we have presented project setup, workflow, and findings on how to set up wireless communication channels for real-time use cases. We elaborated on what affects communication channels and how to alleviate those issues when using the LoRaWAN stack.

Depending on the environment, weather and obstacles we can model communication channels to suit our needs. LoRa showed great results with a high spread factor with the limited amount of data in densely packet areas and predictable results during long-range testing where it excelled. Picking a good placement for the antenna and the right one for the job is crucial to make everything work.

References

1. Ccs.neu.edu. https://www.ccs.neu.edu/home/rraj/Courses/6710/S10/Notes/Antennas.pdf. Accessed 17 Aug 2021
2. Zaidi, A., Athley, F., Medbo, J., Gustavsson, U., Durisi, G., Chen, X.: 5G Physical Layer: Principles, Models and Technology Components, 1st edn. Academic Press, Inc., Orlando

3. Lorawan. https://www.thethingsnetwork.org/docs/lorawan/what-is-lorawan/. Accessed 19 Aug 2021
4. MIT Lecture notes. https://ocw.mit.edu/courses/electrical-engineering-and-computer-science/6-007-electromagnetic-energy-from-motors-to-lasers-spring-2011/lecture-notes/MIT6_007S11_lec32.pdf. Accessed 19 Aug 2021
5. Lorawan. https://www.thethingsnetwork.org/docs/lorawan/modulation-data-rate/. Accessed 19 Aug 2021
6. Lorawan. https://www.thethingsnetwork.org/docs/lorawan/spreading-factors/. Accessed 19 Aug 2021
7. Antenna test. https://www.coredump.ch/2017/04/30/lorawan-868mhz-antenna-test-part-2/. Accessed 22 Aug 2021

Generating Artificial Biscuit Tile Images

Jana Dukić, Ivan Aleksi(✉), Tomislav Matić, and Filip Sušac

Faculty of Electrical Engineering, Computer Science and Information Technology Osijek, Kneza Trpimira 2B, 31000 Osijek, Croatia
jana.dukic@etfos.hr, {ivan.aleksi,tomislav.matic1, filip.susac}@ferit.hr

Abstract. This paper addresses the problem of creating a large image set with artificially inserted biscuit tile surface defects. Those defects commonly occur in real-life ceramic tile production. However, the occurrence rate of a *defected* tile is neglectable to the rate of a *non-defected* tile. To overcome the gap in the size of the two sets of images, *with defects* and *without defects*, artificial image generation increases the number of images *with defects*. This paper enables the future use of machine learning algorithms, which require a relatively equal size of the two image sets. The analysis of surface defects pointed to a similarity of the surface defect to an inverted Gaussian curve. The Gaussian distribution model is proposed for modeling a biscuit tile surface defect. Captured images with surface defects are compared with corresponding artificially created images. Image set *without defects* is used to create the image set *with defects*.

Keywords: Ceramic tile · Augmentation · Artificial · Defects · Detection

1 Introduction

Production of ceramic tiles is a complex process consisting of several stages, and all stages must satisfy certain quality constraints to obtain a good quality of a final product. At the beginning, the raw material is grinded and mixed to get the blended material. After drying, a press machine is used to create the crude tile. Subsequently, a thin layer of glazing is applied to create the biscuit tile. Finally, after firing and cooling, the ceramic tile is created. A simplified view of the ceramic tile production process is depicted in Fig. 1.

Two intermediate products are created during ceramic tile production: a crude tile and a biscuit tile. The final product is a ceramic tile. It is visually inspected and sorted into three classes. The two intermediate products, crude and biscuit tiles, must also be of good quality for an excellent ceramic tile quality. A visual inspection is usually done only in the ceramic tile stage, while in the crude and biscuit tile stage is neglected. To save energy spent on the firing of tiles, inspection of biscuit tiles is also proposed [3].

Because of the highly automated production stages in a ceramic tile production line, the number of defects in the final product is negligible and sparse [3]. This is also true for the two intermediate products. Additionally, different types of defects can appear. All defects can be divided into two main categories. The first one is the edge/corner defects

H. Glavaš et al. (Eds.): OTO 2021, LNNS 369, pp. 378–389, 2022.
https://doi.org/10.1007/978-3-030-92851-3_28

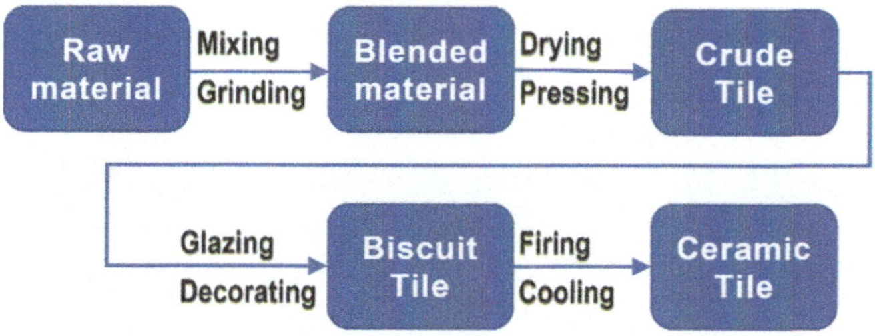

Fig. 1. Ceramic tile production stages.

(a) (b) (c)

Fig. 2. Edge and corner defect examples (a)–(c).

that appear as mechanical damage on a tile's body around its edges and corners due to the press and drying processes, cf. Fig. 2. The second is the surface defects that appear as spots, cracks, and texture anomalies caused by glazing and decoration, cf. Fig. 3. For the crude tile most common are edge and corner defects, while for the biscuit and ceramic tile surface defects are most common. In all stages number of defected products is much less than non-defected products.

Most research papers deal with surface defect detection where the imbalance of non-defected and defected tiles is apparent [1, 4, 5]. Data augmentation is commonly used to try the resolve the problem of imbalanced datasets. Image augmentation techniques include geometric transformations, color space augmentations, kernel filters, mixing images, and similar [6, 7]. Most common augmentation technique is image rotation where, e.g., tile surface image with defects is rotated 90° three times. From one defected tile surface image the result is 4 four images.

In this paper, we propose a model for image augmentation based on artificial defect generation. First, we selected spot surface defects and tested the method on real biscuit tile images. Next, a Gaussian model [8] of spot defect is generated based on real defected biscuit tile images. Then, the developed model is applied to non-defected biscuit tile images to generate additional defected images and thus create a balanced dataset for machine learning methods and their application for defect detection.

Rest of the paper is organized as follows. Section 2 gives an overview of defects modeled in this paper and the proposed Gaussian model of the biscuit tile surface defect.

Evaluation of the model is given in Sect. 3. Finally, Sect. 4 concludes the paper and gives remarks for future work.

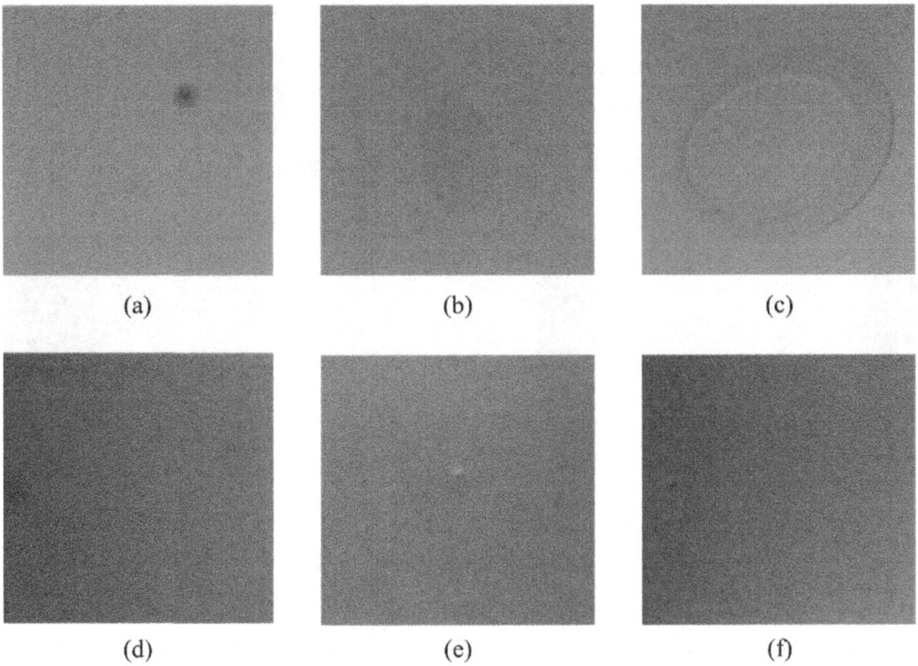

Fig. 3. Surface defect examples (a)–(f).

2 Artificial Biscuit Tile Images

2.1 Biscuit Tile Image Segmentation

Captured biscuit tile image has three distinct areas: the dark-colored background, two dark grey colored conveyor belt lines, and a tile surface itself, cf. Fig. 4a. First, the image area of the tile surface is extracted using the image segmentation method. Then, after image segmentation is done, a tile surface image is obtained, without its surrounding background, cf. Fig. 4b.

In this work, image segmentation is done with the thresholding method. The method uses a threshold level to distinguish background and foreground image areas. In this work, the input image is an 8-bit grayscale image with pixels ranging from 0 to 255. The threshold level is 80. Therefore, pixels with an intensity value less than 80 are labeled as background, while others are foreground. Background pixels are assigned a value of 0, while foreground pixels are assigned a value of 255. Subsequently, the contour tracking method is applied to extract a polygon boundary between tile surface and background pixels. This step is essential for inserting artificial biscuit tile surface defects since they must be inserted only on a tile surface image.

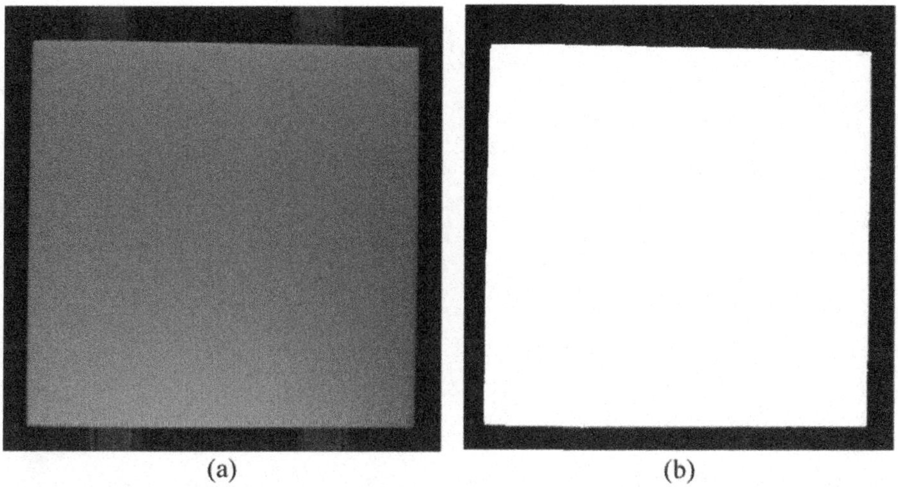

Fig. 4. Biscuit tile: captured image (a); image after applied segmentation method (b).

After the image segmentation method is applied, a tile surface image has $N \times M$ pixels. Parameters N and M represent the number of pixels in the x and y direction, i.e., the width and the height of the tile surface image, respectively. Finally, an artificial biscuit tile surface defect can be inserted into the tile surface image.

2.2 Biscuit Tile Surface Defect

Several types of tile surface defects are known in literature [4, 5, 9]. A common type of surface defect on a biscuit tile is depicted in Fig. 3. The defect has a shape of a reverse Gaussian curve, cf. Fig. 4. The biscuit tile surface defects have a circular shape, and in some cases, an elliptical shape. The defect is characterized by an intensity drop ranging from −25 to −126, relatively to the tile surface.

The dataset used in this paper has 100 images. There are 90 of them without tile surface defects and 10 of them with tile surface defects.

2.3 Gaussian Model of a Biscuit Tile Surface Defect

This paper uses the Gaussian model to model a surface defect on a set of biscuit tile images. Gaussian distribution function with an arbitrary amplitude is defined with

$$f(x, y) = A \cdot \exp\left\{ -\frac{(x-\mu_x)^2}{2\sigma_x} - \frac{(y-\mu_y)^2}{2\sigma_y} \right\}, \quad (1)$$

where A represents an arbitrary amplitude value and the maximum value of a function. Parameters $\mu_x = 1, \ldots, N$ and $\mu_y = 1, \ldots, M$ represents a 2D pixel in the x,y-plane in which the function has the maximum intensity A. Values $x = 1, \ldots, N$ and $y = 1, \ldots, M$ are representing a grid of pixels in the x,y-plane. Variance parameters σ_x and

Fig. 5. Biscuit tile image on a production line with red color marked surface defect.

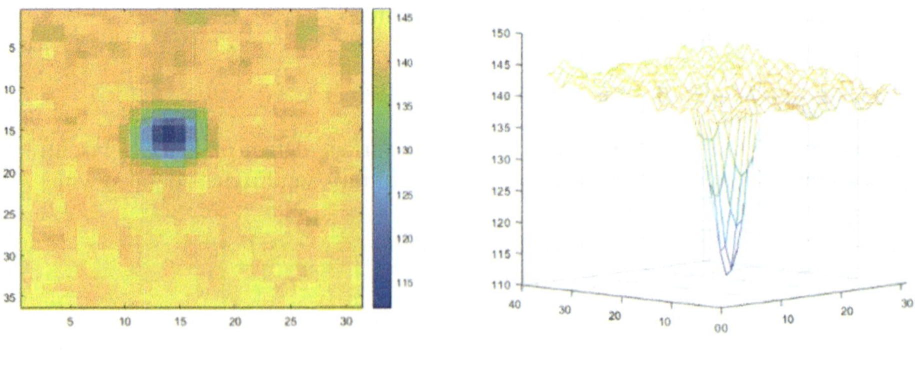

(a) (b)

Fig. 6. Magnified image with the biscuit tile surface defect: 2D (a); 3D (b).

σ_y are defining the width of the Gaussian bell in the x and y direction, respectively. With increasing the variance value in the corresponding x or y direction, the function will become broader in the corresponding direction. When variances are equal in both directions, i.e., when $\sigma_x = \sigma_y$, the Gaussian function has a spherical shape. In the case when the two variances have different values $\sigma_x \neq \sigma_y$ the Gaussian function has an elliptical shape. Additionally, the Gaussian distribution function can have a rotation by the angle $\phi \in [0, 2\pi$. Therefore, the Eq. (1) is generalized with

$$f\left(x_\phi, y_\phi\right) = A \cdot \exp\left\{-\frac{x_\phi^2}{2\sigma_x} - \frac{y_\phi^2}{2\sigma_y}\right\}, \tag{2}$$

$$x_\phi = (x - \mu_x)\cos(\phi) - \left(y - \mu_y\right)\sin(\phi), \tag{3}$$

$$y_\phi = (x - \mu_x)\sin(\phi) + \left(y - \mu_y\right)\cos(\phi). \tag{4}$$

Gaussian model parameters suitable for creating an artificial biscuit tile surface defect are proposed in Table 1. Parameter A has a negative value to obtain a reverse shape of a Gaussian curve. Parameters μ_x and μ_y are in the image of a tile surface, which is created after the image segmentation method is applied. Subsequently, a random number generator is used to create an instance of the surface defect.

Table 1. Gaussian model parameters applied for creating biscuit tile surface defects.

	A	μ_x	μ_y	σ_x	σ_y	$\phi[°]$
Minimum	−126	1	1	2.000	2.000	0.0
Maximum	−25	N	M	6.000	6.000	90.0

A set of three instances of biscuit tile surface defects is generated to demonstrate the result of the proposed Gaussian model for generating biscuit tile surface defects. Parameters of generated instances are presented in Table 2, while their corresponding images are illustrated in Fig. 5.

Table 2. Gaussian model parameters for 3 instances of generated biscuit tile surface defects.

Image	A	μ_x	μ_y	σ_x	σ_y	$\phi[°]$
1	−125	274	690	5.526	5.262	0.0
2	−37	198	675	2.068	4.320	77.6
3	−87	429	328	2.128	4.520	32.6

Fig. 7. Magnified biscuit tile images with artificial surface defect instances, cf. Table 1.

Let F be the image with $N \times M$ pixels, with calculated Gaussian distribution described with Eq. (2). Let I_T be the tile surface image without a surface defect, with $N \times M$ pixels. Finally, from image I_T, an artificial image I_A is created using

$$I_A = I_T + F. \tag{5}$$

In the Eq. 5, the values in F are negative, cf. Table 2, and are representing the defect with the shape of a reverse Gaussian curve. The random number generator is used to create a surface defect with the depth ranging from -37 to -125, as depicted in Table 2. When the negative values from the image F are added to the tile surface image I_T, the defect is created in the resulting image I_A. In this way, the surface defect is created with values like the example that is illustrated in the Fig. 6.

The proposed method is also usable in the scenario when the camera acquisition system has uneven lighting. The tile image area I_T has a visible trend in intensity, cf. Fig. 3. The proposed method adds the defect with the original image, Eq. (5), and the trend in the intensity levels remains unchanged.

3 Implementation and Evaluation

Proposed method is developed and tested using the MATLAB programming environment. Two Graphical User Interfaces (GUI) implementations are developed. The first one is for creating biscuit tile surface defects with manual insertion of each defect, while the second one does this procedure automatically for a large dataset.

3.1 Manual Artificial Biscuit Tile Defects Insertion

First GUI implementation enables user to select (x,y) coordinates on the non-defected tile image where the artificial defect will be generated. Tile edge is segmented (green line in Fig. 8) and the user can select coordinates only on the tile surface, cf. Fig. 8.

Fig. 8. Example of user input for first GUI implementation.

After user coordinates selection, interface generates an artificial defect using the method given in Sect. 2. Parameters for the Gaussian model of the surface defect are randomly selected between the min and max parameter values given in Table 1. New image is formed and saved with the same size and resolution as the input image with an artificial generated spot defect. Result example is depicted in Fig. 9. The procedure can be repeated for the same image to generate multiple spot defects on different coordinates with different random parameters of the proposed model.

Fig. 9. Example of the generated artificial defect based on user input.

3.2 Automatic Artificial Biscuit Tile Defects Insertion

Automatic generation of artificial defects is applied on the batch of biscuit tile images. Input images are images without biscuit tile surface defects. For each input image, the algorithm selects a random number of generated artificial defects. The number of generated defects is in the range defined by the user at the beginning of the procedure. The thresholding image segmentation method is applied on each input image, as it is explained in the Sect. 2. After segmentation, the minimum bounding box algorithm is applied. The bounding box is illustrated with red color in Fig. 10. Subsequently, coordinates of artificial defects are generated with a random number generator. Generated coordinates are inside the bounding box. Finally, the image with inserted artificial defects is stored into the dataset of images with surface defects (Fig. 11).

Fig. 10. Bounding box (red color) detected by the thresholding method.

(a) (b)

Fig. 11. Automatically generated artificial images with: 4 defects (a); 5 defects (b).

4 Evaluation and Discussion of Proposed Method

4.1 Evaluation

To conduct a subjective evaluation of the proposed method, the expert in visual inspection of ceramic tiles was asked to evaluate the result of artificially generated surface defects. A comparison is done between the original surface defect, presented in Fig. 3a, and the artificial surface defect illustrated in Fig. 7. The proposed method is validated as suitable for its application since the original, and the artificial defects are similar, so one cannot distinguish them. The expert had to classify images with a tile surface defect between real and artificially generated images. In the experiment, there were 10 real and 10 artificial images. The expert had a success rate of 55% for correctly classifying real and 45% for correctly classifying artificial images. Therefore, it is hard to distinguish between real and artificial images (Fig. 12).

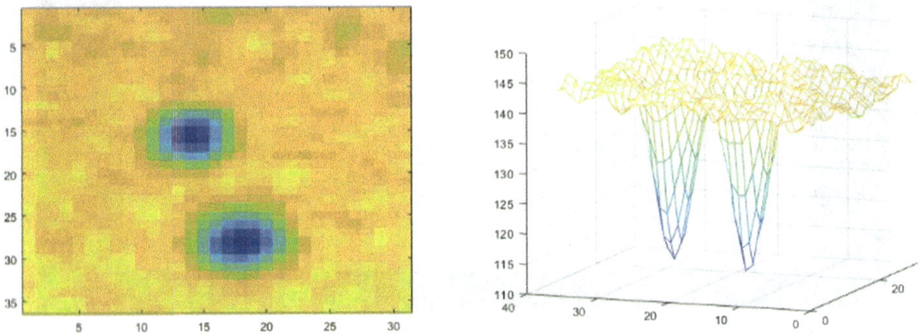

(a) (b)

Fig. 12. Magnified image with real and artificial surface defect: 2D (a); 3D (b).

4.2 Discussion

To the best of our knowledge, there is no similar work in literature that deals with generating artificial tile images. Therefore, it is not possible to compare the proposed method with another one in the literature. Further on, the interactive image generation with the tile surface defects generates a single image with artificially inserted tile surface defects. On the other hand, the automated method generates a large set of images. The proposed interactive method serves as a proof of concept, while the proposed automated method generates the required number of images in the dataset with the tile surface defect. There is a drawback to the proposed method. It can produce only a single type of surface defect. However, the tile surface defect considered in this work is a commonly occurring defect in the ceramic tile industry [4, 5, 9]. In future work, one can model different types of tile surface and edge/corner defects. Finally, the application of the proposed method is in balancing the image dataset required for machine learning algorithms. It is yet unknown how the morel will behave.

5 Conclusion

It is more often to capture an image without defects than with defects in the ceramic tile production line. However, there is a need for a relatively equal number of images with and without the defect to apply machine learning algorithms. Therefore, image augmentation based on artificial defect generation is presented in this work. The Gaussian model for biscuit tile surface defect is presented. The proposed model is used to generate a large set of artificial images with the inserted surface defect. The application of the proposed method is to create two balanced datasets. One dataset has biscuit tile images with no defects, and the other has images with defects. It is done so that the image without defects is used to generate the image with defects. Each image is processed using a thresholding image segmentation algorithm to find the polygon surrounding the tile surface area of the image. Subsequently, the defect is added using the Gaussian model. Captured images of surface defects are compared with artificially created images with defects. The artificial defects are so like the real ones that the expert was not able to distinguish them. The proposed method is not affected by the trend in the uneven lighting environment of the camera acquisition system, and it does not require any image processing of the original biscuit tile image except the image segmentation.

References

1. Zorić, B., Matić, T., Hocenski, Ž.: Classification of biscuit tiles for defect detection using Fourier transform features. ISA Trans. (2021, in press)
2. Matić, T., Aleksi, I., Hocenski, Ž, Kraus, D.: Real-time biscuit tile image segmentation method based on edge detection. ISA Trans. **76**, 246–254 (2018)
3. Hocenski, Ž., Matić, T., Vidović, I.: Technology transfer of computer vision defect detection to ceramic tiles industry. In: Žagar, D., Martinović, G., Rimac Drlje, S., Miličević, K. (eds.) International Conference on Smart Systems and Technologies (SST), pp. 301–305. IEEE, Osijek, Croatia (2016)
4. Hanzaei, S.H., Afshar, A., Barazandeh, F.: Automatic detection and classification of the ceramic tiles' surface defects. Pattern Recogn. **66**, 174–189 (2017)
5. Karimi, M.H., Asemani, D.: Surface defect detection in tiling industries using digital image processing methods: analysis and evaluation. ISA Trans. **53**(3), 834–844 (2014)
6. Shorten, C., Khoshgoftaar, T.M.: A survey on image data augmentation for deep learning. J. Big Data **6**(1), 1–48 (2019). https://doi.org/10.1186/s40537-019-0197-0
7. Mikołajczyk, A., Grochowski, M.: Data augmentation for improving deep learning in image classification problem. In: Hołub, M., Paplicki, P., Stawicki, K. (eds.) International Interdisciplinary PHD Workshop 2018 (IIPHDW 2018), pp. 117–122. IEEE, Świnoujście, Poland (2018)
8. Uvais Qidwai, C.H.C.: Digital Image Processing: An Algorithmic Approach with MATLAB. CRC Press, Boca Raton (2009)
9. Boukouvalas, C., et al.: ASSIST: automatic system for surface inspection and sorting of tiles. J. Mater. Process. Technol. **82**, 179–188 (1998)

The Importance of Employee Education and Information and Communication Tools for Maintenance and Measuring the Performance of Micro-enterprises

Katarina Štavlić[1] (✉), Igor Štavlić[2], and Krešimir Lacković[3]

[1] Polytechnic in Požega, 34000 Požega, Croatia
kstavlic@vup.hr
[2] IGOR, Lifelong Learning Centre, 34000 Požega, Croatia
igorstavlic@gmail.com
[3] University North Center, 48000 Koprivnica, Croatia
klackovic@unin.hr

Abstract. Due to the growing number of micro-enterprises and their contribution to the national economy, there is a need to research the human and IT impact on their performance and maintenance. The aim and purpose of this paper are to emphasise the importance of the successful business of micro-enterprises by presenting the results of empirical research on the presence of continuous education of employees and their use of modern information technology in maintaining and measuring the success of microenterprises. There is a comparison of two surveys, one from 2020 on 121 micro-enterprises and the other from 2016 on a sample of 112 micro-enterprises. The companies included in both surveys are from Slavonia and Baranja and three activities: production, construction and agriculture, fishing and forestry. By applying appropriate statistical and mathematical methods, it was determined that in 2020, compared to 2016, there was an increase in investment in training and education of employees and an increase in monitoring financial performance indicators.

Keywords: Business performance measurement · Micro enterprises · Information and communication tools

1 Introduction

In recent years, micro-enterprises have become increasingly essential stakeholders in the economic system in the Republic of Croatia, contributing to the growth of employment, productivity and total revenues of the national economy. According to the Small Business Development Promotion Act, which complies with Commission Recommendation 2003/361/EC of 6 May 2003, represent natural or legal persons that yearly employ less than ten workers realise a total annual income in the amount equivalent to EUR 2,000,000.00, or have total assets if they are profit taxpayers, i.e. fixed assets if they are

H. Glavaš et al. (Eds.): OTO 2021, LNNS 369, pp. 390–402, 2022.
https://doi.org/10.1007/978-3-030-92851-3_29

income taxpayers in the amount equal to EU 2,000,000.00 [1]. Small and medium-sized enterprises in the Republic of Croatia have been recorded continuous growth and development for the last 20 years. Micro enterprises have a growing share in the total number of companies in the Republic of Croatia; in 2019, compared to 2018, they recorded an increase in the total share of employment, total revenues and exports [2]. Due to their growing contribution, there is a growing need to investigate the impact of employee education and the application of new technologies on maintenance and measuring the success of their business.

Employees are a vital resource for achieving various strategic goals. Satisfied employees are more productive, innovative, loyal and better cooperate in the team, while dissatisfied employees are more passive, inefficient and unmotivated for further professional development. One of the ways employers show their employees that they are valued and essential to the company is by investing in their education. Micro-enterprises in the Republic of Croatia have an average of 3.6 employees, and in a situation of limited human resources, it is necessary to strategically plan the development of employees in the direction of a more successful business. There is no systematic strategic framework for encouraging education for entrepreneurial competencies in Croatia, and micro-entrepreneurs opt for formal, non-formal and informal education.

This paper will analyse continuous investment in the education and training of employees of micro-enterprises, their use of information technology in maintenance and monitoring business performance. The first part of the paper provides a theoretical overview of the importance of measuring business performance and continuing education of employees and their use of new information technologies in the business processes of micro-enterprises. The second part of the paper presents comparative results of empirical research.

2 Business Performance - The Importance of Employee Education and the Application of Information and Communication Tools in Measuring Performance

2.1 Measuring the Performance of Micro-enterprises

Business success can be seen by achieving set goals that can be different - increasing profits, increasing market share, introducing new products, and others. Monitoring the realisation of set goals is possible if the set goals are measurable and achievable. However, the realisation of the set goals does not need to imply successful business. Therefore, each company should determine in its strategic documents the ways of measuring its performance, which can be measured by financial and non-financial indicators or quantitative and qualitative business indicators.

In micro-enterprises, it is necessary to balance owners/entrepreneurs' needs, expectations and interests with the set business goals and accordingly choose the appropriate way to measure business performance. The owners and managers of small businesses are aware of the importance of the existence and monitoring of adequate cash flow for continuous operations and survival. However, many do not appreciate the importance of measuring business performance (seen in the survey results). In addition to cash flow, efficiency and effectiveness are equally crucial for a successful business based on existing resources [3].

Below are some of the selected recommendations of the authors who researched the implementation of a sound performance measurement system in small and medium enterprises [4], and applicable are in micro-enterprises:

- performance indicators should be derived from the strategy,
- performance indicators should provide fast, timely and accurate feedback,
- performance indicators should relate to specific objectives,
- performance indicators should be aimed at improving business,
- performance indicators should be objective, not based on opinion and assessment;
- the performance measurement system should be designed and implemented with the help of external experts,
- performance indicators should use data that is automatically collected whenever possible,
- the performance measurement system should consist of qualitative and quantitative indicators, and others.

It is evident from the above recommendations that their application requires expertise, knowledge and appropriate information technologies, which indicates the need for continuous education of employees in order to understand the purpose and method of measuring business performance.

According to Sikavica et al. [5], business performance indicators can be grouped according to their: nature, the area for which they are set and the approach to measuring organisational performance. The primary division of indicators is into quantitative and qualitative, and given their nature, and there are three primary groups of indicators:

- quantitative value indicators: they are divided into cost indicators (e.g. costs per unit of production, labour costs, employment costs), capital indicators (e.g. fixed assets turnover indicator, return on investment) and income indicators (e.g. sales revenue);
- quantitative physical indicators: these are quantity indicators (e.g. number of products/services sold), quality indicators (e.g. error rate), time indicators (e.g. production time) and frequency indicators (e.g. turnover ratio);
- qualitative indicators: these are non-financial and descriptive indicators and are divided into the target, strategic and programmatic (e.g. customer perceptions, customer satisfaction, employee satisfaction).

In the last 30 years, the field of enterprise performance measurement systems has undergone numerous changes due to the advocacy of some authors to use non-financial indicators in measuring efficiency and effectiveness in addition to traditional financial indicators [3, 6, 7].

Measuring business performance is a complex process in which it is necessary to include several factors that affect business performance. Therefore, entrepreneurs need to choose a set of interrelated key performance indicators whose measurement and monitoring can lead to better management, control, and business improvement. That can be achieved by applying an appropriate performance measurement system. In addition to measurements, it is necessary to compare specific common values to perceive the achieved level of performance better.

2.2 The Importance of Employee Education and the Use of Information Technology in Companies in the Republic of Croatia

Employees are a vital resource for a company's successful business. Drucker [8] states that the most valuable asset of organisations in the 20th Century was production technology and that the most valuable asset of organisations in the 20th Century was the knowledge of workers and their productivity. Therefore, it is crucial in modern business to educate employees to strengthen their knowledge and skills. Pološki Vokić and Grizelj [9] believe that successful world organisations are focused on educating and developing their employees to use their knowledge and skills to improve business, development and growth. Achieving organisational strategies is possible only if employees have the knowledge and skills needed to implement these strategies, so education and development become a strategic imperative of modern organisations. In particular, investing in employee education and development increases value in the value chain, greater efficiency, a better quality of products/services, increased customer/consumer satisfaction, and the like, all of which ultimately lead to tremendous success and competitiveness. To modern organisations, investing in employees by improving their knowledge and skills and using this knowledge and skills pays off many times over, and this has been proven by numerous empirical studies [9].

Croatian micro-entrepreneurs have access to various training sources to improve their knowledge and skills of business management and business performance monitoring with the use of information and communication tools. Formal, non-formal and informal education is available. Due to lower costs and shorter duration, micro-entrepreneurs most often choose specialised courses and training related to work processes to improve their business. However, they need to follow modern technologies and their use and be educated following changes in market requirements and new technologies.

Information technology is changing the ways people live and work, changing the structure and way of doing business of modern businesses worldwide. The use of information technology is a significant segment of every company's business. The Internet and other network technologies enable the connection of business units within the company and the integration of business processes that contribute to more efficient business through collecting, processing, and using information. Information technology is necessary for accurate and timely information storage and its efficient search and use in methods and models that enable quality decision-making. According to Čerić and Varga [10], information technology combines microelectronics, computers, telecommunications and software, which enable the entry, processing and distribution of information. It is one of the vital generic technologies because it penetrates all spheres of economy, science, social and private life and introduces radical changes in them.

Entrepreneurship in the Republic of Croatia began to develop significantly through the Internet, whereby entrepreneurs were given new ways to communicate to facilitate market representation, conduct business, and facilitate internal and external interaction. Continuous positive trends in the application of information and communication technologies in the business of Croatian companies are visible from 2007 to 2019 [11] through the application and access to the Internet, strengthening e-commerce, performing financial and banking operations, owning a website, using administrative services, and others. Croatian companies continued to adapt to the rapid development of information technologies in 2020, and most of them lag behind the world. The percentage of social media use was only 65%, and cloud use was 47% [11]. The lag of micro and small enterprises in the application of information and communication technologies is evident, which may be due to several factors: lack of knowledge about their application, lack of financial resources for the purchase of modern information technologies, low interest in new markets if the company has stable operations, the insufficient workforce in the company to dedicate itself to this business segment or even unsuccessful business and lack of vision and knowledge of the change. Regardless of the reasons, it is necessary to join a new virtual market that offers the opportunity to globalise the business of micro and small enterprises.

3 Research Methodology

Previous scientific research and expert analyzes on micro-enterprises have not fully analyzed and determined the possible factors, causes and reasons for the successful or unsuccessful operation of micro-enterprises. Certainly one of the success factors is the awareness of the internal limitations and external challenges in which they operate and the awareness of their strengths and potentials. This paper focuses on human resources and their education, as well as the use of modern information technologies in maintaining, monitoring and measuring the company's performance. The primary research question is: What is the level presence of continuous education of employees and their use of modern information technology in monitoring the measurement of micro-enterprises business performance?

Some results of the empirical research in 2020 were used for the needs of previous works of the authors (see reference 15), whereby the research methodology was explained in detail. Online survey was aimed at determining the presence of a continuous training of employees and their use of modern information technology in monitoring performance measurement microenterprises. The research included entrepreneurs from the area of Slavonia and Baranja in the fields of agriculture, forestry and fishing, processing and construction. To find the required sample, the database of business entities in the Republic of Croatia in the Digital Chamber was used, which is available on the website of the Croatian Chamber of Commerce. The classification of micro enterprises according to the Small Business Development Act [1] was used, taking into account the quantitative criteria of the number of employees and the amount of income generated.

The research population is micro-enterprises according to the following defined criteria:

- active companies that submitted the annual financial report for 2018
- the number of employees at the end of 2018 (from 0 to 9),
- total annual revenues generated in 2018 (up to HRK 15 million),
- their registered office is in one of the five counties of the region of Slavonia and Baranja, and
- include activities A, C and F according to the National Classification of Activities [12].

2474 micro-companies met these criteria in the region of Slavonia and Baranja and only 445 had a valid e-mail address. One hundred twenty-one entrepreneurs answered the survey, and the obtained answer rate was 27.19%. That is a satisfactory rate because the rate of return of 20% is often set in the world as a marginal rate of return [13]. LimeSurvey software was used for creating and conducting the online survey. Questions related to the presence of continuous education of employees and their use of modern information technology were taken from the survey questionnaire from the research on the success factors of micro-enterprises in the Republic of Croatia (2016) [4]. An online survey was conducted from 27 February 2020 to 15 July 2020. The limitations of the applied methodology are reflected in the more difficult availability of micro-entrepreneurs online contact data, but despite that there were a high percentage of complete responses. The research results on the success factors of micro-enterprises were used for comparing the data with the results from 2016 [4]. Appropriate statistical and mathematical methods for generating and interpreting research results were used.

4 Results and Discussion

Descriptive statistics give the insight into the regional specifics of the number and activity of micro-enterprises and their business performance in five counties. The descriptive statistics of the survey sample and the share of micro-enterprises surveyed by the county are presented in Fig. 1.

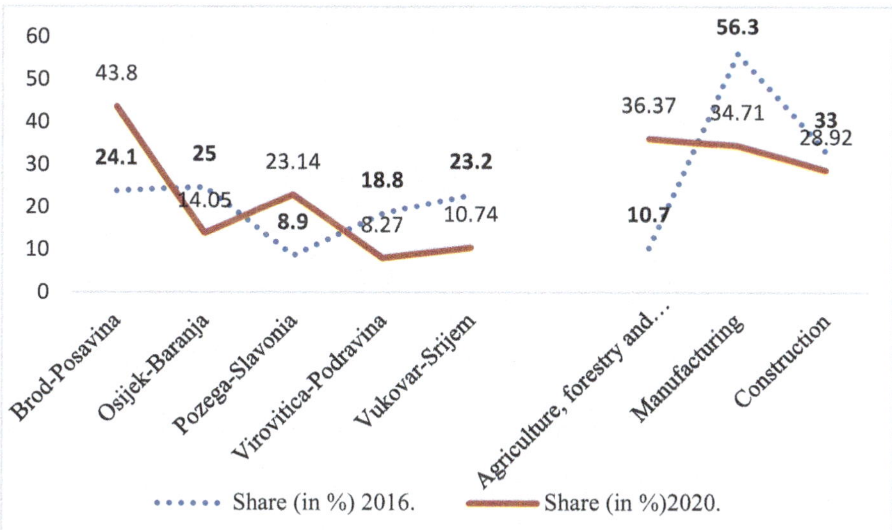

Fig. 1. Descriptive statistics.

There are larger deviations in the distribution of micro-entrepreneurs in 2020 compared with those who responded to the survey in 2016. A higher response was in Brod-Posavina County and from micro-enterprises from the processing industry (in 2016), while in 2020 there is an increase in the number of respondents from agriculture. Due to the insufficient number of micro-enterprises operating at a loss in the sample for 2020, it is not possible to compare all results of the both surveys and perform nonparametric statistical tests to prove the correlation of the given variables with the maintence of the company and business performance.

In order to identify the success factors of micro-enterprises, the questionnaire asked micro-entrepreneurs/owners/managers to assess the importance and impact of the proposed factors on the operations of their enterprises (1-least significant/least impact on success, 5-most important/most significant impact on success). Figure 2 shows the obtained answers shown through the values of the arithmetic mean.

Data in Fig. 2 shows that innovation (arithmetic mean is 4.39) and application of modern technologies in business processes (arithmetic mean is 4.31) are considered the most important factors influencing a company's business success in 2016. In 2020, formal and informal employee education and training (4.13) and application of modern technologies in business processes (4.13) are the most important factors of business success. Modern management methods are considered as the least important factor in both studies.

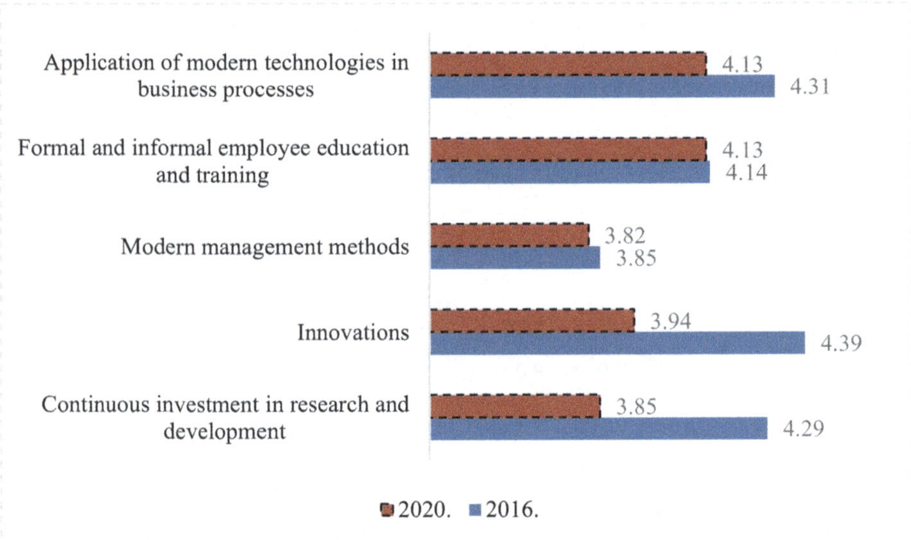

Fig. 2. The importance and influence of individual factors on business operations – comparative analyses.

In order to analyse the potential impact of employee education on monitoring business performance measurement using information and communication tools, the questionnaire asked questions related to areas of investment in the last three years, the application of knowledge and skills by employees, and the level of use of appropriate performance indicators (Fig. 3, Fig. 4, Fig. 5, Fig. 6).

The Fig. 3 presents the answers on the question related to the investment level in the past three years in various activities that influence business performance and maintenance of the company. Respondents were offered activities in which they could invest and rate on a scale of 1 to 5 (a score of 1 meaning that no investment was made at all and a score of 5 meaning that there was a significant investment). Figure 3 shows the result of the comparative analysis presented by frequencies of individual responses, as well as their arithmetic means.

According to the data in the following Fig. 3, it can be seen that the surveyed micro-enterprises, in 2016, invested the most in the purchase of machinery, equipment, software or licenses. Investments follow this in improving organisational or business processes, investments in company reputation and branding, while the lowest investment in product and service design ($\overline{X} = 2.73$) and investment in employee training and development ($\overline{X} = 2.88$). Micro-enterprises in 2020 said that they invested the most in procuring machinery, equipment, programs or licenses. While interesting is that they had the least investments in research and development. From the compared data, it can be seen that there is an increase in the arithmetic mean value in investments in product and service design (16.84%) and employee training and education (6.59%).

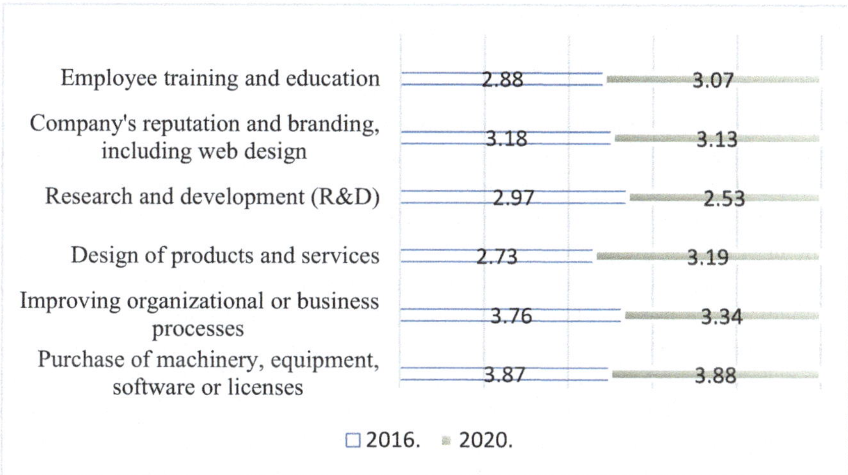

Fig. 3. Types of investments in micro-enterprises – comparative analyses.

For determining the presence of applying the information and communication tools in micro-enterprises as the criterion was used the application and use of knowledge and skills by micro-enterprises employees. The results are shown in Fig. 4. They present the answers to the question relating to human resources and applying employees' different knowledge and skills in their micro-enterprises (1 - never used, 5 - regularly used). The most commonly used skills in 2016 were project management skills, sales skills and computer skills. There are slight differences in 2020 when the sales skills, IT and computer skills and negotiation skills were most commonly used. Compared data shows that the implementation and uses of analytical skills, project management, and knowledge of modern methods for improving business processes have decreased in 2020.

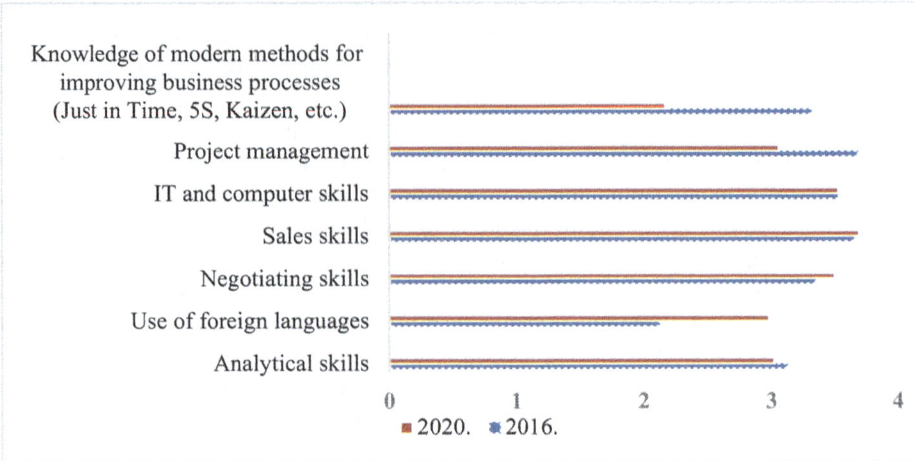

Fig. 4. The application of knowledge and skills by employees of micro-enterprises – comparative analyses.

Surprisingly, there was a decrease in the use of modern methods for improving business processes and analytical skills, even though in the analysed period there was, as described above, an increase in investment in education and training of employees and the purchase of equipment software and licenses.

In the next part of the questionnaire, respondents were asked whether they use specific indicators, tools, and business methods for performance measurement and strategic management. Response results are presented in the following figures.

Figure 5 below represents the comparative analyses of the responses to whether the employees analyse and monitor non-financial performance indicators to measure business performance. The comparative results show a greater focus of micro-enterprises towards monitoring customer satisfaction and complaints, while they are less focused on monitoring indicators related to internal business processes and human resources. Nevertheless, in 2020 compared with 2016, there is an increase in usage of non-financial indicators for measuring employee satisfaction (13.92%), production scrap (146.05%) and product delivery (60.02%). These results indicate an increasing application of knowledge and new technologies in better monitoring and maintenance of business operations in the direction of innovative and sustainable development.

From the results shown in Fig. 6, it is evident that in 2020, compared to 2016, there was an increase in monitoring business performance using liquidity, indebtedness, activity, profitability indicators by 30.41%.

Human resources are the driver and generator of product quality, the application of new technologies, devising a customer-oriented approach and marketing activities, and decision-making considering their financial consequences. According to previous research, the greatest internal strengths and success factors of micro-enterprises are,

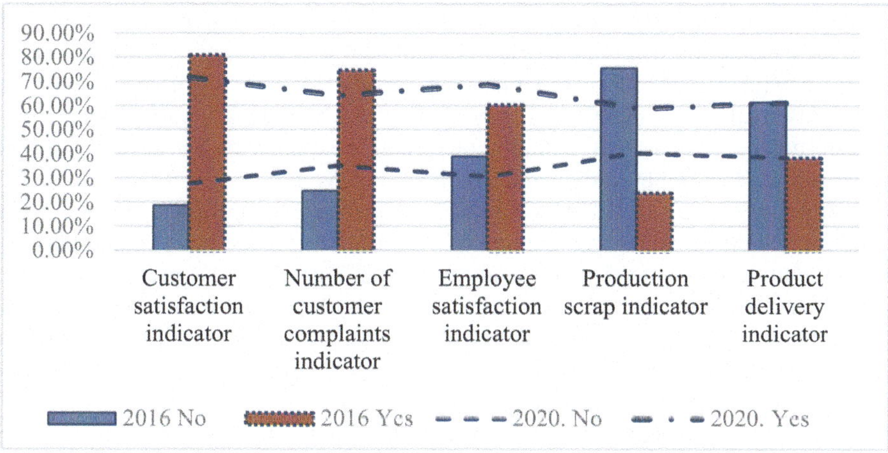

Fig. 5. Use of non-financial performance indicators in micro-enterprises – comparative analyses.

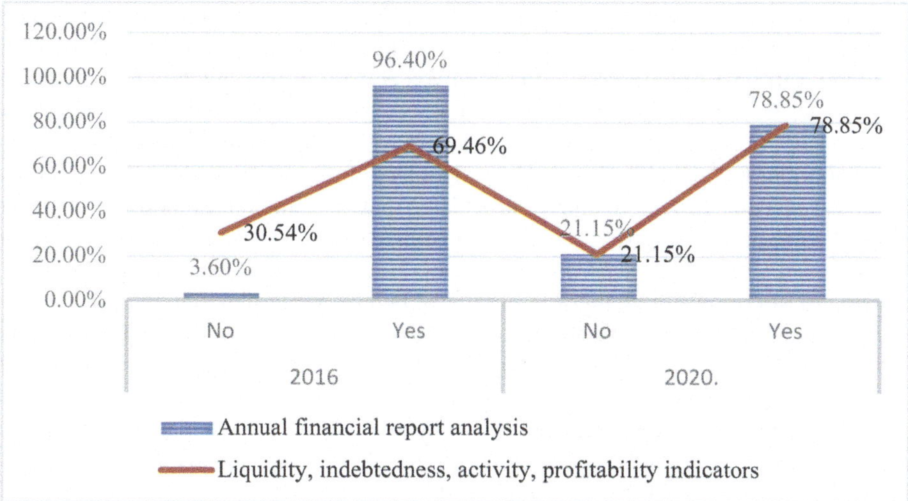

Fig. 6. Use of non-financial performance indicators in micro-enterprises – comparative analyses.

among others, human resources and their knowledge, skills and competencies. Hawkins [16] believes that human capital and human resources can be generated and measured through the level of formal and non-formal education, work experience (emphasis on work in the profession) and specific knowledge and entrepreneurial skills. The results obtained in this paper point to the fact that in the analyzed period there was a reduction in the application of certain knowledge and skills of employees of micro enterprises, which has a possible impact on their business. The impact on business has not been identified, but it represents potential as a topic for new research.

The obtained results regarding the success factors of micro-enterprises partially confirmed the assumption that human resources and their knowledge, skills and competencies are a key factor in the success of micro-enterprises. But these results are similar to the results of Benzing, Chu and Kare [17], according to which the key variables that affect the performance of small and medium enterprises are quality of service and customer service, entrepreneurial honesty and friendliness behavior.

Continuous education of employees and their application of information technologies should result in their greater application of acquired knowledge and skills in business. One of the areas of application of the above is to monitor business performance using non-financial performance indicators. Such business performance monitoring requires strategic planning of micro-entrepreneurs, which is still not sufficiently present in micro-enterprises [18, 19], although it has a growing application which can lead to better business maintenance and performance.

In order to continuously improve IT and other skills and innovate business, it is necessary to act preventively in the education sector by educating students, as potential future entrepreneurs, about e-leadership, knowledge management and the application of IT technologies. The authors Carbonaro [20] and Pereira, Siqueira, Nunes and Dietze [21] write about the need for education on these topics. Given that micro-entrepreneurs consider innovation to be one of the key success factors in this research, it is inevitable to continue to invest continuously in employee education and the application of new technologies in business in order to ensure innovation-based business. This is supported by the results of research by Kay [22] and Porter [23], according to which innovation is an important factor in business success. Holders of all business activities are human resources, especially in micro enterprises. Given that the sample for this study consisted of micro-enterprises from rural areas, the results can be compared with the results of Varghese and Chinnu [24] who point out that continuous training and employee development are important for changes in employee behavior, but that employees in micro-enterprises do not they can dream of high quality training due to financial constraints. They believe that internal training of employees is sustainable in accordance with market trends and requirements.

The limitations of the research in this paper stem from the small territorial and numerical sample and the unavailability of financial data of the surveyed micro enterprises. The limitations of this research are the basis for some future determination of the connection between employee education and the application of new technologies with business success and sustainable business, which should be conducted on a larger territorial and numerical sample. Additional problems in this research arose due to the size of the company, i.e. it is about micro companies in which a small number of employees in the surveyed companies (10% of surveyed companies have no employees, only the owner; 34% of surveyed companies have 1 employee, 15% has 2 employees) and their lack of time and knowledge affected the lower response to filling out surveys and obtaining even more relevant data.

5 Conclusions and Recommendations

Based on the presented research results, several conclusions are possible.The presented research results show that in 2020, compared to 2016, there were changes in the maintenance and way of managing and monitoring the business performance of micro-entrepreneurs in the Republic of Croatia. Namely, in 2020, micro-entrepreneurs believe that formal and informal employee education and training and application of modern technologies in business processes are the most important business success factors. In 2020, compared to 2016, these factors were considered key to business success, which can be linked to increased investment in training and education of employees 2020 by 6.59% compared to 2016. The application of modern technologies in business processes is considered as one of the critical factors of business success in 2020. In both 2016 and 2020, micro-entrepreneurs invested the most in the purchase of machinery, equipment, software and licenses, while the decrease in investment in 2020 was seen in research and development (decrease in the value of the arithmetic mean by 14.81%), improvement of organizational and business processes (decrease in the value of the arithmetic mean by 11.17%) and branding and web design (decrease in the value of the arithmetic mean by 1.57%). Due to that, it can be concluded that increased investment in product and service design, employee education, and increased monitoring of non-financial performance indicators resulted in the more practical application of information and communication tools in the business of micro-enterprises and measuring their performance. Also, the increase in the application of sales skills, the use of foreign languages, negotiation skills and IT and computer skills in 2020 may be a consequence of market changes, and the internationalization and digitalization of business.

The obtained comparative results indicate minor changes in human resource management and the application of new technologies in the business of micro-entrepreneurs. There is no doubt that micro-entrepreneurs, regardless of limited resources, need to mobilise themselves in the direction of training employees for new market trends and the role of new information and communication tools in managing business performance. This directs them to maintain an innovation-oriented business.

References

1. Small Business Development Encouragement Act: National newspapers, Narodne novine no. 29/02, 63/07, 53/12, 56/13, 121/16, Art. 3 (2016). https://www.zakon.hr/z/527/Zakon-o-poticanju-razvoja-malog-gospodarstva. Accessed 15 Sept 2020
2. CEPOR www.cepor.hr: http://www.cepor.hr/wp-content/uploads/2021/01/Izvjesce-2020-HR-web.pdf. Accessed 12 May 2021
3. Laitinen, E.K., Chong, G.: How do small companies measure their performance. Probl. Perspect. Manag. **4**(3), 49–68 (2006). http://businessperspectives.org/journals_free/ppm/2006/PPM_EN_2006_03_Laitinen.pdf. Accessed 11 May 2021
4. Štavlić, K.: Success factors of micro enterprises in the Republic of Croatia, Čimbenici uspješnosti mikro poduzeća u Republici Hrvatskoj, Ph.D. dissertation, Faculty of Economics, Josip Juraj Strossmayer University of Osijek, Croatia (2016)
5. Sikavica, P., Bahtijarević-Šiber, F., Pološki Vokić, N.: Temelji menadžmenta. Školska knjiga, Zagreb (2008)

6. Bititci, U.S., Carrie, A.S., McDevitt, L.: Integrated performance measurement systems: a development guide. Int. J. Oper. Prod. Manage. http://dx.doi.org/https://doi.org/10.1108/01443579710167230 (1997). http://www.emeraldinsight.com/doi/pdfplus/10.1108/01443579710167230. Accessed 11 May 2021
7. Kaplan, R.S., i Norton, D.P.: The balanced scorecard: translating strategy into action. Harv. Bus. Rev. **78**(5), 167–176 (1996)
8. Drucker, P.F.: Management Challenges for the 21st Century. Harper Collins Publishers, New York (1999)
9. Pološki Vokić, N., Grizelj, H.: Obrazovanje i razvoj zaposlenika u hrvatskim organizacijama. Ekonomski pregled **58**(12), 851–880 (2007)
10. Čerić, V., Varga, M.: Informacijska tehnologija u poslovanju. Element, Zagreb (2004)
11. Croatian Bureau of Statistics. www.dzs.hr. Accessed 16 May 2021
12. https://narodne-novine.nn.hr/clanci/sluzbeni/2007_06_58_1870.html. National Occupational Classification – NKD 2007, NN 58/2007. Accessed 05 May 2021
13. Pološki Vokić, N.: Human resources management in large Croatian companies. Econ. Rev. **55**(5–6), 455–478 (2004). https://hrcak.srce.hr/16299. Accessed 16 May 2021
14. Štavlić, K.: Analysis of the innovativity of Croatian micro-entrepreneurs. In: Katalinic, B. (ed.) Proceedings of the 7th International Conference "Vallis Aurea" Focus on: Research & Innovation, Požega, Croatiav, 24–26 September 2020, pp. 603–611. Polytechnic in Pozega, DAAAM International Vienna, Požega, Vienna (2020). ISSN 1847-8204, ISBN 978-3-902734-26-6
15. Stavlic, K.: Comparative analysis of the presence of the new and modern technologies in micro enterprises. In: Katalinic, B. (ed.) Proceedings of the 31st DAAAM International Symposium, Vienna, Austria, pp. 0936–0942, Published by DAAAM International (2020). https://doi.org/10.2507/31st.daaam.proceedings.130. ISBN 978-3-902734-29-7, ISSN 1726-9679
16. Hawkins, W.D.: Exploring success factors of female entrepreneurs in the texas child care market (Order No. 3569855). (2013). ProQuest Dissertations & Theses A&I. (1418478828). http://search.proquest.com/docview/1418478828?accountid=44407. Accessed 06 May 2021
17. Benzing, C., Chu, H.M., Kara, O.: Entrepreneurs in Turkey: a factor analysis of motivations, success factors, and problems. J. Small Business Manage. **47**(1), 58–91 (2009). https://doi.org/10.1111/j.1540-627X.2008.00262.x. http://www.csfstudy.com/uploads/Benzing_2009.pdf. Accessed 20 Apr 2021
18. Greenbank, P.: Micro-business Start-ups: challenging normative decision Making? Mark. Intell. Plan. **4**(4) (2000)
19. Nayak, A., Greenfield, S.: The Use of Management Accounting Information in Managing Micro-Businesses. Finance and the Small Firm. Routledge, London (1994)
20. Carbonaro, A., Breen J.M. (ur): Effective Leadership for Overcoming ICT Challenges in Higher Education. Esmerald Publishing Limited (2021). ISBN 978-1-83982-307-7
21. Pereira, K., Siqueira, S.W.M., Nunes, B.P., Dietze, S.: Linked data in education: a survey and a synthesis of actual research and future challenges. IEEE Trans. Learn. Technol. **11**(3) (2018)
22. Kay, J.: Foundations of Corporate Success: How Business Strategies Add Value. Oxford University Press, New York (2001) http://www.amazon.co.uk/Foundations-Corporate-Success-Business-Strategies/dp/019828781X. Accessed 20 July 2021
23. Porter, M.E.: Competitive Strategy: Techniques for Analyzing Industries and Competitors. The Free Press, New York (1998)
24. Varghese, D., Chinnu, P.B.: HR practices in rural micro enterprises – an empirical investigation. J. Rural Ind. Dev. **7**(2), 01–14 (2019). https://www.researchgate.net/publication/347575493_HR_Practices_in_Rural_Micro_Enterprises_-_An_Empirical_Investigation

Real Fluid Simulation for Determination of Engine Oil Viscosity

Jana Dukić and Dina Jukić(✉)

Faculty of Electrical Engineering, Computer Science and Information Technology Osijek, 31000 Osijek, Croatia
dina.jukic@ferit.hr

Abstract. Viscous characteristics of a Newtonian fluid is described by the dynamic (absolute) and the kinematic viscosity coefficients. The dynamic viscosity coefficient expresses the required applied force per unit area of the one laminar plane of the fluid in order to move it with respect to the adjacent, at an unit velocity. The Kinematic viscosity coefficient is the ratio of dynamic viscosity coefficient and the density of the fluid. This paper presents two simple software procedures for determination of the kinematic viscosity coefficient of real fluids. In addition, the laboratory experimental measurements of some oils' viscosities were done to confirm the proposed procedures, and are presented in the paper as well. The authors intended for the procedures to be used for educational purposes at the college level entry.

Keywords: Engine oil · Real fluid · Simulation · Viscosity

1 Introduction

All fluids are real fluids, as an ideal fluid is an imaginary non-viscous substance which does not display surface tension. In the nature, no such fluid exists, however there are fluids with higher or lower viscosity coefficients.

Viscosity is the measure of the diffusion of momentum due to the microscopic molecular nature of the real fluids [1]. The effect of the momentum is production of resistance to the shearing motions within the fluid. Therefore, this resistance is a frictional force with the origins in interactions among molecules building the fluid. The net frictional force dissipates organized, macroscopic forms of energy, which is kinetic energy in the flow field, and converts it into disorganized, microscopic form of energy, which is heat.

It was shown by Stokes that the frictional force on a sphere moving with constant velocity in a fluid, depends on the sphere's diameter, velocity, and the shear viscosity. The dependence is obtained by solving the equations of motion for the translational flow of fluid around a rigid sphere assuming stick boundary conditions. If slip boundary conditions are assumed, the frictional force is smaller [2]. This is known as the Stokes law, and the frictional force is often called the drag force, to distinguish it from the mechanical friction.

H. Glavaš et al. (Eds.): OTO 2021, LNNS 369, pp. 403–413, 2022.
https://doi.org/10.1007/978-3-030-92851-3_30

This contribution is limited to the Newtonian fluids, which are fluids that do not change their viscous characteristics when a force is applied [3]. All oils are Newtonian fluids. However, viscous characteristics depend on time and temperature. Since the aim of this study is to present procedures for real fluid model or simulation in order to determine viscosity coefficients of engine oils and oils of similar consistencies, the viscous characteristics of analyzed oils weakly depend on both, the time and temperature parameters.

Furthermore, the analyzed fluids are assumed to flow in a laminar motion, which means that the particles of the fluid are grouped and move in regular parallel paths [4]. These paths are not interrupted by any constant flows in the perpendicular direction. That said, the instantaneous deformations of the laminar planes caused by the passing of the object between the planes are small turbulences which do not change the overall direction of the flow of the fluid.

Viscous characteristics of a Newtonian fluid is described by the dynamic (absolute) and the kinematic viscosity coefficients. The dynamic viscosity coefficient expresses the required applied force per unit area of the one laminar plane of the fluid in order to move it with respect to the adjacent, at an unit velocity. The Kinematic viscosity coefficient is the ratio of dynamic viscosity coefficient and the density of the fluid. The usual notation of the engine oil viscosity is of the form: XW-NN, where X indicates the flow rate at zero degrees, whereas NN indicates the kinematic viscosity coefficient at 100 °C.

This paper presents two simple software procedures for determination of the kinematic viscosity coefficient of real fluids, oils. First procedure is a model done in the MATLAB software, and the second procedure is a simulation done in the FlowVision software. Both software procedures were developed in accordance with the principles of the Ostwald viscosimeter [5]. The Ostwald viscosimeter is a manual apparatus for determination of viscosities of solutions of known densities. It is a glass U-tube with two bulbs connected by a capillary. One measures the time of the flow of a solution of unknown viscosity between two markings, and does so again for a standard sample. Solving the equations of motions for the unknown solution and the standard sample gives the ratio of the viscosities equal to the ratio of the densities times the ratio of flow times [6]. That said, manual measurements could result with an error due to individual readings, so recently the Ostwald viscosimeter was modified into digital electro-viscosimeter [2]. Furthermore, [8] proposed the Ansys software solution of the manual Ostwald viscosimeter. However, in this contribution a mathematically simpler approach is proposed. This approach relies on the same base principle of the Ostwald viscosimeter, but does not require the sample solution for comparison. By modelling a closed ball-beaker system, in which the ball is moving thru a liquid, the ball's movement allows for determination of the viscosity of a fluid, due to the principle of relative motion. In addition, the laboratory experimental measurements of oils' viscosities were done to confirm the results of the procedures, and are presented in the paper as well. To eliminate manually measuring the flow times, which could result with an error, an Arduino probe was programmed to measure the time parameter in the laboratory setup. The authors intended for the procedures to be used for educational purposes at the college level entry, but can be used as an initial step of the maintenance of a vehicle as well, as proposed in [9].

2 The Model

To determine the kinematic viscosity coefficient, a simple setup was imagined: A small homogeneous rounded rigid body, the ball, falling in a glass tube filled with fluid. The width of the tube relative to the radii of the ball is such that the effects of the tube walls are neglectable. In a slip model, the motion of the ball is just translatory from the top of the tube to the bottom of the tube. In a stick model, the ball rotates in between lamellar planes of the fluid, changing the direction of the spin, while moving translatory from the top to the bottom of the tube.

In the following analysis of the real fluid, the forces acting on the ball are considered immediately after the ball is set in motion.

2.1 Assuming Slip Boundary Conditions

The forces acting on the ball of the mass m and radii r_b moving in the fluid are: The gravitational force $\vec{F}_G$ directed downwards, the buoyant force $\vec{F}_B$ directed upwards, and the drag force $\vec{F}_D$ directed upwards. The equation of motion considering the magnitudes of the acting forces is then:

$$ma = F_G - F_B - F_D \tag{1}$$

By definition, the gravitational force equals mg, where g is gravitational constant. The buoyant force depends on the density of the fluid ρ_f and the volume of the ball V_b as $\rho_f V_b g$. The Stokes law for the real fluid, assuming slip boundary conditions, gives the drag force on the ball as $6\pi \eta r_b v$, where η is the kinematic viscosity coefficient and v the ball's velocity. Therefore, Eq. (1) becomes:

$$ma = mg - \rho_f V_b g - 6\pi \eta r_b v \tag{2}$$

To simplify the Eq. (2), all constants are written as $\alpha = F_G - F$ and $\beta = 6\pi \eta r_b$. Now, the Eq. (3) is of the form:

$$ma = \alpha - \beta v \tag{3}$$

To obtain the velocity of the ball in dependence of time, the further is applied:

$$m\frac{dv}{dt} = \alpha - \beta v \ \underrightarrow{yields}\ \frac{dv}{dt} = \frac{1}{m}(\alpha - \beta v) \tag{4}$$

Solutions of the Eq. (4) were obtained in MATLAB software using the procedure shown in Fig. 1.

The duration of the procedure was set on $t = 2\,\text{s}$. When the magnitude of the gravitational force equals the sum of the magnitudes of the buoyant and the drag force acting on the ball, the ball stops accelerating and further has constant velocity called *terminal velocity* v_T. While the ball is moving at v_T, it is possible to determine the kinematic viscosity coefficient η. For paraffin oil, common oil and glycerol, the results of the procedure are presented in Table 1, where t_T is the time required to achieve the terminal velocity v_T.

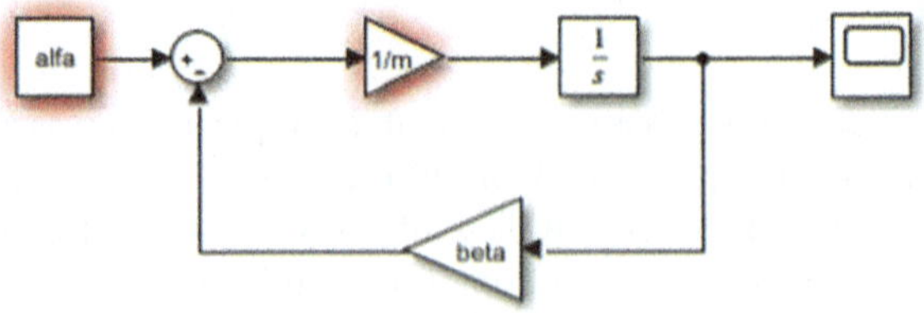

Fig. 1. Block diagram of the ball's motion in the fluid, while assuming slip boundary conditions (slip model)

Table 1. Kinematic viscosity coefficients of real fluids obtained with the slip model

Fluid	v_T (m/s)	t_T (s)	η (Pas)
Paraffin oil	0.650	0.1926	0.13350
Common oil	1.400	0.4712	0.04870
Glycerol	0.074	0.0061	0.98700

2.2 Assuming Stick Boundary Conditions

In a real fluid the ball rotates while overall moving translatory. The rotation of the ball is caused by interaction of the ball with the lamellar planes of the fluid (as if the ball is falling in between window curtains). These interactions cause the changing spin of the ball's rotation as well. As a reaction, the lamellar planes also deform slightly and shortly, which is commonly known as *turbulence*. Therefore, the Stokes law for the real fluid assuming stick boundary conditions, gives the drag force to be:

$$F_D^{'} = 6\pi\eta r_b v + \frac{1}{2}C\rho_b S v^2 \tag{5}$$

where C is a drag coefficient and for a small homogeneous rigid rounded object $C = 0.5$, and S is the surface area of the ball.

Inserting Eq. (5) in the Eq. (1), gives:

$$ma = mg - \rho_f V_b g - 6\pi\eta r_b v + \frac{1}{2}C\rho_b S v^2 \tag{6}$$

Let the $\gamma = \frac{1}{2}C\rho_b S$ as all parameters are constants and it simplifies the writing. Following the same steps done for the slip model, one obtains:

$$\frac{dv}{dt} = \frac{1}{m}(\alpha - \beta v - \gamma v^2) \tag{7}$$

Solutions of the Eq. (7) were obtained in MATLAB software using the procedure shown in Fig. 2.

The duration of the procedure was set on $t = 2\,\text{s}$. The results of the procedure are presented in Table 2.

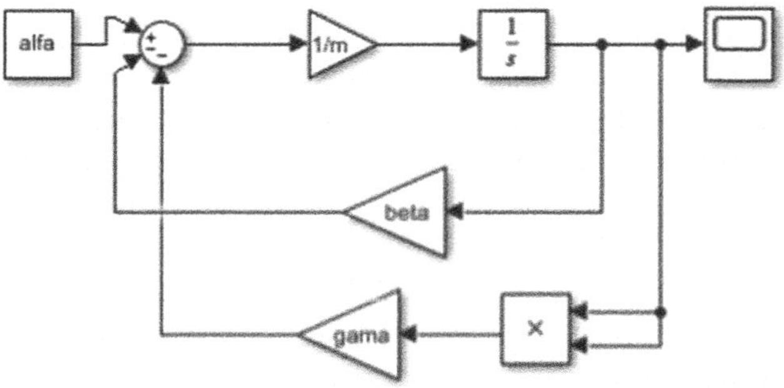

Fig. 2. Block diagram of the ball's motion in the fluid, while assuming stick boundary conditions (stick model)

Table 2. Kinematic viscosity coefficients of real fluids obtained with the stick model

Real fluid	v_T(m/s)	t_T (s)	η (Pas)
Paraffin oil	0.160100	0.4500	0.13400
Common oil	0.248000	0.6500	0.04880
Glycerol	0.006083	0.0745	0.98758

3 The Simulation of the Real Fluid

To simulate the real fluid two softwares were used, the FlowVision and the Solidworks. The SolidWorks software was used to create a domain where the ball is moving. This domain was created using real life parameters of a glass beaker which was later used in a laboratory setup (presented in the Sect. 4). On the other hand, the FlowVision software allows simulation of the real fluid movement where parameters are adjusted for the movement of the observed ball, and separately for fluid behavior. Note that the simulation assumes stick boundary conditions.

First step was to set a physical model by adding a hydrostatic layer and specifying the following parameters: gravity vector, g-point, and stratum (thickness and density). The second step was to create a substance, which is the real fluid, by setting parameters: aggregative state, molar mass, density, viscosity, and specific heat capacity. All the parameters that were set correspond to the real fluids used in the later experiment. The next step was to determine fluid behavior. FlowVision enables description of physical model of motion with the Navier–Stokes theorem which is embedded in the software. The turbulence can be realized by setting in the "Physical processes" window: "Motion" to "Navier–Stokes model" and "Turbulence" to the desired model. The following step refers to the moving ball and adjusting parameters of radii, subdivisions, mass, and translation. In the FlowVision software, the ball can move in any direction of the 3D space, so the remaining settings were set in order to control the direction of the ball's motion. In the end, the ball was positioned in the Cartesian coordinate system at the very top of the previously inserted graduated beaker, as shown in Fig. 3.

Upon setting all mentioned parameters, simulation was started. Results of the simulation for the three real fluids, paraffin, common oil and glycerol, are presented in Table 3. Additionally, the kinematics of the ball was inspected. The time dependencies of the ball's velocities in the three analyzed fluids are shown in Fig. 4, Fig. 5 and Fig. 6. As seen in the figures, the ball is accelerating for the time ds until it reaches definite terminal velocity v_T, which is constant for the rest of the simulation. This indicates that the simulation reflects the physical situation well.

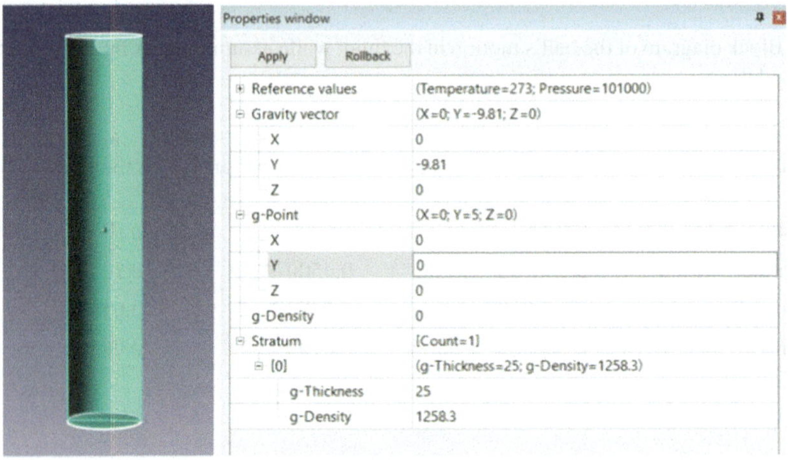

Fig. 3. FlowVision simulation of the ball moving in a glass beaker filled with a real fluid

Table 3. Kinematic viscosity coefficients of real fluids obtained by the simulation

Fluid	Molar mass	Specific heat (J/kg K)	η (Pas)
Paraffin oil	0.4368	2140	0.13400
Common oil	0.8760	535	0.04880
Glycerol	0.0920	4320	0.98660

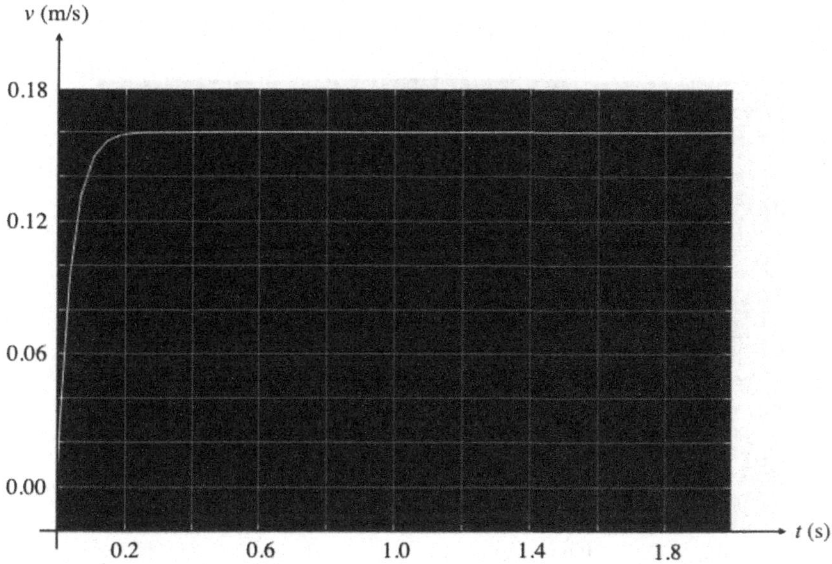

Fig. 4. Time dependence of the velocity of the ball in the paraffin oil

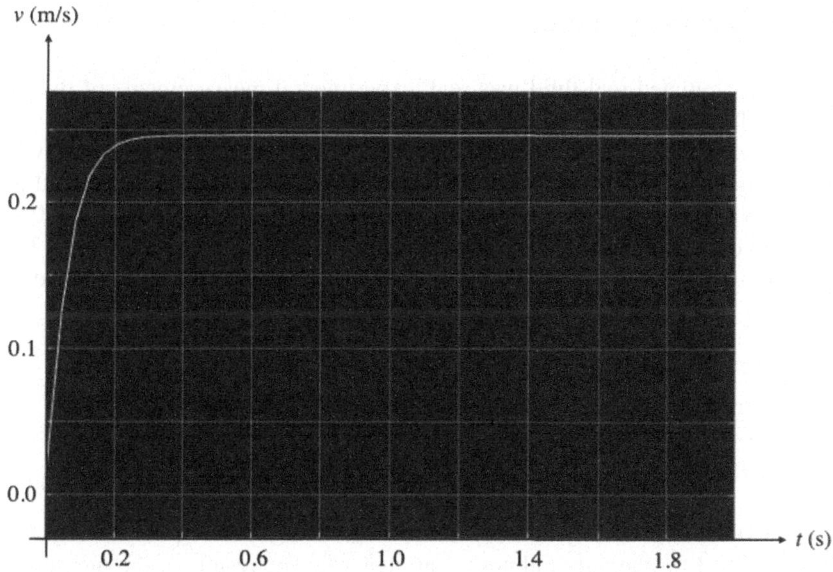

Fig. 5. Time dependence of the velocity of the ball in the common oil

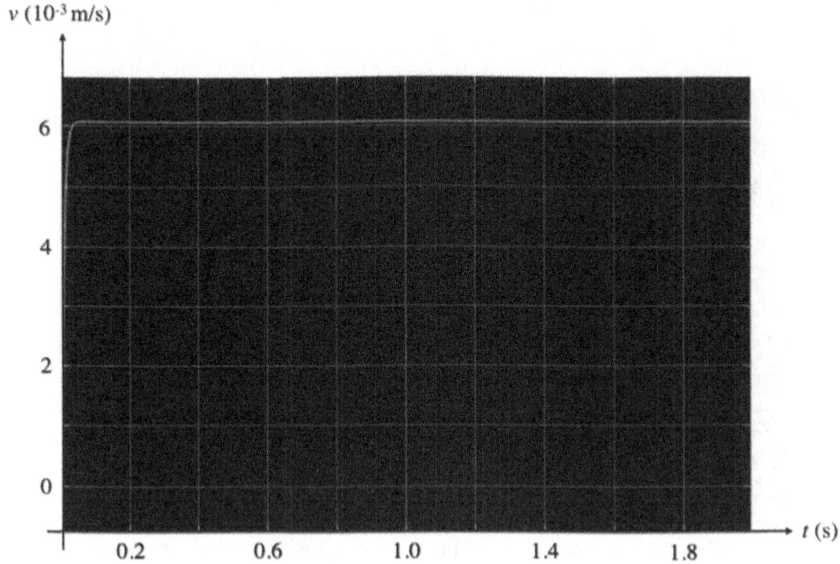

Fig. 6. Time dependence of the velocity of the ball in the glycerol

4 The Laboratory Setup

To test if the simulated real fluid was real, experimental measurements of the kinetic viscosity coefficients were done. The laboratory setup, shown in Fig. 7, followed the simulated situation: a glass beaker of radii $R = 0.024$ m and height H was filled with the real fluid. The time parameter $\mathrm{d}t$ was measured with the Arduino programmed probe. The probe gave the time required for the ball to pass the fraction of the path $\mathrm{d}s = 0.025$ m of the overall distance H. In addition, when the $\mathrm{d}t$ became constant, it indicated that the ball was moving with constant velocity. The terminal velocity v_T was equal to distance over a sum of times. Characteristics of the ball were known and were:

- Density $\rho_b = 1380$ kg/m^3,
- Mass $m = 6.244 \cdot 10^{-4}$ kg,
- Radii $r_b = 4.762 \cdot 10^{-3}$ m,
- Surface area $S = 7.126 \cdot 10^{-5}$ m^2.

Once the terminal velocity was obtained, all parameters were plugged into Eq. (7), which was solved for the only unknown parameter η. The results of the measurements are presented in Table 4.

Fig. 7. The laboratory setup for experimental measurements of kinetic viscosity coefficients

Table 4. Kinematic viscosity coefficients of real fluids measured experimentally

Fluid	v_T (m/s)	$t_T = \sum dt$ (s)	η (Pas)
Paraffin oil	0.105300	0.4512	0.13860
Common oil	0.248000	0.6500	0.04995
Glycerol	0.005905	0.0738	0.98888

5 Comparison of Results Obtained by the Model, the Simulation and Experimentally

Kinetic viscosity coefficients of three real fluids were determined by three means, by modelling, by simulation, and experimentally. As the same setup was used in all three methods, it is justified to make a comparison of the results obtained. Absolute differences among the results are presented in Table 5, whereas relative differences among the results are presented in Table 6, where Δ_{M-S} indicates the difference between the model and the simulation, Δ_{M-E} indicates the difference between the model and the experiment, and the Δ_{S-E} indicates the difference between the simulation and the experiment. For better visualization of the results obtained by the three methods, they are shown in Fig. 8.

As was expected, the model done in the MATLAB Software and the simulation done in the FlowVision gave almost identical results, the exception being kinetic viscosity coefficients of the glycerol, with the absolute difference of, that is relative difference of only 0.1%. Kinetic viscosity coefficients of all tested fluids in laboratory setup are slightly higher than the coefficients obtained by the model and the simulation. This is due to the additional turbulence caused by placing the ball in the fluid. However, their relative differences are all less than 5%, as seen in the Table 6, which is to say that the modeled and the simulated fluid was done real.

Table 5. Absolute comparison of the results obtained by the three methods

Fluid	MATLAB model	FlowVision simulation	Experiment	Relative differences		
	η_M (Pas)	η_S (Pas)	η_E (Pas)	Δ_{M-S}	Δ_{M-E}	Δ_{S-E}
Paraffin oil	0.13400	0.13400	0.13860	0.00000	0.00460	0.00460
Common oil	0.04880	0.04880	0.04995	0.00000	0.00115	0.00115
Glycerol	0.98758	0.98660	0.98888	0.00098	0.00130	0.00130

Table 6. Relative comparison of the results obtained by the three methods

Fluid	MATLAB model	FlowVision simulation	Experiment	Relative differences		
	η_M (Pas)	η_S(Pas)	η_E(Pas)	Δ_{M-S}	Δ_{M-E}	Δ_{S-E}
Paraffin oil	0.13400	0.13400	0.13860	0.00	3.31%	3.31%
Common oil	0.04880	0.04880	0.04995	0.00	2.30%	2.30%
Glycerol	0.98758	0.98660	0.98888	0.10%	0.13%	0.23%

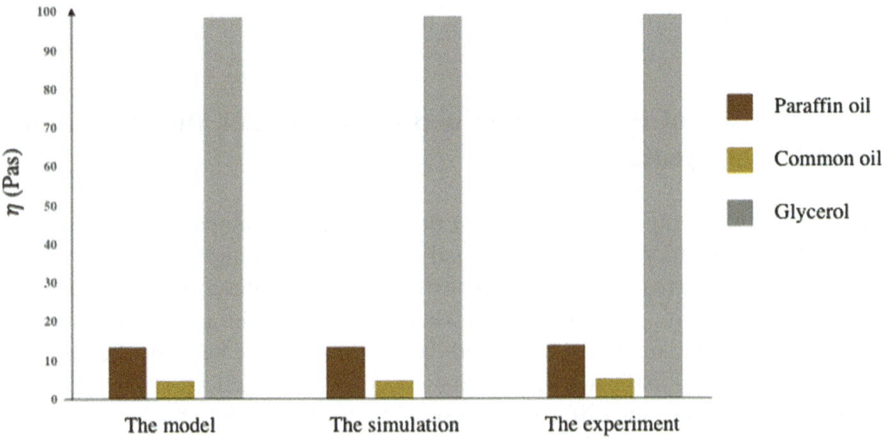

Fig. 8. Results given by the model, simulation and experiment

6 Conclusions

Kinetic viscosity coefficients of three real fluids were determined by three means. In all three methods, the same setup was used: a ball falling in the tube filled with viscous fluid. Firstly, the model of the real fluid was done in the MATLAB software. Secondly, the real fluid was simulated in the FlowVision, upon building the domain in the Solidworks. Finally, the coefficients were determined experimentally.

Relative differences among the results revealed the following: The kinetic viscosity coefficients obtained by the model and by the simulation statistically do not differ, which indicates that both approaches give solid results. Furthermore, the relative difference between the results obtained experimentally and the results obtained by the model, as well as the relative difference between the results obtained experimentally and the results obtained by the simulation, is less than statistically significant (<0.05). Nevertheless, this small difference exists due to the fact that additional turbulence was caused while dropping the ball in the glass beaker while performing the experiment in the laboratory setup, which was not included in the model/simulation.

However, it can be concluded that the simple model or simulation presented in this contribution is well to be used to determine the viscosity coefficients of the real fluids which viscose characteristics do not strongly depend on time.

References

1. Doering, C.R., Gibbon, J.D.: Applied Analysis of the Navier-Stokes Equations. Cambridge University Press, London (1995)
2. Walser, R., Mark, A.E., van Gunsteren, W.F.: On the validity of Stokes' law at the molecular level. Chem. Phys. Lett. **303**(5–6), 583–586 (1999)
3. Papanastasiou, T., Georgiou, G., Alexandrou, A.N.: Viscous Fluid Flow. CRC Press, New York (2021)
4. Fjørtoft, R.: Application of integral theorems in deriving criteria of stability for laminar flows and for the baroclinic circular vortex. Grøndahl and søns boktr, I kommisjon hos Cammermeyers boghandel, Oslo (1950)
5. Janakiraman, D.: Wilhelm Ostwald, the father of physical chemistry. Resonance **17**(5), 454–466 (2012)
6. Langton, N.H., Vaughan, P.: The effects on flow time of tilting an Ostwald viscometer. J. Sci. Instr. **42**(5), 317 (1965)
7. Saxena, I., Kumar, V.: Automatic version of Ostwald viscometer for conductive liquids. Instr. Exp. Tech. **64**(2), 327–330 (2021)
8. Azziz, H.N., Hussein, E.Q., Abdulhameed, M.A.: Analysis of fluid flow states in viscometers in ansys software. In: IOP Conference Series: Materials, Science and Engineering, vol. 1067, no. 1, pp. 012091-012116 (2021)
9. Glavaš, H., Karakašić, M., Kljain, M., Desnica, E.: Essential preventive automobile maintanance during a pandemic. Tehnički vjesnik **6**(28), 2190–2199 (2021)

Winery Monitoring System: A Semi-automated Solutions for Product Quality Maintenance and Production Assistance

Ivana Kovačević, Tomislav Keser(✉), and Robert Miling

Faculty of Electrical Engineering, Computer Science and Information Technology Osijek, J.J. Strossmayer University of Osijek, Kneza Trpimira 2B, 31000 Osijek, Croatia
{ivana.kovacevic,tomislav.keser}@ferit.hr

Abstract. This paper provides a conceptual overview of the existing solutions for monitoring the conditions for wine production and product quality maintenance in the winery environment. The biggest challenge for developing such monitoring system is to measure the essential parameters of wine during and after production. Monitoring systems are designed to allow winemakers monitoring and control the wine transformational processes, in real-time. Also, it enables and helps to make right constructive decision during the period when parameters changing have effect on a production. This article provides a comprehensive summary of the evolution of winery monitoring systems that helps to maintain wine quality during its production. In particular, the paper reviews different IoT-based monitoring systems along with required communication technologies. This paper investigates and contrasts the numerous monitoring systems for the same purpose thus giving comprehensive overview to such system conceptual composition and giving emphasis to the maintenance-product quality relations.

Keywords: Monitoring system · Wine production · Quality maintenance

1 Introduction

Viticulture, like other agriculture activities, is risky by nature. Over time, the conditions of wine production have changed, so there is a growing need for a system that will allow winemakers and oenologists to monitor the entire course of production from grape ripening and harvesting to the final product bottled wine. Many authors have suggested low-cost solutions to this using wireless temperature monitoring systems. However, a correct winemaking monitoring needs to control many other processes and consequently many other parameters in addition to temperature. Effectively managing the complexity of risk enables producers to ensure long-term business success by simultaneously achieving financial goals and conserving human and natural resources. All in all, the conclusion is that monitoring tasks need to be anticipated and intensified. So this initiative is triggered because the good news is that winemakers can adopt methods to reduce liability and risk. Although wine production still counts on a significant and rooted traditional component, integrating Information and Communication Technologies (ICT) in

H. Glavaš et al. (Eds.): OTO 2021, LNNS 369, pp. 414–424, 2022.
https://doi.org/10.1007/978-3-030-92851-3_31

this field can improve productivity, increase competitiveness and facilitate batch processes. The Internet of Things (IoT) paradigm has progressed substantially in the last few years thanks to the continuous advances in many areas, such as miniaturization and electronic technology. Nowadays, the IoT is a flagship to move the real world into a digital one, and it is present in many areas, from houses to cities. The advanced systems integrated into wine production to obtain online monitoring and control must be understood as a means to carry out wine production considering a comprehensive knowledge of the whole process variables in real-time. The development of new technologies leads to a change in traditional agriculture. Network sensor technology helps monitor environmental conditions and essential parameters. In agriculture, a sensor network technology can monitor crop growth, control irrigation, control wine fermentation conditions, and more.

Several implementations of various methods for different sensors have been studied. However, it is not easy to implement winery monitoring sensor nodes for high-quality wine production because wine cellars are mostly places with limited communication between sensor nodes. In addition, there are various technical problems in measuring the parameters inside the barrels in which the fermentation of the wine takes place.

This paper aims to measure various parameters important in wine production and their monitoring to avoid possible errors in production. In this paper, we propose a new architecture of sensor nodes that will enable the maintenance of product quality and production assistance.

2 Related Work

Previous technologies for monitoring fermentation environmental conditions, such as temperature and humidity, have worked by placing measuring sensors on the walls of the winery's cellar. When the parameters are measured, the winery workers manually control the measurement and record the measured data manually. This way of managing the winery is not sustainable because there is a possibility of data loss that can later lead to fatal failures.

Several semi-automated solutions have been developed to measure and monitor various parameters during wine production to avoid this problem. Mandela and his collaborators developed IoT multiplatform networking to monitor and control wineries and vineyards. In his paper [1], he claims that it is mandatory to measure temperature and humidity through every production stage. The proposed solution is divided into four main areas. The first group refers to wireless devices with different sensors for measuring parameters distributed throughout the vineyard whose task is to collect data and interact with other nodes, creating a network and sending data wirelessly via various interfaces. The second area implies assembling sensors information. There is an aggregation point in most locations that helps sensors collect data and send it via a single interface, acting as a gateway. The central server performs the essential control tasks and collects data delivered to it by the deployed devices. Also, the central server is responsible for automatically sending information and in real-time activated alarms defined in advance by the user. The fourth area relates to user interfaces that monitor and control measurement parameters using various mobile devices. ZigBee is used for communication technology. It is a low-cost, low-power wireless technology designed to robustly transfer small amounts of data, making it suitable for critical environments such as winemaking. The architecture of this proposed system is shown in Fig. 1 below.

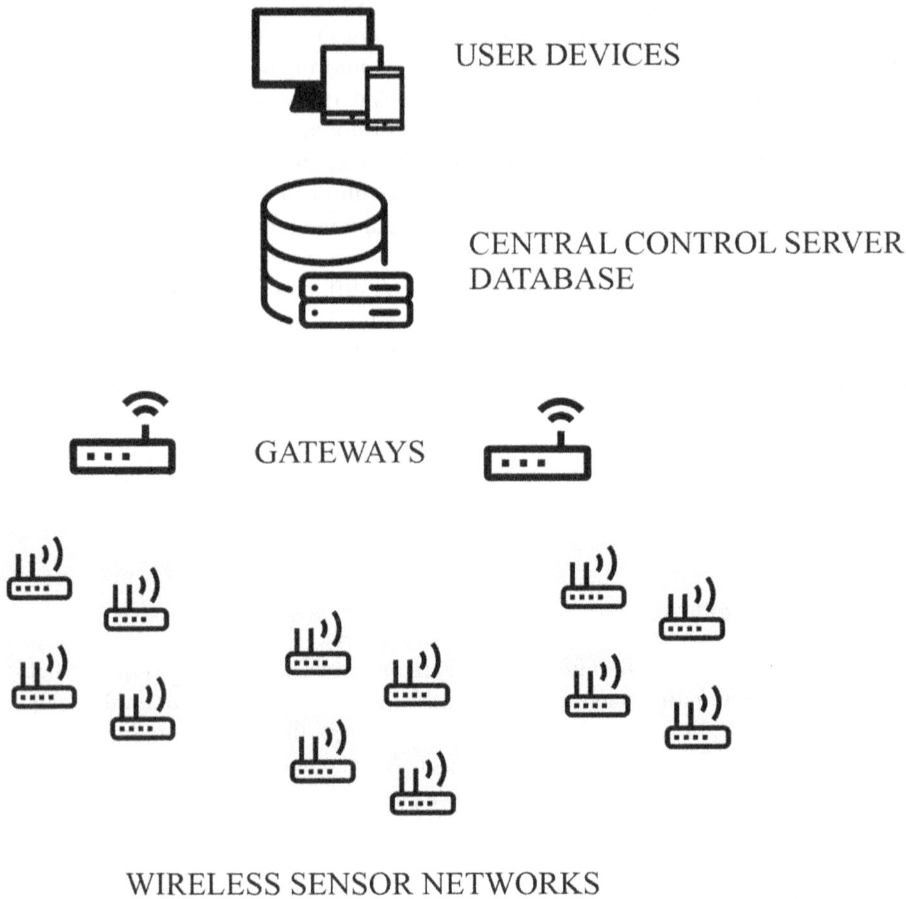

Fig. 1. System architecture [1].

In [2], a real-time monitoring system based on smart cork is proposed to monitor the casks' structural health, the room temperature, the light level and balance between the wine and air inside the barrels. The winery monitoring system consists of several parts. The "smart plug" is used to measure essential parameters, the communication architecture used to transfer data from the barrel to the cloud, and the cloud platform for storing and displaying data obtained by measurements and barrels to end-users in an understandable way. A sensor node for measuring essential parameters is built into the barrel stopper. The system uses sensors to measure humidity, temperature, distance, light level. The ZigBee protocol is used for communication between the sensor and the microcontroller. The "smart plug" integrates the entire system through which information is obtained from each container (barrel). All this information is sent to the cloud, stored, and graphically displayed in real-time to winemakers. Figure 2 below shows the system architecture of the Smart Winery system.

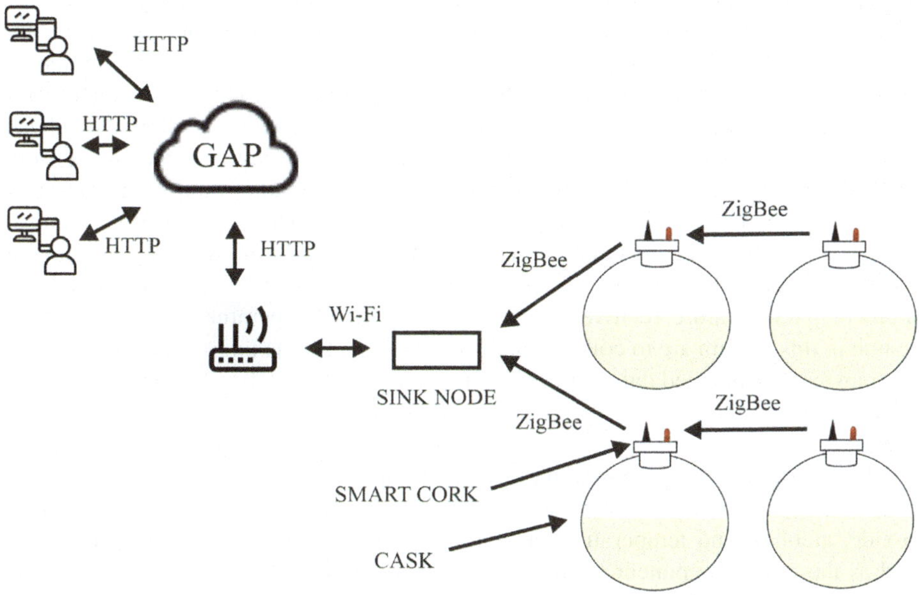

Fig. 2. System architecture [2].

The authors of [3] propose a system that brings progress in researching and monitoring conditions within the winery. As carbon dioxide is denser than air, a higher concentration of carbon dioxide will be found on the winery floor than at the measuring points, mostly placed at a certain height inside the winery. Due to the higher concentration of carbon dioxide on the floor, there is a need for a more sensory system. The system uses IEEE 802.15.4 communication protocol to create a star network. The main task of this system is to implement a network such that it allows data to be routed through multiple paths. The specified network contains a coordinator, a router, and an end node. The system consists of two types of nodes; the first type includes sensors for measuring temperature and humidity, while the second type measures the carbon dioxide concentration and the listed parameters. All nodes use a microcontroller with an integrated RF transceiver. The nodes are implemented according to a low-power architecture that allows complete shutdown of the microcontroller and sensor between measurements. For communication with the carbon dioxide sensor, the microcontroller uses the UART protocol, while for communication with the sensors for measuring temperature and humidity, it uses I2C. Fledge, open-source cloud software used for IoT applications, is used for data transfer. Fledge collects and forwards data from the ZigBee coordinator and forwards it to the database, where data is stored and displayed in a user-friendly way.

The example of a winery management system described in [4] measures the internal and external factors that influence the wine aging process. The development and implementation of a remote distributed system for monitoring parameters that impact the wine aging process are presented. The monitoring process was performed using an RS-485 industrial network. Inside each barrel, sensors are placed to measure the parameters that are important during the aging process of the wine: pH, redox potential, the amount of wine's dissolved oxygen, and temperature. Also, two environmental parameters were measured, temperature and humidity. For external/Internet communication, the hub has a built-in GPRS modem. Wine parameters are measured using two specific sensors. PHEHT sensor measures pH, redox potential, and temperature, and OPTOD sensor measure the amount of released dioxide and temperature. These sensors communicate via the RS-485 bus and are suitable for constant contact with the wine. Environmental conditions in the winery are measured with an HMT310 sensor, which enables the measurement of temperature, relative humidity, condensation temperature. The main tasks of the hub in this system are to collect data from sensors, provide temporary local storage, and transfer the collected data to the IoT remote mySENSE platform. The application's graphical interface allows the system's local configuration and checks the status of the system.

The sensor network proposed in [5] monitors the temperature and relative humidity in the winery and the wine fermentation process by tracking the concentration of carbon dioxide, alcohol, and temperature in the tanks. The PIC16F877A microcontroller is used as the main component of the system. This component processes the data from the sensor, translates it from analog to digital, and sends it to the server via an Ethernet module. A sensor node placed inside the barrel monitors and collects sensor data and sends the specified data via the Internet in real-time. The sensor network consists of a single computer that collects data from all sensors mounted on each wine tank in the winery. The FT232BM component is used to connect the storage server and each sensor module installed in the tank. The microcontroller communicates with the FT232BM part via the UART protocol. The collected data can be analyzed to reveal possible links between environmental conditions in wineries and wine quality itself.

The wireless sensor system for monitoring the winery environment described in [6] is based on indoor air quality testing. The primary factors affecting indoor air quality are temperature, humidity, suspended particles, volatile organic compounds, and various bioaerosols and chemicals. Measuring all of the above parameters that determine air quality over a long period allows winemakers and technologists to review previous environmental conditions based on which, in the future, they can make decisions based on existing information. For the needs of this system, the so-called COTS (commercial off-the-shelf) sensors are integrated into IoT modules for measuring multiple parameters throughout the building and wireless transmission of the same data accessed from a remote location. Temperature, relative humidity, carbon dioxide, volatility of organic compounds, and the concentration of particles in the air are measured during the characterization of the air in the winery. Measurements are taken every 15 min. The wireless node uses the IEEE 802.15.4 communication protocol, enabling two-way wireless communication to transmit small data packets. A microcontroller is used to communicate with the various system components connected to the board. The microcontroller has a built-in RTC (real-time clock), which reads information from the sensor at certain intervals, communicates with the radio module, and controls all electronic components' states. Temperature, relative humidity concentration, and carbon dioxide are measured with a COZIR sensor. UART and I2C are used for communication between microcontrollers and sensors. This system differs from existing solutions in that it can display real-time data in one convenient location.

In [7], the authors present a solution to the fermentation environment monitoring system for different wineries. This system measures the appropriate parameters inside the barrel. The entire monitoring system is installed in a stopper that closes the barrels in the wine ferments to prevent air from entering the barrel. Inside the cork is a built-in sensor node consisting of external wireless communication and various sensors that monitor the inside of the barrel. The sensor node contains temperature sensors that measure temperature depending on the type of alcoholic liquid, sensors for controlling environmental conditions, sensors for measuring CO2 (carbon dioxide), pH (acidity), tartaric acid, and a sensor for measuring acid volatility. The measured data is sent via a wireless transceiver using the IEEE 802.15.4 communication protocol.

Wine Smart Barrel System (Wine-SBS) [8] is a new proposed system architecture and HTTP communication mechanism for monitoring fermenting conditions in the winemaking process of Debina varietal sparkling wine. Wine-SBS consists of a wine fermentation vessel constructed either by stainless steel or a wooden barrel and includes several different interconnected sensors and actuators. The system provides a smart cask, shown in Fig. 3 below, to monitor the wine fermentation of the Debina variety of semi-sparkling wine. The system, installed inside each barrel, continuously monitors aspects of the wine such as color, transparency, pH, pressure, temperature, and alcohol levels. The development makes use of Wi-Fi, which considerably increases the power consumption of the IoT device. All measurements are uploaded to a web server, which enables the monitored data to be accessed. High level architectural diagram of the Smart Barrel is shown in Fig. 3 below.

The WineDuino system proposed in [9] consists of nodes integrated into the silicone bungs, allowing direct monitoring of enological parameters during wine aging in barrels. Nodes collect data and provide a wireless transmission to a coordinator unit, collecting and forwarding data to a remote server. The nodes have been developed and built with Arduino open-source technology. The board node gains data from the sensors at four hourly frequencies, and after readings, data are sent to the coordinator unit through the Xbee S1 radio module. WineDuino System is used to measure temperature and pH for monitoring MLF (Malolactic Fermentation) in wine. The system uses IEEE 802.15.4 networking protocol for communication between node unit to coordinator unit. Transmitting data from the coordinator unit to the remote server database is enabled using Wi-Fi technology. Data are stored in a MySQL database, which many clients can query to provide textual data tables, graphical visualization plots, and export data function in.csv format. The user can view data remotely via a browser or a dedicated app.

In [10], the main objective was to develop and manufacture a novel, cost-effective, automated system for real-time monitoring of wine maceration-fermentation during the production chain. The system includes a liquid sampler, a preparation unit, an analysis unit, a data logger linked to an alarm, and several other modules. The modular analysis unit includes several classic physicochemical sensors (i.e., temperature, pressure, carbon dioxide), a highly effective and robust biosensor, and a sensor capable of detecting

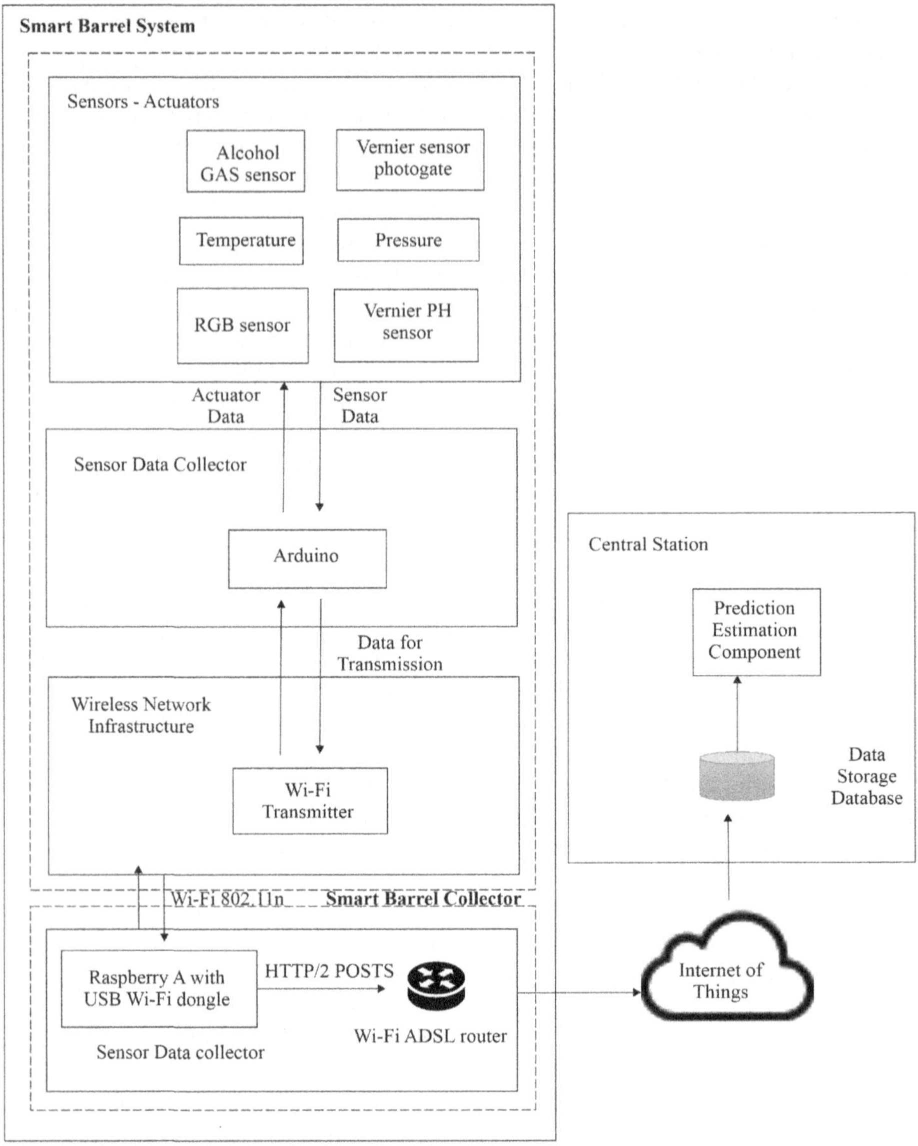

Fig. 3. High level architectural diagram of the smart barrel [8].

chromatic characteristics, polyphenols, and glucose. Several embedded measurement systems are integrated using a LabWindows/CVI software which forms a complex monitor and control system. In the automated monitoring system, alarm capabilities are implemented, signaling when critical parameters are reached. The process parameters can be controlled locally, through a touch-screen or LCD near the fermentation tank, or remotely.

Table 1 below shows a brief overview of existing solutions with measurement parameters and used communication technologies. All these listed parameters are important in wine production.

Table 1. Examples of existing solutions with measurement parameters and communication technologies

Related work	Measuring parameters	Communication technologies
IoT multiplatform networking to monitor and control wineries and vineyards	Temperature, humidity, environmental conditions, density, sugar, color, pH, wine temperature	ZigBee
Smart winery: a real-time monitoring system for structural health and ullage in fino style wine	Room temperature, humidity, light level, balance between wine and air inside the barrel	ZigBee, Wi-Fi
Wireless sensor network with mesh topology for carbon dioxide monitoring in a winery	Carbon dioxide, temperature, humidity	IEEE 802.15.4, UART, I2C, ZigBee
Distributed monitoring system for precision enology of the Tawny Port wine aging process	pH, redox, potential, wine's dissolved oxygen, wine temperature, environment temperature and humidity	RS-232C, RS-485
Sensor module for monitoring wine fermentation process	Temperature, relative humidity, carbon dioxide, alcohol, temperature	Ethernet, UART
Remote monitoring of winery and creamery environments with a wireless sensor system	Temperature, humidity, suspended particles, volatile organic compounds, bioaerosols and chemicals, carbon dioxide	IEEE 802.15.4, UART, I2C
A new method for winery environment monitoring using wireless sensor networks	Temperature, environmental conditions, carbon dioxide, pH, tartaric acid, acid volatility	IEEE 802.15.4
IoT architecture for monitoring wine fermentation process of debina variety semi-sparkling wine	Color, temperature, transparency, pH, pressure, alcohol	HTTP. Wi-Fi
An open-source and low-cost monitoring system for precision enology	Temperature, pH	Xbee S1 radio module, IEEE 802.15.4, Wi-Fi
Remote monitoring and control of winemaking parameters	Temperature, pressure, carbon dioxide, polyphenols, glucose	RS-232C, Wi-Fi

3 A Proposal of a System for Monitoring Qualitative Parameters in the Wine Production Process

In a winemaking process, a new type of characteristics has been discovered throughout the development, consequently paving the way to new winemaking techniques and a variety of new kinds of wines. Because of this and the recent climate-changing factors, there is a rising necessity for a conceptual monitoring system that can survey and track the number of different winemaking parameters making the whole production process somewhat easier for the end-user. The proposed system is a type of wireless monitoring system is based on the Internet-of-Things paradigm. The wine production process can be divided into several stages, as show in Fig. 4.

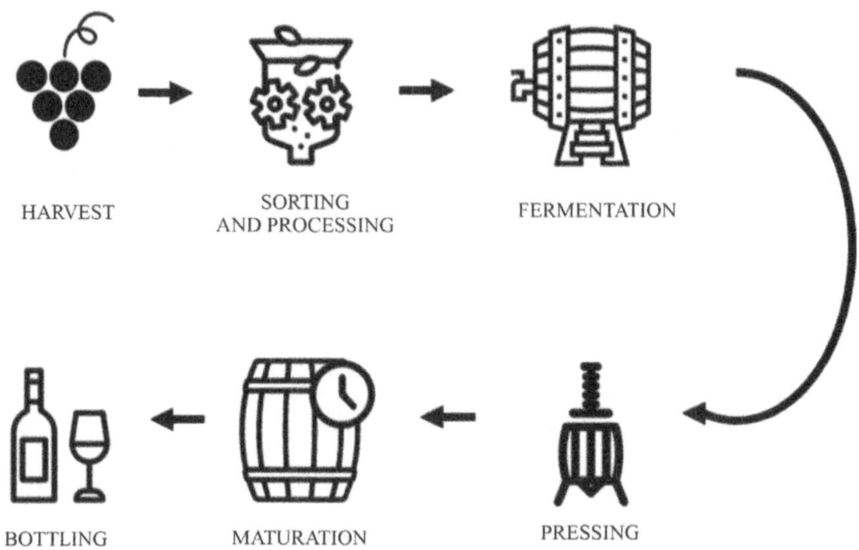

Fig. 4. Winemaking process.

Measurements are carried out before and during harvest, when the must is in barrels, during fermentation, ripening, pouring, forging, and storage. Each phase contains specific measurements according to which further procedures are determined. The proposed system measures the parameters in the must, fermentation, maturation, and storage phase. During other phases, parameters cannot be measured with the equipment we currently have, and these are still performed in specially equipped laboratories. Table 2 shows the measurement parameters in each phase together with the limit values.

As shown in Fig. 5 below, several wireless monitoring nodes will be placed throughout the winery to measure specific parameters and transmit the measured data to a server through a wireless network. This transmitted data will provide critical information for winery workers and oenologists so that the appropriate measures can be made in time to prevent and reduce errors. Unlike previous proposals, which mostly placed measuring sensors on each barrel, our system takes a wine sample from each barrel. During sampling from the barrel, specific parameters (temperature and flow) are measured. During

Table 2. Winemaking phases with measurements parameters

Phase	Parameter	Value
Must	Cellar temperature	10–15 °C
	Cellar humidity	max 80%
	Sulfur	30 g/l for red wine 40 g/l for white wine
	CO_2	
Fermentation	Must temperature	18–20 °C
	Sulfur	30 g/l for red wine 40 g/l for white wine
Maturation	Wine temperature	15 °C
	Sulfur	30 g/l for red wine 40 g/l for white wine
	CO_2	
Storage	Cellar temperature	15–20 °C
	Cellar humidity	50–70%

sampling from the barrel, specific parameters (temperature and flow) are measured. The samples are brought to a central measurement unit where measurements are made. The central measurement unit can measure temperature, sugar, flow, and SOx.

The measured data can be stored locally and on a remote server. End-users have access to measured data via a web interface accessed via various mobile devices and a local user interface located on the central measurement unit. The usage of aerial communications enables data transmitting. Also, the monitoring system contains an expert system that, based on the measured parameters, can identify an error in the wine production process and inform the winery workers of unfavorable conditions to prevent a more significant error in the future.

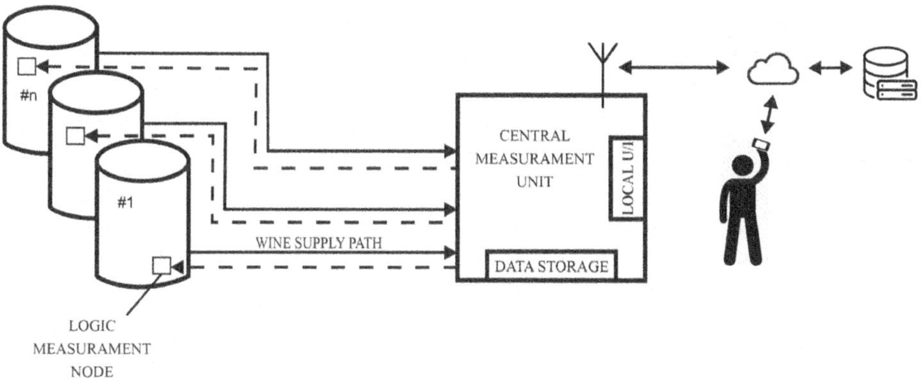

Fig. 5. Principal structural diagram.

4 Conclusion

A new semi-automated solution for product quality maintenance and production assistance enables end-users to control wine production. This system contains many wireless monitoring nodes for specific parameters placed through the winery for transmitting the data they collect to a server through a wireless network. This transmitted data will provide crucial info for the staff in the winery so that the correct measures can be made in time to prevent and dispose of errors.

Acknowledgment. This work results from implementing research activities on the project ESIA – Expert System for Intelligent Agriculture (KK.01.1.1.07.0036). The stated project is funded by European Regional Development Fund.

References

1. Mendela, A., Cendón, B., González, L., Crespo, R., Nevares, I.: IoT multiplatform networking to monitor and control wineries and vineyards. In: International Information Management Corporation, pp. 1–10. Future Network & Mobile Summit, Lisbon, Portugal (2013)
2. Cañete, E., Chen, J., Martín, C., Rubio, B.: Smart winery: a real-time monitoring system for structural health and ullage in fino style wine casks. Sensors **18**(3), 803 (2018)
3. Nelson, J., et al.: Wireless sensor network with mesh topology for carbon dioxide monitoring in a winery. In: IEEE Topical Conference on Wireless Sensors and Sensor Networks (WiSNeT), pp. 30–33, San Diego, USA (2021)
4. Morais, R., Peres, E., Boaventura-Cunha, J., Mendes, J., Cosme, F.: Distributed monitoring system for precision enology of the Tawny Port wine aging process. Comput. Electron. Agric. **145**, 92–104 (2018)
5. Angelkov, D., Martinovska Bande, C.: Sensor module for monitoring wine fermentation process. In: 1st International Conference on Applied Physics, System Science and Computers (APSAC2016), pp. 28–30, Dubrovnik, Croatia (2018)
6. Madrid, N., Boulton, R., Knoesen, A.: Remote monitoring of winery and creamery environments with a wireless sensor system. Build. Environ. **119**, 128–139 (2017)
7. Kim, B.W., Kang, S., Bang H.: A new method for winery environment monitoring using wireless sensor networks. In: 7th International Conference on Computing and Convergence Technology (ICCCT), pp. 98–101, Seoul, South Korea (2012)
8. Tomtsis, D., Kontogiannis, S., Kokkonis, G., Zinas, N.: IoT architecture for monitoring wine fermentation process of debina variety semi-sparkling wine. In: SouthEast European Design Automation, Computer Engineering, Computer Networks and Social Media Conference, pp. 42–47, Kastoria, Greece (2016)
9. Di Gennaro, S.F., Matese, A., Mancin, M., Primicerio, J., Palliotti, A.: An open-source and low-cost monitoring system for precision enology. Sensors **14**(12), 23388–23397 (2014)
10. Epure, P., Ursutiu, D., Samoila, C.: Remote monitoring and control of winemaking parameters. In: 5th Experiment International Conference, pp. 214–219 Madeira, Portugal (2019)

Maintenance of Malted Grain Quality – A Five-Year-Long Study

Kristina Habschied[1(✉)], Alojzije Lalić[2], Kresimir Mastanjevic[1], Daniela Horvat[2], Ivan Abičić[2], and Gordana Šimić[2]

[1] Faculty of Food Technology Osijek, University of J.J. Strossmayer, F. Kuhača 18, 31000 Osijek, Croatia
khabschi@ptfos.hr.com

[2] Agricultural Institute Osijek, Južno predgrađe 17, 31000 Osijek, Croatia

Abstract. To ensure the demanded quality for malting barley, each season a number of quality indicators have to be determined and evaluated. Depending on the variety, location and growing conditions (precipitation, temperature, soil type, and agro-technical measures), the achieved quality indicators have to fulfil a certain theoretical value to set the variety into an end-purpose group. End purpose can be designated as feed or malting. However, based on their intended end-use and in respect to their characteristics, barley varieties in Croatia can also be classified as dual, namely malting-feed varieties. Most important parameters that have to be analyzed every year include total protein content, soluble proteins, moisture, malt extract, extract difference, friability, viscosity, Kolbach index, wort color, wort pH. For malting barley varieties, these values have to be within the set and recommended values, taken from different literature sources. The aim of this work was to present a five-year long, continuous examination of all mentioned quality control indicators. All indicators are prone to variations due to the variety, location and growing conditions, so a five-year continuous study can provide mean values for the tested indicators important for growers, maltsters and brewers. Expressed variations can be noted in protein content. Namely, 2018 resulted with the lowest protein content in comparison to other years, for almost all varieties.

Keywords: Malting · Barley · Quality maintenance

1 Introduction

As one of the earliest cultivated plants for human and animal nutrition, barley was also recognized as a potential raw material for fermentation. Abreast with the cultivation it was utilised as a source of fermentable extract for the production of beer and spirits [1]. So far, over 300.000 barley varieties are categorized as two-row and six-row [2]. Barley is now used not only as a feed material but its varietal properties are being carefully selected to be utilized in malting, brewing and distilling. Quality requests set by masters and the brewing and distilling industries in general, require just the opposite to the requests of the food and feed industries [1]. Efforts of the Institute of Agriculture Osijek

H. Glavaš et al. (Eds.): OTO 2021, LNNS 369, pp. 425–434, 2022.
https://doi.org/10.1007/978-3-030-92851-3_32

in creating new varieties that conform with the malting industry requests are continuous. They actively developed new varieties of spring and winter barley intended for brewing and whisky malt (WM) production [1].

The brewing industry prefers the two-row spring varieties [3, 4]. Some varieties are referred to as standards (Vanessa, Tiffany, Casanova, etc.) and serve for monitoring the impact of various agro-technical and environmental conditions on the crop (climate specifics of the season, different cultivation methods, location changes, etc.) as well as for orientation when introducing new varieties in the assortment [1]. Plant breeding lasts for 10–15 years from selection to the point where the variety can be put on the market [5, 6]. Barleys appropriate for malting, after the varietal recognition takes place, are then tested for their technological indicators preferably during a statistically relevant period of 3–5 years [1]. Agronomic quality indicators, as the micromalting process and the determination of technologically relevant indicators are carried out in the renowned institutes specialized for malting and brewing. Countries with developed malting and brewing industry provide annual reports for all the standard and other most abundant cultivars in the breeding [1]. The aim of this paper is to present which basic quality indicators regarding malt quality are necessary to insure the malt quality maintenance over the years, preferably during a statistically relevant period.

2 Materials and Methods

2.1 Sampling and Analysis

Samples. Fifteen barley varieties were grown in the fields at location Osijek. The experiment was conducted in randomized block designs (RCBD) with six replications; plot size was 7.56 m^2. Sampling (5 kg per sample) was performed on cleaned and processed barley grains (according to the EBC 3.3.1. method) [7] and the barley samples were kept refrigerated in sterile dry containers. Grain samples (5 kg per sample) were collected as untreated and conditioned grain, scaled and packed into in double-walled paper bags (1 kg). Until micromalting, the material was stored in sterile dry containers for two months in a dry and cool place (18–20 °C) to overcome the post-harvest grain dormancy.

Micromalting. Micro-malting of barley was carried out in a laboratory micromalting plant, Joe White Malting Systems (Pty. Limited East Melbourne, Victoria, Australia; Automatic Micro Malt Unit, 10 kg capacity). The malting scheme is presented in Table 1. Four samples of 2 × 1 kg of each barley variety were malted after which the kiln took place, as shown in Table 1. Degermination of dry malt was conducted after the kilning. After the micromalting and kilning, 500 g of each sample was weighed and stored in paper bags for one month in order for moisture content to stabilize prior analysis.

Barley and Malt Quality Analysis. Malt quality indicators were determined according to the European Brewing Convention (EBC) [7] and Middle European Brewing Analysis Commission (MEBAK) [8] methods (Table 2).

Table 1. The applied micro malting scheme of barley samples

Steeping	Micro malting stage	Air flow (%)	T (°C)	t (h)
	Immersion steeping	-	14	5
	Dry steeping	100	14	19
	Immersion steeping	-	14	4
	Dry steeping	100	14	20
	Immersion steeping	-	14	2
	Dry steeping	moisture correction to 44.5% by spraying with water		
Germination	Germination parameters	75	14	96
	Turning over time: 2 Number of rotation during turn over: 3			
Kilning	First phase	100	50	16
	Second phase	100	60	1
	Third phase	90	70	1
	Fourth phase	40	80	1

Table 2. Used the Middle European Brewing Analysis Commission (MEBAK) and European Brewery Convention (EBC) methods for the analysis of malt

		Method	
	Unit	MEBAK®	EBC®
Micromalting		2.5.3.1	
Moisture	%		3.2/4.2
Fine extract content	% d. wt.	4.1.4.2.2	
Extract difference	%	4.1.4.2.10	
Friability			4.15
Total proteins			4.3.1
Soluble proteins	mg/L		4.9.1
Kolbach index	%		
Wort colour	EBC	4.1.4.2.8.2	
Viscosity	mPas. 8.6%e	4.1.4.4.2	
pH	-	4.1.4.2.7	

3 Results and Discussion

Tables 2, 3, 4, 5, 6 show the results of continuous quality monitoring over a five-year period (2016–2020). Most important quality indicators that insure the malt quality have been monitored continuously for five years. In order to maintain the malt quality many of these quality indicators have to be within the recommended limits for malting barley. For example, protein content is one of the most important quality indicators and the maximum amount of protein content in malt should not go over 11.5% [9]. Over the years a significant difference can be noted between the same variety. In 2016 maximum value for protein content was determined in variety 13 (12.5%) while majority of varieties did not go over 11%. In 2017, all tested varieties showed an increase in protein content, where 13 again showed the highest value, 13.65%. In contrast, in 2018, almost all varieties showed a decrease in protein content.

Malt moisture was also decreased in 2018, with the highest value being 5.72% for sample 15. For all other years the moisture was above 6%. Maximal value was noted for sample 1 in 2020, 7.30%. Preferably values range 4.5–5% [10].

Ideally, malt extract values should be around 80% [11], and this is pretty much the case for all samples during our five-year study.

Extract difference, should be <2.5% [8], which is not the case for most of the samples. None of the varieties showed consistence in this property through the years, and some showed values between the recommendations, depending on the year. In 2016 and 2017 none of the varieties showed satisfactory values for this indicator. In 2018 three variety showed extract difference values below 2.5% (samples 2,3,6) and in 2020 only one (11), while in 2019 eleven varieties were below 2.5% for malt extract difference.

Soluble proteins should not go over 4.4–5.6% [11], which is mostly the case for all years. There is only one sample (3) in 2018 that showed an increased vale for tis quality indicator.

Malt friability should be >80% [8]. In 2016, 2017 and 2020 this was not achieved in any of the samples. 2019 was noted as the best year for this property since 13 out of 15 varieties showed appropriate results.

Wort viscosity should be below 1.8 mPa × s [8]. This property especially follows the literature recommendations in 2019 while in all other years a certain discrepancies appear between the varieties.

Kolbach index values conform the recommended values (<42% [11]) in 2016, 2017 and 2020. In 2018 and 2019 some of the varieties show increased values for this property.

Wort color is for the most years satisfactory (3–5 EBC [8]) for all varieties, but shows increased values in 2019 for sample 10 with the maximal value of 8.3 EBC.

pH values should be between 5.9 and 6.1 [8]. Only in 2020 the values were within the recommendations (Table 7).

Table 3. Malt quality indicators for 2016

Year/Sample 2016	Malt protein (%)	Moisture (%)	Malt extract (%)	Extract difference (%)	Soluble protein (%)	Malt friability (%)	Malt viscosity (mPas)	Kolbach index (%)	Wort color (EBC units)	Wort pH
1	10.65	6.41	79.53	6.74	3.9	46.56	1.944	36.62	4.16	5.52
2	10.55	6.84	79.42	5.13	4.1	57.28	1.69	38.86	4.14	5.48
3	10.45	6.59	79.8	6.55	3.95	46.4	1.911	37.8	3.31	5.48
4	11.4	6.36	79.07	5.41	4.05	48.96	2.019	35.53	4.07	5.5
5	10.7	6.54	80.06	4.61	4.1	52.7	1.685	38.32	4.09	5.64
6	10.35	6.79	78.86	6.75	3.9	45.94	1.899	37.68	3.62	5.6
7	10.7	6.76	80.17	4.11	4.15	59.34	1.657	38.79	4.34	5.62
8	11	6.7	80.42	4.32	4.1	64.66	1.524	37.27	4.12	5.55
9	10.5	6.6	80.74	3.7	4.25	72.94	1.566	40.48	3.49	5.53
10	10.98	6.53	78.92	6.74	4.25	42.82	1.85	38.69	4.24	5.61
11	10.9	6.55	78.84	6.03	3.9	41.64	1.965	35.78	4.55	5.56
12	11.05	6.34	79.15	6.63	4	44.1	1.934	36.2	4.4	5.6
13	12.5	6.8	78.66	3.39	4.55	54.74	1.774	36.4	3.63	5.63
14	10.45	7.03	79.6	3.61	4.25	56.26	1.657	40.67	3.51	5.56
15	10.6	6.96	78.81	5.23	4.1	52.46	1.713	38.68	3.97	5.61

Values are means of triplicate

Table 4. Malt quality indicators for 2017

Year/Sample 2017	Malt protein (%)	Moisture (%)	Malt extract (%)	Extract difference (%)	Soluble protein (%)	Malt friability (%)	Malt viscosity (mPas)	Kolbach index (%)	Wort color (EBC units)	Wort pH
1	13.40	6.68	80.61	4.93	4.45	51.18	1.754	33.21	3.48	5.60
2	10.90	6.84	83.99	3.12	4.20	68.52	1.816	38.53	3.84	5.65
3	12.70	6.22	79.14	9.64	3.85	21.18	2.366	30.31	3.62	5.62
4	12.20	6.52	80.77	7.37	4.10	38.02	2.381	33.61	3.59	5.57
5	11.50	6.74	80.77	7.18	3.90	34.24	2.115	33.91	3.52	5.61
6	10.95	6.73	80.76	5.55	3.90	50.36	2.184	35.62	3.54	5.63
7	13.30	6.20	79.43	5.51	4.70	38.94	2.256	35.34	3.52	5.56
8	13.50	6.73	80.76	6.06	4.40	31.76	2.299	32.59	3.41	5.62
9	11.90	6.81	80.84	6.78	4.15	32.58	2.663	34.87	3.62	5.61
10	13.40	6.90	80.41	4.12	4.40	43.48	1.902	32.84	3.67	5.54
11	13.10	6.73	81.80	6.28	4.30	35.84	2.51	32.82	3.65	5.59
12	12.10	6.87	80.79	8.21	4.05	24.12	2.601	33.47	3.84	5.61
13	13.65	6.84	79.42	5.85	4.70	40.9	2.184	34.43	3.68	5.58
14	12.00	6.38	80.53	4.31	4.10	54.46	1.879	34.17	3.93	5.59
15	12.95	6.90	81.24	7.00	4.25	20.02	2.924	32.82	4.01	5.66

Values are means of triplicate.

Table 5. Malt quality indicators for 2018

Year/Sample number	Malt protein (%)	Moisture (%)	Malt extract (%)	Extract difference (%)	Soluble protein (%)	Malt friability (%)	Wort viscosity (mPas)	Kolbach Indeks (%)	Wort color (EBC units)	Wort pH
2018										
1	11.05	5.40	80.21	7.98	4.40	45.3	1.948	39.82	4.56	5.53
2	10.25	5.44	81.78	1.12	4.60	80.3	1.589	44.88	4.45	5.63
3	10.15	5.36	82.83	2.15	5.75	79.64	1.588	56.65	4.72	5.66
4	9.90	5.08	80.92	5.26	4.00	51.8	2.206	40.40	3.98	5.63
5	9.75	5.25	84.26	2.87	4.25	84.72	1.596	43.59	4.54	5.59
6	9.70	5.27	83.25	1.64	4.40	90.34	1.534	45.36	4.07	5.64
7	10.4	5.49	81.01	7.00	4.20	52.46	1.952	40.38	3.86	5.58
8	9.30	5.03	79.96	5.75	4.00	52.04	2.342	43.01	4.15	5.62
9	9.95	5.44	81.47	6.40	4.25	48.78	2.528	42.71	4.61	5.65
10	9.70	5.35	80.88	4.57	4.30	63.56	2.190	44.33	3.78	5.67
11	9.60	5.33	81.57	5.69	4.15	51.26	2.038	43.23	3.94	5.61
12	9.75	5.44	79.84	7.08	4.10	49.1	1.920	42.05	4.17	5.57
13	10.15	5.66	82.30	2.76	4.50	72.94	1.660	44.33	4.61	5.61
14	9.55	5.34	82.50	3.87	4.20	80.42	1.591	43.98	4.27	5.64
15	9.35	5.72	84.21	3.90	4.10	76.86	1.712	43.85	3.89	5.65

Values are mears of triplicate.

Table 6. Malt quality indicators for 2019

Year/Sample 2019	Malt protein (%)	Moisture (%)	Malt ex-tract (%)	Extract differ-ence (%)	Soluble protein (%)	Malt friabil-ity (%)	Malt viscosity (mPas)	Kolbach index (%)	Wort color (EBC units)	Wort pH
1	10.8	6.34	81.42	2.06	4.60	86.30	1.505	42.59	3.90	5.71
2	10.45	5.43	80.44	3.86	4.35	94.20	1.563	41.63	5.71	5.90
3	10.25	6.67	82.57	0.62	4.55	94.04	1.433	44.39	5.68	5.64
4	10.75	6.55	83.49	1.45	4.50	94.04	1.477	41.86	5.16	5.64
5	11.15	5.95	81.35	2.15	4.60	78.92	1.515	41.26	6.07	5.74
6	11.15	6.35	83.08	0.93	4.50	97.58	1.417	40.36	4.59	5.75
7	10.50	5.89	83.45	1.03	4.50	91.24	1.498	42.86	4.59	5.76
8	10.20	6.20	82.52	2.37	4.40	86.50	1.52	43.14	3.95	5.77
9	11.75	6.57	81.95	1.45	4.55	94.20	1.432	38.72	4.21	5.66
10	10.25	5.59	81.21	2.04	4.35	84.98	1.51	42.44	8.34	5.72
11	10.75	6.13	82.04	1.44	4.75	87.00	1.487	44.19	5.41	5.71
12	11.60	6.15	81.44	2.77	4.45	81.00	1.519	38.36	4.72	5.79
13	11.70	6.40	82.51	1.45	4.70	94.08	1.432	40.17	4.72	5.66
14	11.60	6.63	85.55	1.35	4.70	94.60	1.487	40.52	4.45	5.68
15	11.20	6.20	81.80	2.78	4.45	70.32	1.553	39.73	3.71	5.75

Values are means of triplicate.

Table 7. Malt quality indicators for 2020

Year/Sample 2020	Malt protein (%)	Moisture (%)	Malt extract (%)	Extract difference (%)	Soluble protein (%)	Malt friability (%)	Malt viscosity (mPas)	Kolbach index (%)	Wort color (EBC units)	Wort pH
1	12.85	7.30	80.07	5.06	4.30	36.16	1.674	35.53	3.56	5.85
2	10.85	6.39	79.61	5.21	4.00	51.38	1.899	36.87	3.63	5.94
3	10.85	6.59	81.25	6.47	4.05	45.72	2.007	37.33	3.41	5.96
4	10.90	6.57	80.71	4.93	4.25	51.22	1.976	38.99	3.56	5.96
5	12.30	6.68	80.92	3.71	4.35	61.22	1.593	35.37	3.63	5.89
6	12.75	6.51	81.27	4.83	4.40	48.34	1.664	34.51	3.10	5.97
7	11.15	6.28	79.71	7.34	4.00	35.14	2.058	35.87	3.01	5.97
8	11.80	6.95	80.56	5.66	4.30	42.20	1.952	36.44	2.98	5.89
9	11.50	6.76	79.76	7.68	3.90	24.60	2.075	33.91	3.24	5.91
10	12.55	6.99	80.60	4.64	4.25	38.36	1.998	33.86	3.07	5.92
11	10.90	6.35	82.25	1.86	4.10	73.44	1.552	37.61	3.96	5.95
12	11.50	6.72	81.37	3.71	4.05	46.90	1.969	35.22	4.08	6.00
13	12.00	6.67	81.43	4.22	4.35	55.54	1.748	36.25	3.69	6.00
14	11.10	6.69	81.76	3.41	3.85	59.30	1.654	34.68	4.72	6.04
15	11.80	6.56	80.19	5.13	4.10	44.78	2.000	34.75	3.68	5.98

Values are means of triplicate.

4 Conclusions

Continuous quality monitoring is an important parameter of quality maintenance. Barley, and other cereals as well, is currently exposed to dramatic climatic changes which influence the quality of final products, malt being one of them. Some of the variations in malt quality parameters can be easily noted during this period of five years. The quality of malt is very important for maltsters, but also affects the market price. Quality monitoring of malt means that continuous efforts should be implemented every season in order to get an insight in main quality properties and achieve a better yield and quality by predicting the weather influence and soil needs for fertilization in a sense of agrotechnical measures.

References

1. Habschied, K., August, H., Mastanjević, K., Šimić, G., Lalic, A., Krstanović, V.: Technological quality of new Croatian barley varieties intended for whisky malt production. Curr. Res. Nutr. Food Sci. J. **7**, 783–790 (2019)
2. Trimindis, A.: Single malt whisky is produced from three main ingredients, barley, water and yeast. Why Barley? www.singlemaltlodge.com. Accessed 23 Aug 2021
3. DeClerck, J.: Barley. In: A Textbook of Brewing, vol. 1. Chapman and Hall, London (1957)
4. Dillon, G.: How Scotch Whisky is Made #1: Barley. https://flaviar.com/blog/barley-in-the-whisky-making-process. Accessed 21 Aug 2021
5. Thomason, W.E., Phillips, S.B.: Methods to evaluate wheat cultivar testing environments and improve cultivar selection protocols. Field Crops Res. **99**(2–3), 87–95 (2006)
6. Patterson, H.D., Silvey, V.: Statutory and recommended list trials of crop varieties in the United Kingdom. J. R. Soc. Stat. Soc. Ser. A (Gen.) **143**(3), 219–252 (1980)
7. European Brewery Convention (EBC), Analytica, 5th edn. Fachverlag Hans Carl, D-Nürnberg, Germany (1998)
8. Middle European Brewing Analysis Commission (MEBAK), 3rd edn. Selbstverlag der MEBAK, Freising-Weihenstephan, Germany (1997)
9. Narziss, L.: Die Technologie der Malzbereitung. 7th edn. Stuttgart, Germany (1999)
10. Russel, I.: Whisky technology, production and marketing. Chapter 2: Dolan C.S.T. - Malt whiskies: Raw materials and processing, pp. 27–73 (Institute and Guild of Brewing- Typical malt distilling malt specification IGB (2001)
11. Barley and malt quality specifications. American Malting Barley Association, Inc., Milwaukee, Wisconsin. http://www.ambainc.org/ni/Guidelines.pdf. Accessed 21 Aug 2021

Maintaining the Quality Control of Beer

Kristina Habschied(✉) and Krešimir Mastanjević

Faculty of Food Technology Osijek, University of J.J. Strossmayer, F. Kuhača 18, 31 000 Osijek, Croatia
khabschi@ptfos.hr

Abstract. Today's brewing industries rely on continuous and reliable analytical methods in order to control the quality of produced beer, but also to respect the legislation related to beer quality. To achieve the wanted quality control it is important to bear in mind several things, such as size and age of the brewing industry, process automation and skills of the workers, availability of laboratory and technical evaluation of the laboratory equipment (calibration and general maintenance). Brewing process quality control involves a lot of indicators, but some of the most important are gravity of wort and beer, wort and beer pH, air injection into wort, yeast cell counts, yeast culture and propagation control, yeast viability, dissolved oxygen, carbon dioxide, sulfur dioxide, aerobic and anaerobic bacteria counts, microbiological culturing media, turbidity, color, alcohol, foam quality, sensory evaluation. Considering that beer can be packaged into cans, glass bottles, kegs or PET, many factors should be taken into an account. For example, PET packaging has the shortest storage time, 4 months in comparison to other kinds of packaging, such as kegs, cans or glass bottles. Sensory quality control evaluates the consumers' perception and often beer bought in store is not as the one tested in the brewery. However, the input of efforts and resources involved in beer quality control is continuous and important for the overall quality of beer.

Keywords: Beer · Quality insurance · Brewing industry

1 Introduction

1.1 The Importance of Beer Quality Maintenance

"Making great beer is hard. Making the same great beer every day is harder. Ensuring that your great beer is still great after packaging is harder still. And the hardest thing of all is ensuring that every customer gets the great beer you brewed, whether they drink it in your pub or another establishment, or from a glass, bottle or can…Our collective success is built on quality."

-Ken Grossman, Sierra Nevada Brewing Co.

The production of beer is a well known process. In modern brewing industry, due to the closed system, almost all errors are condensed to a minimum. However, certain mistakes can by-pass even the best maintenance and operational systems [1]. In a world on many choices of beer, uniformity is still the key to success. This means that the

H. Glavaš et al. (Eds.): OTO 2021, LNNS 369, pp. 435–446, 2022.
https://doi.org/10.1007/978-3-030-92851-3_33

consumers rely on certain beer quality characteristic for a chosen beer brand. They want consistency.

Even though quality control analysis seems standard and available in every brewery, this is not the case. In order to insure the consistent quality, breweries have to rely on many analytical methods, that are at times costly (chemicals, instruments) or require a skilled worker to conduct the analysis and interpret the results.

Quality control, its assurance and maintenance are important for several reasons: interest of other brewing companies and stakeholders; and quality tools can be utilized in solving current (potential) problems or improving unit operations [2]. Also, the aim of every corporation, in this case the brewery, is to enhance quality management as a part of the effort to:

- increase profits,
- improve the quality of both products and operations,
- prove environmental responsible behavior.

A quality management program can be implemented via several different systems [2]:

- Quality Management System (QMS) ISO 9001/2;
- Environmental Management System (EMS) ISO 14001;
- Health and Safety Management System (HSMS) OHSAS 18001:2007;
- Food Safety Management Systems (FSMS) ISO 22000:2005;
- Hazard Analysis Critical Control Points (HACCP).

There is a distinction between quality control and quality assurance. Namely, quality control monitors products to uncover any shortcomings or defects in order to allow the management to release or deny the product into sales. Quality assurance, on the other hand, attempts to improve and stabilize production (and associated processes) to avoid, or at least minimize, issues which led to the defect(s) in the first place.

Quality assurance assures the quality of the product (Beer) by systematic monitoring of raw materials, in process product, and end product.

Both should be implemented in: malting and brewing processes.

2 Quality Indicators

Quality maintenance must be identified in every brewery in order to the quality requirements whose standards, specifications and procedures are set by the brewing industry [1] and must follow the country's legislation. They mostly regard safety and quality upon production, packaging and selling [1].

2.1 Usual Quality Control Check-Points

Quality control and assurance should also be implemented in malting industry and usually they:

cover the barley check for suitability for malting and
prevent dead or unfit barley from entering the process.
Some of the typical barley and malt quality analysis involves [1]:

In barley

- color and odor
- % moisture and water sensitivity
- nitrogen content (total nitrogen) and protein
- % foreign seeds and materials
- absence of damage and fungal growth
- germinative capacity and germination energy
- barley type and variety
- barley
- kernel size

In malt

- time
- % kernel moisture
- kernel size
- temperature and flowrate of water and air
- color of malt
- protein, starch and enzymes levels
- sensory evaluation.

Quality maintenance in the brewery depends on several factors [1, 3]**:**

- size and age
- process complexity
- degree of automation
- styles of beer
- workforce size and degree of skill
- available in-house laboratory analyses – physical-chemical, microbiological
- instrument calibration and maintenance.

2.2 Monitoring of Common Quality Properties

There are several aspects of beer quality that should be monitored regularly: foam stability, flavor (in)stability, colloidal stability, microbiological stability.

Foam stability is one of the visual properties that appeal the consumer. Breweries are prone to provide a stabile and retentive foam head as it is one of the main indicators of beer freshness and quality [4]. Rich head of foam (see Fig. 1) is a property of certain types of beer (lager, pilsner, and wheat beer among others), and some do not appreciate foam in their glass, and instead they enjoy in the lacy pattern at the bottom of the finished beer [5]. Lacing or cling can be described as the adhesion of beer foam that occurs to the side of the glass during beer consumption. This is typical for Belgian beers [6]. According to [5], foam quality involves properties such as stability, retention, viscosity, whiteness, bubble size, density. Since the physics behind the foam formation and retention is so complex, many methods have been developed and are currently in use in order to quantify the quality of beer foam [7–23].

a) b)

Fig. 1. Foam stability a) and gushing b) (author's archive).

Flavor (in)stability is the major backbone of the overall perception of beer quality. Achieving the flavor stability for a longer time is a challenging process for the brewers. Due to the variety of beer styles, it is difficult to predict the exact changes in each style, but some general observations can be made based on the empirical data. For example, lagers and strong ales will display different flavor changes during storage. However, at first bitterness will decline, resulting in perceivably harsher flavor. Secondly, a decline in fruity/estery and floral notes occurs and some beers may develop a ribes (blackcurrant buds, tomcat urine) aroma. Others tend to develop a wet paper or cardboard character. Bready, sweet, toffee-like, honey, earthy, straw, hay, woody, winey and sherry-like notes can emerge as well [24–27]. Other important hurdles in delivering a good quality flavored beer are, as for foam stability, packaging, transport and storage. After packaging, the producer is rarely in control of the transporting (temperature fluctuations, agitation) and storage conditions (temperature fluctuations and time) [28]. A radical idea was proposed by Torline et al. [29] who suggested that the aged character should be maximized in beer before it leaves the brewery, on the basis that no further flavor change will occur. Flavor instability can occur in practically every step of the production, from malt to beer.

Colloidal stability is one of the important properties of beer. It refers to a non-biological haze in beer. Clarity is one of the cherished characteristics of beers such as lagers and pilsners. Beer is considered clear and brilliant if no haze forms when chilled to 4 °C or lower [30, 31]. Haziness is often considered as an error in production and as a possible health risk, so many consumers avoid hazy beers. However, some German and

Belgium wheat (white) beers are desirably hazy and the expansion of craft and organic beers on the market pushed the limits of appreciated haziness.

Fig. 2. Hazy beer (author's archive).

The lack of filtration and chemical stabilization in such beers, result in hazy beers (see Fig. 2). However, colloidal stability of beer is still the most important factor in beer quality [32] since colloidal particles significantly affect storage time, and influence its appearance as well. Nowadays, colloidal stabilization includes one or more unit operations that result in colloidal stability of beer. By definition, beer is considered to be colloidal stable if it can be stored for several months at 25 °C without exhibiting any changes in composition or other properties; in other words, beer has to be able to remain clear without any signs of precipitation. The level of colloidal stability depends on the desired storage time and temperature after packaging [32].

Ensuring the microbiological stability in beer is not as difficult as ensuring these other types of stability. As a naturally protected process (ethanol formation, low pH, antiseptic action of hop acids, low nutrient level, low oxygen concentration and carbonation), beer fermentation is not as susceptible to microbiological spoilage as one may think. There are no known human pathogens found in beer, but this should not encourage producers to omit any of the sanitation operations. There are many microorganisms that can contaminate the production so the control of microbial contaminants is an important part in ensuring consistently uniform and high quality beer [28]. This is especially important for craft breweries and homebrewers who omit the filtration and pasteurization. There are many tools available for detection, determination and quantification of microbial contaminates. To avoid or at least minimize the damage, quality control should be done in every step of the production. Recommended cleaning regimes and testing procedures should be followed. However, the best microbiological control is primarily related with [28]:

- a well-designed and maintained plant,
- application of cleaning-in-place,
- effective detergents and sterilants,
- strict microbiological monitoring.

Microorganisms should be completely absent or present in very low numbers in [28]:

- raw materials
- beer
- finished product
- strategic surfaces of process machinery (filler heads, pipes, pipelines, etc.).

Constant awareness of threats, simple hygiene and good production practice are still the most effective methods in reducing the microbiological risk.

Many associations all over the world prescribe, develop and recommend methods for beer analysis [3], and some promote brewing, especially craft entrepreneurship.

- American Society of Brewing Chemists (ASBC)
- American Homebrewers Association (AHA)
- American Malting Barley Association, Inc. (AMBA)
- Brewers Association (BA)
- Beer Institute (BI)
- Master Brewers Association of the Americas (MBAA)
- European Brewing Convention (EBC)
- Middle European Brewing Analysis Commission (MEBAK).

They all have a common interest in making the brewing quality maintenance more accessible and simple for all brewers.

The most popular in Europe are MEBAK and EBC [33, 34]. The flowing text describes some of the most important physical-chemical, microbiological, sensory and sanitation indicators of quality maintenance.

Important physical-chemical quality analysis that indicate good quality maintenance [33–36].

Protein – beer proteins play an important role in foam quality and retention. They contribute to the mouthfeel, but are also responsible for haze and undesirable bitterness in beer. Methods determine the protein content of beer (% by weight) by the Kjeldahl method, by combustion, and by spectrophotometer.

Specific gravity of wort and beer – this is the ratio of the density of a substance to that of a standard substance.

Apparent extract – apparent extract is a measure of the solids dissolved in a fermenting liquid without correction for ethanol content.

Real extract – determined from the residue in the distilling flask after determination of alcohol. It corresponds to a method used for alcohol determination.

Wort and beer pH – even the slight shifts in pH values can be detected and impair the beer quality. Every beer style has its optimal pH, and this is why it should be monitored regularly.

Air injection into wort – air, or better to say oxygen form the air can act detrimental to beer causing off-flavors by oxidation. However, air is injected into wort in order to promote the yeast cell growth. Aeration is very important in the wort stage but can ruin the finished beer.

Dissolved oxygen – is usually quantified by colorimetric determination in beer where oxygen reacts with reduced indigo carmine (disodium indigo disulfonate). It is suitable for use in pale beers containing up to 2 mg/L dissolved oxygen and is recommended for calibration of dissolved oxygen analyzers.

Carbon dioxide – is a by-product of fermentation and dissolved CO_2 can be found in tanks, bottles, and cans. It is dependent on establishment by agitation of partial gas pressures in the headspace above beer in a container at a particular temperature. It is very hard to determine and quantify gaseous compounds, but these methods are sufficiently precise for the control of unit operations in breweries.

Sulfur dioxide – can be determined using the *p*-rosaniline method, Monier-Williams method, and to a limited extent, flow injection analysis, pulse polarography and ion chromatography.

Turbidity – beer clarity is an important property. Turbidity can be present due to the processing factors, as a result of aging or due to the mishandling (chilling) of the packaged product. Formazin turbidity standards have proved to provide a reference scale. This method allows the reporting of beer turbidity in formazin turbidity units.

Bitterness – the European Brewery Convention uses the EBC Bitterness Units as a uniform unit for expressing the bitter flavor of beer. There are several methods prescribed by different associations but the BU method (prescribed by ASBC) gives satisfactory similar results for bitter flavor, regardless of whether the beer was made with fresh or old hops. The IAAs (iso-α-acids) of beer brewed with old or poorly stored hops, and with certain special hop extracts, can be significantly lower than the BU figure.

Color – beer color is an essential component of the overall sensory perception of the product. Color is commonly quantified by using a spectrophotomer in order to determine the absorbance or transmittance of the sample.

Alcohol – can be determined in beer or distillate volumetrically and gravimetrically, refractometrically, by gas chromatography, low alcohol concentrations by an enzymatic method; alcohol and original extract content by using a relationship of absorbance at near-infrared wavelengths.

Viscosity – viscosity of wort and beer can indicate the content and degradation state of contributory factors, such as β-glucan, derived from malt and wort. Commonly this is enzymatic method.

Foam Quality – there are many described methods for foam stability and quality quantification, but the rate of foam collapse by the Sigma value method (modified Carlsberg method) and the foam flashing method are often used in the industry.

Important microbiological quality analysis that indicate good quality maintenance [33–35]. Here are several indicators of microbiological quality of beer.

Yeast cell counts
Yeast culture and propagation control
Yeast viability
Aerobic and anaerobic bacteria counts
Microbiological culturing media

Sensory Analysis. Beer flavor and aroma are very complex especially if we consider that it originates from raw materials used for brewing: water, barley malt, hops and yeast. Standard beer consists of about 94% water, 1–2% residual sugars, 4% ethanol and 0.1% of various flavor compounds. 0.1% seems like a minimal amount but these compounds are responsible for the unique flavor of beer. More than 1000 different flavor compounds have been identified in beer originating directly from the raw materials. Some compounds form during processing, lagering, maturation or packaging. Evolution, changes of these compounds happens during storage so the beer that has left the brewery is newer the same as the one on the consumer's table [37]. Beer aroma is related to chemical volatile compounds of the barley malt (thermal treatment during malting), hops and yeast metabolism (development of beer during fermentation and aging).However, several different volatile compounds, divided in five groups, can affect the final flavor quality of beer [38]:

- from ingredients, such as barley malt and hops,
- from roasting malt and boiling wort,
- as yeast metabolism by-products during fermentation,
- from microorganism contamination,
- from inappropriate storage conditions, such as oxygen and sunlight exposure.

The volatile compounds affect beer's organoleptic profile and are composed mainly of aliphatic and aromatic alcohols, esters, organic acids, aldehyde, carbonyl compounds, and terpenic substances. Although the raw materials are the same to majority of beer styles, some aromas and flavors are inherent for the traditionally produced beers. Many of the sensory characteristics are related to yeast metabolism during fermentation and aging. There are many reports about the (bio)chemistry of beers' sensory properties, especially regarding the composition and structure of volatile compounds. They usually vary depending on the brewing process, yeast strain, raw-material (wheat vs. barley malt), etc. [39–50].

Sanitation [51]. Often the most important quality management can be done by simple proper sanitation in the brewing industry. Stainless steel (grade 304 or 316) [52] is the best material for brewing equipment, it is easy to clean, highly resistant to corrosion, and it is relatively ductile. Special attention should be payed for cleaning the welded conjunctions due to possible bacteria build up in them. Cleaning can be performed in different ways:

- sterilization - destroys any form of life
- disinfection - kill the microorganisms that are of concern
- sanitization - lowering the number of contaminants to the acceptable level
- cleaning - removing the visible and larger scale dirt [53].

High temperatures and the addition of hops (antibacterial agent) during brewing process inhibit the possible bacterial contamination. Low pH values, low oxygen and nutrients on one hand and concentration of ethanol on the other act preventative to contaminants. Cleaning in Place (CIP) is a common practice for the industry. COP (Cleaning Out of Place), is sometimes used too due to the easier visual inspection. However, it does take up a lot more time. Proper cleaning requires the following steps:

- wetting of the equipment
- reaction between the soil and the water/chemical used
- removal of both
- prevention of reoccurrence [51].

Disinfection can prevent microbial regrowth and some disinfectants used in industry include peracetic acid, hydrogen peroxide, and chlorine [53]. An important issues that should be addressed is the use of environmentally and economically friendly cleaning agents.

3 Conclusions

Maintaining a good quality product is very hard, especially with a medium such as beer. The fluctuations in raw material quality and process parameters make it hard to deliver a standardized and uniform beer to the consumer. Packaging, transport and storage make this job even harder since many variables such as temperature fluctuations, oxygen and storage time significantly influence the beer quality. Therefore, continuous and dedicated quality control by applying the required manufacturing and laboratory practice, should result in quality beer on the tables of the consumers. Breweries should implement a good quality policy, including procedures for managing the quality of its brands and raw materials, packaging materials and final products. The best way to implement quality is to acknowledge some of the QMS (Quality Management System) or FSMS (Food Safety Management System). The implementation of ISO 9001 standard and the HACCP standard should result in quality improvement. Outsourcing of food safety and quality experts, and regular monitoring of customers satisfaction are key factors in insuring the quality maintenance during beer production process.

References

1. Brewing Quality Control and Quality Management. https://www.brewer-world.com/brewing-quality-control-and-quality-management/. Accessed 19 Aug 2021
2. Vrellas, C.G., Tsiotras, G.D.: Quality management in the global brewing industry. Int. J. Qual. Reliab. Manag. **32**(1), 42–52 (2015)
3. Amad, M.J.: Aspects of Brewing Quality Control. https://cdn.ymaws.com/www.acil.org/resource/resmgr/annual_meeting/2016_am/Aspects_of_Brewing_Quality_C.pdf. Accessed 19 Aug 2021
4. Nyarko, E.K., Glavaš, H., Habschied, K., Mastanjević, K.: Determination of foam stability in lager beers using digital image analysis of images obtained using RGB and 3D cameras. Fermentation **7**(2), 46 (2021)
5. Bamforth, C.W.: The foaming properties of beer. J. Inst. Brew. **91**, 370–383 (1985)
6. BJCP, Beer Style Guidelines. Edited by Gordon Strong, Kristen England. www.bjcp.org. Accessed 24 Aug 2021
7. Hackbarth, J.J.: Multivariate analyses of beer foam stand. J. Inst. Brew. **112**, 17–24 (2006)
8. Prins, A., van Marle, J.T.: Foam formation in beer: some physics behind it. In: Monographs of the European Brewing Convention, vol. 27, pp. 26–36 (1999)
9. Ronteltap, A., Hollemans, M., Bisperink, C.G.J., Prims, A.R.: Beer foam physics. Tech. Q. Master Brew. Assoc. **28**, 25–32 (1991)
10. Fisher, S., Hauser, G., Sommer, K.: Influence of dissolved gases on foam. In: Monographs of the European Brewing Convention, vol. 27, pp. 37–46 (1999)
11. Evans, D.E., Sheehan, M.C.: Don't be fobbed off: the substance of beer foam—a review. J. Am. Soc. Brew. Chem. **60**(2), 47–57 (2002)
12. ASBC. Methods of Analysis. Method Beer-22. Foam Collapse Rate. Approved 1962, rev. 1975; American Society of Brewing Chemists: St. Paul, MN, USA (2018)
13. Klopper, W.: Foam stability and foam cling. In Proceedings of the Congress on European Brewery Convention. Elsevier, Amsterdam, The Netherlands (1964)
14. Rudin, A.: Measurement of the foam stability of beers. J. Inst. Brew. **63**, 506–509 (1957)
15. Rasmussen, J.N.: Automated analysis of foam stability. Carlsberg Res. Commun. **46**(1–2), 25–36 (1981). https://doi.org/10.1007/BF02906195
16. Jackson, G., Bamforth, C.: The measurement of foam-lacing. J. Inst. Brew. **88**, 378–381 (1982)
17. Constant, M.: A practical method for characterizing poured beer foam quality. J. Am. Soc. Brew. Chem. **50**, 37–47 (1992)
18. Vundla, W., Torline, P.: Steps toward the formulation of a model foam standard. J. Am. Soc. Brew. Chem. **65**, 21–25 (2007)
19. Amerine, M.A., Martini, L., Mattei, W.D.: Foaming properties of wine. Ind. Eng. Chem. **34**, 152–157 (1942)
20. Wilson, P., Mundy, A.: An improved method for measuring beer foam collapse. J. Inst. Brew. **90**, 385–388 (1984)
21. Evans, D.E., Surrel, A., Sheehy, M., Stewart, D.C., Robinson, L.H.: Comparison of foam quality and the influence of hop α-acids and proteins using five foam analysis methods. J. Am. Soc. Brew. Chem. **66**, 1–10 (2008)
22. Evans, D.E., Oberdieck, M., Red, K.S., Newman, R.: Comparison of the Rudin and NIBEM methods for measuring foam stability with a manual pour method to identify beer characteristics that deliver consumers stable beer foam. J. Am. Soc. Brew. Chem. **70**, 70–78 (2012)
23. Smith, R.J., Davidson, D., Wilson, R.J.: Natural foam stabilizing and bittering compounds derived from hops. J. Am. Soc. Brew. Chem. **56**, 52–57 (1998)

24. Drost, B.W., Eerde, P., Van Hoekstra, S., Strating, J.: Fatty acids and the staling of beer. In: Proceedings of the 13th European Brewery Convention Congress, Estoril, vol. 9, pp. 451–458 (1971)
25. Meilgaard, M.: Stale flavor carbonyls in brewing. Brew. Dig. **47**, 48–57 (1972)
26. Whitear, A.L.: Factors affecting beer stability. In: EBC Flavor Symposium, Monograph VII, pp. 203–210 (1981)
27. Dalgliesh, C.E.: Flavour stability. In: Proceedings of the 16th European Brewery Convention Congress, Amsterdam, pp. 623–659 (1977)
28. Bamforth, C.W.: Beer: A Quality Perspective, 1st edn. Academic Press Elsevier, New York (2009)
29. Torline, P., Dercksen, A., Axcell, B.: An electronic look at beer flavor instability. Master Brew. Assoc. Am. Tech. Q. **36**, 289–292 (1999)
30. Briggs, D.E., Boulton, C.A., Brookes, P.A., Stevens, R.: Brewing: Science and Practice. Woodhead Publishing Limited, Cambridge (2004)
31. Lewis, M.J., Young, T.W.: Brewing, 2nd edn. Kluwer Academic and Plenum Publishers, New York (2002)
32. Mastanjević, K., Krstanović, V., Lukinac, J., Jukić, M., Vulin, Z., Mastanjević, K.: Beer-the importance of colloidal stability (non-biological haze). Fermentation **4**(4), 91 (2018)
33. Middle European Brewing Analysis Commission (MEBAK), 3rd edn. Selbstverlag der MEBAK, Freising-Weihenstephan, Germany (1997)
34. European Brewery Convention (EBC), Analytica, 5th edn. Fachverlag Hans Carl, D-Nürnberg, Germany (1998)
35. ASBC methods. https://www.asbcnet.org/Methods/BeerMethods/Pages/default.aspx. Accessed 21 Aug 2021
36. Guido, L.F.: Sulfites in beer: reviewing regulation, analysis and role. Scientia Agricola **73**(2), 189–197 (2016)
37. Parker, D. K.: Beer: production, sensory characteristics and sensory analysis. In: Alcoholic Beverages, pp. 133–158 (2012)
38. Humia, B.V., Santos, K.S., Barbosa, A.M., Sawata, M., Mendonça, M.D.C., Padilha, F.F.: Beer molecules and its sensory and biological properties: a review. Molecules **24**(8), 1568 (2019)
39. Holt, S., Mukherjee, V., Lievens, B., Verstrepen, K.J., Thevelein, J.M.: Bioflavoring by non-conventional yeasts in sequential beer fermentations. Food Microbiol. **72**, 55–66 (2018)
40. Mastanjevic, K., et al.: From malt to wheat beer: a comprehensive multi-toxin screening, transfer assessment and its influence on basic fermentation parameters. Food Chem. **254**, 115–121 (2018)
41. Kobayashi, M., Shimizu, H., Shioya, S.: Beer volatile compounds and their application to low-malt beer fermentation. J. Biosci. Bioeng. **106**, 317–323 (2008)
42. Dack, R.E., Black, G.W., Koutsidis, G., Usher, S.J.: The effect of Maillard reaction products and yeast strain on the synthesis of key higher alcohols and esters in beer fermentations. Food Chem. **232**, 595–601 (2017)
43. Rodman, A.D., Gerogiorgis, D.I.: Dynamic optimization of beer fermentation: Sensitivity analysis of attainable performance vs. product flavor constraints. Comput. Chem. Eng. **106**, 582–595 (2017)
44. Stefanuto, P.H., et al.: Advanced method optimization for volatile aroma profiling of beer using two-dimensional gas chromatography time-of-flight mass spectrometry. J. Chromatogr. A **1507**, 45–52 (2017)
45. Toh, D.W.K., Chua, J.Y., Liu, S.Q.: Impact of simultaneous fermentation with *Saccharomyces cerevisiae* and *Torulaspora delbrueckii* on volatile and non-volatile constituents in beer. LWT Food Sci. Technol. **91**, 26–33 (2018)

46. Ocvirk, M., Mlinaric, N.K., Kosir, I.J.: Comparison of sensory and chemical evaluation of lager beer aroma by gas chromatography and gas chromatography/mass spectrometry. J. Sci. Food Agr. **80**, 3627–3635 (2018)
47. Quifer-Rada, P., et al.: A comprehensive characterisation of beer polyphenols by high resolution mass spectrometry (LC–ESI-LTQ-Orbitrap-MS). Food Chem. **169**, 336–343 (2015)
48. Da Silva, G.C., et al.: Method development by GC–ECD and HS-SPME–GC–MS for beer volatile analysis. Food Chem. **167**, 71–77 (2015)
49. Bogdan, P., Kordialik-Bogacka, E.: Alternatives to malt in brewing. Trends Food Sci. Technol. **65**, 1–9 (2017)
50. Gagula, G., Mastanjević, K., Mastanjević, K., Krstanović, V., Horvat D., Magdić, D.: The influence of packaging material on volatile compounds of pale lager beer. Food Packag. Shelf Life **24**, 100496 (2020)
51. Deraney, N., Manning, J., Morse, L., Palmer, J.: Efficient Brewery Sanitation. https://web.wpi.edu/Pubs/E-project/Available/E-project-042215-141239/unrestricted/Efficient_Brewery_Sanitation_MQP__Deraney_Manning_Morse_Palmer.pdf. Accessed 21 July 2021
52. The Alloys of Brew. https://www.reliance-foundry.com/blog/brewing-metallurgy. Accessed 21 Sept 2021
53. Cluett, J., Rowlands, D., Khanyile, D., Hulse, G.: Principles of Hygiene in the Beverage Industry, 1st edn. The Institute of Brewing and Distilling (IBD)–Africa Section, Supreme Printers, Hout Bay, South Africa (2003)

Maintaining the Quality of Homemade Kulen (Dry Croatian Sausage) During a Three-Year Period

Krešimir Mastanjević(✉), Dragan Kovačević, Ana Domaćinovć, Snježana Keleković, and Kristina Habschied

Faculty of Food Technology Osijek, Josip Juraj Strossmayer University of Osijek, F. Kuhača 20, 31000 Osijek, Croatia
kmastanj@ptfos.hr

Abstract. The objective of this research was maintenance of the quality parameters of homemade kulen during a three-year period. Homemade kulen is a dry fermented sausage with specific sensorial characteristic (smell and taste) which originates from the use of the high quality pig meat, specific spices, smoking (use of open fire) and long ripening time (at least four months). The production of this indigenous dry sausage take place on small farms and it is conditioned by the climatic conditions. Namely, to produce a good quality kulen it is necessary to start the production during winter months, when the slaughter and production take place. After the smoking, fermentation and drying, the ripening takes place. The physical-chemical parameters (pH, a_W, moisture content, fat content, protein content and collagen content) of 25 samples of homemade kulen were determined each year. The current Croatian legislative prescribes only the moisture content (40%) and protein content (22%) for this kind of dry sausage. The results of this study showed variation in investigated parameters between the samples and between the production year. Variations can be contributed to different recipes used by producers, different climatic conditions and to different production technologies (smoking with different wood, ripening time, pig meat quality, etc.).

Keywords: Homemade kulen · Traditional production · Quality · Maintenance · Physical-chemical parameters

1 Introduction

Homemade kulen (Fig. 1) is dry sausage produced in Croatian region Slavonija and Baranja. The production is still in small scale, so there are many producers, which use different recipes and different production procedure. Homemade kulen manufacturing is a seasonal character and beings in late fall and last until the late spring (approximately six months), due to the favorable weather conditions (temperature and relative humidity) [1]. The peculiar aroma of this dry sausage manly originates from long-term ripening (more than 3 months), smoking, lactic acid bacteria activity, and special spices used in its production (sweet paprika powder, hot paprika powder and garlic) [2, 3]. According to

H. Glavaš et al. (Eds.): OTO 2021, LNNS 369, pp. 447–456, 2022.
https://doi.org/10.1007/978-3-030-92851-3_34

the actual Croatian Legislation [4] homemade kulen is a dry sausage made from coarsely chopped pork and pork back fat, salt, red spicy peppers and garlic, with the possibility of adding sugar, other spices or spice extracts, starter cultures, additives and smoke aromas. They should not contain more than 40% of moisture and must contain at least 22% of meat proteins [4].

The sausage is produced from the parts of the pig with the highest commercial value, the meat from the leg and loin with the addition of approximately 20% of shoulder and neck meat. The minced meat with the addition of max. 10% of pork back fat is seasoned with 2% of rock salt. 0.6% of sweet paprika powder, 0.6% of hot paprika powder and 0.3% of minced fresh garlic. This mixture is then stuffed in natural casing – pig appendix. The raw homemade kulen is then smoked, dried and ripened until it receives the satisfactory texture and aroma.

Fig. 1. Homemade kulen.

Dry fermented sausage are usually made from pork, or combination of pork and beef, mutton rarely, horse meat and chicken. Minced meat is mixed with pork fat, salt, spices (paprika, pepper, garlic, etc.) and, especially in industrial production, nitrate/nitrite, ascorbate, sugars, etc., in different mass fraction (different recipes). Specific sensory properties of certain dry sausage in addition to being associated with the technology of production (primary the smoking) and added substances (primarily spices) are a result of fermentation and maturation (proteolysis and lipolysis due to the activity of endogenous and exogenous enzymes) or activity of technological microflora that is specific for each microclimatic location [5–7]. The dry sausages from different European countries (Croatia, Spain, Italy, Greece and Bosina and Hezegovina) were investigated for its physico-chemical properties with the purpose of standardization of its production [6, 8–16]. These study showed large variations in chemical composition and water activity and pH [17].

The small scale produces still use the traditional production without the possibility of controlled conditions (temperature and relative humidity) in smoking, drying and ripening stage. The production conditions are dependent on the weather condition in the area of production. All mentioned results in high variability in chemical composition, of homemade kulen between different producers.

2 Material and Methods

2.1 Material

The samples of homemade kulen were collected at the traditional homemade kulen competition organized by The Chamber of Commerce in Osijek, Croatia. In year 2019. 47 samples were collected, in the years 2020 and 2021, 25 samples were collected. The lower number of samples collected in years 2020 and 2021 is probably related to the COVID-19 pandemic.

In the production of homemade kulen the pig meat from the pigs of at least 12 month old and over 150 kg live weight must be used. After the slaughter, dissection and deboning the meat is cooled to temperatures of 0–4 °C. The met pH at this stage should be from 5.6 to 5.9. The meat is then grinded trough a grinding plate with holes of 8 to 12 mm in diameter. The meat is them seasoned with 0.7% red paprika, 0.6% of hot paprika and garlic in the amount of 0.2%. Before this the salt is added in amount of 2%. After the mixing the mixture is stuffed into a pig's appendix (lat. *intestinum caecum*).

Raw homemade kulen is then smoked with open fire using dry hard wood (hornbeam, beech and its sawdust) approximately six to seven times during a 14 days period. After the smoking, homemade kulan is dried and ripened. After that the homemade kulen is ready for consumption.

2.2 Physico-Chemical Analysis

After discarding the casing, the samples were cut into small pieces and homogenized in a knife mill Gridomix GM 200 (Retsch, Germany). The FoodScan Meat Analyser (Foos, Sweden) was used to determine moisture, total protein, total fat and collagen according to the AOAC 2007 04.

The pH level was measured in a homogenate of the sample mixed with distilled water (1:10) with pH/Ion 510 – Bench pH/Ion/mV Meter (Eutech Instruments Pte Ltd/Oakton Instruments, USA) according to ISO 2917:1999 (HRN ISO 2917:2000) and manufacturer instructions (pH/Ion 510 Instruction Manual).

Water activity (a_w) was determined using a Rotronic Hygrolab 3 (Rotronic AG, Switzerland) according to manufacturer instructions, at room temperature (20 ± 2 °C). All measurements were conducted in three parallel.

3 Results and Discussion

In the Table 1 the general composition pH and a_w of the 47 samples of homemade kulen were presented. The moisture were from 25.92 to 44.13%.

Relatively large differences are related to the defend stages ripening of homemade kulen. According to the official Croatian legislative the maximal prescribed value of moisture content for this kind of sausage is 40%. The six samples of homemade kulen

Table 1. General composition, pH and a_w of homemade kulen in year 2019.

Sample	Fat (%)	Moisture (%)	Protein (%)	Collagen (%)	pH	a_w
HK1	30.23	34.22	29.13	2.08	5.41	0.89
HK2	21.65	33.99	35.42	3.61	5.95	0.87
HK3	14.4	44.13	33.56	2.17	5.34	0.89
HK4	14.71	39.68	36.49	2.39	5.47	0.86
HK5	13.22	41.27	36.27	2.27	5.49	0.88
HK6	13.34	41.44	35.89	2.54	5.46	0.88
HK7	23.15	25.97	38.48	3.31	5.86	0.80
HK8	21.25	36.76	34.02	2.93	5.24	0.88
HK9	23.61	35.83	34.78	2.66	5.02	0.88
HK10	20.98	36.55	34.43	2.91	5.24	0.88
HK11	22.05	32.26	35.55	2.38	5.43	0.86
HK12	23.04	35.14	33.24	2.95	5.35	0.87
HK13	23.87	32.05	34.43	3.43	5.12	0.86
HK14	20.87	30.91	36.41	3.11	5.96	0.83
HK15	21.3	37.31	33.37	2.74	5.57	0.89
HK16	27.56	34.14	30.19	3.15	5.65	0.86
HK17	27.87	28.93	33.27	4.96	5.81	0.84
HK18	24.64	25.92	36.28	4.39	6.09	0.81
HK19	27.63	30.58	33.64	3.01	5.74	0.86
HK20	28.17	28.96	34.37	4.26	5.62	0.84
HK21	23.24	30.53	34.85	4.83	5.53	0.83
HK22	17.14	40.03	34.99	2.79	5.33	0.89
HK23	23.2	32.31	34.44	3.34	5.46	0.84
HK24	15.86	37.11	36.16	3.03	5.58	0.87
HK25	19.3	33.94	36.26	3.22	5.2	0.87
HK26	17.65	43.54	32.71	2.23	5.24	0.91
HK27	17.27	42.98	33.01	1.89	5.23	0.89
HK28	22.93	32.66	34.77	3.01	5.73	0.87
HK29	32.52	27.47	34.04	4.11	5.29	0.82
HK30	22.01	30.03	35.23	4.26	5.4	0.84
HK31	33.3	31.22	28.05	3.33	5.4	0.88
HK32	36.27	30,03	26.07	1.5	5.05	0.86

(continued)

Table 1. *(continued)*

Sample	Fat (%)	Moisture (%)	Protein (%)	Collagen (%)	pH	a_w
HK33	28.04	30.67	29.88	2.95	5.23	0.82
HK34	27.51	26.67	28.11	0.27	5.49	0.86
HK35	26.01	27.83	34.46	3.62	5.51	0.85
HK36	18.68	35.98	34.39	2.69	5.53	0.87
HK37	25.63	29.64	34.54	3.74	5.12	0.86
HK38	24.41	34.92	29.51	0.75	5.74	0.88
HK39	20.37	31.52	36.48	3.47	5.41	0.85
HK40	19.6	33.11	36.31	3.21	5.35	0.86
HK41	21.4	33.79	34.47	2.49	5.59	0.86
HK42	22.68	32.11	33.61	2.7	5.39	0.85
HK43	26.46	30.64	33.63	2.5	5.57	0.86
HK44	25.56	29.17	35.34	2.91	5.52	0.84
HK45	22.96	32.67	34.53	2.93	5.61	0.86
HK46	28.96	32.65	30.5	1.78	5.38	0.87
HK47	27.75	27.67	34.52	4.25	5.47	0.84

were not in agreement with prescribed values (HK3, HK5, HK6, HK22, HK26 and HK27).

The fat content ranged from 13.22 to 36.27% with the average value of 23.04%. Protein content of examined samples were from 26.07 to 38.48%, collagen content ranged from 0.27 to 4.96%. pH values were in interval from 5to 6.09 and a_w 0.80 to 0.91.

In the year 2020 two samples had a higher moisture content then prescribed (HK2 and HK3) (Table 2). Fat content ranged from 11.99 to 38.41%. Mass fraction of protein ranged from 26.94 to 40.15% and collagen from 1.10 to 5.13%. pH ranged from 5.09 to 6.19 and a_w 0.78 to 0.89.

Basic chemical composition, pH and a_w of 25 samples of homemade kulen in year 2021 are presented in Table 3. The moisture content varied from 25.35 to 41.51%. Four samples showed higher values then legally prescribed level. Mass fraction of fat were in range from 12.22 to 38.09%. Protein content of investigated samples were 24.08 to 37.72% and collagen from 0.85 to 4.01%. pH ranged from 5.12 to 6.15 and a_w 0.81 to 0.90.

In Table 4 the average, minimum and maximum values of examined physical-chemical parameters for each production year are presented. Average values of fat content were from 22.85 to 23.15% and showed little variations. Moisture content varied from 32. 02 to 35.52%. Protein contents were from 33.26 to 34.26. Ambrosiadis et al. [18] reported higher values for moisture (38.74%), fat (40.75%) content and a_w (0.931) for Greek dry sausage. Also, the protein content (16.41%) and pH (4.88) were lower.

Table 2. General composition, pH and a_W of homemade kulen in year 2020.

Sample	Fat (%)	Moisture (%)	Protein (%)	Collagen (%)	pH	a_W
HK1	16.99	37.2	33.13	1.11	6.19	0.84
HK2	12.63	43.81	35.2	2.01	5.32	0.88
HK3	11.99	47.55	33.34	1.21	5.31	0.89
HK4	22.87	34.12	35.26	2.26	5.5	0.87
HK5	23.16	35.07	33.26	2.29	5.52	0.86
HK6	24.23	29.97	36.26	2.87	5.24	0.84
HK7	24.61	30.62	35.03	3.37	5.28	0.85
HK8	24.82	24.31	40.15	3.44	5.56	0.78
HK9	26.72	35.59	29.01	2.05	5.54	0.88
HK10	26.72	35.57	29.08	1.5	5.6	0.89
HK11	21.74	27.86	38.78	3.09	5.32	0.81
HK12	22.11	32.36	35.54	2.83	5.09	0.85
HK13	21.36	29.88	38.34	3.55	5.09	0.84
HK14	22.41	28.14	35.45	5.13	5.27	0.79
HK15	31.36	27.64	31.11	4.13	5.4	0.83
HK16	25.87	31.18	34.72	1.58	5.76	0.85
HK17	32.71	27.03	33.48	4.86	5.41	0.82
HK18	38.41	28.17	26.94	2.37	5.31	0.82
HK19	19.85	35.12	35.03	3.45	5.35	0.86
HK20	23.02	30.45	35.69	4.08	5.54	0.83
HK21	20.56	35.59	34.41	3.08	5.51	0.88
HK22	24.09	34.62	32.32	2.05	5.35	0.88
HK23	21.76	34.12	34.36	1.95	5.94	0.87
HK24	20.77	36.52	33.41	3.44	5.51	0.87
HK25	18.08	35.47	37.22	2.94	5.55	0.87

Similar as Greek dry sausages the Spanish dry sausage Iberian chorizo showed lower values for moisture content (18.6–19.42%), protein content (27.22–28.68%) and a_W (0.77–0.88), but higher values for fat content (33.50–35.98%) [9, 19]. pH values in the investigation by [18] (5.25–5.80) were similar to values presented in Table 4.

De Nobile et al. [20] reported higher protein (38.52%) and fat (27.98) content but lower moisture content (26.01) for Italian salami.

Croatian dry sausage *Istarska kobasica* showed higher fat content (49.01%), but lower protein (23.36%) and moisture content (22.31%) [12]. Other Croatian dry sausage such as Slavonska kobasica and *Kulenova seka* showed similar values of physical-chemical parameters as homemade kulen [21].

Table 3. General composition, pH and a_w of homemade kulen in year 2021.

Sample	Fat (%)	Moisture (%)	Protein (%)	Collagen (%)	pH	a_w
HK1	17.84	38.64	34.99	2.66	5.26	0.88
HK2	12.22	40.07	37.72	2.56	5.34	0.88
HK3	21.74	33.74	35.86	2.25	5.22	0.86
HK4	21.29	34.9	35.39	2.4	5.19	0.87
HK5	22.03	33.56	35.72	2.78	5.12	0.86
HK6	19.03	39.42	33.74	2.19	5.55	0.89
HK7	21.82	35.63	33.42	3.15	5.67	0.88
HK8	22.34	35.04	33.96	3.1	5.25	0.88
HK9	22.53	34.02	34.36	3.19	5.23	0.87
HK10	19.41	40.14	33.38	1.41	5.75	0.89
HK11	21.87	32.61	35.35	2.82	5.56	0.86
HK12	19.87	41.51	32.36	1.74	5.49	0.90
HK13	36.46	29.05	27.57	2.49	5.6	0.85
HK14	30.11	25.35	34.28	4.01	5.41	0.81
HK15	38.09	31.65	24.08	1.64	5.23	0.87
HK16	21.33	38.57	32.61	0.85	5.59	0.89
HK17	24.81	37.15	30.9	1.64	5.78	0.86
HK18	19.53	40.47	33.2	1.74	5.89	0.89
HK19	24.23	35.63	32.7	1.92	5.29	0.88
HK20	18.23	39.69	33.63	2.92	5.33	0.89
HK21	22.09	35.9	32.96	2.73	5.58	0.87
HK22	20.93	36.22	32.88	2.23	5.37	0.86
HK23	26.49	33.98	32.72	1.91	6.15	0.88
HK24	22.65	31.85	34.85	2.86	5.53	0.86
HK25	24.47	33.14	32.95	2.02	6.05	0.83

Portuguese Chorizo had higher pH of 5.78, a_w 0,94, higher mass fraction of water 50.40% and fat 25.60%, but lower protein content 19.10% [22].

Traditional Spanish dry sausage Fuet had similar moisture content 33.09% higher fat content 26.83% and lower mass fraction of protein 29.14%. Values of pH were similar 5.44 but water activity showed lower values (0.763) [23].

Martin et al. reported lower moisture content of 26.04%, lower water activity 0.785 but higher pH of 5.08 for Spanish dry sausage salchichón [24].

And Turkish-style fermented sausage, also known as sucuk had higher fat content (41.55%) but lower mass fraction of water (20.01%) and protein (34.77%). Sucuk showed lower pH value of 4.82 [25].

Table 4. Average, minimum and maximum values of general composition, pH and a_W of homemade kulen in years 2019–2021.

Year	2019			2020			2021		
	Average	Min.	Max.	Average	Min.	Max.	Average	Min.	Max.
Fat (%)	23.20	13.22	36.27	23.15	11.99	38.41	22.85	12.22	38.09
Moisture (%)	33.33	25.90	44.13	32.02	24.31	47.55	35.52	25.35	41.51
Protein (%)	33.83	26.07	38.48	34.26	26.94	40.15	33.26	24.08	37.72
Collagen (%)	2.96	0.27	4.96	2.82	1.11	5.13	2.37	0.85	4.01
pH	5.47	5.01	6.09	5.46	5.09	6.19	5.50	5.12	6.15
a_W	0.861	0.801	0.907	0.850	0.783	0.895	0.871	0.810	0.904

Jahic and Pracic [11] reported higher values of moisture of 43.58% but lower fat content (17.10%) and protein content (31.47%) for dry sausage Bosnian sudžuk.

This discrepancies between the homemade kulen and other European dry sausages may be related to smoking (most of the mentioned sausages are not smoked) and the by longer ripening stage of homemade kulen [1, 5].

4 Conclusions

Fat, moisture, collagen, protein content, pH and water activity of homemade kulen showed large variation between producers for all investigated years. This shows that producers still use very different recipes in production and that the samples homemade kulen were in different stage of production (different ripening stage).

The work on the standardization of production is needed. This can be achieved by using the controlled condition (temperature 15–17 °C, and relative humidity around 75%) of homemade kulen production. The small scale producers should use the automatic chambers for smoking, drying and ripening.

References

1. Kovačević, D., Mastanjević, K., Šubarić, D., Jerković, I., Marijanović, Z.: Fizikalno-kemijska svojstva, boja i tekstura slavonskog kulena – tradicionalne hrvatske trajne kobasice. MESO Prvi hrvatski časopis o mesu **XII**, 275–275 (2010)
2. Jerković, I., Kovačević, D., Šubarić, D., Marijanović, Z., Mastanjević, K., Suman, K.: Authentication study of volatile flavour compounds composition in Slavonian traditional dry fermented salami "kulen." Food Chem. **119**, 813–822 (2010). https://doi.org/10.1016/j.foodchem.2009.07.024
3. Karolyi, D., Salajpal, K., Đikić, M., Kostelić, A., Jurić, I.: Psysicochemical characteristics of slavonian kulen. MESO Prvi hrvatski časopis o mesu **VII**, 35–37 (2005)
4. Pravilnik o mesnim proizvodima. https://narodne-novine.nn.hr/clanci/sluzbeni/2018_07_62_1292.html. Accessed 26 Aug 2021

5. Kovačević, D., Mastanjević, K., Šubarić, D., Suman, K.: Physico-chemical and colour properties of homemade Slavonian sausage. MESO Prvi hrvatski časopis o mesu **11**, 280–284 (2009)
6. Mastanjević, K., Kovačević, D., Frece, J., Markov, K., Pleadin, J.: The effect of autochthonous starter culture, sugars and temperature on the fermentation of Slavonian Kulen. Food Technol. Biotechnol. **55**, 67–76 (2017). https://doi.org/10.17113/ftb.55.01.17.4688
7. Kovačević, D., Suman, K., Lenart, L., Mastanjević, K., Šubarić, D., Frece, J.: Smanjenje udjela soli u domaćoj slavonskoj kobasici: utjecaj na sastav, fizikalno-kemijska svojstva, boju, teksturu, senzorska svojstva i zdravstvenu ispravnost. MESO Prvi hrvatski časopis o mesu **13**, 244–249 (2011)
8. Gonzalez-Fandos, E., Vazquez de Castro, M., Martinez-Laorden, A., Perez-Arnedo, I.: Behavior of Listeria monocytogenes and other microorganisms in sliced Riojano Chorizo (Spanish dry-cured sausage) during storage under modified atmospheres. Microorganisms **9**, 1384 (2021). https://doi.org/10.3390/microorganisms9071384
9. Tejada, L., et al.: Development of an Iberian chorizo salted with a combination of mineral salts (seawater substitute) and better nutritional profile. Front. Nutr. **8**, 66 (2021). https://doi.org/10.3389/fnut.2021.642726
10. Olivares, A., Navarro, J.L., Flores, M.: Effect of fat content on aroma generation during processing of dry fermented sausages. Meat Sci. **87**, 264–273 (2011). https://doi.org/10.1016/j.meatsci.2010.10.021
11. Jahić, S., Pračić, N.: The influence of raw materials and different fermentation intervals on quality parameters of the traditionally produced Bosnian sudžuk. Technologica Acta Sci. Prof. J. Chem. Technol. **11**, 11–16 (2018)
12. Bratulić, M., et al.: Quality of traditionally produced Istrian sausage and identification of autochthonous lactic acid bacteria strains as potential functional starter cultures. Eur. Food Res. Technol. **247**(11), 2847–2860 (2021). https://doi.org/10.1007/s00217-021-03835-6
13. Girolami, A., Napolitano, F., Faraone, D., Di Bello, G., Braghieri, A.: Image analysis with the computer vision system and the consumer test in evaluating the appearance of Lucanian dry sausage. Meat Sci. **96**, 610–616 (2014). https://doi.org/10.1016/j.meatsci.2013.08.006
14. Operta, S., Čorbo, S., Tahmaz, M.R.J., dipl vet Dževdetbegović, M., dipl ing tehn Šehović, A.: Utjecaj sirovine na fizikalno-kemijska i senzorna svojstva bosanskog sudžuka proizvedenog u kontroliranim uvjetima. MESO Prvi hrvatski časopis o mesu **XIV**, 472–479 (2012)
15. Mastanjević, K., Kovačević, D., Frece, J., Markov, K., Pleadin, J.: Utjecaj dodatka autohtone bakterijske starter kulture i šećera, te temperature na fermentaciju Slavonskog kulena. Food Technol. Biotechnol. **55**, 67–76 (2017). https://doi.org/10.17113/ftb.55.01.17.4688
16. Kovacevic, D., Mastanjevic, K., Pleadin, J., Frece, J.: Physicochemical, microbiological, and colour attributes of horse salami established during the ripening period. Ital. J. Food Sci. **28**, 96–106 (2016). https://doi.org/10.14674/1120-1770/ijfs.v463
17. Simunović, S., et al.: Estimation of fat content in fermented sausages by means of Computer Vision System (CVS). Sci. J. "Meat Technol." **62**, 27–32 (2021). https://doi.org/10.18485/2021.62.1.3
18. Ambrosiadis, J., Soultos, N., Abrahim, A., Bloukas, J.G.: Physicochemical, microbiological and sensory attributes for the characterization of Greek traditional sausages. Meat Sci. **66**, 279–287 (2004). https://doi.org/10.1016/S0309-1740(03)00100-1
19. Trejo, A., Martín, M.J., Gómez-Quintana, A., Cava, R., García-Parra, J.J., Ramírez, M.R.: Effect of high-pressure treatment and storage temperature on top-quality (Montanera) Iberian dry-cured pork sausages (chorizo). J. Food Sci. **86**, 1963–1978 (2021). https://doi.org/10.1111/1750-3841.15698
20. Del Nobile, M.A., Conte, A., Incoronato, A.L., Panza, O., Sevi, A., Marino, R.: New strategies for reducing the pork back-fat content in typical Italian salami. Meat Sci. **81**, 263–269 (2009). https://doi.org/10.1016/j.meatsci.2008.07.026

21. Kovačević, D., Mastanjević, K., Šakić, I., Frece, J., Pleadin, J.: Utjecaj dodatka različitih šećera na proces fermentacije hrvatske trajne kobasice kulenove seke. MESO Prvi hrvatski časopis o mesu **XVI**, 324–328 (2014)
22. Martín, M., et al.: Comparative effect of high hydrostatic pressure treatment on Spanish and Portuguese traditional chorizos and evolution at different storage temperatures. J. Food Process. Preserv. **45**, 15082 (2020). https://doi.org/10.1111/jfpp.15082
23. Martínez-Zamora, L., Peñalver, R., Ros, G., Nieto, G.: Substitution of synthetic nitrates and antioxidants by spices, fruits and vegetables in clean label Spanish chorizo. Food Res. Int. **139**, 109835 (2021). https://doi.org/10.1016/j.foodres.2020.109835
24. Martín, I., Rodríguez, A., Sánchez-Montero, L., Padilla, P., Córdoba, J.J.: Effect of the Dry-Cured fermented sausage "Salchichón" processing with a selected Lactobacillus sakei in Listeria monocytogenes and microbial population. Foods **10**, 856 (2021). https://doi.org/10.3390/foods10040856
25. Sarıçoban, C., Unal, K.: Influence of pre-treated bitter orange albedo on the physicochemical, textural and sensory properties of fermented sausages (sucuk). J. Food Sci. Technol. **1**, 9 (2021). https://doi.org/10.1007/s13197-021-05158-6

Mobile Application for Mentoring Graduation Theses

Ivica Lukić[✉], Matej Peršić, and Mirko Köhler

Faculty of Electrical Engineering, Computer Science and Information Technology Osijek, 31000 Osijek, Croatia
ivica.lukic@ferit.hr

Abstract. In this paper mobile version of the system for mentoring graduation theses (MAK) is presented. The goal of application is to enable the Students, Mentors and Reviewers to do their tasks inside of the MAK system. After successful authentication application sends token to user. Token and user name are used for further requests to decide if the user is a Student or Mentor. With that in mind, the user gets the appropriate interface. Inside every activity the user can do various activities in the MAK system, where users input is the input to the system. After an action, the user gets a Toast message which is used as an output from the system. The messages show the state of the procedure or an error which is related to the state of the Student's paper, and disables the navigation to the next activity. The application lifecycle management is proposed and explained in further chapters.

Keywords: Kotlin · MAK system · Mobile application · Retrofit

1 Introduction

In this paper is presented mobile application development process for already existing Web application for mentoring final papers and graduate theses MAK. This application will ensure easier access in MAK system using mobile devices. Students are enabled to add and monitor progress of their theses. Student can track progress of his thesis and have different versions of thesis, according to Mentor's guidance and recommendations. Mentors are enabled to lead their Students step by step, demand changes in theses, grade theses etc. After each change Student can upload new version of thesis and MAK system enables Mentors to track changes.

Chapter one contains a short summary and the main idea of this paper. Second chapter contains the review of the field where used technologies are listed and a brief review of the field is made. Third chapter contains a review of the system MAK where the user roles inside of the systems are described. Fourth chapter contains a short review of the Kotlin programming language. Fifth chapter describes the problem of logging in and signing the users out of the MAK system, and determining which activity the user sees when navigating through the application regarding his role in the system.

For a successful program solution, it is needed to use the knowledge of Kotlin and HTTP requests. Using them and other technologies, it is needed to develop a valid solution.

H. Glavaš et al. (Eds.): OTO 2021, LNNS 369, pp. 457–466, 2022.
https://doi.org/10.1007/978-3-030-92851-3_35

For the application development and implementation application lifecycle management (ALM) is used, because ALM is very important part of every application used by many end users. ALM is not only software development lifecycle (requirements, design, coding, testing, project and change management), ALM continues after development. The end of ALM is after application is no longer in use. That is why it is important to keep in mind how the application will be managed even after production.

2 Review of the Field

As technology rapidly progressed, mobile devices became a common pocket accessory. In addition to calling and texting, they were capable for more complex tasks, ranging from writing notes to even playing games, such as Snake on Nokia phones [1]. Those were first mobile applications, but they were only developed in the company which produced mobile device.

When the first iPhone appeared, it revolutionized the whole idea of mobile devices. Today Apple and Google are dominating the mobile market and there is an increasing demand of high-quality mobile applications that can be developed very fast and simple as possible [2].

The tool used for development of applications for the Apple devices is called XCode. Also programming technologies like UI-Kit and Swift-UI are widely used iOS application. Despite high prices and relatively small portion on the market which iPhones have, Apple is gaining almost a third of the all profit in the whole mobile industry [3].

The MAK application which is presented in this paper is made for Google's platform called Android. For Android platform the tool called Android Studio is most widely used, and programming languages for Android development are Java, and newer languages Kotlin and Flutter [4].

On the Faculty of Electrical Engineering, Computer Science and Information Technology Osijek (FERIT) are used similar programming solutions to our MAK application. These solutions are using mobile application for the Merlin system, for Student's class schedule on FERIT. Both solutions are using the system of AAI identities and as such provide the appropriate interface on successful authentication of a user.

2.1 Android

Android is an operating system used in mobile devices such as smart-phones, tablets, and even smart watches and automobiles. The first version of Android is launched in 2008. and its newest version is Android 12. Today it's the most widely spread operating system due to its accessibility and affordable price, so it can be used in devices of all price ranges [5].

2.2 Used Technologies

In this paper programming language Kotlin is used for the MAK mobile application. Kotlin is the most popular programming language for Android applications, has easier reading and better workflow in comparison to Java. Next to Kotlin a library FileAndFolderPicker for easier work with data received from the user's devices is used. Retrofit for working with HTTP requests is also used [6].

3 MAK System for Mentoring Graduation Theses

Information system for mentoring graduation theses (cro. Informacijski sustav za mentorske aktivnosti) – MAK, is a system used for easier maintaining of the procedures related to writing and grading of final papers and graduate theses at the Faculty of Electrical Engineering, Computer Science and Information Technology Osijek. It covers a part of the procedure from the moment the Graduate Committee approves the topic assignment of final paper or graduate thesis until the President of the Graduate Committee, or Secretary of the Graduate Committee approves the passing grade.

The whole MAK procedure has 9 different steps which are show in Fig. 1. The requirements for ALM are shown here. In first step initial version of Student's paper is uploaded. In the second step Mentor approves initial version or demands changes. This step is repeated until Mentor is satisfied with initial version. The third step is where Student sets final version of thesis. Consequently, in the fourth step Mentor confirms final version or demands changes. After four step final version is prepared and Student in fifth step confirms that his thesis is ready for Defense Commission and register exam in ISVU system. In the sixth step mentor grades thesis or final paper for undergraduate Students. Undergraduate Student's paper is assigned to Reviewer which approves or disapproves Mentor's grade. For graduate Students Mentor selects Defense Commission. Defense Commission is approved or disapproved by Secretary of the Graduate Committee in step seven. Step eight is thesis defense for graduate Students and Reviewer reading for undergraduate Students. In final step, Secretary of the Graduate Committee confirms thesis Defense Commission procedure and Reviewer's opinion. Steps for Secretary of the Graduate Committee are made in Web application instead of in mobile application.

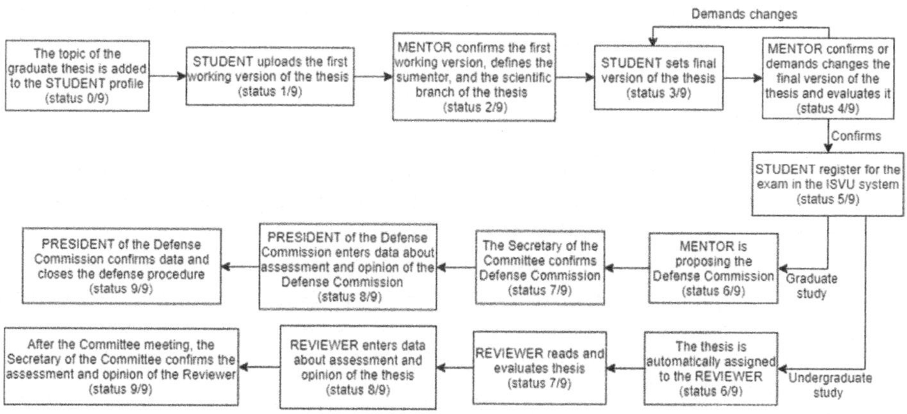

Fig. 1. Activity workflow of the MAK system.

3.1 Roles Inside the Mobile Application

Inside the system MAK there are numerous roles. According to them, the user sees the appropriate part of the application and has according authorizations. The roles that are used in this paper are Student, Mentor and Reviewer.

Student is a role inside the system who adds his final paper or graduate thesis into the system and can track the procedure and status of his final paper or graduate thesis.

Mentor is a person inside the system who can monitor his Students and approve versions of his Student's final paper or graduate thesis.

Reviewer is a person inside the system who gets the Student's final paper or graduate thesis for reading and provides feedback about the paper. For undergraduate students there is a panel of thesis Reviewers which are assigned to read the thesis. These Reviewers are recognized as academic experts or authorities in the area in which the research was conducted. Thesis Defense Commission is organized for graduate students and has three members. First one is a President of Defense Commission, second is Mentor and third member is selected as recognized as academic experts or authority in the area in which the research was conducted.

President of the Defense Commission a person inside the system who gets the Student's final paper or graduate thesis for reading and provides feedback about the paper.

4 Solution

4.1 Kotlin Programming Language

Kotlin is a general-purpose programming language. It is designed to interoperate fully with Java, and the Java Virtual Machine (JVM) version for Kotlin's standard library should be same version as the Java Class Library. Today it is popular for Android development. Also, Retrofit library is used for working with HTTP requests. FileAndFolderPicker library is used for solving the problem of loading files from the user's devices [7].

4.2 Application Lifecycle Management

As mentioned in chapter one, MAK mobile application used Application Lifecycle Management divided is six phases [8]. ALM is divided in phases such as: requirements, design, implementation, testing, deployment, project and change management. Authors gathered data from different type of users, search requirements for application and made analysis of the requirements. One of the authors developed Web application for MAK system and all requirements for different types of user are taken from Web application. In design phase authors proposed Form, Reports, Queries and API (Application Interface) design. Next phase was coding implementation using Kotlin Programming Language according to designed Forms and Queries in previous phase. After coding and testing Mobile application is set in production phase. After production application is in change management and maintenance phase. In future Work authors will try to improve application and explain change management.

The problem of user authentication is solved using an existing system of authentication and authorization infrastructure called AAI@EduHR [9]. This system is used for science and higher education in the Republic of Croatia, and which stores the user's electronic identities. Users are logged in using their username and given password as shown in Fig. 2. After successful authentication system is able to determine if the user is a Student or Mentor, and application sends token to user. Token and user name are used for further requests to decide if the user is a Student or Mentor. If user is Student, the part of application related to Student's activities will be displayed. If user is Mentor, the user will see the part of the application related to Mentor's and Reviewer's activities. While using the app, it is needed to keep track of the user via generated token and his username. When the user is done with his activities, he has an option to log out of the system.

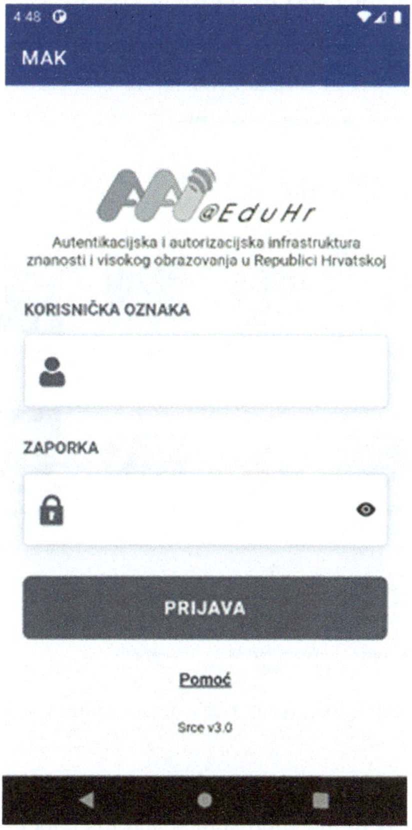

Fig. 2. Login to the MAK system.

4.3 Solution Procedure

The first problem that needs to be solved is loading users token and username in the MainActivity of mobile application after successful authentication. MainActivity reads API provided in Web application for mentoring final papers and graduate theses MAK. This API sends JSON document as response. If the JSON document text contains a suffix student, the current user is a Student, so Student's main activity is launched. Otherwise, the activity for Mentors and Reviewers is launched.

Student Activities

In StudentsMainActivity a String is loaded from MainActivity that contains the Student's username and token. With them in store, the Student can do various tasks. Student is presented with a Tab Layout, in which he sees 3 fragments. First fragment called StudentFinalPaperInfo contains information about the Student and a submit button. If the paper is in valid procedure status, the Student can upload his initial (status 1/9) or final version (status 3/9) of the paper as shown in Fig. 3.

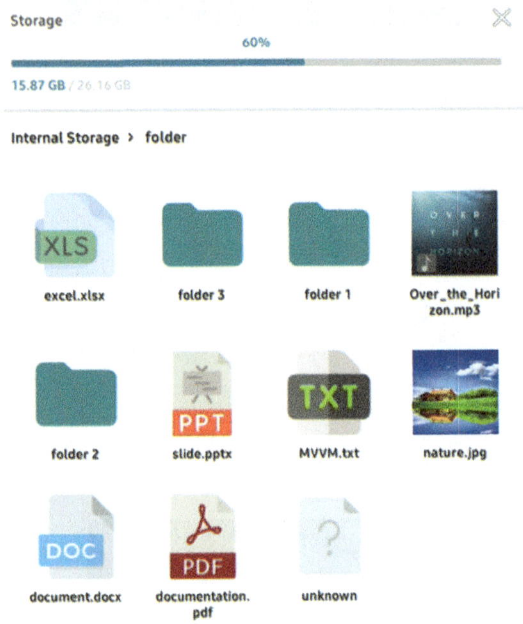

Fig. 3. A document upload from user's device using FileAndFolderPicker.

In the second fragment called StudentConfirmsFinalPaper, Student can press a button on which if the paper is in the valid status, and his paper is registered in the ISVU system, the procedure will advance to status 5/9. In the final fragment called StudentLogOutFragment Student can log out and go to the initial state of the application in the MainActivity, where he can log in.

Mentor Activities

After successful authentication Mentor can see three different buttons displayed in the OtherUserMainActivity.

When clicking on the Mentor button, user is presented with MentorsRecyclerActivity, where his mentoring Students are displayed. After selecting desired Student, Mentor is presented with MentorChoosesActivity. In this activity Mentor is enabled to advance to activities in which he can confirm the initial (status 2/9), final version of the paper (status 4/9) or choose the defense committee of the paper as shown in Fig. 4.

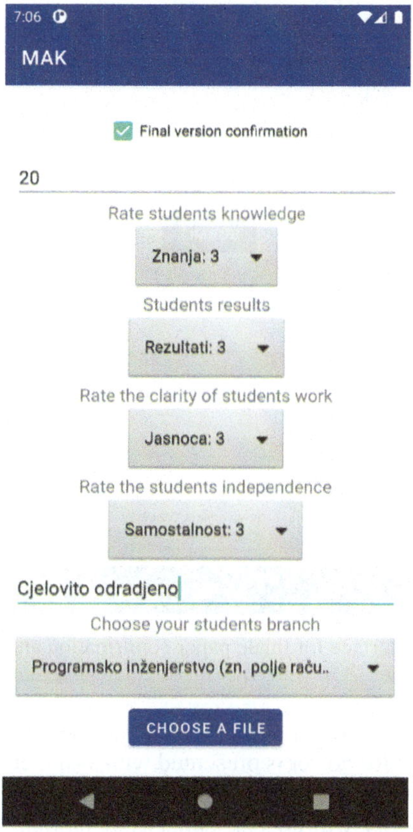

Fig. 4. Interface for finale paper confirmation and grading.

Reviewer Activities

Reviewer activities are used for undergraduate Students. After Mentor grades final paper of undergraduate Student, final paper is assigned to proper Reviewer to read final paper and approve (status 8/9) or disapprove Mentor's grade as shown in Fig. 5.

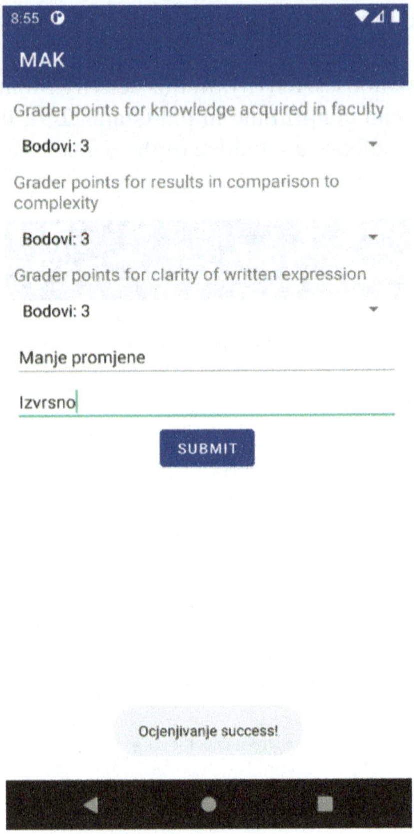

Fig. 5. Interface for finale paper confirmation and grading.

After clicking on the Reviewer button in the OtherUserMainActivity, just like in MentorsRecyclerActivity, Reviewer is presented with assigned Students. When clicking on a Student, if the status is valid and Student is in undergraduate study program, PreddiplomskiOcjenjivacActivity will appear. Otherwise, ObranaRadaActivity will appear for graduate Student and procedure for thesis defense is displayed (status 8/9). The last button in OtherUserMainActivity enables the user to log out of the system (Fig. 6).

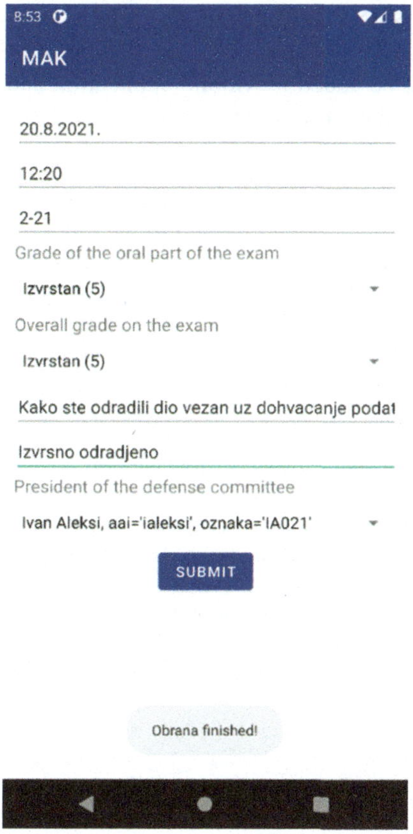

Fig. 6. Interface for thesis defense.

5 Conclusion

In this paper authors explained analysis, design and implementation of MAK mobile application in which the users of MAK system can perform their activities using their mobile devices. The solution has been implemented for mobile devices running Android, in the Kotlin programming language.

A simple application enables Students, Mentors and Reviewers to complete complex tasks through a simple interface. The flow of the application is simple and the user can complete various tasks just like in the web application of MAK system.

Kotlin is a modern programming language that makes the development of Android applications easier then Java, with minimal losses. The most appealing property of Kotlin is less writing and easier reading then Java.

The created application enables Students to update the versions of their paper and simple control of their procedures. Mentors and Reviewers are enabled to easily monitor their Students and work with their papers, all from their mobile devices.

Any future updates, like adding more roles to MAK system, new design for user interface, new properties and etc., will be made and tracked using SVN (Sub versioning) tools in Android Studio. SVN tool is used for application version control and change management that is the last phase of ALM. For each time as Web application is updated the same changes should be added to Mobile application and it is recommended that both application use the same ALM system.

References

1. Snake. http://www.springer.com/lncs. Accessed 14 Aug 2021
2. Galloway, S.: The Four: The Hidden DNA of Amazon, Apple Facebook and Google. Penguin Putnam Inc., New York (2017)
3. Profit in the mobile market. https://www.forbes.com/sites/johnkoetsier/2019/12/22/global-phone-profits-apple-66-samsung-17-everyone-else-unlucky-13/. Accessed 8 Aug 2021
4. Smyth, N.: Android 9, Kotlin i Android Studio 3.2 u jednoj knjizi, Kompjuter biblioteka, Beograd (2018)
5. Android operating system. https://www.androidauthority.com/history-android-os-name-789433/. Accessed 8 June 2021
6. Retrofit. https://square.github.io/retrofit/. Accessed 8 Aug 2021
7. FileAndFolderPicker. https://android-arsenal.com/details/1/8131#!description. Accessed 8 June 2021
8. Software Product lifecycle. https://milenasoftware.com/?page_id=208. Accessed 17 Aug 2021
9. AAI@EduHr. https://www.aaiedu.hr/. Accessed 17 Aug 2021

Maintaining the Descriptive Geometry's Design Knowledge

Ivan Grgić[1](✉), Mirko Karakašić[1], Željko Ivandić[1], and Hrvoje Glavaš[2]

[1] Mechanical Engineering Faculty in Slavonski Brod, University of Slavonski Brod, Trg I. B. Mažuranić 2, 35000 Slavonski Brod, Croatia
igrgic@unisb.hr

[2] Faculty of Electrical Engineering, Computer Science and Information Technology Osijek, Josip Juraj Strossmayer University of Osijek, Kneza Trpimira 2B, 31000 Osijek, Croatia

Abstract. One of the most important parts of product lifetime management is maintaining the knowledge that was implemented during product design. However, the problem occurs when the design knowledge has been lost or it is left written in the old form. In this work, the design knowledge is presented by the traditional drawing of an object using descriptive geometry. The aim is to present an approach to refresh the drawing with traditional CAD software and preserve the design knowledge. The line segment and the pentagon pyramid were drawn on paper using the traditional drawing technique with a pencil considering descriptive geometry rules. As a part of the traditional drawing method, they were never processed by conventional CAD software. They are imported as an image file, scaled, and processed with the aim to extract the written knowledge for further maintenance in a new CAD form. With such an approach, the design knowledge is refreshed and ready to be used again. This approach can help save time in much more complex applications such as in a redesign or reverse engineering process of products and their life-cycle management overall.

Keywords: Computer-Aided Design (CAD) · Design knowledge · Descriptive geometry · Maintenance

1 Introduction

Design knowledge is a set of visions, proposals, tools, and reflections. This knowledge has to be explicit (to be clearly expressed by whoever produces it), discussable (to permit the exchange of opinions among many interested interlocutors), transferrable (to be applicable by other designers) and possible to accumulate (to form a reservoir of design knowledge that could be the starting point for producing further knowledge by other researchers) [1]. The knowledge implemented in developed products should be saved in order to be reanalyzed and reused when becomes necessary to adapt the outdated product to the new customer needs. In the product development process, the knowledge constantly circulating among participants changing its forms frequently [2]. The forms can be tables, graphs, reports, technical drawings, 3D models, etc.

H. Glavaš et al. (Eds.): OTO 2021, LNNS 369, pp. 467–481, 2022.
https://doi.org/10.1007/978-3-030-92851-3_36

Losing such knowledge would significantly extend the time of the new product development and it will be a big step backward, which is not desirable. Therefore, a focus should be on how to preserve the design knowledge from the loss when it is presented by the old handmade technical drawings. However, many mechanical engineering drawings of operational products still exist only in this form and are needed for the maintenance of existing systems and/or as a basis for designing the next generation of the product the model of which they describe.

Various approaches were presented to create CAD models from 2D sketches [3–5], and the approaches dealing with automatic object recognition from images in [6–11]. However, in the recently published study presented in [12] the application of CAD combined with computer image processing technology in the mechanical drawing is analyzed. It was emphasized that the application of CAD combined with computer image processing technology in mechanical drawing has applications in descriptive geometry (18%), applications in reading assembly views (5%), applications in sectional views (29%), applications in assembly drawings (21%), and other aspects (27%). The focus fell on descriptive geometry (DG).

Descriptive Geometry as a branch of geometry is of vital importance in mechanical engineering. It helps engineers to represent three-dimensional objects on a plane, using a set of well-defined rules. Basic of the DG are taught through secondary education, mostly in technical schools, and through the course of the mathematics in other types of schools, but the special emphasis of DG teaching was given particularly in all branches of engineering at the university's level. So, could spatial perception be learned? Could teachers remind us of it? Does gender affect it? Could virtual and augmented reality help? A lot of studies have been conducted to answer such and similar questions [13–18]. Locke in his book [19] pointing out the 3D modeling technology offer great potential to make the solution of DG problems directly in a 3D environment; help engineering students to become more efficient future engineers. The same author offers a free online textbook for teaching DG with AutoCAD and Inventor. Bokan et al. [20] have discussed teaching methods in an undergraduate-level course of DG and they have pointed that it is useful to have an interactive tool for 3D object spatial representation in order to develop student's spatial intuition. In their work, the reasons for using AutoCAD in teaching DG have been explained and they have presented some student's projects in terms of software development. Another software development has been presented in [21] for DG practicing. Besides the student learning, the software also makes easy the evaluation of the student work by the instructors. An example of using AutoCAD for DG exercises has been presented in [22]. A very useful online place for learning DG with definitions and animations can be seen in [23]. Furthermore, the authors in [24] presented the methodology suitable for free form geometries, comprising the generation and processing of a 3D voxel image obtained from a hand drawing. Authors in [25] presented the creation of a set of 3D curves fitting the voxel image and the automatic generation of surface patches on the resulting curve network. Elias and Kebisek in [26] propose a systematic tool that allows the reconstruction of a 3D pseudo-wireframe starting from a 2D vectorial input. Such a tool is discussed in detail and has been implemented into the MatLab environment in order to validate and test the procedures. Furferi et al. [27] presented an overview and analysis of known methods for 3D model reconstruction from 2D orthographic views. These methods deal with the automatic processing of technical drawings of parts in vector format. Authors in [28] carried out a systematic clear and concise step-by-step procedure for 3D retrieval starting from wireframe models. Since the intent is to afford as clear as possible, guided, a procedure for 3D reconstruction, mathematical description is limited to the simplest case of polyhedral objects. The case of medicine application and an example in the field of paintings and heritage is shown in [29] and [30].

In general, the CAD software is too generic and they do not have specific tools or interfaces related to a specific subject in DG. From that point of view, it is a disadvantage, on the other, it is a benefit because it constantly forces its participants to think about how to relate traditional drawing with the CAD software and get the same solution and even 3D presentation of the same solution. At the same time, the designers getting knowledge from and about the object, useful for further processing. The approach described in this paper aims to transfer the knowledge left in the traditional drawing to 3D CAD software to help in its maintenance. The Solidworks software was chosen for conducting this study. The objects presented by the traditional drawing together with the simple DG problem are used as representatives. This study can be useful for reverse engineering and in the life-cycle management of products.

The paper is organized as follows: Sect. 1 presents the basics of the descriptive geometry knowledge implemented in the technical drawing. Section 2 presents the utilization of the SolidWorks software to DG tasks. Results and the discussion are presented in Sect. 3, and the conclusion is made in Sect. 4.

2 Position and Metric Relations of a Point, Line, and Line Segment

The space in a paper can be organized through the x, y, and z axes resulting in three projection planes representing the orthogonal projection planes such as front, top, and side view, Fig. 1a). Such space is also divided into eight octants denoted with Roman numerals starting from I to VIII. The plane Π_1 is determined with the axes x and y (positive and/or negative) and it is also called the 1st projection plane and represents the top view of an object. Analogously the Π_2 (front view) and Π_3 (side view). Let us explain the basics of the projecting of an object by using the point in space.

Point T (x,y,z) belongs to the 1st octant, Fig. 1b). The orthogonal projections of a T point are as follows: T'(x,y)-top view; T''(x,z)-front view, and T'''(y,z)-side view. The same solution was drawn by traditional drawing using a ruler set and a pencil and it is presented in Fig. 2.

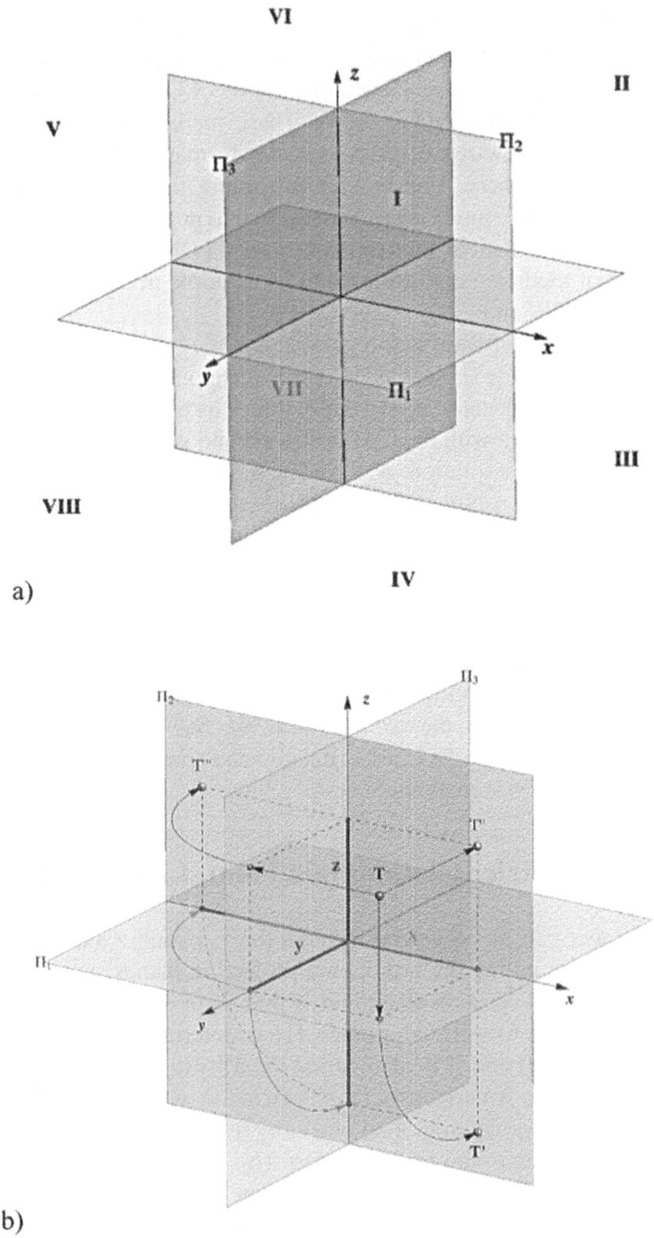

Fig. 1. a) The planes Π_1, Π_2, and Π_3 divide the space into eight octants; b) The projected point T-spatial perception [23].

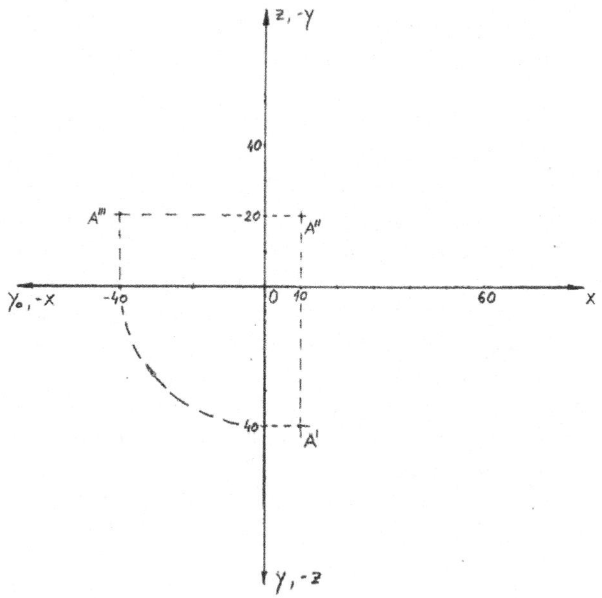

Fig. 2. The projected point A-traditional drawing on a paper.

2.1 Projection of a Line Segment

The projecting procedure of a line segment is analogous to the procedure earlier explained by determining the projections of a point. If we add another point in space and connect it with the existing one, the line segment will appear. An example of the line segment AB is shown below in Fig. 3.

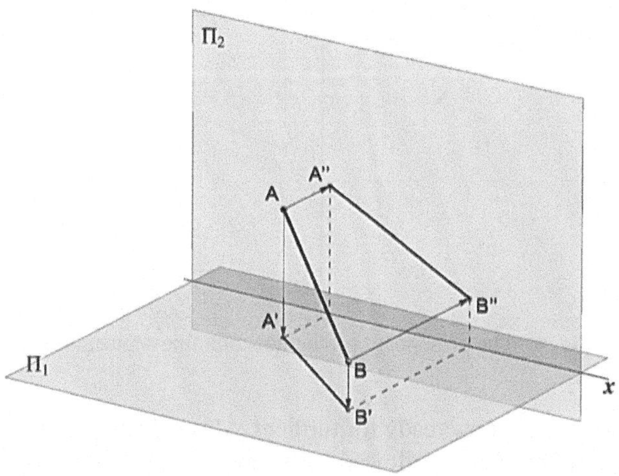

Fig. 3. The projected line segment AB [23].

The case study is presented by the orthogonal projections of a line segment determined with the two points A (10, 40, 20) and B (60, 20, 40). The implemented knowledge is as follows:

Considering given coordinates, the line segment AB is located in the space, and in-plane it is determined by its orthogonal projections. The in-plane position of point A is determined by the x, y, and z coordinate axes. According to Fig. 1, it can be seen that point A is positioned in the first octant (all axis coordinates are positive numbers). The projection of the front view of point A to the front plane Π_2 is determined by the x and z coordinate axes and is denoted as A´´. The projection of the top view of point A to the front plane Π_1 is determined by the coordinate axes x and y and is denoted as A´, and finally, the projection of the side view of point A to the front plane Π_3 is determined by the coordinate axes x and y_o and is denoted as A´´´. This is a method of parallel projection in which the projecting rays fall perpendicular to the objects and transmit their projections further to the projection planes in their original form. Here, the projection object is located between the designer and the surface on which it is projected.

The result is shown in Fig. 4. In nature, the line segment is drawn on the paper but for the clear view, it is converted to the black and white figure.

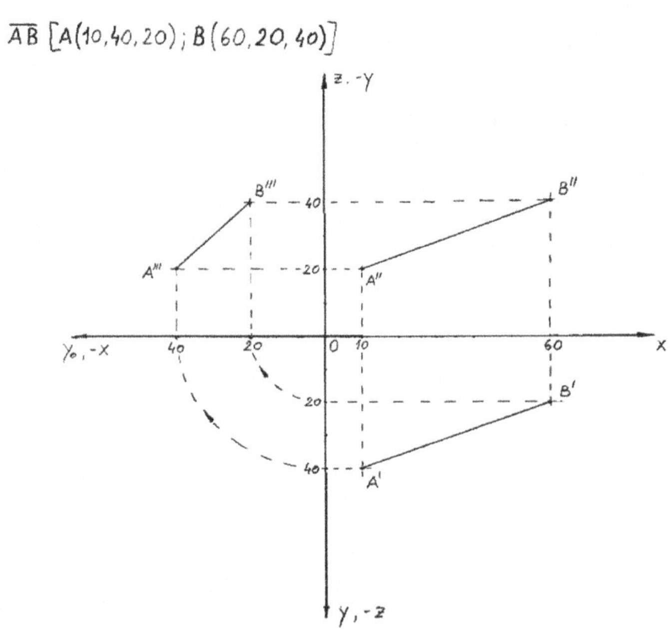

Fig. 4. The orthogonal projections of a line segment AB.

As it is, the drawing is now ready for further CAD analysis. The proposed method is presented step by step in Fig. 5 followed by its application to the two case studies, one presenting the line segment and the other presenting the pentagonal pyramid.

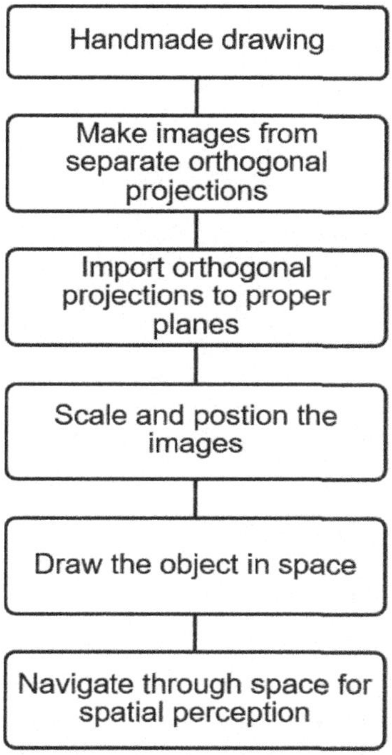

Fig. 5. Approach to extract the DG design knowledge.

2.2 Utilization of the CAD Software to DG Problem – Line Segment

The utilization requires several steps to do. It forces the user to think about what to do in each step. In precise, in each step it is necessary to relate orthogonal projections with images, axes (positive/negative), octants according to axes, proper planes in the software and finally get the full spatial perception and relation between 2D drawing and 3D model of the line segment. For acquiring images, it is possible to do this with any camera device. The basic skills of SolidWorks software are necessary such as 2D/3D sketching in planes, editing objects (scale, rotate, move), and navigating through space. As a part of the knowledge, the user must be able to recognize and separate the orthogonal projections as independent drawings because they are and will represent the views when inserted in software 3D space. The user can do this step either to draw separate views or use some photo editor if it is possible. It is important to make images of it as a next step.

When the part is opened in SolidWorks for sketching, it is recommended to make planes visible to relate them with eight octants presented in Fig. 1a). An example of the plane's visibility is shown in Fig. 6.

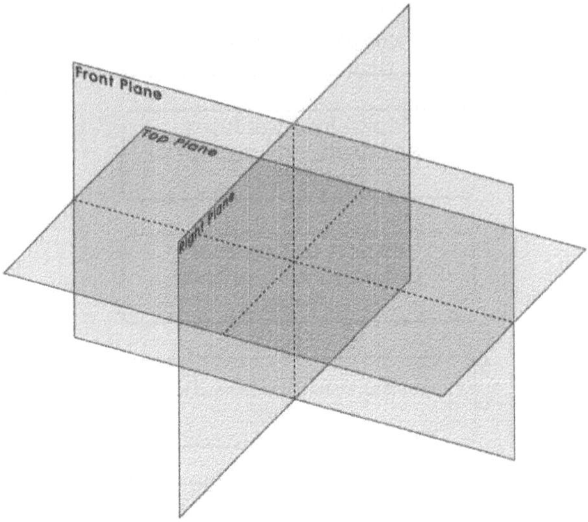

Fig. 6. The planes in SolidWorks.

The next step is inserting images in related planes. The image representing the front view of an object must be properly placed on a front plane in software, Fig. 7. The user must be aware of proper octant and axes determination. This will determine the correct image positioning in the software space. The image scaling is easy because the image contains marked numbers on axes and the scaling can be done using any of it. The scaled and positioned image is shown in Fig. 8 and Fig. 9. Analogously the top and the side view have to be done.

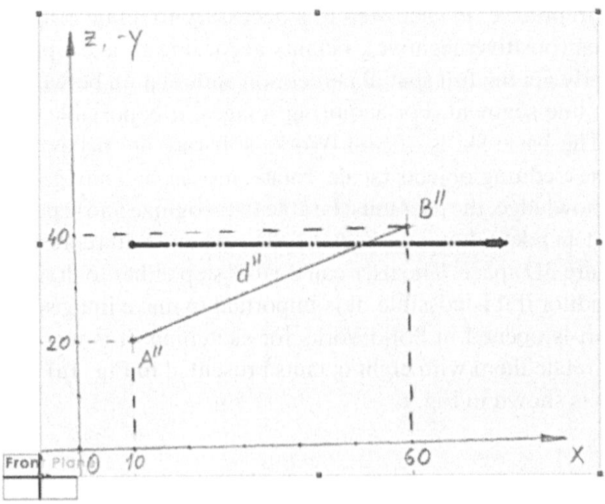

Fig. 7. Front view of a line segment AB inserted on a front plane prepared for positioning and scaling.

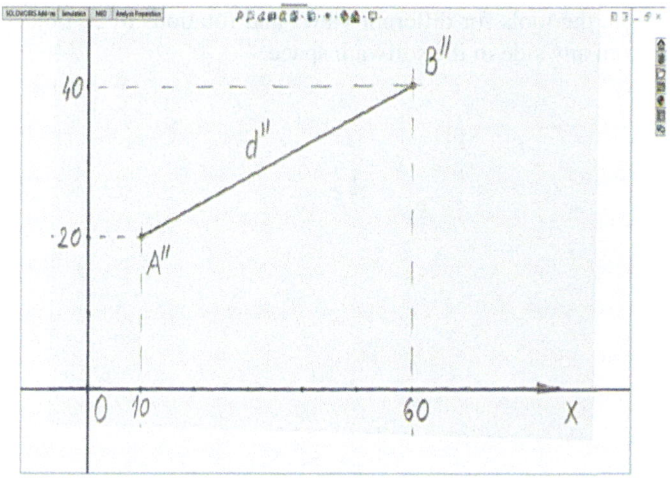

Fig. 8. Scaled and aligned image with axes.

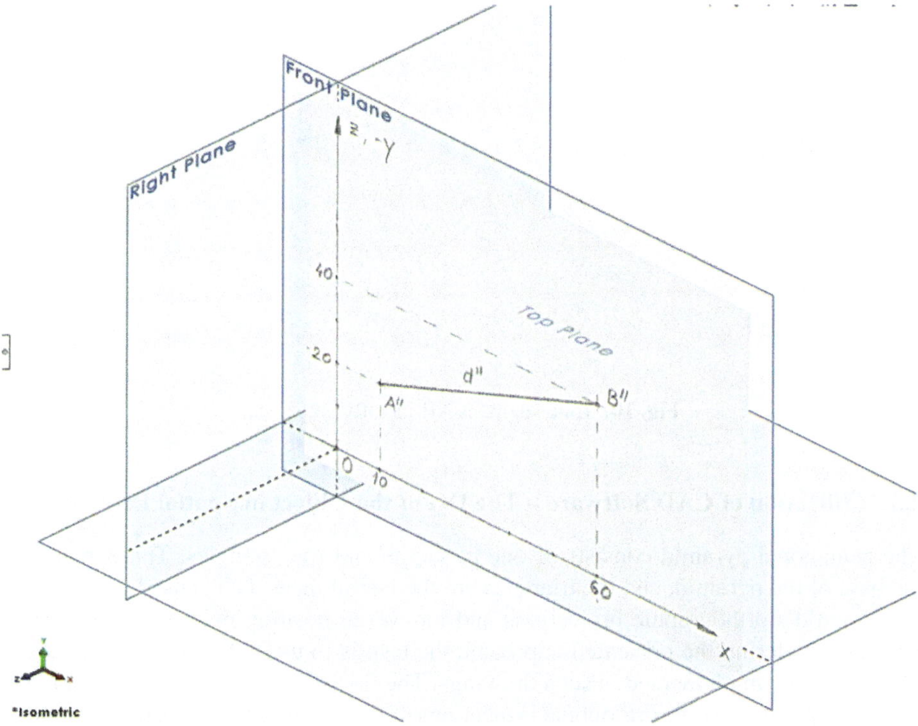

Fig. 9. Scaled and aligned image with axes in the software space.

The sketching step is the next and the basic knowledge about 2D and 3D sketching in software is necessary. The combination of points, dashed and full lines will lead the user to a last solution which is a 3D view of line segment AB, Fig. 10. As a better spatial perception, the tools for different views and rotations of an object can be used for viewing it from any side in the software space.

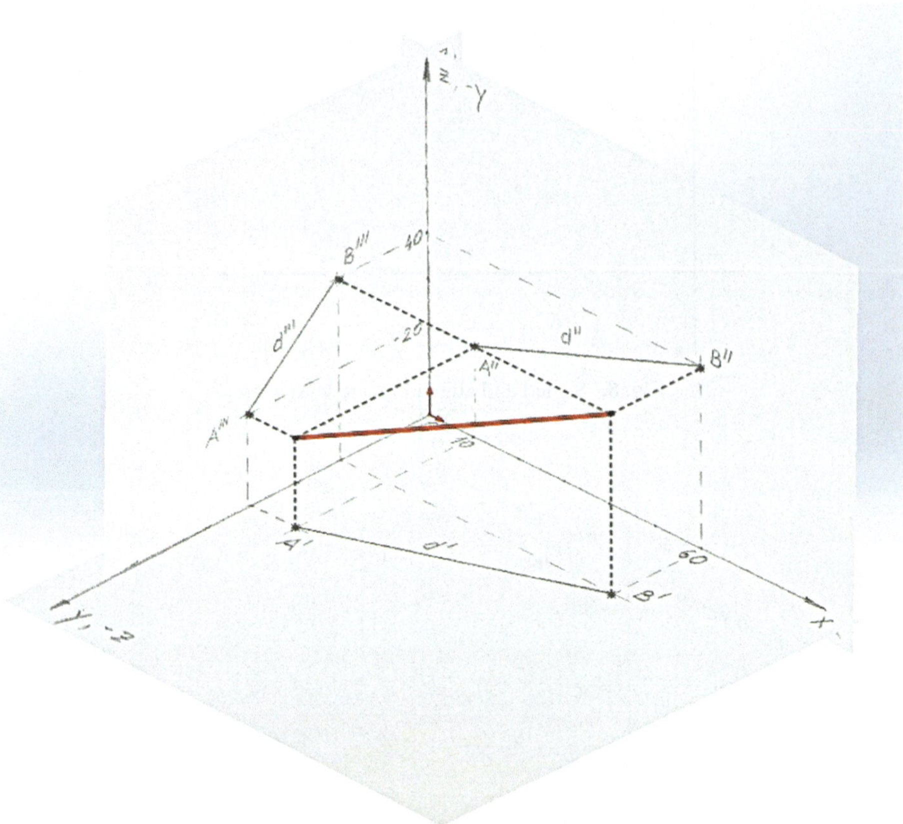

Fig. 10. Line segment AB in software space.

2.3 Utilization of CAD Software – The DG of the Object in Spatial Perception

The pentagonal pyramid consists of one pentagon and five triangles. The pentagon is the base of the pyramid, and the triangles are the lateral faces. DG is used to determine the pyramid's orthographic projections and its overall position in space drawn on the paper. Considering the presented approach, the task is to use CAD software to extract the knowledge implemented in such drawings. The first step is presented in Fig. 11. The second step is to separate orthogonal projections of the pyramid and an example is shown in Fig. 12. The pentagonal pyramid with its base lies in the plane R presented by R_x, R_y, and R_z coordinates. Line R_x-R_z is the trace of the plane R over the front-view plane, line R_x-R_y is the trace of the plane R over the top-view plane, and R_z-R_{yo} is the trace of the plane R over the side-view plane. The third, fourth, and fifth steps were presented in detail in Fig. 7, 8, 9, 10, hence Fig. 13 is the solution for the case of the pyramid. After inserting the projecting rays the base is modeled on the R plane and finally, the object can be created (Fig. 14). To see the relations between the object, projections, and planes it is recommended to navigate through the CAD space.

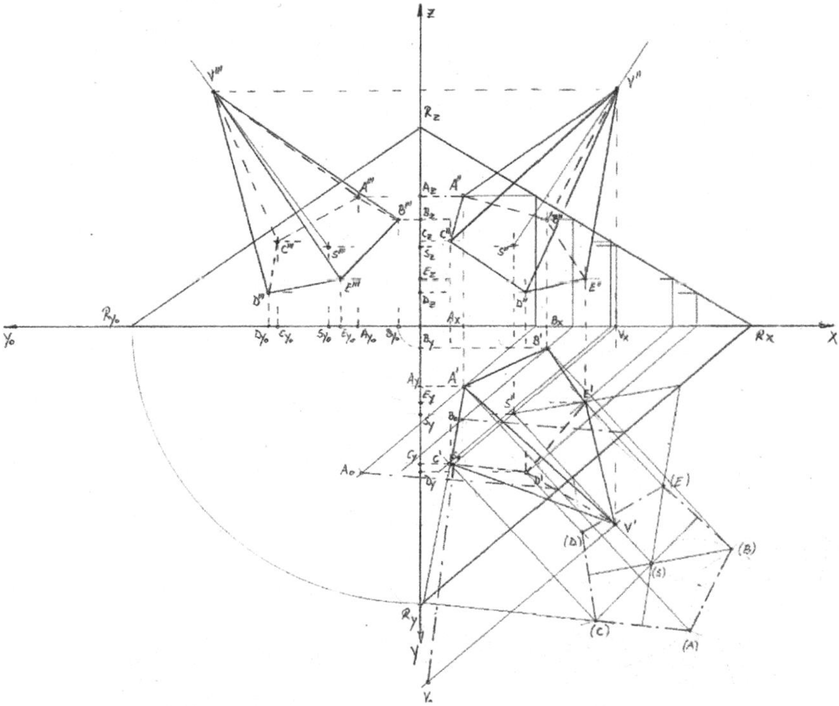

Fig. 11. DG used for pentagonal pyramid perception.

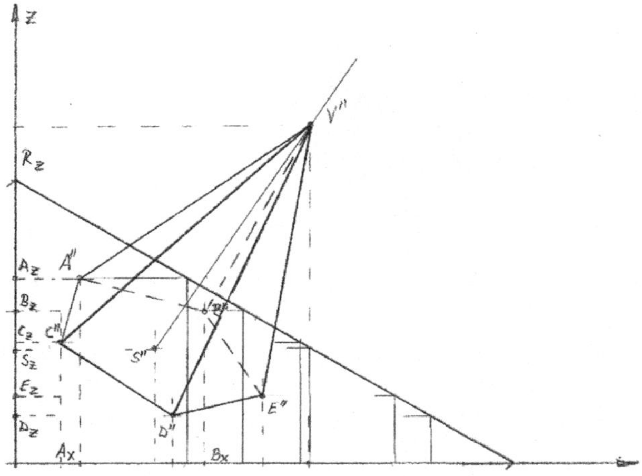

Fig. 12. The front view of the pentagonal pyramid.

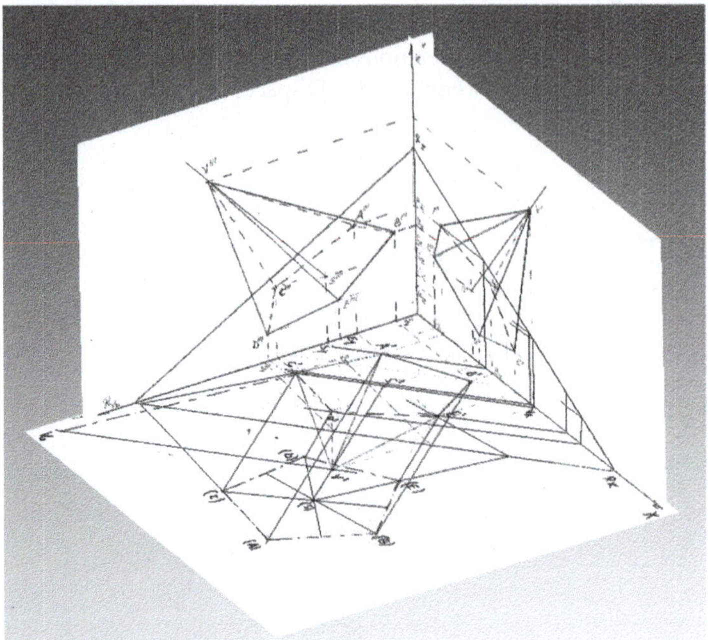

Fig. 13. Images inserted, scaled, and positioned in software; front, top, and a side view of a pentagonal pyramid.

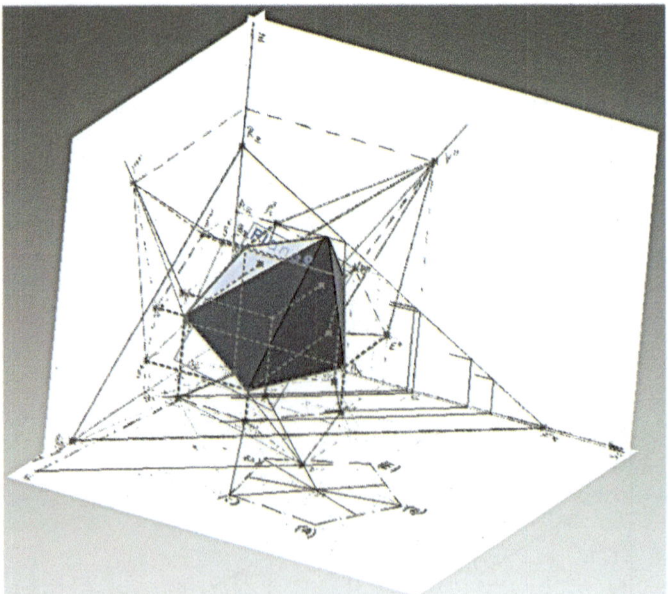

Fig. 14. The pentagonal pyramid in CAD space.

3 Discussion

The approach consisted of several steps to follow, has been successfully applied to the two case studies. It is not based on the automatic reconstruction of the drawings to generate a new CAD file like the ones presented in [6–11]. The aim is to force the user to be focused in each step to recognize every feature presented by the drawing. This is the knowledge left on the paper at some point in the past and has to be refreshed and maintained. The life-cycle management of a product depends on it especially in the case of reverse engineering and redesign. The approach presented in this study is not limited to specific products/objects, however, it can be applied as it is, it can end at any stage or be extended depends on the user's arbitrarity.

In the case of both case studies, the same steps were followed. For acquiring images, it is possible to do this with any camera device or scanner. The basic skills of CAD software are necessary such as 2D/3D sketching in planes, editing objects (scale, rotate, move), and navigating through space. As a part of the learning process, the user must be able to recognize and separate the orthogonal projections as independent drawings because they are and will represent the views when inserted in software 3D space. This can be done either to draw separate views or use some photo editor if it is possible.

In each step it is necessary to relate orthogonal projections with images, axes (positive/negative), octants according to axes, proper planes in the software and finally get the full spatial perception and relation between 2D drawing and 3D model. It is recommended to make planes visible to relate them with eight octants presented in Fig. 1. The image representing the front view of an object must be properly placed on a front plane in software, etc. The user must be aware of proper octant and axes determination. This will determine the correct image positioning in the software space. The image scaling is easy because the image contains marked numbers on axes and the scaling can be done using any of it. Furthermore, sketching over the inserted planes can help in feature extraction and recognition. This is not just clicking over the animated view and observe how points, planes, lines, and objects appear "flying" to their right position in relation to the available tools reported in [21–23].

Considering reported studies such as [26] the programming is not necessary to be used for this approach.

It is necessary to draw attention to the following:

- Handmade drawing accuracy is desirable
- A special effort is needed for image preparation
- The quality of the final solution depends on the image positioning accuracy
- The approach should be tested in other CAD software, not just SolidWorks.

4 Conclusion

Using 3D possibilities of CAD software with a little effort it is possible to explore descriptive geometry innovatively. This approach offers 2D and 3D view relation for the same object. With this approach, the handmade drawings can be inserted into CAD software and generate 3D objects from them simultaneously generating knowledge about the object and its further maintenance in a modern way. With this approach, during refreshing the knowledge implemented to the paper, we want the user to be involved in each step while generating the CAD solution, and not just to be an observer of some pre-solved animation.

We believe that this method can be used with every CAD software that can generate a 3D object.

In the future, with this method, we will try to extract the knowledge related to ordinary technical handmade drawings which include tolerances, manufacturing data, dimensions, cross-sections, details, etc., and for sketch drawings as well.

References

1. Manzini, E.: New design knowledge. Des. Stud. **30**(1), 4–12 (2009). https://doi.org/10.1016/j.destud.2008.10.001
2. Ognjanovic, M.: Creativity in design incited by knowledge modeling, In: Lindemann, U., Birkhofer, H., Meerkamm, H., Vajna, S. (eds) International Conference on Engineering Design, ICED 99, Design Society (1999)
3. Cheon, S.-U., Kim, B.C., Mun, D., Han, S.: A procedural method to exchange editable 3D data from a free-hand 2D sketch modeling system into 3D mechanical CAD systems. Comput. Aided Des. **44**(2), 123–131 (2012). https://doi.org/10.1016/j.cad.2011.10.003
4. Kim, D.H., Kim, M.-J.: A new modeling interface for the pen-input displays. Comput. Aided Des. **38**(3), 210–223 (2006). https://doi.org/10.1016/j.cad.2005.10.007
5. Yang, C., Sharon, D., Panne, M.: Sketch-based modeling of parameterized objects. In: The Proceedings of 2nd Eurographics Workshop on Sketch-Based Interfaces and Modeling (2005)
6. Dori, D., Tombre, K.: From engineering drawings to 3D CAD models: are we ready now? Comput. Aided Des. **27**(4), 243–254 (1995). https://doi.org/10.1016/0010-4485(95)91134-7
7. Ansary, F.T., Vandeborre, J.-P., Daoudi, M.: A framework for 3D CAD models retrieval from 2D images. Annales Des. Télécommunications **60**, 1337–1359 (2005). https://doi.org/10.1080/16864360.2005.10738335
8. Intwala, A.: Image to CAD: feature extraction and translation of raster image of CAD drawing to DXF CAD format. In: Nain, N., Vipparthi, S.K., Raman, B. (eds.) CVIP 2019. CCIS, vol. 1147, pp. 205–215. Springer, Singapore (2020). https://doi.org/10.1007/978-981-15-4015-8_18
9. Khalifa, I., Moussa, M., Kamel, M.: Range image segmentation using local approximation of scan lines with application to CAD model acquisition. Mach. Vis. Appl. **13**, 263–274 (2003). https://doi.org/10.1007/s00138-002-0083-0
10. Hetzel, G., Leibe, B., Levi, P., Schiele, B.: 3D Object Recognition from range images using local feature histograms. In: Proceedings of the 2001 IEEE Computer Society Conference on Computer Vision and Pattern Recognition. CVPR 2001, IEEE Xplore (2001)
11. Lowe, G.D.: Three-dimensional object recognition from single two-dimensional images. Artif. Intell. **31**(3), 355–395 (1987). https://doi.org/10.1016/0004-3702(87)90070-1
12. Fan, N.: Application of CAD combined with computer image processing technology in mechanical drawing. In: Sugumaran, V., Xu, Z., Zhou, H. (eds.) MMIA 2020. AISC, vol. 1234, pp. 319–324. Springer, Cham (2021). https://doi.org/10.1007/978-3-030-51556-0_46
13. Astur, S.R., Troop, J., Sava, S., Constable, R.T., Markus, J.E.: Sex differences and correlations in a virtual Morris water task, a virtual radial arm maze, and mental rotation. Behav. Brain Res. **151**, 103–115 (2004). https://doi.org/10.1016/j.bbr.2003.08.024

14. Cutmore, R.H.T., Hine, J.T., Maberly, J.K., Langford, M.N., Hawgood, G.: Cognitive and gender factors influencing navigation in a virtual environment. Int. J. Hum. Comput. Stud. **53**(2), 223–249 (2000). https://doi.org/10.1006/ijhc.2000.0389
15. Waller, D., Hunt, E., Knapp, D.: The transfer of spatial knowledge in virtual environment training. Presence Teleoperators Virtual Environ. **7**(2), 129–143 (1998). https://doi.org/10.1162/105474698565631
16. Oman, M.C., Shebilske, L.W., Richards, T.J., Tubré, C.T., Beall, C.A., Natapoff, A.: Three dimensional spatial memory and learning in real and virtual environments. Spatial Cogn. Comput. **2**(4), 355–372 (2002). https://doi.org/10.1023/a:1015548105563
17. Allen, G., et al.: Virtual environments and the enhancement of spatial behavior: towards a comprehensive research agenda. Presence Teleoperators Virtual Environ. **9**(6), 593–615 (2000). https://doi.org/10.1162/105474600300040402
18. Diaz, D.D., Sims, K.V.: Augmenting virtual environments: the influence of spatial ability on learning from integrated displays. High Ability Stud. **14**(2), 191–212 (2003). https://doi.org/10.1080/1359813032000163915
19. Locke, E.: Engineering Descriptive Geometry A Collection of Teaching and Learning Modules for the Dummies with Autodesk AutoCAD and Inventor/California State University Los Angeles (2007). https://suniseaproducts.weebly.com/free-online-descriptive-geometry--sheet-metal-design-textbook.html. Accessed 21 July 2021
20. Bokan, N., Ljucović, M., Vukmirović, S.: Computer-aided teaching of descriptive geometry. J. Geometry Graph. **13**(2), 221–229 (2009)
21. Jacques, J.J., Teixeira, G.F.: Development of CAD software for descriptive geometry practice. In: Proceedings - International Conference on Geometry and Graphics, Salvador (2006)
22. Jacobson, E.C.: Using AutoCAD for descriptive geometry exercises in undergraduate structural geology. Comput. Geosci. **27**, 9–15 (2001)
23. GeomTeh3D. http://www.grad.hr/geomteh3d/radne_eng.html. Accessed 21 July 2021
24. Governi, L., Furferi, R., Palai, M., Volpe, Y.: 3D geometry reconstruction from orthographic views: a method based on 3D image processing and data fitting. Comput. Ind. **64**, 1290–1300 (2013). https://doi.org/10.1016/j.compind.2013.02.003
25. Furferi, R., Governi, L., Palai, M., Volpe, Y.: From 2D orthographic views to 3D pseudo-wireframe: an automatic procedure. Int. J. Comput. Appl. **5**(6), 18–24 (2010). https://doi.org/10.5120/918-1296
26. Elias, M., Kebisek, M.: An overview of methods for 3D model reconstruction from 2D orthographic views. In: Proceedings of the 3rd International Workshop "Innovation in Information Technologies - Theory and Practice", 6th–10th September, Dresden, Germany (2010)
27. Furferi, R., Governi, L., Palai, M., Volpe, Y.: 3D model retrieval from mechanical drawings analysis. Int. J. Mech. **5**(2), 91–99 (2011)
28. Yan, Q.-W., Chen, C.L.P., Tang, Z.: Efficient algorithm for the reconstruction of 3D objects from orthographic projections. Comput. Aided Des. **26**(9), 699–717 (1994). https://doi.org/10.1016/0010-4485%2894%2990020-5
29. Budak, I., et al.: Comparative analysis of 3D digitization systems in the field of dental prosthetics. Tech. Gaz. **20**(2), 291–296 (2013). https://hrcak.srce.hr/100249
30. Bernik, A., Vusić, D., Kober, D.: Implementation of augmented reality application and computer graphics: the case of the stolen paintings. Tech. Gaz. **26**(6), 1570–1575 (2019). https://doi.org/10.17559/TV-20181015160248

Author Index

H. Glavaš et al. (Eds.): OTO 2021, LNNS 369, pp. 483–484, 2022.
https://doi.org/10.1007/978-3-030-92851-3

The manufacturer's authorised representative in the EU is Springer
Nature Customer Service Centre GmbH, Europaplatz 3, 69115 Heidelberg,
Germany. If you have any concerns regarding our products, please
contact ProductSafety@springernature.com

Printed and bound by CPI Group (UK) Ltd, Croydon, CR0 4YY
30/09/2026
02225251-0015